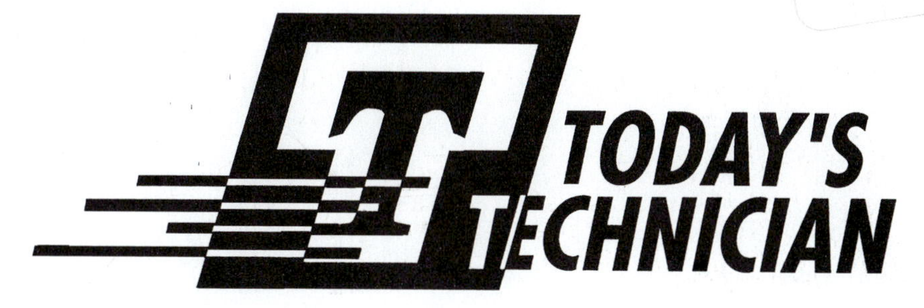

Shop Manual for

Basic Automotive Service and Systems

Second Edition

Online Services

Delmar Online

To access a wide variety of Delmar products and services on the World Wide Web, point your browser to:

http://www.delmar.com
or email: info@delmar.com

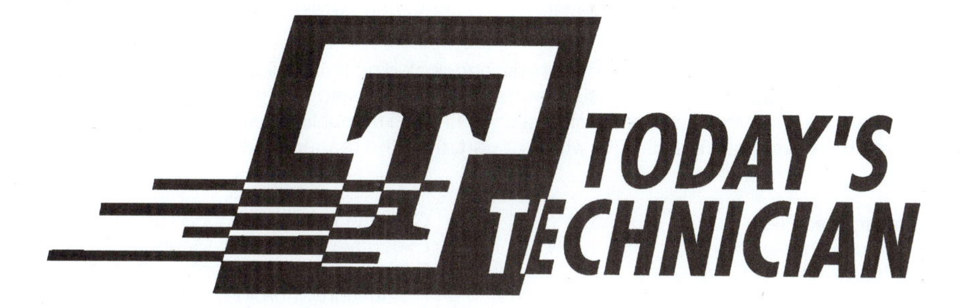

Shop Manual for
Basic Automotive Service and Systems
Second Edition

Jay Webster
California State University
Long Beach, California

Clifton E. Owen
Griffin Technical Institute
Griffin, Georgia

Jack Erjavec
Series Advisor
Professor Emeritus, Columbus State Community College
Columbus, Ohio

Africa • Australia • Canada • Denmark • Japan • Mexico • New Zealand • Philippines
Puerto Rico • Singapore • Spain • United Kingdom • United States

NOTICE TO THE READER

Publisher does not warrant or guarantee any of the products described herein or perform any independent analysis in connection with any of the product information contained herein. Publisher does not assume, and expressly disclaims, any obligations to obtain and include information other than that provided to it by the manufacturer.

The reader is expressly warned to consider and adopt all safety precautions that might be indicated by the activities herein and to avoid all potential hazards. By following the instructions contained herein, the reader willingly assumes all risks in connection with such instructions.

The Publisher makes no representation or warranties of any kind, including but not limited to, the warranties of fitness for particular purpose or merchantability, nor are any such representations implied with respect to the material set forth herein, and the publisher takes no responsibility with respect to such material. The publisher shall not be liable for any special, consequential, or exemplary damages resulting, in whole or part, from the readers' use of, or reliance upon, this material.

Delmar Staff:
Business Unit Director: Alar Elken
Executive Editor: Sandy Clark
Acquisitions Editor: Vernon R. Anthony
Development Editor: Catherine Wein
Editorial Assistant: Bridget Morrison
Executive Marketing Manager: Maura Theriault
Channel Manager: Mona Caron
Executive Production Manager: Mary Ellen Black
Project Editor: Christopher Chien
Production Coordinator: Karen Smith
Art/Design Coordinator: Cheri Plasse
Cover Image by David Kimball

Printed in the United States of America
5 6 7 8 9 10 XXX 05

For more information, contact Delmar, 3 Columbia Circle, PO Box 15015, Albany, NY 12212-0515; or find us on the World Wide Web at http://www.delmar.com

Asia
Thomson Learning
60 Albert Street, #15-01
Albert Complex
Singapore 189969
Tel: 65 336 6411
Fax: 65 336 7411

Australia/New Zealand
Nelson/Thomson Learning
102 Dodds Street
South Melbourne, Victoria 3205
Australia
Tel: 61 39 685 4111
Fax: 61 39 685 4199

Canada
Nelson/Thomson Learning
1120 Birchmount Road
Scarborough, Ontario
Canada M1K 5G4
Tel: 416-752-9100
Fax: 416-752-8102

International Headquarters
Thomson Learning
International Division
290 Harbor Drive, 2nd Floor
Stamford, CT 0692-7477
Tel: 203-969-8700
Fax: 203-969-8751

Japan
Thompson Learning
Palaceside Building 5F
1-1-1 Hitotsubashi, Chiyoda-ku
Tokyo 100 0002 Japan
Tel: 813 5218 6544
Fax: 813 5218 6551

Latin America
Thomson Learning
Seneca, 53
Colonia Polanco
11560 Mexico D. F. Mexico
Tel: 525-281-2906
Fax: 525-281-2656

South Africa
Thomson Learning
Zonnebloom Building
Constantine Square
526 Sixteenth Road
P.O. Box 2459
Halfway House, 1685
South Africa
Tel: 27 11 805 4819
Fax: 27 11 805 3648

Spain
Thomson Learning
Calle Magallanes, 25
28015-MADRID
ESPANA
Tel: 34 91 446 33 50
Fax: 34 91 445 62 18

UK/Europe/Middle East
Thomson Learning
Berkshire House
168-173 High Holborn
London
WC1V 7AA United Kingdom
Tel: 44 171 497 1422
Fax: 44 171 497 1426

Thomas Nelson & Sons LTD
Nelson House
Mayfield Road
Walton-on-Thames
KT 12 5PL United Kingdom
Tel: 44 1932 2522111
Fax: 44 1932 246574

Library of Congress Cataloging-in-Publication Data

Webster, Jay.
Basic automotive service and systems / Jay Webster, Clifton Owen.
— 2nd ed.
p. cm. — (Today's technician)
Includes index.
ISBN 0-8273-8544-7 (alk. paper)
1. Automobiles—Maintenance and repair. I. Owen, Clifton.
II. Title. III. Series.
TL152.W38 1999
629.28'72—dc21

99-35324
CIP

CONTENTS

Photo Sequences

Job Sheets

PREFACE

Thanks to the support the *Today's Technician* series has received from those who teach automotive technology, Delmar Publishers is able to live up to its promise to provide new editions every three years. We have listened to our critics and our fans and present this new, revised edition. By revising our series every three years, we can and will respond to changes in the industry, changes in the certification process, and to the ever-changing needs of those who teach automotive technology.

The *Today's Technician* series, by Delmar Publishers, features textbooks that cover all mechanical and electrical systems of automobiles and light trucks. Principal titles correspond with the eight major areas of ASE (National Institute for Automotive Service Excellence) certification. Additional titles include remedial skills and theories common to all of the certification areas and advanced or specialized subject areas that reflect the latest technological trends.

Each title is divided into two manuals: a Classroom Manual and a Shop Manual. Dividing the material into two manuals provides the reader with the information needed to begin a successful career as an automotive technician without interrupting the learning process by mixing cognitive and performance-based learning objectives.

Each Classroom Manual contains the principles of operation for each system and subsystem. It also discusses the design variations used by different manufacturers. The Classroom Manual is organized to build upon basic facts and theories. The primary objective of this manual is to allow the reader to gain an understanding of how each system and subsystem operates. This understanding is necessary to diagnose the complex automobile systems.

The understanding acquired by using the Classroom Manual is required for competence in the skill areas covered in the Shop Manual. All of the high priority skills, as identified by ASE, are explained in the Shop Manual. The Shop Manual also includes step-by-step instructions for diagnostic and repair procedures. Photo Sequences are used to illustrate many of the common service procedures. Other common procedures are listed and are accompanied with drawings and photographs that allow the reader to visualize and conceptualize the finest details of the procedure. The Shop Manual also contains the reasons for performing the procedures, as well as when that particular service is appropriate.

The two manuals are designed to be used together and are arranged in corresponding chapters. Not only are the chapters in the manuals linked together, the contents of the chapters are also linked. Both manuals contain clear and thoughtfully selected illustrations. Many of the illustrations are original drawings or photos prepared for inclusion in this series. This means that the art is a vital part of each manual.

The page layout is designed to include information that would otherwise break up the flow of information presented to the reader. The main body of the text includes all of the "need-to-know" information and illustrations. In the side margins are many of the special features of the series. Items such as definition of new terms, common trade jargon, tools list, and cross-referencing are placed in the margin, out of the normal flow of information so as not to interrupt the thought process of the reader.

Jack Erjavec

Highlights of this Edition—Shop Manual

The Shop Manual has been updated to correlate with the new content of the Classroom Manual. More information has been added on securing a job, technical and legal certification for the technician, entry level tasks and procedures, inspection and maintenance checks, supplement restraint systems and climate control operation. Some of the tasks described require close supervision to be performed correctly, but the entry-level technician can be expected to assist in the repair or perform the repair on their own within a few months of employment. Those tasks are presented as part of an inspection routine or part of a larger repair.

Job Sheets have been added to the end of each chapter. The Job Sheets provide a format for students to perform some of the tasks covered in the chapter. In addition to walking a student through a procedure, step-by-step, these Job Sheets challenge the student by asking why or how something should be done, thereby making the students think about what they are doing.

Highlights of this Edition—Classroom Manual

The Classroom Manual content and organization has been based on ASE testing areas. Other content updates include information on duties and responsibilities in an automotive repair shop, locating and applying for a position with a repair facility, safety inspection, repair orders, the use of computers in the repair shop, and electronic devices. Electronic devices and systems that are expected to be standard equipment in a few years are introduced.

The manual is arranged so that the basic theories and general information chapters are followed with a description and operation of the various systems of a typical car or light truck.

Classroom Manual

To stress the importance of safe work habits, the Classroom Manual dedicates one full chapter to safety. Included in this chapter are common safety practices, safety equipment, and safe handling of hazardous materials and wastes. This includes information on MSDS sheets and OSHA regulations. Other features of this manual include:

Cognitive Objectives

These objectives define the contents of the chapter and define what the student should have learned upon completion of the chapter.
Each topic is divided into small units to promote easier understanding and learning.

Marginal Notes

New terms are pulled out and defined. Common trade jargon also appears in the margin and gives some of the common terms used for components. This allows the reader to speak and understand the language of the trade, especially when conversing with an experienced technician.

A Bit of History

This feature gives the student a sense of the evolution of the automobile. This feature not only contains nice-to-know information, but also should spark some interest in the subject matter.

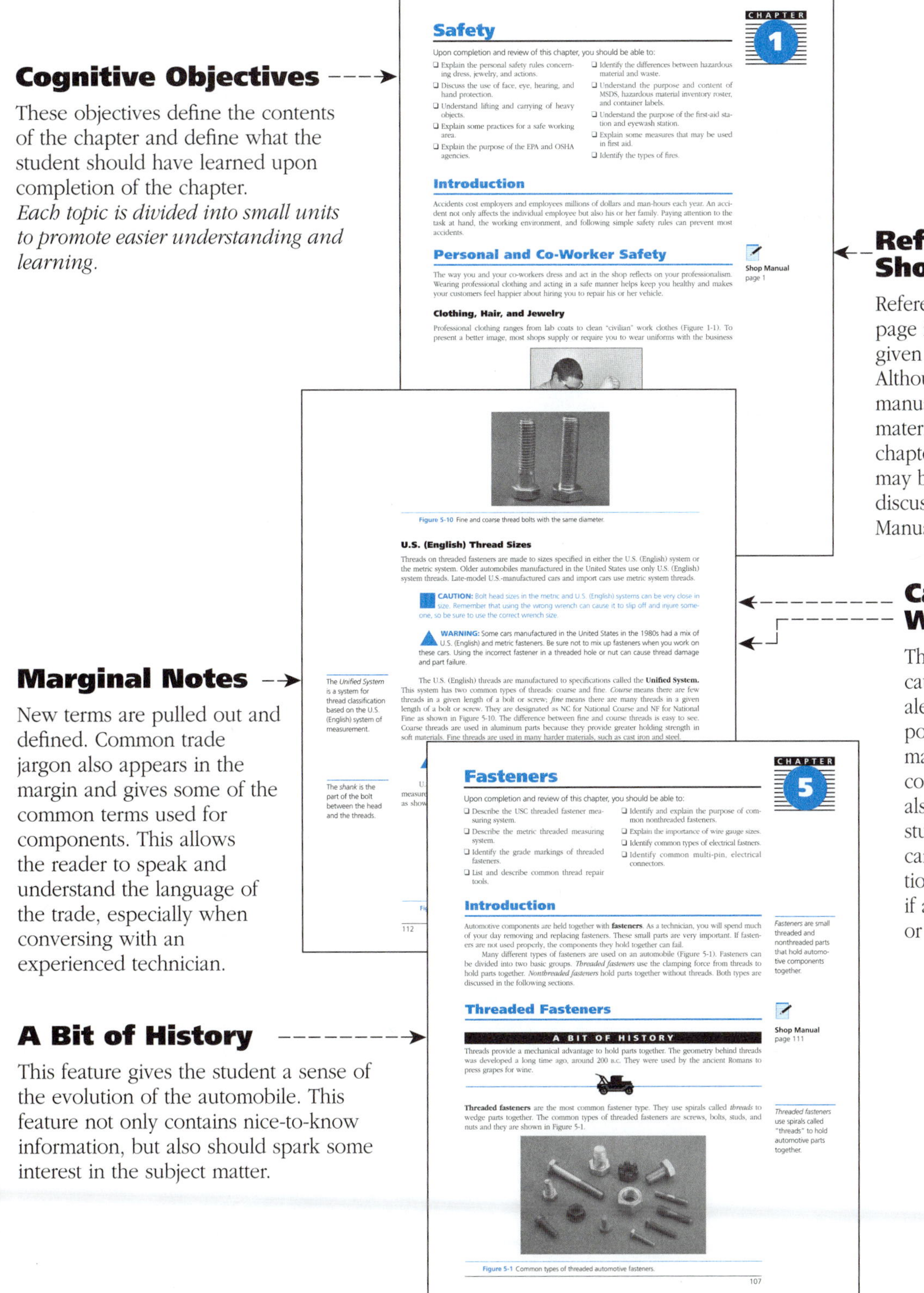

CHAPTER 1

Safety

Upon completion and review of this chapter, you should be able to:

- ❑ Explain the personal safety rules concerning dress, jewelry, and actions.
- ❑ Discuss the use of face, eye, hearing, and hand protection.
- ❑ Understand lifting and carrying of heavy objects.
- ❑ Explain some practices for a safe working area.
- ❑ Explain the purpose of the EPA and OSHA agencies.
- ❑ Identify the differences between hazardous material and waste.
- ❑ Understand the purpose and content of MSDS, hazardous material inventory roster, and container labels.
- ❑ Understand the purpose of the first-aid station and eyewash station.
- ❑ Explain some measures that may be used in first aid.
- ❑ Identify the types of fires.

Introduction

Accidents cost employers and employees millions of dollars and man-hours each year. An accident not only affects the individual employee but also his or her family. Paying attention to the task at hand, the working environment, and following simple safety rules can prevent most accidents.

Personal and Co-Worker Safety

Shop Manual page 1

The way you and your co-workers dress and act in the shop reflects on your professionalism. Wearing professional clothing and acting in a safe manner helps keep you healthy and makes your customers feel happier about hiring you to repair his or her vehicle.

Clothing, Hair, and Jewelry

Professional clothing ranges from lab coats to clean "civilian" work clothes (Figure 1-1). To present a better image, most shops supply or require you to wear uniforms with the business

Figure 5-10 Fine and coarse thread bolts with the same diameter.

U.S. (English) Thread Sizes

Threads on threaded fasteners are made to sizes specified in either the U.S. (English) system or the metric system. Older automobiles manufactured in the United States use only U.S. (English) system threads. Late-model U.S.-manufactured cars and import cars use metric system threads.

CAUTION: Bolt head sizes in the metric and U.S. (English) systems can be very close in size. Remember that using the wrong wrench can cause it to slip off and injure someone, so be sure to use the correct wrench size.

WARNING: Some cars manufactured in the United States in the 1980s had a mix of U.S. (English) and metric fasteners. Be sure not to mix up fasteners when you work on these cars. Using the incorrect fastener in a threaded hole or nut can cause thread damage and part failure.

The *Unified System* is a system for thread classification based on the U.S. (English) system of measurement.

The U.S. (English) threads are manufactured to specifications called the **Unified System.** This system has two common types of threads: coarse and fine. *Course* means there are few threads in a given length of a bolt or screw; *fine* means there are many threads in a given length of a bolt or screw. They are designated as NC for National Coarse and NF for National Fine as shown in Figure 5-10. The difference between fine and course threads is easy to see. Coarse threads are used in aluminum parts because they provide greater holding strength in soft materials. Fine threads are used in many harder materials, such as cast iron and steel.

The *shank* is the part of the bolt between the head and the threads.

112

CHAPTER 5

Fasteners

Upon completion and review of this chapter, you should be able to:

- ❑ Describe the USC threaded fastener measuring system.
- ❑ Describe the metric threaded measuring system.
- ❑ Identify the grade markings of threaded fasteners.
- ❑ List and describe common thread repair tools.
- ❑ Identify and explain the purpose of common nonthreaded fasteners.
- ❑ Explain the importance of wire gauge sizes.
- ❑ Identify common types of electrical fastners.
- ❑ Identify common multi-pin, electrical connectors.

Introduction

Fasteners are small threaded and nonthreaded parts that hold automotive components together.

Automotive components are held together with **fasteners**. As a technician, you will spend much of your day removing and replacing fasteners. These small parts are very important. If fasteners are not used properly, the components they hold together can fail.

Many different types of fasteners are used on an automobile (Figure 5-1). Fasteners can be divided into two basic groups. *Threaded fasteners* use the clamping force from threads to hold parts together. *Nonthreaded fasteners* hold parts together without threads. Both types are discussed in the following sections.

Threaded Fasteners

Shop Manual page 111

A BIT OF HISTORY

Threads provide a mechanical advantage to hold parts together. The geometry behind threads was developed a long time ago, around 200 B.C. They were used by the ancient Romans to press grapes for wine.

Threaded fasteners use spirals called "threads" to hold automotive parts together.

Threaded fasteners are the most common fastener type. They use spirals called *threads* to wedge parts together. The common types of threaded fasteners are screws, bolts, studs, and nuts and they are shown in Figure 5-1.

Figure 5-1 Common types of threaded automotive fasteners.

107

References to the Shop Manual

Reference to the appropriate page in the Shop Manual is given whenever necessary. Although the chapters of the two manuals are synchronized, material covered in other chapters of the Shop Manual may be fundamental to the topic discussed in the Classroom Manual.

Cautions and Warnings

Throughout the text, cautions are given to alert the reader to potentially hazardous materials or unsafe conditions. Warnings are also given to advise the student of things that can go wrong if instructions are not followed or if a nonacceptable part or tool is used.

Terms to Know

A list of new terms appears next to the Summary. Definitions for these terms can be found in the Glossary at the end of the manual.

Summaries

Each chapter concludes with summary statements that contain the important topics of the chapter. These are designed to help the reader review the contents.

Review Questions

Short answer essays, fill-in-the-blanks, and multiple-choice type questions follow each chapter. These questions are designed to accurately assess the student's competence in the stated objectives at the beginning of the chapter.

Figure 5-43 The seal is trapped between the mating connectors when they are plugged together.

Terms to Know

Bolt
Butt connector
Dies
Fasteners
Helicoil
Keys
Nuts
Pins
Pitch gauge
Rivets
Screws
Shank
Snap rings
Society of Automotive Engineers
Solder
Splines
Stud
Taps
Tensile strength
Terminal connectors
Thread-restoring files
Threaded fasteners
Thread chasers
Unified System
Washers
Weather-tight

Summary

- ❑ Threaded fasteners use threads to hold automotive parts together. Common threaded fasteners include bolts, screws, studs, nuts, and washers.
- ❑ Threads are measured and classified according to the U.S. (English) and metric systems. U.S. (English) threads are made to standards of the Unified System.
- ❑ Unified System threads are classified according to the bolt shank diameter, length, and number of threads per inch. Threads in this system can be classified as fine or coarse.
- ❑ Metric system threads are classified according to shank diameter, length, and pitch.
- ❑ Both metric and U.S. (English) fasteners have grade markings to show fastener strength. U.S. (English) fasteners use marks on the bolt head to indicate grades. Metric bolts use property numbers to indicate grades.
- ❑ A pitch gauge can be used to determine the size of a fastener.
- ❑ Fasteners must be torqued to ensure they are not over- or undertightened. Overtightened fasteners loose their strength.
- ❑ Damaged threads may be repaired with a tap, die, or by installing a helicoil.
- ❑ Many automotive parts are held together with nonthreaded fasteners. Common nonthreaded fasteners include keys, snap rings, rivets, splines, and pins.
- ❑ Electrical system repair involves the use of automotive wire and electrical connectors. Wire is sized according to American Wire Gauge sizes. The smaller the cross section of the wire core, the larger the AWG number. Replacement wires must be the same gauge and should be the same color.
- ❑ Electrical terminal connectors fit on the end of the wire. Butt connectors join two wires together. Connectors must be the correct shape and wire gauge size.
- ❑ Weather-tight connectors are used to protect the interior of the connector from moisture.

Review Questions

Short Answer Essays

1. Describe the difference between a bolt and a screw.
2. Explain the difference between a bolt and a stud.
3. Describe how the length of a bolt is measured.
4. Explain how to tell the strength of a U.S. (English) bolt.

Review Questions

Short Answer Essays

1. Explain the measurement markings on a USC micrometer.
2. List and describe the different components on a micrometer.
3. Explain how stepped or go/no-go feeler gauges are machined.
4. Explain how to convert 3 inches to millimeters.
5. Discuss the components of a dial caliper.
6. Explain how pressure and vacuum are used to move air into the engine's cylinders.
7. List and discuss the differences between a USC and a metric dial caliper.
8. Describe the standard for measuring vacuum.
9. Explain how liquid can be used to transfer force.
10. Explain the metric system of measurement.

Fill-in-the-Blanks

1. The calipers on a dial caliper are moved by a(n) __________ __________.
2. Some micrometers have an additional scale called a(n) __________ __________.
3. One rotation of the spindle on a USC micrometer moves the spindle __________ __________.
4. A 3- to 4-inch micrometer can measure __________ inch.
5. A typical dial caliper will measure up to __________ inch(es) in __________-inch segments.
6. A stepped feeler gauge may be known as a(n) __________ gauge.
7. Large bores may be measured with a micrometer and__________ or a(n) __________.
8. A dial indicator may have a(n) __________ or a(n) __________ scale.
9. A micrometer has a(n) __________ to prevent an over torque of the spindle.
10. The vernier scale is used to read __________ of an inch.

158

Shop Manual

To stress the importance of safe work habits, the Shop Manual also dedicates one full chapter to safety. Other important features of this manual include:

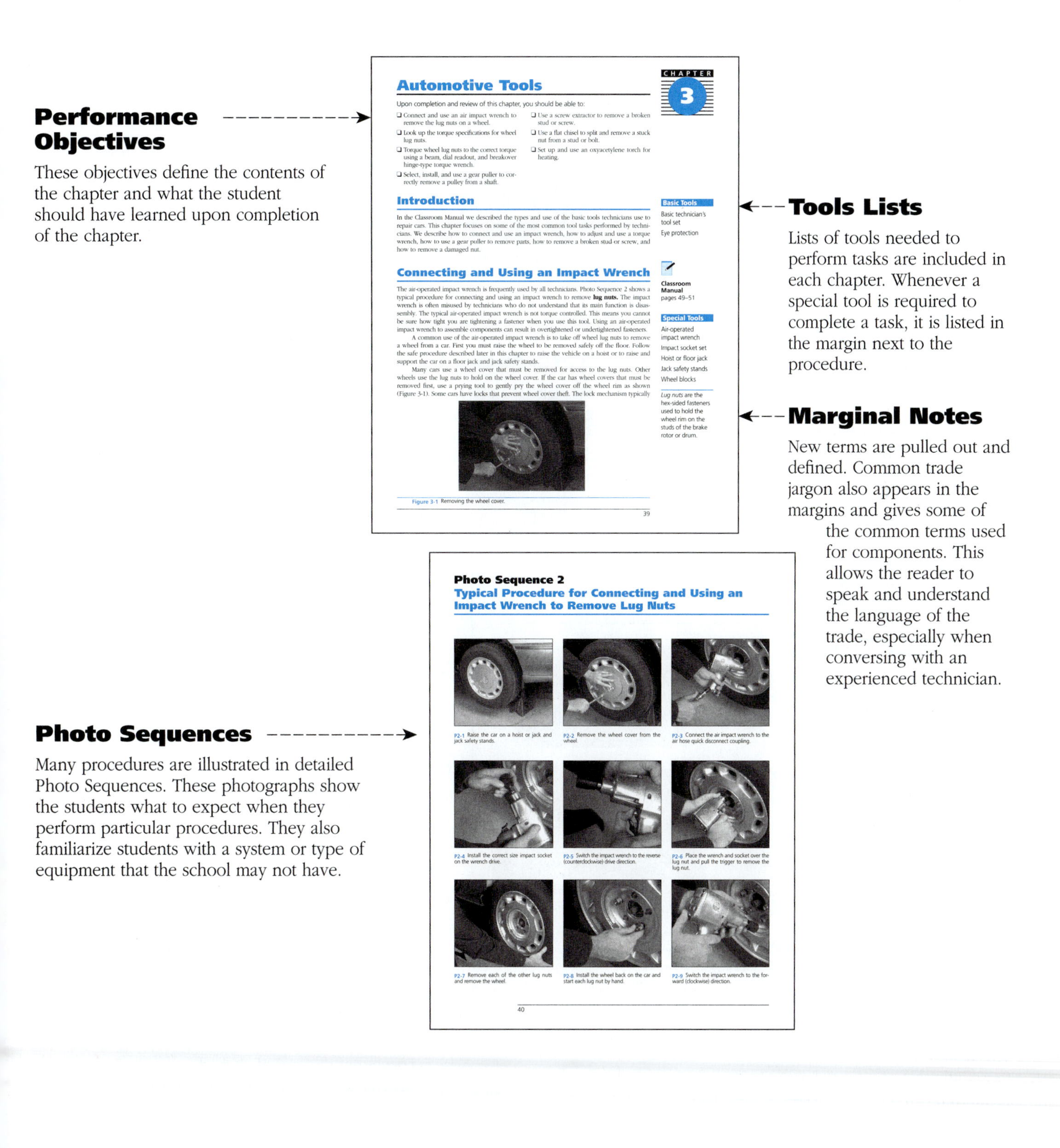

CHAPTER 3

Automotive Tools

Upon completion and review of this chapter, you should be able to:

- ❑ Connect and use an air impact wrench to remove the lug nuts on a wheel.
- ❑ Look up the torque specifications for wheel lug nuts.
- ❑ Torque wheel lug nuts to the correct torque using a beam, dial readout, and breakover hinge-type torque wrench.
- ❑ Select, install, and use a gear puller to correctly remove a pulley from a shaft.
- ❑ Use a screw extractor to remove a broken stud or screw.
- ❑ Use a flat chisel to split and remove a stuck nut from a stud or bolt.
- ❑ Set up and use an oxyacetylene torch for heating.

Introduction

In the Classroom Manual we described the types and use of the basic tools technicians use to repair cars. This chapter focuses on some of the most common tool tasks performed by technicians. We describe how to connect and use an impact wrench, how to adjust and use a torque wrench, how to use a gear puller to remove parts, how to remove a broken stud or screw, and how to remove a damaged nut.

Basic Tools

Basic technician's tool set

Eye protection

Connecting and Using an Impact Wrench

The air-operated impact wrench is frequently used by all technicians. Photo Sequence 2 shows a typical procedure for connecting and using an impact wrench to remove **lug nuts.** The impact wrench is often misused by technicians who do not understand that its main function is disassembly. The typical air-operated impact wrench is not torque controlled. This means you cannot be sure how tight you are tightening a fastener when you use this tool. Using an air-operated impact wrench to assemble components can result in overtightened or undertightened fasteners.

A common use of the air-operated impact wrench is to take off wheel lug nuts to remove a wheel from a car. First you must raise the wheel to be removed safely off the floor. Follow the safe procedure described later in this chapter to raise the vehicle on a hoist or to raise and support the car on a floor jack and jack safety stands.

Many cars use a wheel cover that must be removed for access to the lug nuts. Other wheels use the lug nuts to hold on the wheel cover. If the car has wheel covers that must be removed first, use a prying tool to gently pry the wheel cover off the wheel rim as shown (Figure 3-1). Some cars have locks that prevent wheel cover theft. The lock mechanism typically

Classroom Manual pages 49–51

Special Tools

Air-operated impact wrench

Impact socket set

Hoist or floor jack

Jack safety stands

Wheel blocks

Lug nuts are the hex-sided fasteners used to hold the wheel rim on the studs of the brake rotor or drum.

Figure 3-1 Removing the wheel cover.

39

Photo Sequence 2

Typical Procedure for Connecting and Using an Impact Wrench to Remove Lug Nuts

P2-1 Raise the car on a hoist or jack and jack safety stands.

P2-2 Remove the wheel cover from the wheel.

P2-3 Connect the air impact wrench to the air hose quick disconnect coupling.

P2-4 Install the correct size impact socket on the wrench drive.

P2-5 Switch the impact wrench to the reverse (counterclockwise) drive direction.

P2-6 Place the wrench and socket over the lug nut and pull the trigger to remove the lug nut.

P2-7 Remove each of the other lug nuts and remove the wheel.

P2-8 Install the wheel back on the car and start each lug nut by hand.

P2-9 Switch the impact wrench to the forward (clockwise) direction.

40

Performance Objectives

These objectives define the contents of the chapter and what the student should have learned upon completion of the chapter.

Tools Lists

Lists of tools needed to perform tasks are included in each chapter. Whenever a special tool is required to complete a task, it is listed in the margin next to the procedure.

Marginal Notes

New terms are pulled out and defined. Common trade jargon also appears in the margins and gives some of the common terms used for components. This allows the reader to speak and understand the language of the trade, especially when conversing with an experienced technician.

Photo Sequences

Many procedures are illustrated in detailed Photo Sequences. These photographs show the students what to expect when they perform particular procedures. They also familiarize students with a system or type of equipment that the school may not have.

Service Tips

Whenever a short-cut or special procedure is appropriate, it is described in the text. Generally, these tips are things commonly done by experienced technicians.

Nut steel wheel

Nut aluminum wheel cast or stamped

Chamfer

Figure 3-2 Types and parts of the lug nuts. (Courtesy of DaimlerChrysler Corporation)

fits over one of the lug nut studs. Use the key to remove the lock then pry off the cover. Be careful not to bend or distort the wheel cover or it will not fit back on properly.

The lug nuts hold the wheel onto studs. Passenger cars typically use either four or five lug nuts to hold on the wheel. These studs are attached to the brake rotor or brake drum. There are several designs of lug nuts (Figure 3-2). Lug nuts that hold on steel wheels are often different than those that hold on aluminum wheels. One side of the lug nut are chamfered. The chamfered side always goes on toward the wheel rim. The chamfer fits in a matching chamfer on the wheel and centers the wheel over the studs as it is tightened.

The lug nuts are hex shaped or six sided to fit standard wrench sizes. The wrench sizes are different for different cars. Select a six-point impact socket (Figure 3-3) that fits the hex sides of the lug nut snugly.

SERVICE TIP: You should make match marks on the wheel rim and one of the studs with chalk or grease pencil. Always replace the wheel in the same position in case it has been balanced on the car.

WARNING: Using a loose fitting impact socket on a wheel lug nut can cause the corners of the hex to be rounded off and not fit a wrench properly.

CAUTION: Always use impact sockets with an impact wrench. Standard sockets can fracture and explode when used on an impact wrench. Always wear eye protection when using an impact wrench.

Figure 3-3 A six-point impact socket is used with an impact wrench.

Warnings and Cautions

Throughout the text, cautions are given to alert the reader to potentially hazardous materials or unsafe conditions. Warnings are also given to advise the student of things that can go wrong if instructions are not followed or if a nonacceptable part or tool is used.

WARNING: Store brake fluid in a sealed, clean container. Brake fluid will absorb water and its boiling point will be lowered. Exposure to petroleum products such as grease and oil may cause deterioration of brake components when the contaminated fluid is installed.

WARNING: Do not use petroleum-based (gasoline, kerosene, motor oil, transmission fluid) or mineral-oil-based products to clean brake components. These types of liquids will cause damage to the seals and rubber cups and decrease braking efficiency.

CAUTION: Wear eye protection when dealing with brake fluid because it can cause permanent eye damage. If brake fluid gets in the eyes, flush with cold, running water and see a doctor immediately.

Diagnosing Brake Systems

Classroom Manual pages 315–323

Brake system failures can be classed as either *hydraulic* or *mechanical* failures. Both will create symptoms that can be readily connected to a component or assembly. Mechanical failures will be discussed by common symptoms and their probable causes.

There are two conditions that constitute hydraulic failure: *external leaks* and *internal leaks*. Although mechanical conditions will create the leaks, they will be addressed as hydraulic problems to clarify the operational symptoms. An external leak is defined as "brake fluid exiting the system completely." An internal leak is defined as "a fluid leak within the system that does not exit the system."

External Leaks

A driver may complain that when the brakes were applied, the pedal dropped drastically and it took much more distance to stop the vehicle. The most common cause of this problem is an external leak somewhere in the system. A complete leak where the fluid exits the system results in a complete loss of pressure. The pedal will drop much lower than normal or go to the floorboard immediately upon brake application. If a single-piston master cylinder is being used, the pedal will hit the floorboard and cause a complete loss of all brakes. Pumping the pedal in an effort to regain control only compounds the problem. On a dual-piston master cylinder, the pedal will drop much lower than normal, but the driver will have brakes on one of the split systems and can stop the car. Again, pumping the pedal will not improve braking. An external leak of this type is noticeable when performing a visual inspection. The brake fluid level will also drop in the reservoir.

If the master cylinder is leaking externally, brake fluid will be leaking between the rear of the master cylinder and power booster or firewall (Figure 11-3). Usually, an external master

Figure 11-3 Brake fluid between the power booster and mast... of the master cylinder. (Reprinted with permission)

References to the Classroom Manual

Reference to the appropriate page in the Classroom Manual is given whenever necessary. Although the chapters of the two manuals are synchronized, material covered in other chapters of the Classroom Manual may be fundamental to the topic discussed in the Shop Manual.

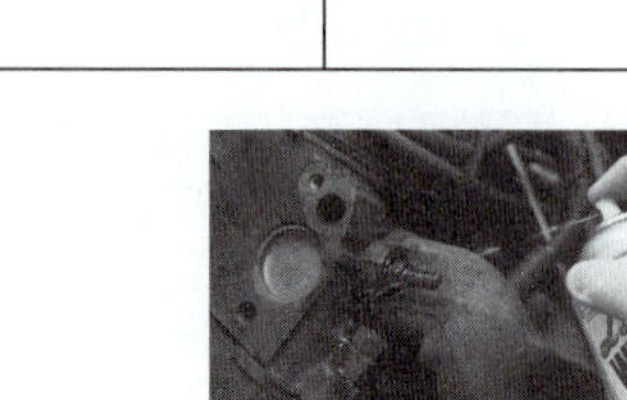

Figure 3-33 Penetrating oil can free a rusted nut.

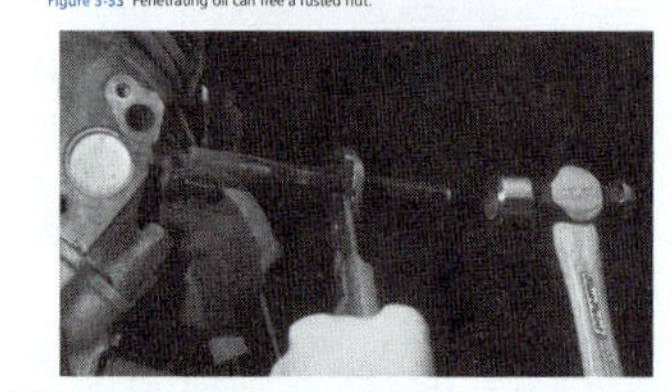

Figure 3-34 Splitting a nut with a flat chisel.

the bolt or stud. Usually one cut will free the nut. Sometimes you will have to cut it on opposite sides to get it off.

A tool called a **nut splitter** is available to split a nut (Figure 3-35). It is positioned over the stuck nut. A forcing screw driven by a wrench causes a cutter to penetrate and split the nut.

Fasteners used in areas of the car subject to heat and corrosion often seize over time. An example is the fasteners used under the car on the exhaust system. Seizing is also a problem when steel fasteners are used in aluminum parts. A good prevention technique is to coat these fasteners with an antiseize compound when they are assembled. This will make your job much easier if you have to remove the parts later. An example of an antiseize compound is shown in Figure 3-36.

A *nut splitter* is a tool used with a wrench to split a stuck nut for removal.

CUSTOMER CARE: When a little extra time is taken to do a professional job, like using a torque wrench or coating fasteners with antiseize compound, the repair job will last longer. A technician's reputation as a craftsperson will pay off in the long run when customers want her or him to work on their cars.

56

Job Sheet 13

13

Name ______________________ Date ____________

Removing, Inspecting, and Replacing Spark Plugs and Wires

Upon completion of this job sheet, you should be able to replace the spark plugs and wires.

Tools Needed

Hand tool set
Service manual

Procedures

1. Locate and record the specifications for the engine.
   Engine ID (type, size, valve)____________
   Spark plug type ____________
   Spark plug torque____________
   Spark plug gap____________
   Spark plug wire type ____________
2. Determine the best method of reaching the spark plugs.
3. Remove the wire from the first plug to be removed.
4. Remove the spark plug and inspect it. Record all findings.
   Cyl #1 ________ Cyl #2 ________
   Cyl #3 ________ Cyl #4 ________
   Cyl #5 ________ Cyl #6 ________
   Cyl #7 ________ Cyl #8 ________
5. Set the air gap on the new plug and install into the cylinder head.
   What type of tool was used to measure the gap?____________
6. Select the new spark plug wire by matching it to the old one.
7. Route the new wire alongside the old wire to the distributor cap or coil pack. Replace the old wire at the cap or coil terminal.
8. Connect the wire to the spark plug.
9. Start the engine to check the installation of this spark plug.
10. Repeat steps 3 through 10 until all plugs and wires are replaced.
11. Recheck the placement of the wires and their connections at the plug and distributor/coil pack.

215

Customer Care

This feature highlights those little things a technician can do or say to enhance customer relations.

Job Sheets

Located at the end of each chapter, the Job Sheets provide a format for students to perform procedures covered in the chapter. A reference to the ASE Task addressed by the procedure is referenced on the Job Sheet.

Case Studies

Case Studies concentrate on the ability to properly diagnose the systems. Each chapter ends with a Case Study in which a vehicle has a problem, and the logic used by a technician to solve the problem is explained.

Terms to Know

Terms in this list can be found in the Glossary at the end of the manual.

ASE Style Review Questions

Each chapter contains ASE Style Review Questions that reflect the performance objectives listed at the beginning of the chapter. These questions can be used to review the chapter as well as to prepare for the ASE certification exam.

Diagnostic Charts

Chapters include detailed diagnostic charts linked with the appropriate ASE task. These charts list common problems and most probable causes. They also list a page reference in the Classroom Manual for better understanding of the system's operation and a page reference in the Shop Manual for details on the procedure necessary for correcting the problem.

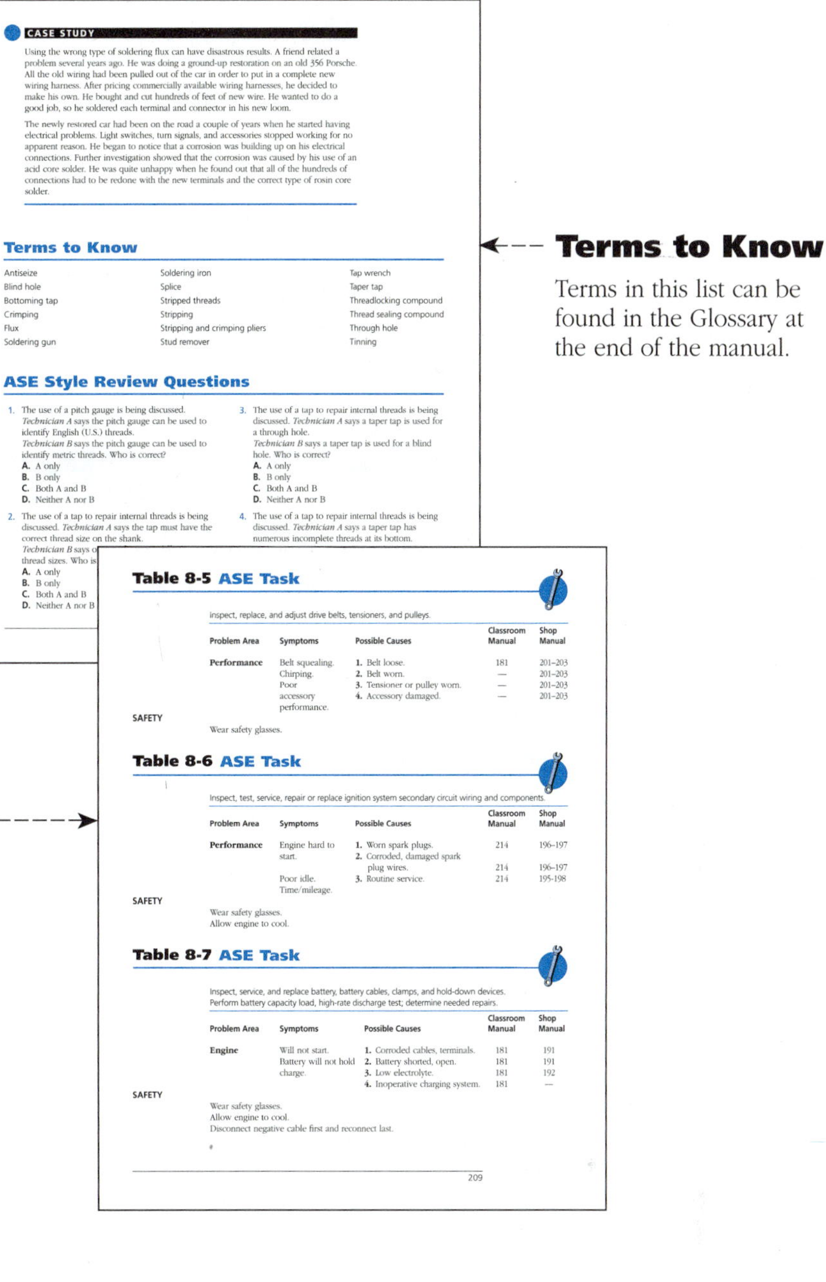

CASE STUDY

Using the wrong type of soldering flux can have disastrous results. A friend related a problem several years ago. He was doing a ground-up restoration on an old 356 Porsche. All the old wiring had been pulled out of the car in order to put in a complete new wiring harness. After pricing commercially available wiring harnesses, he decided to make his own. He bought and cut hundreds of feet of new wire. He wanted to do a good job, so he soldered each terminal and connector in his new loom.

The newly restored car had been on the road a couple of years when he started having electrical problems. Light switches, turn signals, and accessories stopped working for no apparent reason. He began to notice that a corrosion was building up on his electrical connections. Further investigation showed that the corrosion was caused by his use of an acid core solder. He was quite unhappy when he found out that all of the hundreds of connections had to be redone with the new terminals and the correct type of rosin core solder.

Terms to Know

Antiseize
Blind hole
Bottoming tap
Crimping
Flux
Soldering gun
Soldering iron
Splice
Stripped threads
Stripping
Stripping and crimping pliers
Stud remover
Tap wrench
Taper tap
Threadlocking compound
Thread sealing compound
Through hole
Tinning

ASE Style Review Questions

1. The use of a pitch gauge is being discussed. *Technician A* says the pitch gauge can be used to identify English (U.S.) threads. *Technician B* says the pitch gauge can be used to identify metric threads. Who is correct?
   A. A only
   B. B only
   C. Both A and B
   D. Neither A nor B
2. The use of a tap to repair internal threads is being discussed. *Technician A* says the tap must have the correct thread size on the shank. *Technician B* says o… thread sizes. Who is…
   A. A only
   B. B only
   C. Both A and B
   D. Neither A nor B
3. The use of a tap to repair internal threads is being discussed. *Technician A* says a taper tap is used for a through hole. *Technician B* says a taper tap is used for a blind hole. Who is correct?
   A. A only
   B. B only
   C. Both A and B
   D. Neither A nor B
4. The use of a tap to repair internal threads is being discussed. *Technician A* says a taper tap has numerous incomplete threads at its bottom.

Table 8-5 ASE Task

Inspect, replace, and adjust drive belts, tensioners, and pulleys.

Problem Area	Symptoms	Possible Causes	Classroom Manual	Shop Manual
Performance	Belt squealing.	1. Belt loose.	181	201–203
	Chirping.	2. Belt worn.	—	201–203
	Poor accessory performance.	3. Tensioner or pulley worn.	—	201–203
		4. Accessory damaged.	—	201–203

SAFETY

Wear safety glasses.

Table 8-6 ASE Task

Inspect, test, service, repair or replace ignition system secondary circuit wiring and components.

Problem Area	Symptoms	Possible Causes	Classroom Manual	Shop Manual
Performance	Engine hard to start.	1. Worn spark plugs.	214	196–197
		2. Corroded, damaged spark plug wires.	214	196–197
	Poor idle. Time/mileage.	3. Routine service.	214	195-198

SAFETY

Wear safety glasses.
Allow engine to cool.

Table 8-7 ASE Task

Inspect, service, and replace battery, battery cables, clamps, and hold-down devices. Perform battery capacity load, high-rate discharge test; determine needed repairs.

Problem Area	Symptoms	Possible Causes	Classroom Manual	Shop Manual
Engine	Will not start.	1. Corroded cables, terminals.	181	191
	Battery will not hold charge.	2. Battery shorted, open.	181	191
		3. Low electrolyte.	181	192
		4. Inoperative charging system.	181	—

SAFETY

Wear safety glasses.
Allow engine to cool.
Disconnect negative cable first and reconnect last.

209

Reviewers

Special thanks to the following instructors for reviewing this material:

Charles Capsel, Antelope Valley College
Rick Curlee, Hill College Technical
Donald R. Deal, Red River AVTS
Dan Encinas, East Los Angeles College
Thomas J. Fitch, Monroe Community College
Earl J. Friedell, Jr., DeKalb Technical Institute
Jon Gerdy, New Castle School of Trades
James Helmle, Milwaukee Area Technical College
Bob Klauer, Metro Tech
Norris Martin, Texas State Technical College – Waco
Don Moseley, Monterey Peninsula College
Alan Penuela, Ventura College
Michael Stiles, Indian River Community College

Contributing Companies

I would also like to thank these companies who provided technical information and art for this edition:

A & E Manufacturing Company
Actron Manufacturing Company
American Honda Motor Co., Inc
Brake Parts, Inc.
Breton Publishers
C. Thomas Olivo Associates
Central Tools, Inc.
Cooper Automotive/Champion Spark Plug Company
Chicago Rawhide
CRC Industries, Inc.
DaimlerChrysler Corporation
Danaher Tools
Davis
Deere & Company
Delco-Remy Co.
DuPont Automotive Finishes
EIS Brake Parts
Federal-Mogul Corporation
Fel-Pro, Inc.
Ford Motor Company
General Motors Corporation, Service Operations
Gray Automotive
Griffin Technical Institute
Goodson Shop Supplies
Hunter Engineering Company
Klein Tools
The L.S. Starrett Co.
Lisle Tools Corp.
MATCO Tool Company
Mazda Motor of America, Inc.
Mitsubishi Motor Sales of America, Inc.
Moog Automotive, Inc.
Mitchell International
© National Institute for Automotive Service Excellence (ASE)
National Safety Council
Nissan North America
National Lubricating Grease Institute
Owatonna Tool Company
POP Fasteners
Proto Tools
Sealed Power Corp.
Sears Industrial Tools
Siebe North, Inc.
Society of Automotive Engineers, Inc.
Snap-on Tools Company
Spalding Lincoln-Mercury-Mazda-Dodge
©1998 Stanley Tools, a Product Group of The Stanley Works, New Britain, CT
The Timken Co.
U.S. Navy
Volvo Car Corporation
Western Emergency Equipment
York Technical College

Safety

Upon completion and review of this chapter, you should be able to:

- ❑ Understand the obstacles to avoid when moving a vehicle in and out of a shop and how to arrange a bay for best use.
- ❑ Perform a shop safety inspection.
- ❑ Practice safe procedures during the cleaning of parts.
- ❑ Read and interpret an MSDS and hazardous material inventory roster.
- ❑ Store hazardous material and waste.
- ❑ Inspect and use a Class B fire extinguisher.

Introduction

Personal safety habits reflect on the technician's professionalism and work. Some people think that small cuts, bruises, or other injuries are part of the automotive business. While that may be true to a very small extent, the problem is that even the most minor injury is usually caused by a lack of attention to detail. A cut may be small, but the vehicle's owner does not want blood on the upholstery. Also, the technician loses time and money locating and applying a bandage. Paying attention to the small items will help prevent bigger accidents from happening. This chapter will discuss some of the small details and some of the larger issues.

Shop Arrangement

Classroom Manual
page 5

The owner, architect, and builder design the layout of an automotive shop. A properly designed **shop** has office space for management, parts, a customer waiting area, and the work bays (Figure 1-1). Even though a technician does not have much input during the design and construction phase, he or she is directly affected by the placement and size of the work bay.

National or regional chain *shops* such as Precision Tune are built to the same plan. Dealerships are built based on a single design that can be adjusted according to the size of the dealership and expected business.

The technician is responsible for moving the vehicles in and out of the shop and bay. On the first day at the shop, every technician should take a good look at the traffic pattern in and out of the work bay. Most of the time, it is straightforward without obstruction. Other times, skill and attention are required to move a vehicle. Study the area around the bay. A pillar or other permanent structure may block easy entrance into the bay (Figure 1-2). It may be better to enter the shop from a different door and enter the bay from a different angle. Just remember that the structure will still be there when the vehicle is moved from the bay.

Trash cans, parts washers, and other moveable objects should be moved if possible. Sometimes the size of the shop dictates the positioning of machines and toolboxes.

The bay is not only a place where the technician performs work, it is also where personal tools are stored. Younger technicians may be able to fit their tools in a small box. However, the older technicians will probably have top and boxes that may measure six feet high by six feet long (Figure 1-3). These take a great deal of space. Technicians must arrange the bay to accommodate their tools for storage and ease of use. The bay must be set up to accommodate work areas around the vehicle without blocking access to the tools or blocking an easy exit in an emergency.

Many larger shops are installing **computer terminals** in some or all work bays (Figure 1-4). Terminals can be awkward to move and easily abused. Once a terminal is installed in the bay, everything else, including storage and work, is performed around the terminal's position. Computer keyboards and processing units are not good places to set drinks, loose tools, or fasteners. The terminal is another tool for the technician and must be treated as such.

A *computer terminal* is a computer link to a main computer or server. It allows several persons to access or enter data on the server at the same time.

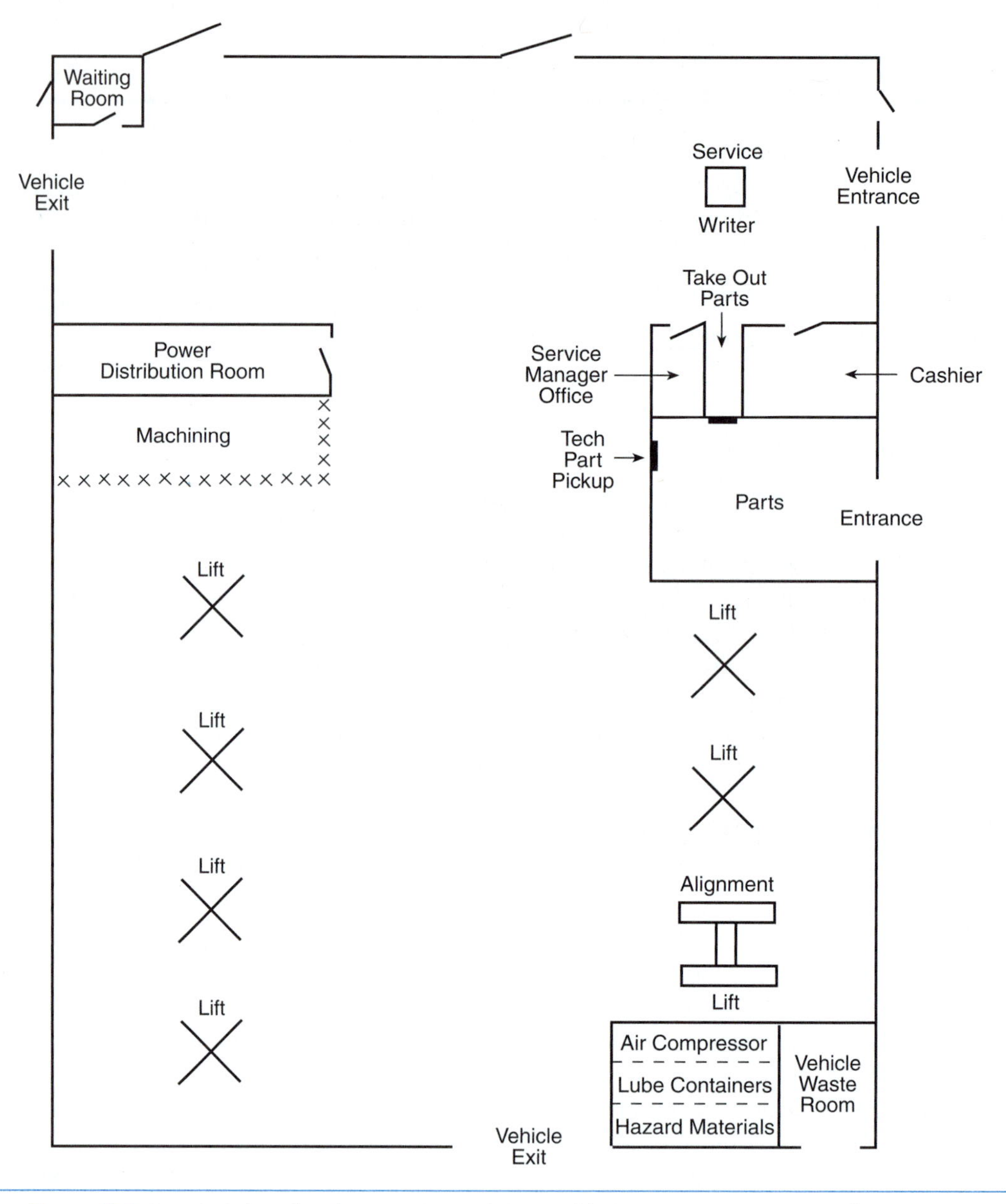

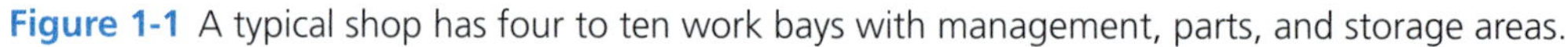

Figure 1-1 A typical shop has four to ten work bays with management, parts, and storage areas.

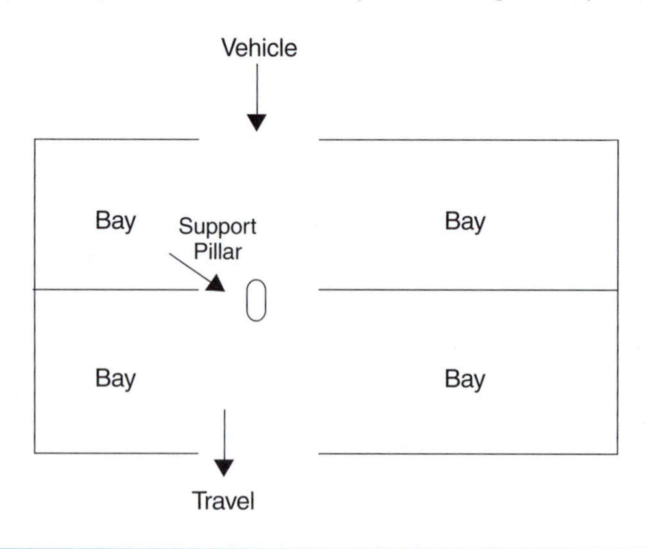

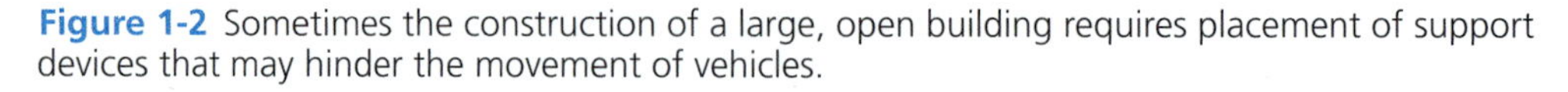

Figure 1-2 Sometimes the construction of a large, open building requires placement of support devices that may hinder the movement of vehicles.

Figure 1-3 Toolboxes can be large, pretty, and take up a great deal of space in the work bay. (Courtesy of MATCO Tool Company)

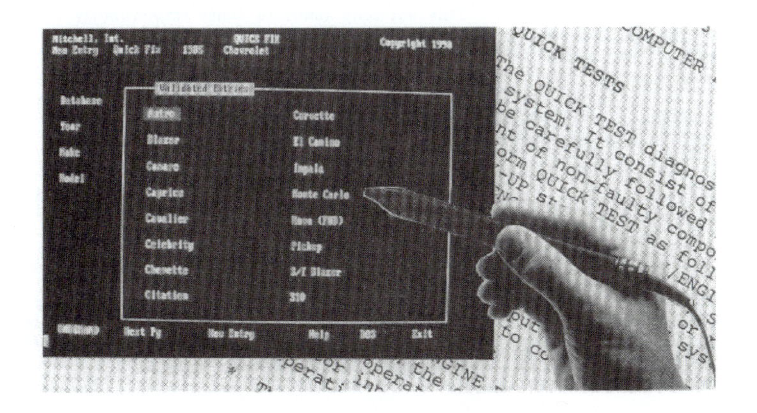

Figure 1-4 Terminals in the work bays allow the technician to retrieve and enter data quickly. (Courtesy of Mitchell International)

Shop Inspection

Classroom Manual
page 5

A newly-hired technician should walk around the shop to locate the first-aid station, the eye wash station, and the machines and equipment needed to repair a vehicle. Location of hazardous material and waste storage areas and special tools should also be located. In most cases, someone from shop management or another employee will escort a new technician through the shop to point out features that will be needed or avoided. On the tour, make a note of possible hazards that may exist.

Most businesses have some type of **safety inspection** that is conducted on a periodic basis. The inspection includes storage of hazardous materials and waste, fire extinguishers, storage of tools, and equipment condition. A typical inspection sheet is shown in Figure 1-5. Many items on the checklist may be unfamiliar to the reader. Special tools and equipment are covered in a later chapter.

A *safety inspection* is a comprehensive inspection of the shop following a very detailed checklist.

Although the inspection sheet shown is specifically designed for a school, it provides a good list of items that should be checked. Note that it includes items such as fire extinguishers, location of exits, and **signage** (Figure 1-6). Machining equipment should be surrounded by some type of protection device to prevent anyone from accidentally coming in contact with an operating machine. Many shops do not have this protection, but instead place this type of equipment out of the normal working area. It is up to the technicians to observe all safety precautions when working in the machining area and keep it isolated from any customers. The inspection sheet, such as the example shown in Figure 1-5, can be easily modified to fit a shop as needed.

Signage is the term used to designate all signs at a location.

NATIONAL STANDARD SCHOOL SHOP SAFETY INSPECTION CHECK LIST

Prepared by the Joint Safety Committee of the
AMERICAN VOCATIONAL ASSOCIATION — NATIONAL SAFETY COUNCIL

____ Date____________________

____

INTRODUCTION

A safe environment is an essential part of the school shop safety education program. The safe environment will exist only if hazards are discovered and corrected through *regular* and *frequent* inspections by school personnel—administrators, teachers and students. Safety inspections are to determine if everything is satisfactory.

Inspections may be made at the request of the board of education, the school administration or upon the initiative of the teacher. Some communities have drawn upon the cooperative service of professional safety engineers, inspectors of state labor departments, insurance companies and local safety councils to supplement and confirm inspections by school personnel.

The National Standard School Shop Safety Inspection Check List, recommended by the President's Conference on Industrial Safety is an objective inspection procedure for the school shop.

DIRECTIONS

WHO INSPECTS?

This will depend upon local policies. It is recommended, however, that shop teachers, and students—the student safety engineer and/or student safety committee—participate in making regular inspections. This not only tends to share responsibility but stimulates a broader interest in the maintenance of a safe school shop.

WHEN TO INSPECT?

As a minimum, a safety inspection should be made at the beginning of every school term or semester. More frequent inspections may be advisable.

HOW TO INSPECT?

Inspections should be well planned in advance.

Inspections should be systematic and thorough. No location that may contain a hazard should be overlooked.

Inspection reports should be clear and concise, but with sufficient explanation to make each recommendation for improvement understandable.

FOLLOW-UP

The current report should be compared with previous records to determine progress. The report should be studied in terms of the accident situation so that special attention can be given to those conditions and locations which are accident producers.

Each unsafe condition should be corrected as soon as possible in accordance with accepted local procedures.

A definite policy should be established in regard to taking materials and equipment out of service because of unsafe conditions.

The inspection report can be used to advantage as the subject for staff and class discussion.

CHECKING PROCEDURE

Draw a circle around the appropriate letter, using the following letter scheme:

S — Satisfactory (needs no attention)
A — Acceptable (needs some attention)
U — Unsatisfactory (needs immediate attention)

Recommendations should be made in all cases where a "U" is circled. Space is provided at the end of the form for such comments. Designate the items covered by the recommendations, using the code number applicable (as B-2).

In most categories, space is provided for listing of standards, requirements or regulations which have local application only.

A. GENERAL PHYSICAL CONDITION

1. Machines, benches, and other equipment are arranged so as to conform to good safety practices........................ S A U
2. Condition of stairways............................ S A U
3. Condition of aisles .... S A U
4. Condition of floors .... S A U
5. Condition of walls, windows, and ceiling............ S A U
6. Illumination is safe, sufficient, and well placed....... S A U
7. Ventilation is adequate and proper for conditions..... S A U
8. Temperature control .... S A U
9. Fire extinguishers are of proper type, adequately supplied, properly located and maintained........................... S A U
10. Teacher and pupils know location of and how to use proper type for various fires.................................... S A U
11. Number and location of exits is adequate and properly identified .... S A U
12. Proper procedures have been formulated for emptying the room of pupils and taking adequate precautions in case of emergencies S A U
13. Lockers are inspected regularly for cleanliness and fire hazards. S A U
14. Locker doors are kept closed....................... S A U
15. Walls are clear of objects that might fall............ S A U
16. Utility lines are properly identified................. S A U
17. Teachers know the procedure in the event of fire including notification of the fire department and the evacuation of the building. S A U
18. Air in shop is free from excessive dust, smoke, etc.... S A U
19. ____________________________ S A U
20. ____________________________ S A U
21. ____________________________ S A U
22. ____________________________ S A U
23. Evaluation for the total rating of A. GENERAL PHYSICAL CONDITION .... S A U

Figure 1-5 The safety inspection checklist provides a means to perform a systematic inspection of the shop. (Courtesy of National Safety Council)

Figure 1-6 All exits must be marked in letters large enough to be read from across the room or shop.

As you inspect, look for items that may not be specifically listed on the inspection form. Empty or unopened cans of cleaner in the trash; oily rags not properly stored; or a lack of gloves at the cleaning vat may not be listed as such on the form, but are definitely safety hazards. A good inspection involves common sense.

Classroom Manual
pages 2, 3, 6, and 7

Parts Cleaning

WARNING: Always wear eye protection and chemical gloves when using caustic cleaning solvents. If possible, wear long-sleeved shirts during the cleanup. Caustic cleaning solvents can cause damage to the eyes and exposed skin.

A *caustic* chemical is one that can burn or corrode. Battery electrolyte (acid) is a caustic chemical.

A *work apron* is usually made of heavy material that is resistance to caustic chemicals.

Part washers may be small and fairly portable or large enough to automatically wash an entire engine or transmission at once. The large washers use steam, heat, and pressure to clean the component.

During vehicle repairs, there will be a time when parts must be cleaned. The technician needs to wear protective gloves and face protection while cleaning parts (Figure 1-7). The cleaning material is usually **caustic** and will be harmful to flesh, particularly the face and eyes. It may also be advisable to wear a **work apron** if larger parts are being cleaned.

A typical **parts washer** is a tub mounted over a drum of chemical cleaner (Figure 1-8). The cleaner is pumped through a hose, normally to a hollow brush. The part is brushed while the cleaner is flowing. Even though the system is low pressure and low volume, the cleaner can splash.

After donning the gloves and face protection, place the part low in the tub. Switch on the unit and use the brush to clean the part. Keep the brush in contact with the part to help prevent splashing. As the part is cleaned, it will probably be necessary to rinse it at times to survey the work completed. Hold the part low in the tub and at arm length. This will keep the splashing chemicals contained in the tub and away from the eyes and body.

Sometimes a part may be cleaned directly on the vehicle using a spray-on substance like carburetor cleaner. The chemicals in some spray-on cleaners violate EPA standards for released hazardous materials. Read the container label for information on use of all cleaners and containment of their waste (Figure 1-9). Spray-on cleaners splash much more violently, and face/eye protection must be worn.

Figure 1-7 Virile gloves should always be worn when cleaning parts. Protective clothing may also be needed.

Figure 1-8 Part washers are usually self-contained units. The waste and used chemicals are treated as hazardous waste. (Courtesy of Snap-on Tools Company)

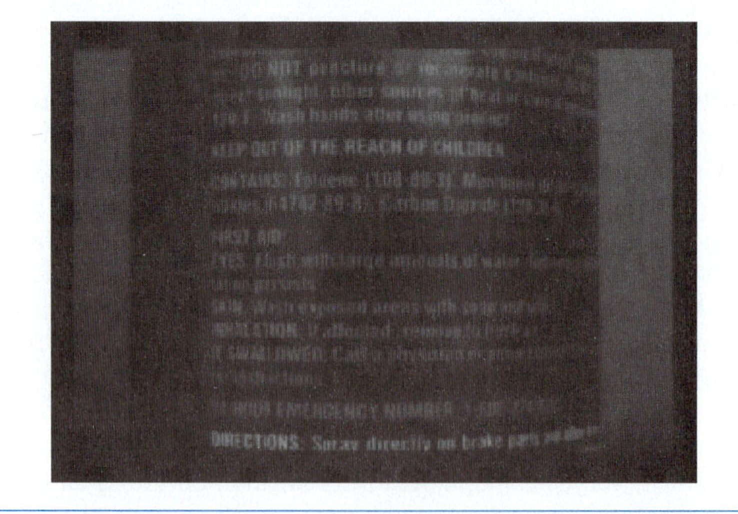

Figure 1-9 Container labels list the chemical properties of the contents and first aid treatment.

Face, eye, and hand protection should *always* be worn when cleaning automotive parts. Every cleaning operation produces some measure of health risk either from the cleaner, the material being removed, or both. Remember that most automotive cleaners are considered to be hazardous materials and once used must be treated as hazardous waste.

Hazardous Material and Waste

Classroom Manual pages 6–10

Right-to-know laws place a great deal of responsibility on employers to ensure that employees know about all hazardous materials they may be using. It is the employee's responsibility to utilize that training.

Material Safety Data Sheets

The *Material Safety Data Sheet (MSDS)* is most important to you as an employee. In Figure 1-10 the chemical listed is a common brake cleaner (Sections 1 and 2). Search the sheet until you find the safety hazards that may cause harm to humans and the environment (Sections 3 and 4). Using that information, you should be able to decide what personal protection measures must be taken to protect yourself. Go to Section 5 on Figure 1-10 and compare your measures with the items listed there. In case of accident, analyze the first-aid treatment that may be needed. Survey your shop and locate the items you may need to prevent injury and the items you may need for first aid for this chemical. The minimum would be an eyewash station and a place to wash any flesh that may be heavily exposed to this chemical.

Hazardous Material Inventory Roster

Since hazardous materials are used almost daily, they need to be stored close to the work area, usually inside the shop. Use the roster to inspect the shop for possible storage areas. In some instances, the container must be labeled in accordance with the most hazardous material stored there. All types of containers should be closed and labeled. Look for small containers that may be refilled from a larger storage unit. The label on the container should match the one on the larger unit. Check throughout the shop and its immediate exterior for large storage containers or flammable liquid storage areas. There may be more than one hazardous materials storage area depending on the shop size and amount of material on hand.

The storage device should be a closed, metal cabinet of some type that is marked with fire warnings (Figure 1-11). Some materials, such as fuel, should be *stored outside the building* in approved metal containers. Inspect the work area to see if any hazardous materials have been left on the workbenches or floor. Remember, the storage and safe use of hazardous materials is one of the safety requirements of the shop and employees. Anyone using a hazardous material must use it safely and store it properly. Any material remaining in the container after use must be returned to the storage area or disposed of properly.

Hazardous *waste* is a different matter from hazardous *materials*. Waste is something to be discarded; it is not reuseable and must be stored in marked and completely sealed containers. Look at the shop again for hazardous waste. In most cases, oil and antifreeze are the most common hazardous waste in an automotive shop. The EPA does not consider most used materials as hazardous waste until they are picked up for recycling or disposal.

A closed, empty *container* in the trashcan or dumpster can build internal pressure on a warm day. When the EPA inspector presses the spray button, the pressure is released, hence the container is hazardous waste.

Extract of MATERIAL SAFETY DATA SHEET

SECTION 1

BRAKE FLUID - DOT 3 BRAKE FLUID

Item Name: DOT 3 BRAKE FLUID

SECTION 2

Ingredients/identity Information

Proprietary: NO

Ingredient: VISCOSITY ENHANCER

Ingredient: LUBRICITY ADDITIVE

Ingredient: MODIFIER/COUPLERS

Ingredient: CORROSION INHIBITOR

SECTION 3

Special Fire Fighting Proc: WATER MAY BE INEFFECTIVE BUT CAN BE USED TO COOL CONTAINERS. WATER/FOAM MAY CAUSE FROTHING, ESPECIALLY IF SPRAYED INTO CONTAINERS OF HOT, BURNING LIQUID.

Reactivity Data

Stability: YES

Cond To Avoid (Stability): HEAT, SPARKS, OPEN FLAMES, OR OTHER SOURCES OF IGNITION

Materials To Avoid: STRONG OXIDIZING AGENTS

Hazardous Decomp Products: DENSE SMOKE, CO, C02, & OTHER OXIDES

Hazardous Poly Occur: No

SECTION 4

Health Hazard Data

Route of Entry - inhalation; YES

Route of Entry - Skin: YES

Route of Entry - Ingestion: YES

IRRITATION, INHALATION: RESPIRATORY TRACT IRRITATION/INTOXICATION/ FATAL/HARMFUL/KIDNEY, LIVER, GASTROINTESTINAL, TESTICULAR, & CENTRAL NERVOUS SYSTEM EFFECTS/BLADDER STONES/NAUSEA/VOMITING.

Signs/Symptoms Of Overexp: SKIN: IRRITATION/ABSORPTION. EYE: IRRITATION. INHALATION: RESPIRATORY TRACT IRRITATION/INTOXICATION/ FATAL/HARMFUL/KIDNEY, LIVER, GASTROINTESTINAL, TESTICULAR, & CENTRAL NERVOUS SYSTEM EFFECTS/BLADDER STONES/NAUSEA/VOMITING.

Emergency/First Aid Proc: EYES: IMMEDIATELY FLUSH W/LARGE AMOUNTS OF WATER & CONTINUE FLUSHING UNTIL IRRITATION SUBSIDES. SKIN: WASH THOROUGHLY W/SOAP & WATER. INHALATION: REMOVE TO FRESH AIR. INGESTION: GIVE LARGE QUANTITIES OF WATER IMMEDIATELY. INDUCE VOMITING. OBTAIN MEDICAL ATTENTION IN ALL CASES.

Precautions for Safe Handling and Use: Steps If Matl Released/Spill: CONTAIN IMMEDIATELY. DON'T ALLOW TO ENTER SEWER/WATERCOURSES. REMOVE ALL IGNITION SOURCES. ABSORB W/APPROPRIATE INERT MATERIAL (SAND/CLAY) PICK UP LARGE SPILLS USING VACUUM PUMPS/SHOVELS/ BUCKETS. PLACE IN DRUMS/SUITABLE CONTAINERS.

Waste Disposal Method: IF SPILLED/DISCARDED, MAY BE A REGULATED WASTE. WASTE MATERIAL MAY BE LANDFILLED OR INCINERATED AT AN APPROVED FACILITY IN ACCORDANCE W/LOCAL, STATE, & FEDERAL REGULATIONS. MATERIALS SHOULD BE RECYCLED IF POSSIBLE. FLAMMABLE & COMBUSTIBLE LIQUID. Precautions-Handling/Storing: DON'T TRANSFER TO UNMARKED CONTAINERS. STORE IN CLOSED CONTAINERS AWAY HEAT, SPARKS, OPEN FLAME, OR OXIDIZING MATERIALS. Other Precautions: AVOID CONTACT W/SKIN & EYES. AVOID INGESTION & INHALATION OF MISTS OR VAPORS. AVOID CONSUMPTION OF FOOD & BEVERAGE IN WORK AREAS.

Section 5

Control Measures

Respiratory Protection: IF VAPOR/MIST IS GENERATED, USE AN ORGANIC VAPOR RESPIRATOR W/DUST & MIST FILTER. ALL RESPIRATORS MUST BE NIOSH-APPROVED. DON'T USE COMPRESSED OXYGEN IN HYDROCARBON ATMOSPHERES.

Ventilation: GENERAL ROOM VENTILATION IS NORMALLY SUFFICIENT TO PREVENT CONCENTRATION BUILDUP. Protective Gloves: IMPERVIOUS SYNTHETIC RUBBER

Eye Protection: SPLASH-PROOF SAFETY GOGGLES

Other Protective Equipment: PLASTIC FACE SHIELD, IMPERVIOUS SYNTHETIC CLOTHING, BOOTS, & APRONS,

Hygienic Practices: REMOVE/LAUNDER CONTAMINATED CLOTHING BEFORE

REUSE. DISCARD LEATHER ARTICLES. AVOID EATING/DRINKING/SMOKING. WASH HANDS.

Figure 1-10 Each section of the MSDS lists specific information about the properties of the material.

Figure 1-11 Hazardous material storage areas should be marked as flammable and then be enclosed. (Courtesy of Snap-on Tools Company)

However, used material such as oil must be stored under certain conditions (Figure 1-12). Each material must have its own container. The container must be labeled with the material's name and be marked as USED.

Another type of hazardous waste is the can used to store propellant-driven material such as a brake cleaner. Once empty with all of its pressure released, the can must be split open before its disposal. The EPA considers a **container** of this type to be hazardous waste if it is not split open. Check the shop to see if there are empty cans that are not split. Be sure to check the trashcans.

Inspection and Use of a Fire Extinguisher

Classroom Manual
pages 11–14

The danger of fire in an automotive shop is present at most times. Fire can be caused by oily rags improperly stored, hazardous materials exposed to heat, fuel leaks, or poor storage of hazardous material or waste. In most cases, the type of fire will not be electrical, paper, or

Figure 1-12 Used material containers must be closed and labeled.

metallic such as magnesium. The fire will probably be a Class B type, which involves flammables such as fuel, oil, and solvents. However, you must understand the type of fire you are combating and whether or not the fire extinguishers in your shop are operable.

The fire extinguisher normally found in an automotive shop is either a Class B or a multipurpose extinguisher that can be used on different types of fires. Look for a "B" enclosed in a square on the fire extinguisher (Figure 1-13). It may have more than one symbol indicating that it is a multipurpose extinguisher. Obviously, inspection of a fire extinguisher cannot be done as you try to stop the fire. Fire extinguishers must be inspected periodically, preferably by the same person. Take time to locate and inspect the fire extinguishers in the shop. There are usually several fire extinguisher points throughout the shop. Look at the **tag** and gauge of each to make sure the extinguisher is charged and ready to work. The safety clip or wire must be in place and there must be no damage to the fire extinguisher. Initial and date the tag if the extinguisher is ready for use. If there are different types of extinguishers available, learn their locations and the proper use of each type during the inspection.

The extinguisher's *tag* is usually a small card tied directly to the extinguisher. The inspector may enter his or her initials, date, and status of the extinguisher.

We will discuss the use of a Class B fire extinguisher since that is the most common type of fire in an automotive shop. Class B fires must have their oxygen supply cut off before they go out. Water or other spreading material is not applicable on Class B fires because it spreads the fuel. When a fire is discovered, sound the alarm and locate a Class B fire extinguisher. Position yourself about eight feet from the base of the fire, **upwind** if possible. Remove the safely clip, aim the nozzle at the base of the fire, and pull the trigger. Continue using the extinguisher until the fire is out or the extinguisher is empty. If necessary, spray the fire area to cool any hot spots that may re-ignite the fire. If you or any of your co-workers feel the fire is too large for your shop's fire extinguishers, leave the area and notify the local fire department.

Upwind means the wind is blowing on the person's back. If the wind is in the fire fighter's face (downwind), the flame and the fire fighting material can be blown toward the fighter.

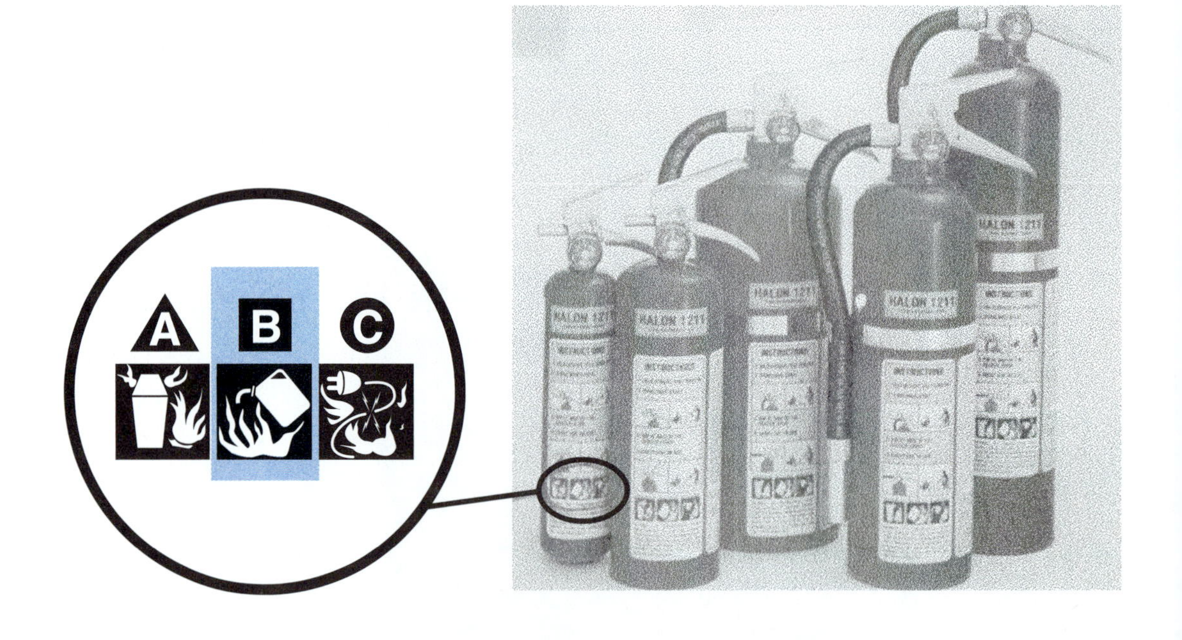

Figure 1-13 The fire extinguisher highlighted is a multipurpose fire extinguisher for use against Class A, B, and C fires.

Photo Sequence 1
Using A Dry Chemical Fire Extinguisher

P1-1 Multipurpose dry chemical fire extinguisher.

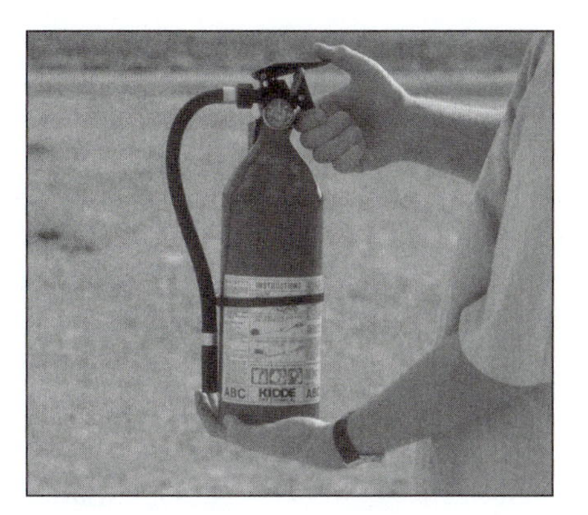

P1-2 Hold the fire extinguisher in an upright position.

P1-3 Pull the safety pin from the handle.

P1-4 Stand 8 feet from the fire. Do not go any closer to the fire.

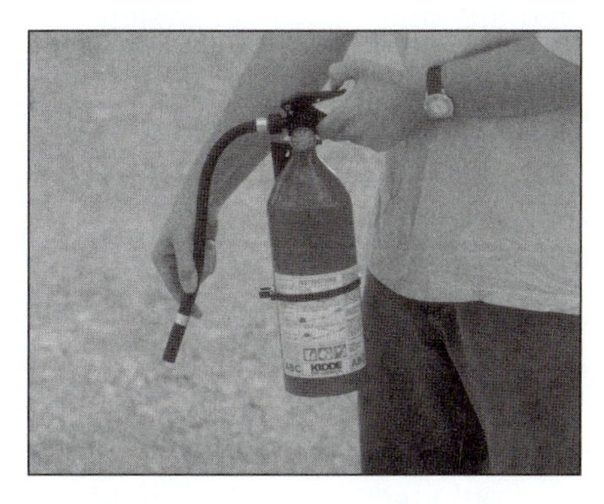

P1-5 Free the hose from its retainer and aim it at the *base* of the fire.

P1-6 Squeeze the lever while sweeping the hose from side to side. Keep the hose aimed at the *base* of the fire.

CASE STUDY

A combination of chemicals and misinformation led to a possible dangerous situation in a tech school a few years ago. A young student was cleaning the posts and cable terminals on a vehicle. After using a battery post and terminal cleaner, he sprayed WD40, a brand name for a liquid lubricant, over the top of the posts and terminals. For some reason he reconnected the cables improperly, negative first followed by the positive. When the positive cable neared its post, a spark arced across the gap setting the WD40 on fire. His shirt sleeve also caught fire. Since the policy in this school was to have a fire extinguisher at the vehicle during battery or fuel system repair, the fire was quickly extinguished and no injury or damage occurred. During the investigation following the accident, the student stated that he had always been told to put oil or grease on the post and terminals to reduce corrosion. Incorrect information, even if given with the best intentions, can lead to accidents resulting in serious injuries and damage.

Terms to Know

Caustic

Safety inspection

Work apron

Signage

ASE Style Review Questions

1. Fire fighting is being discussed. *Technician A* says to use a fire extinguisher that has a circle and a "B" on it to fight a cleaner solvent fire.
   Technician B says no spreading agent is to be used for fighting a gasoline fire. Who is correct?
   A. A only
   B. B only
   C. Both A and B
   D. Neither A nor B

2. Hazardous material and waste are being discussed. *Technician A* says both should be stored the same way.
   Technician B says waste is not considered to be waste until it has been picked up for disposal or thrown into a dumpster. Who is correct?
   A. A only
   B. B only
   C. Both A and B
   D. Neither A nor B

3. The setup of the work bay is being discussed.
   Technician A says a computer terminal should be placed so it will not interfere with the operations within the bay.
   Technician B says computer terminals should be placed where they are not likely to be hit or damaged by routine work. Who is correct?
   A. A only
   B. B only
   C. Both A and B
   D. Neither A nor B

4. Fire extinguisher use is being discussed.
   Technician A says to stand eight feet from the base of a fire and downwind if possible.
   Technician B says to squeeze or pull the trigger and then pull out the safety clip. Who is correct?
   A. A only
   B. B only
   C. Both A and B
   D. Neither A nor B

5. Safety inspections are being discussed. *Technician A* says a good part of an inspection is common sense.
   Technician B says the machining equipment should be isolated. Who is correct?
   A. A only
   B. B only
   C. Both A and B
   D. Neither A nor B

6. Part cleaning is being discussed. *Technician A* says a parts washer usually has its own tub and pump system.
   Technician B says most cleaning of automotive parts results in hazardous waste. Who is correct?
   A. A only
   B. B only
   C. Both A and B
   D. Neither A nor B

7. *Technician A* says the container label of a cleaning material should be read before using the cleaner.
   Technician B says an empty cleaning container should be split before it is thrown away. Who is correct?
   A. A only
   B. B only
   C. Both A and B
   D. Neither A nor B

8. Class B fires are being discussed. *Technician A* says a spreading agent fire extinguisher can be used to separate the burning materials.
   Technician B says any fire extinguisher labeled with a B is to be used on this type fire. Who is correct?
   A. A only
   B. B only
   C. Both A and B
   D. Neither A nor B

9. *Technician A* says gloves should be worn when using caustic cleaning materials.
Technician B says face and eye protection should be worn when cleaning parts. Who is correct?
A. A only
B. B only
C. Both A and B
D. Neither A nor B

10. The hazardous material inventory roster is being discussed. *Technician A* says most hazardous materials on the roster should be stored in open containers.
Technician B says only the material specifically stored inside the shop should be listed. Who is correct?
A. A only
B. B only
C. Both A and B
D. Neither A nor B

Table 1-1 ASE Task

Safety practices and habits.

Problem Area	Symptoms	Possible Causes	Classroom Manual	Shop Manual
			1	1

The first standard in all ASE tasks is the requirement to perform all work and training in a safe environment and in a safe manner.

1 Job Sheet 1

Name ______________________________ Date ______________

Performing a Shop Safety Inspection

Upon completion of this job sheet, you should be able to perform a safety inspection of an automotive repair shop.

Tools Needed

Safety inspection checklist, if available

Procedures

1. Study the checklist to review the items to be checked. If a checklist is not available, use the following steps to complete the inspection.
2. Establish an inspection pattern by starting at one side and one end of the shop and follow the walls. Inspect each area as it is reached and record your results in the appropriate space on the checklist.
3. For areas in the center or set away from the main work area, establish a pattern that will ensure that each area is inspected, but reduces the amount of time needed.
4. In the absence of a shop checklist, inspect and record findings in the following steps.
5. Trash

	YES	NO
A. Trash on floor, workbenches, work area?	______	______
B. Trash containers overfull?	______	______
C. Spills on floor, tables?	______	______
D. Wrong items thrown in trashcans?	______	______
E. Empty pressurized containers not split?	______	______
F. Oily/dirty rags left unattended?	______	______
G. Oily rag container present and not overflowing?	______	______
H. Does oily rag container have automatic closing lid?	______	______

Remarks ____

____

____

6. Tools and equipment

	YES	NO
A. Tools left unstored when not in use?	______	______
B. Tools left in walking area?	______	______
C. Equipment improperly stored or maintained?	______	______
D. Equipment power connections damaged?	______	______
E. Hoses leaking, cords frayed, connections damaged, parts missing?	______	______
F. Equipment missing parts?	______	______
G. Safety shields/guards in place?	______	______
H. Loose mountings?	______	______
I. Grinding wheels or bits damaged?	______	______
J. Belts/drive mechanisms worn?	______	______

K. Other damage or lack of maintenance? ______ ______
L. Machining equipment improperly isolated or stored? ______ ______
M. Equipment stored in separate rooms or barricaded? ______ ______
N. Equipment not readily accessible to customers or other non-employees?
O. Adapters properly stored and clean? ______ ______

Remarks ____

____

____

7. Hazardous materials/waste	**YES**	**NO**
A. Individual containers stored properly?	______	______
B. Containers closed, labeled, and clean?	______	______
C. Storage areas marked properly and controlled?	______	______
D. Storage areas marked with flammable sign, closed, not available to non-employees, and isolated from work areas?	______	______
E. Used materials properly stored and labeled?	______	______
F. Empty containers stored properly?	______	______
G. Any material classified as hazardous waste improperly stored, not labeled, or left out?	______	______
H. Drain pans left unemptied?	______	______
I. Empty pressurized containers not split?	______	______
J. Trashcans and dumpsters checked?	______	______

Remarks ____

____

____

8. Fire extinguishers	**YES**	**NO**
A. Type and number of shop fire extinguishers on hand?	______	______
B. Fire extinguisher locations marked and accessible?	______	______
C. Locations marked with large red signs and directional signs on other walls if necessary?	______	______
D. All signs visible across the shop?	______	______
E. Extinguishers in areas away from possible fires? Easily accessible?	______	______
F. Fire extinguishers serviceable?	______	______
G. Inspection tags updated?	______	______
H. Extinguishers fully charged?	______	______
I. Safety wires in place?	______	______
J. Any extinguishers damaged?	______	______

Remarks ____

____

____

9. First aid and eyewash	**YES**	**NO**
A. All locations marked and clean of debris?	______	______
B. First-aid kit stocked?	______	______
C. Area around stations free of obstructions?	______	______

10. Exits

	YES	NO
A. Exits clearly marked?	______	______
B. All red signs visible from across the room/shop?	______	______
C. Does shop have a red light required in dark areas of the building?	______	______
D. Any exists blocked by trash, equipment, or furnishings?	______	______

Remarks ____

____

____

11. Other possible hazards/remarks ____

____

____

____

____

____

____

____

____

____

____

____

____

____

____

____

____

____

____

____

____

____

____

Instructor's Check ____

____

Shop Operations and the Technician's Role

Upon completion and review of this chapter, you should be able to:

- ❑ Explain the basic duties of key shop personnel.
- ❑ Discuss the technician's basic responsibilities.
- ❑ Discuss the educational requirements for automotive technicians.
- ❑ Discuss the routing of a repair order.
- ❑ Complete a sample repair order.
- ❑ Write a cover letter and resumé.

Introduction

The shop employs personnel in several key positions (Figure 2-1). As stated in the Classroom Manual, the service manager is responsible for the service department. To make the operation run smoothly, he or she hires service writers, accountants, cashiers, parts managers, personnel, and the technicians. We will discuss shop operations that would be typical of dealerships and large independents.

Service Writers and Accountants

Classroom Manual
page 19

An *accountant* may be referred to as a bookkeeper and is responsible for receiving payments, making payments, and filing/tracking all funds for the department.

The service writer is usually the first person to greet the customer and the **accountant** is the last. Both are essential to the shop's operation.

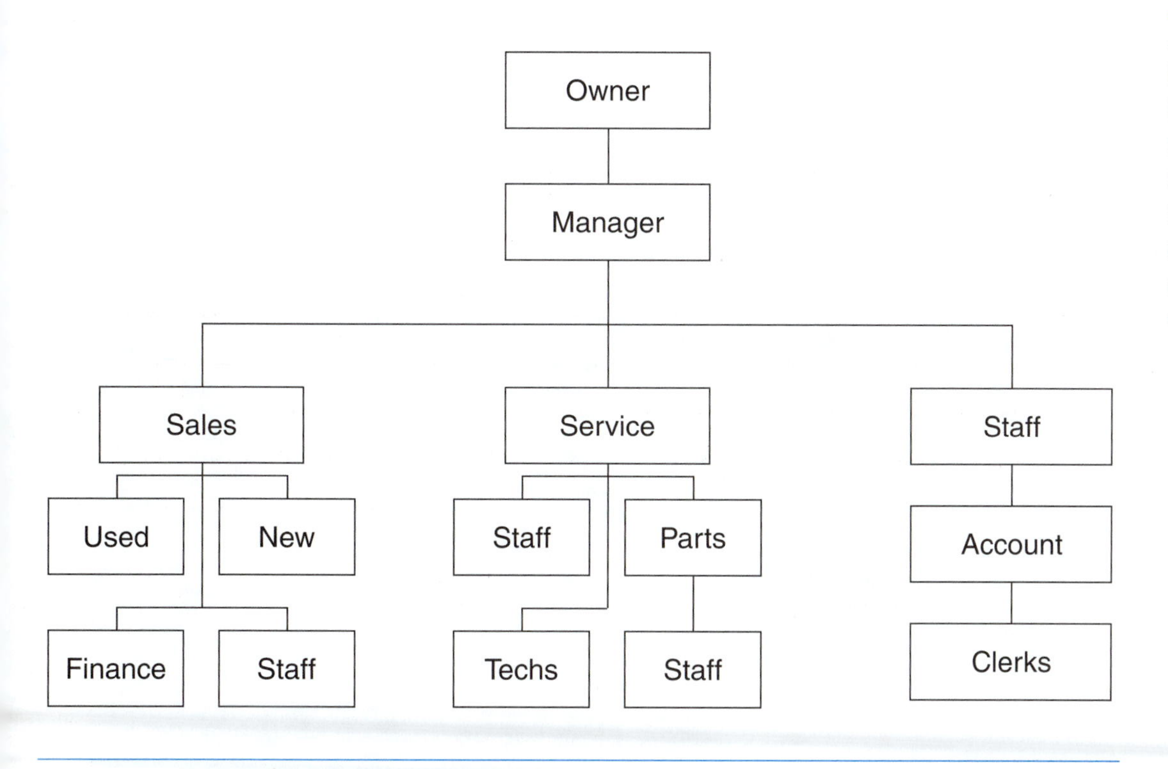

Figure 2-1 Although a business may be separated into different departments, each department must work together to make the business profitable.

Figure 2-2 The service writer initiates the repair order.

The *repair order* is a document that tracks the vehicle from its arrival at the shop for repairs to the time the repairs are paid by the customer.

A *database* is a computerized record system. Information on the customer, vehicle, and repair data can be entered or retrieved quickly.

The service writer greets the customer and initiates the **repair order** (Figure 2-2). She or he must be able to communicate with a wide variety of customers and must understand enough of the vehicle's operating systems to interpret the customer's complaint. The diagnosis goes much quicker if the complaint is clear and concise on the repair order. The service writer may create an estimate of the repair. In many instances, the customer only conducts business with the service writer and the accountant.

The accountant may also act as the cashier for the service department (Figure 2-3). The customer will pick up his or her vehicle keys and pay the repair cost at the cashier's window. However, the accountant does much more. The information on the customer, the vehicle, the repair accomplished, and so on is usually collected and entered into a computerized **database.** The information is used to remind customers of scheduled maintenance and may be used to resolve a customer's complaint. In larger shops, there may be several individuals working in a service department's accounting office.

Classroom Manual
page 22

The *parts manager* usually reports to the service manager. The department is responsible for ordering parts, entering them on repair orders, staffing the pickup window for walk-in customers, maintaining inventory, and ordering out-of-stock items from other dealerships.

Parts include any component needed to repair or maintain a vehicle.

Parts

The **parts manager** usually reports to the service manager. The **parts** department is responsible for ensuring that sufficient routine maintenance items are present (Figure 2-4). Also, parts

Figure 2-3 The accountant acts as the service department's cashier in many shops.

Figure 2-4 The parts department stocks parts and materials that are required on a routine basis.

personnel must order, deliver, pick up, and enter parts on the repair order. The department operates a pickup window for walk-in customers. A running inventory is kept at all times to prevent shortages of routine items. In addition, dealerships are linked by telephone and satellite communications to other dealerships. In this manner, it may be possible to get an out-of-stock part from a dealer on the other side of town instead of waiting several days to get one from the manufacturer's warehouse. The parts are charged against the repair order or a separate invoice, which is sent to the accountant payment collection.

The Technician

Classroom Manual
page 22

Entry-level technicians are usually recent graduates or may still be attending school. This is their first official job as automotive technicians.

The technician performs the basic profit-making portion of the shop's operations: vehicle repairs. The technician may also have the most personal money invested in the position. Most shops will not hire a technician that does not have personal tools (Figure 2-5). A basic tool set may cost over a thousand dollars and it is not unusual for an experienced technician to have over $20,000 worth of tools collected through her or his career.

The technician must follow all regulations concerning automotive environmental hazards and safety. The repair should be performed correctly the first time. The best customer relations possible is a quick, fair, accurate repair to the vehicle.

Entry-level technicians are hired to perform basic services such as oil and filter changes, tires, brakes, and other routine maintenance (Figure 2-6). As the technician gains experience, he or she is given more and more responsibilities. In order to stay abreast of vehicle operations, it is necessary for the technician to attend training at the manufacturer school, parts-vendor-sponsored training, or a technical school.

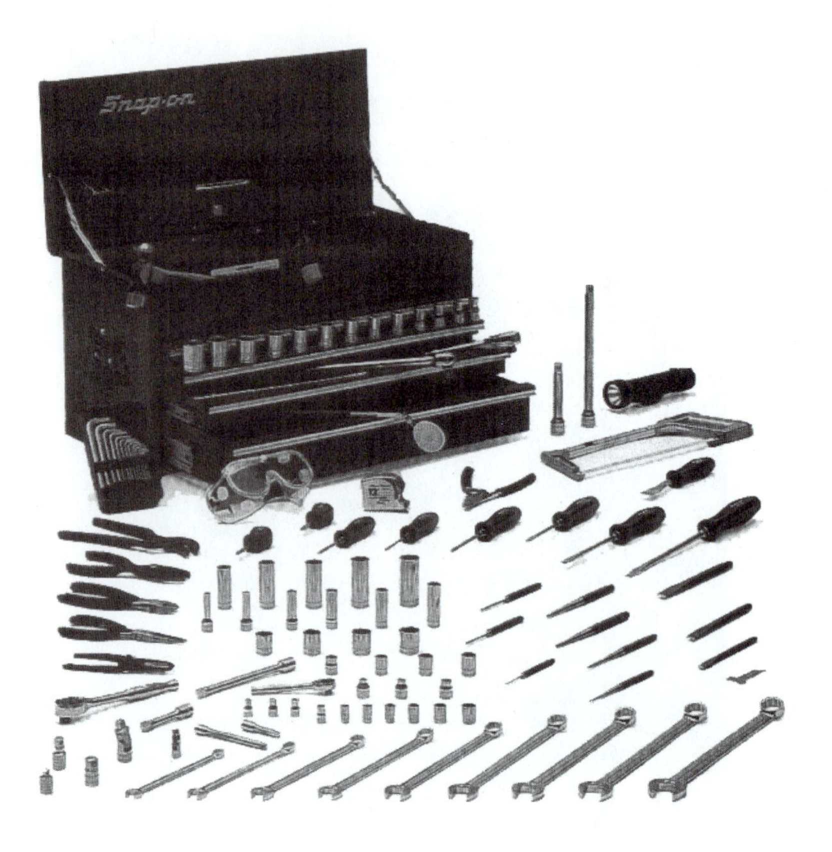

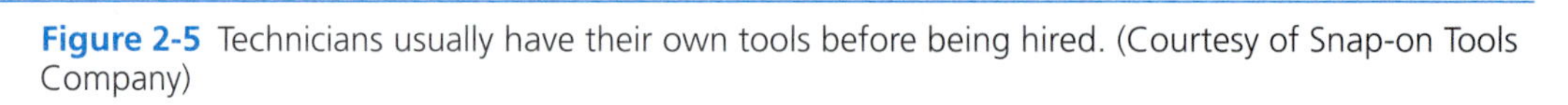

Figure 2-5 Technicians usually have their own tools before being hired. (Courtesy of Snap-on Tools Company)

Figure 2-6 Most entry-level technicians are hired for basic lubrication and other minor services.

Technician Education and Certification

Classroom Manual
pages 23–25

Education usually refers to formal *training* of some sort, however, a great deal of education is done on the job by learning from experienced technicians.

Many technicians receive their initial **training** in the backyard working on family and personal vehicles. Most secondary schools (high school) offer some type of technical or vocation training. The technician may enter the automotive workforce after high school, but may be at a disadvantage when compared to technical school graduates. Depending on the person's individual situation, the high school graduate may attend a technical institution during day or evening hours. This will allow the person to work and attend training. Most technical schools offer a program in automotive technology (Figure 2-7).

Even with a technical school diploma or degree, the automotive technician must continue to attend training. Vehicles change in some way almost every year. Future vehicles may be electric- or natural-gas-powered, combination of both, or something entirely new. With the number of vehicles on the road, government regulations, and customer preference, the automotive technician can look forward to a life of learning.

Technical certification is a means of letting the managers and customers know that the technician is proficient in the field. The technician may receive certification from manufacturer schools, ASE, or other certification agencies. Almost every class attended produces some type of certificate or certification (Figure 2-8). Some companies will pay a bonus for ASE certification or will pay the cost of successfully completing the certification.

The single item that tracks a technician's day-to-day performance is the repair order. The repair order is opened by the service writer and assigned to a technician.

Repair Orders

Classroom Manual
page 22

A completed repair order lists the customer's complaint, the diagnosis, repair parts, labor, and costs of each. Some may include a history of the most recent services performed on the vehicle (Figure 2-9). While the individual form may differ from shop to shop, the same information is normally used.

Figure 2-7 Most technical schools offer additional training in basic automotive technology after high school. (Courtesy of Griffin Technical Institute)

York Technical College

and

The Center for Alternative Energy Transportaton

awards this

Certificate

to

Clifton Owen

for the successful completion of the

Inaugural Electric Vehicle and Maintenance Training Course

from November 18, 1996 to December 6, 1996

Instructor

President

Figure 2-8 Most training courses offer some type of certificate to show successful completion of the course. (Courtesy of York Technical Institute)

Labor time may be based on a labor guide or the actual time the technician spent on the repair.

Upon receipt of the repair order, the technician retrieves the vehicle and performs the diagnosis. Assuming that parts are available, the technician performs the repairs and conducts a test drive when needed. Parts personnel charge the parts to the repair order. The technician includes the **labor time** and other information about the repairs required by the shop. This may include the method or type of testing performed and the diagnosis.

The repair order is routed to the accountant who computes the bill based on information supplied by the technician, parts, and shop policy. Warranty repairs do not cost the customer, but the same information is entered on the order and the customer receives a copy. The repair order information is entered into the database and the order is filed after the customer pays. The manufacturer reimburses the dealership for warranty repairs.

Recall listings can be found on the Internet, at dealerships, or in paper manuals. The National Safety and Transportation Board also keeps a list of recalls.

Another type of repair that is done by the technician and is paid for by the manufacturer is **recall** repair. This is a defect found on a model or make of vehicle after many units have been sold to the public. Customer complaints or manufacturer testing and inspection may find the defect. Sometimes the manufacturer makes a voluntary recall or it is ordered to do so by the federal government after an extensive investigation. The defect may be a safety hazard or some modification needed to increase the life of the vehicle. In either case, each owner of vehicles with the defect are notified by mail and encouraged to take the vehicle to a dealership for repairs at no cost. Manufacturers are required to track the vehicle as it is sold and resold because sometimes the defect may not be found for several years.

RO NUMBER	DATE CLOSED	MILEAGE	W / [illegible]	OP CODE	DESCRIPTION	OP CODE	DESCRIPTION	OP CODE	DESCRIPTION	OP CODE	DESCRIPTION
63539	30MAR98	51509	5364UC	806	USED CAR INSPE	05	REPLACE FRONT	05	CLEAN AND ADJU		
57012	02AUG97	36133	5399PM	1000	TRANS. FLUID L	1000	SAME AS LINE A	01M	OIL AND FILTER	00	NO CHARGE
51886	07FEB97	22202	6564PM	01M	OIL AND FILTER	1000	ALIGNEMENT ON	22M	ROTATE AND BAL	1000	CHECKED ALL FL
48433	04OCT96	12269	3045PM	01	OIL & FILTER C	1000	RESECURE				

SHOWN ABOVE ARE THE MOST RECENT SERVICES PERFORMED ON YOUR VEHICLE.

CURRENT MILEAGE 51557

BASED ON THE CURRENT MILEAGE OF YOUR VEHICLE, THE MANUFACTURER RECOMMENDS THAT THE SERVICES MARKED WITH AN 'X' BE PERFORMED NOW.

LUBRICATION / OIL AND OIL FILTER CHANGE / SAFETY CHECK / CHECK FLUID LEVELS / SERVICE AIR CLEANER / SERVICE EMISSION CONTROL SYSTEM / SERVICE BATTERY AND STARTING SYSTEM / SERVICE COOLING SYSTEM / SERVICE AUTOMATIC TRANSMISSION / REPLACE FUEL FILTER / ROTATE TIRES / WHEEL BALANCE / PACK FRONT WHEEL BEARINGS / FRONT END ALIGNMENT / INSPECT EXHAUST SYSTEM / INSPECT SHOCKS, SUSPENSION, AND STEERING / INSPECT AND SERVICE BRAKES / INSPECT AND ADJUST ALL BELTS / REPLACE WINDSHIELD WIPER BLADES / SERVICE WINDSHIELD WASHER SYSTEM / SCOPE AND SET ENGINE / TUNE UP / CARBURETOR OVERHAUL / SERVICE AIR CONDITIONER / ADJUST PARKING BRAKE / INSPECT AND TEST ELECTRICAL SYSTEM / SERVICE DIFFERENTIAL / DRAIN AND FLUSH COOLING SYSTEM

NOT RESPONSIBLE FOR LOSS OR DAMAGE TO CARS OR ARTICLES LEFT IN CARS IN CASE OF FIRE, THEFT OR ANY OTHER CAUSE BEYOND OUR CONTROL.

MERCURY LINCOLN

SPALDING LINCOLN-MERCURY
MAZDA-DODGE
2570 NORTH EXPRESSWAY
GRIFFIN, GEORGIA 30223
(404) 228-9485

mazda Dodge

EXCLUSION OF WARRANTIES

Any warranties on the parts and accessories sold hereby are made by the manufacturer. The undersigned purchaser understands and agrees that dealer makes no warranties of any kind, express or implied, and disclaims all warranties, including warranties of merchantability or fitness for a particular purpose, with regard to the parts and/or accessories purchased; and that in no event shall dealer be liable for incidental or consequential damages or commercial losses arising out of such purchase. The undersigned purchaser further agrees that the warranties excluded by dealer include, but are not limited to, any warranties that such parts and/or accessories are of merchantable quality or that they will enable any vehicle or any of its systems to perform with reasonable safety, efficiency, or comfort.

DATE: SIGNED: X

CUST. NO. STOCK NO. M3175A TAG NO. T099 COLOR TAN REPRINT PAGE 1 OF 1

DATE	VEHICLE IDENTIFICATION NUMBER	MILEAGE	DELIVER DATE	LICENSE NUMBER	YEAR	MAKE AND MODEL	WRITTEN BY	RO NUMBER
03JUN98	JM1BB1410T0326042	51557	04MAR96		96	MAZDA PROTEGE	7501	65619

I hereby authorize the repair work herein set forth to be done along with the necessary material and agree that you are not responsible for loss or damage to vehicle or articles left in vehicle in case of fire, theft or any other cause beyond your control or for any delays caused by unavailability of parts or delays in parts shipments by the supplier or transporter. I hereby grant you and/or your employees permission to operate the vehicle herein described on streets, highways or elsewhere for the purpose of testing and/or inspection. An express mechanic's lien is hereby acknowledged on above vehicle to secure the amount or repairs thereto. Not responsible for damage from freezing due to lack of antifreeze.

PRELIMINARY ESTIMATE $ ______________

AUTHORIZED BY: X

	DATE	TIME	BY
REVISED ESTIMATE(1) $			
REVISED ESTIMATE(2) $			

I HEREBY ACKNOWLEDGE THAT I WAS NOTIFIED & GAVE ORAL APPROVAL OF THE ABOVE REVISED ESTIMATES:

X ______________
CUSTOMER SIGNATURE

NAME

ADDRESS

CITY/STATE/ZIP

BILL TO: P.O. NO.

TIME PROMISED
17:30 01JUN98

HOME PHONE

BUSINESS PHONE

ENGINE NO. TRANSM. NO. AXLE NO. PROD. DATE

LABOR RATE
63.00

METHOD OF PAYMENT CASH
SELLING DEALER 23464
WARRANTY EXPIRES

LINE	OP CODE	LABOR INSTRUCTIONS AND DESCRIPTION
# A		CUST STATES THAT RADIO IS INOP

GAS/OIL/GREASE	AMOUNT
TOWING	AMOUNT
MISC. CHARGES	AMOUNT
SUBLETS P.O. NO.	AMOUNT

FOREMAN'S SIGNATURE X

TECH COPY

Figure 2-9 This repair shows a recent history (top) and the customer's complaint (center). (Courtesy of Spalding Lincoln/Mercury/Mazda/Dodge)

Vehicle Identification Number (VIN)

Figure 2-10 The VIN has codes that relate to specific information about systems of the vehicle. (Courtesy of Chevrolet Motor Division, General Motors Corporation)

Classroom Manual
page 22

The *vehicle identification number* (VIN) is a number assigned to a particular vehicle. Information on the engine size and type, transmission, year of production, and other vehicle data is encoded into the VIN.

Hourly rate refers to the fee charged by the shop for one hour of labor.

Completing a Repair Order

The following steps are based on the form in Figure 2-9. Most of the information will be included on almost any shop's repair order.

Complete the service writer's portion of the repair order by entering the date, **vehicle identification number (VIN)**, mileage, make and type of vehicle, customer personal information, and complaint (Figure 2-10). Enter the complaint as clearly and concisely as possible. Be sure to secure the customer's signature authorizing the repairs. If requested, provide an estimate of the cost and time of completion to the customer.

Enter the technician's diagnosis based on the complaint. Most shops require the technician to enter a starting and ending time on the repair order to track the actual time the vehicle was in the shop. Enter the labor time and, if not already provided, the **hourly rate**. Once the repair is complete, provide a short, concise statement about the repair that will give the customer an idea of the work performed. The information may be handwritten or printed by a computer terminal based on numerical codes entered by the technician. The technician must enter his or her employee identification in order to receive credit and payment for the repair. Many shops are using computer-generated forms that allow the service writer, technician, parts personnel, and accountant to process a complete repair order by entering codes into a terminal.

Complete the parts department information by entering part numbers, quality, cost per unit, and total cost of parts. The repair order is given to the accountant for final calculations including taxes before the customer pays.

Some shops require a foreman or supervisor to sign orders for major repairs. The supervisor may choose to test drive the vehicle. The repair order should not go to the accountant until the required signature is obtained. The accountant is the last stop for the repair order before filing. With the repair order complete, notify the service writer that the vehicle is ready for the customer to pick up.

Classroom Manual
page 23

An *application* is a paper form designed by a company. It is set up to collect only the applicant information needed by the company for hiring.

Applying for Employment

Becoming employed depends not only on technical skills, but also on the technician's communication skills. Normally the first contact the technician has with a potential employer is the job **application** supplied by the employer. If at all possible, the application should be accompanied by a *cover letter* and *resumé*.

Cover Letters

A cover letter is a short, one-page document that expresses the desire for a particular job (Figure 2-11). It should also include a brief overview of the technician's abilities. Information for a cover letter should appear on only one sheet of paper and be addressed to the person who performs the hiring interview. In some cases, that is the service manager. In other instances, it may be a personnel officer.

SAMPLE
COVER LETTER

Ms. Jane Owner
Owner's Auto Refurnishers
My City, Nebraska 11111

Working Person
123 4th Street
Anywhere, USA 22222

Dear Ms. Owner,

Mr. Service Manager notified me of an open position within your automotive detailing department. I would like to fill that position for your firm and am enclosing a resumé.

During my last two years of high school, I worked part-time at Joe's Auto Flea Market. I have spoken with Mr. Joe about applying with you. He feels I would have better career opportunities with your firm.

I like detailing older vehicles to bring them back to their original condition. My patience has rewarded Mr. Joe with some better-than-average sales. While I was given a portion of the sales price, I need to move to a larger company that can offer me some career opportunities. I feel my willingness to work with other members of your firm will be a great benefit to both of us. I look forward to meeting you and becoming a part of Owner's Auto Refurnishers.

Thank you for your time and consideration.

Signed,

Working Person

Working Person

Figure 2-11 The cover letter should express the applicant's interest in the position and a brief explanation of qualifications.

Resumés

A resumé provides a short history of the technician's education and experience. It is usually written in short phrases to save space. Like the cover letter, most resumés should be completed on one sheet (Figure 2-12).

The heading should include personal information about the technician, including address and telephone number. Work history is next and should start with the current or latest employment. Most employers would like to have the history for about the last five years. Large companies involved in government or sensitive work may require as much as ten or more years of history.

List the employer, the address, and the dates of employment. If the street address in not available, make sure the city is listed. Follow the employer information with a brief description of duties and responsibilities. Do not exaggerate, but do not be shy when providing skill information. It is not required to list reasons for leaving a job on the resumé, but it may be required on the job application. Do not lie about the reason for leaving even if it was for cause (fired). A quick check by the personnel office will find the reason.

List education in the same order as the work history. Ensure that any academic awards or extracurricular activities are noted. This provides the interviewer with some background on the technician's ability to work with other people.

SAMPLE RESUMÉ

Working Person
123 4th Street
Anywhere, USA 22222
Telephone 555-555-5555
SSAN 123456789

WORK EXPERIENCE

June 1997-Present: Joe's Auto Flea Market, 321 North Street, Anywhere, USA. Detailing vehicles for sales and private owners. Performed light preventive maintenance to vehicles before sale. Performed safety inspection of vehicles before sales. Checked trade-in vehicle for any major maintenance or safety defects.

EDUCATION

September 1995-Present: Anywhere High School, Anywhere, USA. Attended general curriculum. Second place winner in state VICA automotive paint and body competition. GPA 3.0.

SKILLS AND HOBBIES

Excellent skill and patience preparing vehicles for painting. Operate paint booth and related painting equipment. Enjoy refurbishing older vehicles.

Figure 2-12 The resumé is intended to give the potential employer an idea of the applicant's experience, education, and skills.

Information on additional skills and hobbies can, and should, include skills that may or may not directly relate to the job and other data that show the technician's personal interest and goals.

CASE STUDY

A young technician employed for over three years with the same company was released for padding a ticket (over charging the customer for parts or work not completed). The technician applied at a dealership in a city about forty miles from his first job. He was hired and had worked for about four months with a good record. By chance, the customer who had been over charged had also moved to the same city. When the customer brought a vehicle in for service, he recognized the technician, refused to allow the shop to perform the work, and stated his reasons. The service manager investigated and released the technician. The service manager stated to the technician that, had he not lied on the application, he could have been hired under probation and allowed to prove himself. The final result was a fairly skilled technician who had to take a lesser job which resulted in loss of pay and benefits.

Terms to Know

Accountant
Application
Cover letter
Database
Labor time
Parts manager
Recall
Repair order
Resumé
Vehicle identification number (VIN)

ASE Style Review Questions

1. The duties of the service writer are being discussed.
   Technician A says the service writer completes the repair order.
   Technician B says the service writer may provide a repair estimate to the customer. Who is correct?
   A. A only
   B. B only
   C. Both A and B
   D. Neither A nor B

2. Technician education is being discussed.
   Technician A says training is life long.
   Technician B says training may be accomplished through manufacturer school or vendor-sponsored training classes. Who is correct?
   A. A only
   B. B only
   C. Both A and B
   D. Neither A nor B

3. The parts department is being discussed.
   Technician A says all parts are sold on the shop's repair orders.
   Technician B says the technician performing the service must charge the parts to the order. Who is correct?
   A. A only
   B. B only
   C. Both A and B
   D. Neither A nor B

4. Technician certification is being discussed.
   Technician A says technicians can highlight their technical knowledge through certifications.
   Technician B says a technician may have certifications from several agencies. Who is correct?
   A. A only
   B. B only
   C. Both A and B
   D. Neither A nor B

5. A resumé is being discussed. *Technician A* says education history should precede work history.
   Technician B says the heading should include an address and telephone number. Who is correct?
   A. A only
   B. B only
   C. Both A and B
   D. Neither A nor B

6. Cover letters are being discussed. *Technician A* says the letter should list the technician's special skills.
   Technician B says the letter may be sent to the shop's personnel officer or the service manager. Who is correct?
   A. A only
   B. B only
   C. Both A and B
   D. Neither A nor B

7. Applying for a job is being discussed. *Technician A* says the job applicant should have good communications skills.
   Technician B says an untruthful application may result in dismissal. Who is correct?
   A. A only
   B. B only
   C. Both A and B
   D. Neither A nor B

8. Repair orders are being discussed. *Technician A* says the technician is responsible for computing the labor cost.
   Technician B says the accountant usually enters the labor time on the order. Who is correct?
   A. A only
   B. B only
   C. Both A and B
   D. Neither A nor B

9. *Technician A* says the service manager is responsible for the company's accounting office. *Technician B* says the parts manager usually reports to the service manager. Who is correct?
   A. A only
   B. B only
   C. Both A and B
   D. Neither A nor B

10. *Technician A* says the service writer initiates the repair order.
    Technician B says the technician enters the VIN onto the repair order. Who is correct?
    A. A only
    B. B only
    C. Both A and B
    D. Neither A nor B

Job Sheet 2

2

Name ______________________ Date ____________

Complete a Cover Letter and Resumé

Upon completion and review of this job sheet, you should be able to prepare and write a cover letter and resumé for a job application.

Instructor's Note: You may wish to coordinate with your local job placement center for the cover letter and resumé formats desired by local businesses.

Tools Needed

None

Procedures

1. Prepare notes on work and education experience. List any special skills, hobbies, and activities that highlight abilities to perform a job well and show communication skills.
2. Determine the person who should receive the letter and resumé. Telephone the business and ask for the name and address of the person if necessary. Check the spelling of the person's name and title.
3. Prepare a cover letter by:
   A. Creating a heading addressed to the person who is doing the hiring.
   B. Keeping all paragraphs three to five sentences long.
   C. Stating the reason for the letter in the first paragraph should, i.e., "I understand from Mr. Soso that a position is opened. . . ."
   D. Stating in the second paragraph an overview of why you should be considered for the job, i.e., "I have worked at Joe's Auto Repair and have experience in. . . ." The second paragraph should also show your desire to work for the company and how you can assist the company, i.e., "I feel my abilities and experience would be of great benefit to your business. . . ."
   E. Expressing thanks in the last paragraph to the individual for his or her time and consideration, and closing with a desire to hear from them, i.e., "I look forward to working with your company. . . ."
   F. Signing the letter with: Sincerely yours,
   Your Signature
4. Prepare the resumé by:
   A. Creating a heading. Include your full name, address, telephone number, and Social Security Number if desired.
   B. Listing your work history, starting with the most recent and extending back at least three to five years. Example:

   1995–Present: Joe's Auto, 1 Any Street, Anywhere, USA. Performed oil changes and routine services; diagnosed and repaired brake systems; machined drums and rotors; diagnosed tire wear; mounted and balanced tire assemblies; replaced drive belts, spark plugs, and filters; and assisted air-conditioning technician on AC repairs.

C. Listing your education history, beginning with the most recent and extending back over last several schools attended. Example:

June 1997—Ford EECV system diagnostics. Ford Training Center.
July 1996–May 1997—Best Technical School, Automotive Technology Diploma. Dean's List last two quarters.

D. Listing any special skills and activities. Example:

Worked with pit crew on mini-stock. Member of church youth group. Coached t-ball team. Had some training in machining during technical school.

5. Proofread your cover letter and resumé and ask someone else to proofread it. An instructor or person who has experience in English and application forms is best suited for this task.

Job Sheet 3

3

Name ________________________________ Date ____________________

Locating an Automotive Technician Position

Upon completion of this job sheet, you should be able to locate available automotive technician positions locally, state- or province-wide, and nationally.

Instructor's Note: This is a general job sheet employing common methods of locating a job. Students may use different methods to find employment information. It is not expected that all students who read this book have or should have the transportation to visit off-campus sites.

Tools Needed

Internet access
Local and state/province-wide newspapers
Local office of labor department

Procedures

1. Use newspapers to locate an available automotive technician position. Record the following information from the newspaper. At least two of the positions should not be local, but should be within a reasonable commute.
2. First employer

   Name of employer ________________________________

   Employer address/telephone ________________________________

   Employment requirement:

   Technical training required? Y N

   If so, what type? ________________________________

   ____

   Experience required? Y N

   If so, what kind? ________________________________

   ____

   Tools required? Y N

   Can application be:

   _____ Mailed _____ E-mailed _____ By appointment

   _____ Labor Department _____ School job placement

   _____ Temporary hire service

3. Second employer

Name of employer ______________________________

Employer address/telephone ______________________________

Employment requirement:

Technical training required? Y N

If so, what type? ______________________________

____

Experience required? Y N

If so, what kind? ______________________________

____

Tools required? Y N

Can application be:

_____ Mailed _____ E-mailed _____ By appointment

_____ Labor Department _____ School job placement

_____ Temporary hire service

4. Third employer

Name of employer ______________________________

Employer address/telephone ______________________________

Employment requirement:

Technical training required? Y N

If so, what type? ______________________________

____

Experience required? Y N

If so, what kind? ______________________________

____

Tools required? Y N

Can application be:

_____ Mailed _____ E-mailed _____ By appointment

_____ Labor Department _____ School job placement

_____ Temporary hire service

5. Fourth employer

Name of employer ____

Employer address/telephone ____________________________________

Employment requirement:

Technical training required? Y N

If so, what type? ____

____

Experience required? Y N

If so, what kind? ____

____

Tools required? Y N

Can application be:

_____ Mailed _____ E-mailed _____ By appointment

_____ Labor Department _____ School job placement

_____ Temporary hire service

6. Fifth employer

Name of employer ____

Employer address/telephone ____________________________________

Employment requirement:

Technical training required? Y N

If so, what type? ____

____

Experience required? Y N

If so, what kind? ____

____

Tools required? Y N

Can application be:

_____ Mailed _____ E-mailed _____ By appointment

_____ Labor Department _____ School job placement

_____ Temporary hire service

7. Based on your search, for which of the job openings are your best qualified?

____

____

8. Explain why you are qualified for the openings you selected.

Instructor's Check

Job Sheet 4

4

Name ______________________ Date ______________

Automotive Business Positions

Upon completion of this job sheet, you should be able to discuss the types of positions, employee duties, and the general employment requirements for a typical automotive business.

Instructor's Note: This job sheet will require the student to make either telephone or personal contact with various individuals in several local automotive businesses. It is suggested that the instructor or school obtain permission from the businesses so the student can visit or conduct a telephone interview. In addition, the student may require access to the Internet or the local labor office.

Tools Needed

Telephone and transportation
Access to Labor Department job description files

Procedures

1. The following subjects could be asked of a service manager or other shop supervisor.

A. Types of positions available within the service department:

____
____
____
____
____
____

Number of technicians __________

Number of parts personnel __________

Number of service writers __________

Number of staff personnel __________

B. Basic requirements for employment as:

Service writer ____
____
____

Parts and parts manager ____
____
____
____

Staff personnel ____
____
____
____
____

Technicians

C. What training is offered or required annually for the following?

Service writer

Parts and parts manager

Staff personnel

Technicians

D. What are some of the major benefits offered to shop employees?

2. The following subjects could be asked of a parts manager or parts personnel.

A. What training is needed to work in the parts department?

B. How many persons work in this parts department?

C. What are their basic duties and responsibilities?

D. How do they advance in position and salary? ______________________________

____

____

3. The following subjects could be asked of a automotive technician.

A. What was your training before beginning your first automotive repair job? ____________

____

____

____

B. What kind of tool set did you have when you began your first automotive repair job?

____

____

C. How does your first tool set compare to the set you have now? ________________

____

____

D. Do you work on any type vehicle or just selected ones? (dealership vs. independent)

____

____

____

E. How do you see the future of automotive technicians? ______________________

____

____

____

____

____

■ **NOTE:** Steps 1–3 may be used to interview personnel from different automotive repair shops.

4. The following information should be obtained for the local Labor Department, local library, school library, or the Internet.

A. The official job description for:

Service manager ____

____

____

____

Accountant ____

____

____

____

Service writer

Automotive repair technician (may be listed as mechanic)

Heavy truck repair technician (may be listed as mechanic)

Small engine repair technician (may be listed as mechanic)

Marine engine repair technician (may be listed as mechanic)

5. Based on the information collected in step 4, what is the major "official" difference(s) between the four job descriptions as defined by the Labor Department?

Automotive Tools

Upon completion and review of this chapter, you should be able to:

- ❑ Connect and use an air impact wrench to remove the lug nuts on a wheel.
- ❑ Look up the torque specifications for wheel lug nuts.
- ❑ Torque wheel lug nuts to the correct torque using a beam, dial readout, and breakover hinge-type torque wrench.
- ❑ Select, install, and use a gear puller to correctly remove a pulley from a shaft.
- ❑ Use a screw extractor to remove a broken stud or screw.
- ❑ Use a flat chisel to split and remove a stuck nut from a stud or bolt.
- ❑ Set up and use an oxyacetylene torch for heating.

Introduction

Basic technician's tool set

Eye protection

In the Classroom Manual we described the types and use of the basic tools technicians use to repair cars. This chapter focuses on some of the most common tool tasks performed by technicians. We describe how to connect and use an impact wrench, how to adjust and use a torque wrench, how to use a gear puller to remove parts, how to remove a broken stud or screw, and how to remove a damaged nut.

Connecting and Using an Impact Wrench

Classroom Manual
pages 49–51

Special Tools

Air-operated impact wrench

Impact socket set

Hoist or floor jack

Jack safety stands

Wheel blocks

Lug nuts are the hex-sided fasteners used to hold the wheel rim on the studs of the brake rotor or drum.

The air-operated impact wrench is frequently used by all technicians. Photo Sequence 2 shows a typical procedure for connecting and using an impact wrench to remove **lug nuts.** The impact wrench is often misused by technicians who do not understand that its main function is disassembly. The typical air-operated impact wrench is not torque controlled. This means you cannot be sure how tight you are tightening a fastener when you use this tool. Using an air-operated impact wrench to assemble components can result in overtightened or undertightened fasteners.

A common use of the air-operated impact wrench is to take off wheel lug nuts to remove a wheel from a car. First you must raise the wheel to be removed safely off the floor. Follow the safe procedure described later in this chapter to raise the vehicle on a hoist or to raise and support the car on a floor jack and jack safety stands.

Many cars use a wheel cover that must be removed for access to the lug nuts. Other wheels use the lug nuts to hold on the wheel cover. If the car has wheel covers that must be removed first, use a prying tool to gently pry the wheel cover off the wheel rim as shown (Figure 3-1). Some cars have locks that prevent wheel cover theft. The lock mechanism typically

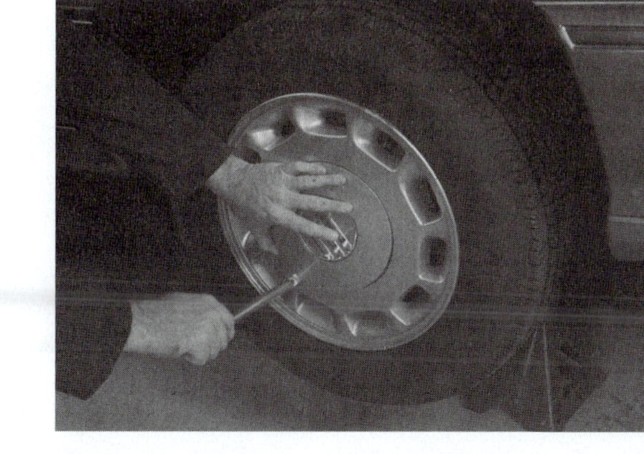

Figure 3-1 Removing the wheel cover.

Photo Sequence 2
Typical Procedure for Connecting and Using an Impact Wrench to Remove Lug Nuts

P2-1 Raise the car on a hoist or jack and jack safety stands.

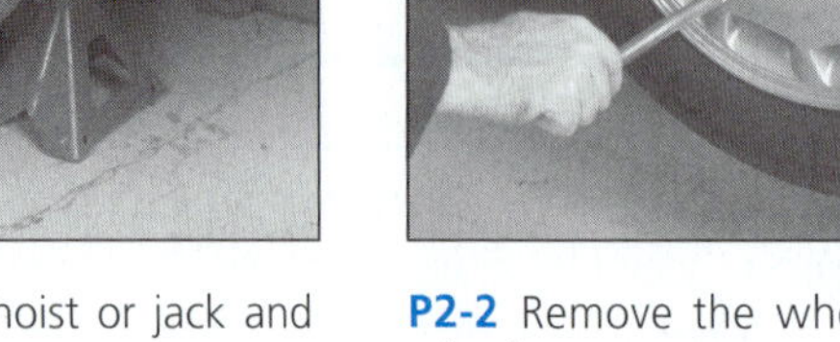

P2-2 Remove the wheel cover from the wheel.

P2-3 Connect the air impact wrench to the air hose quick disconnect coupling.

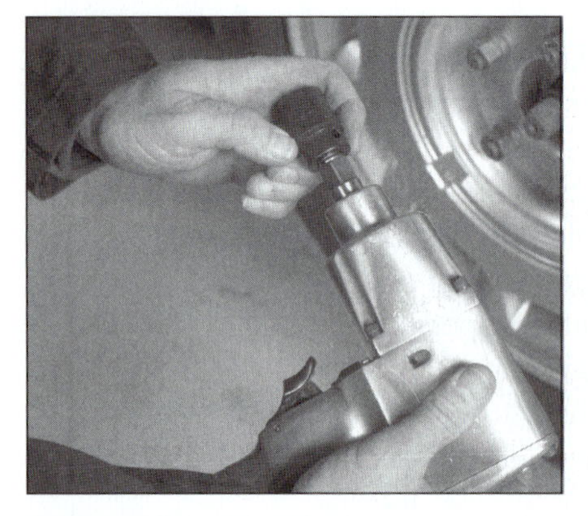

P2-4 Install the correct size impact socket on the wrench drive.

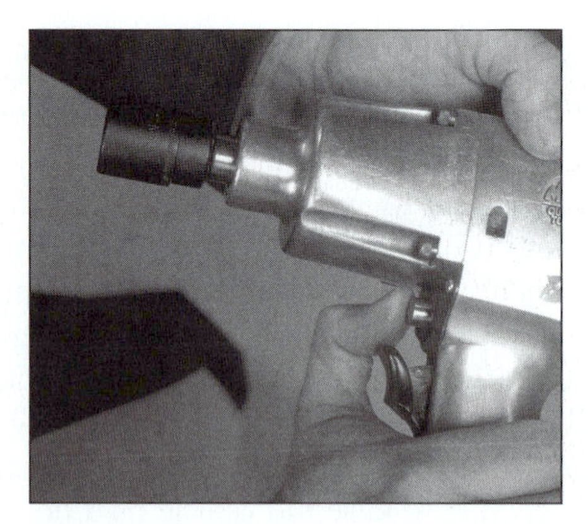

P2-5 Switch the impact wrench to the reverse (counterclockwise) drive direction.

P2-6 Place the wrench and socket over the lug nut and pull the trigger to remove the lug nut.

P2-7 Remove each of the other lug nuts and remove the wheel.

P2-8 Install the wheel back on the car and start each lug nut by hand.

P2-9 Switch the impact wrench to the forward (clockwise) direction.

Typical Procedure for Connecting and Using an Impact Wrench to Remove Lug Nuts (continued)

P2-10 Carefully run the nuts up but do not tighten.

P2-11 Look up the tightening torque and tightening sequence for the lug nuts in a shop service manual.

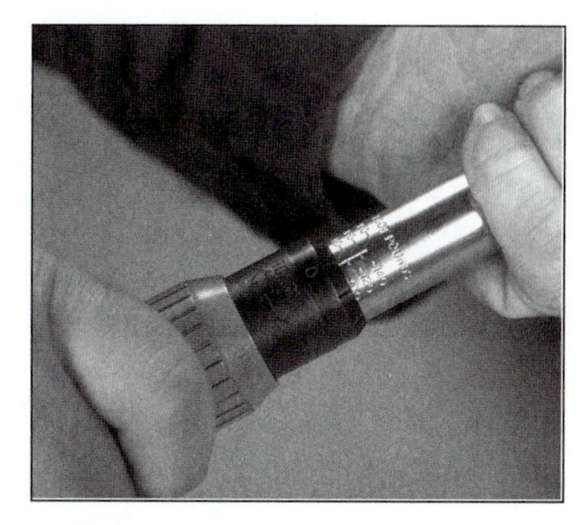

P2-12 Set the torque wrench to the specified torque.

P2-13 Tighten each lug nut in the correct order to the specified torque.

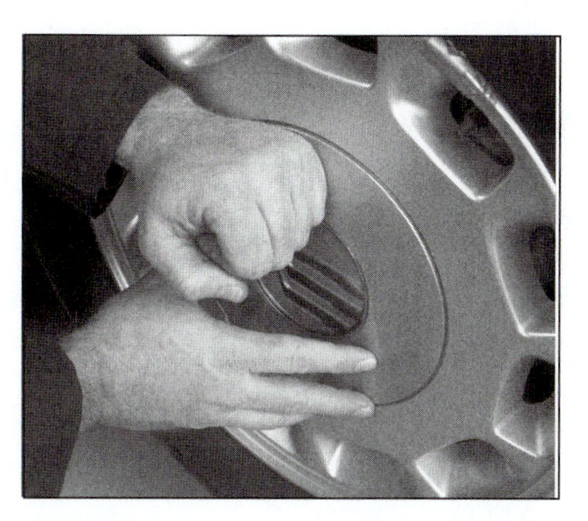

P2-14 Replace the wheel cover and lower the car.

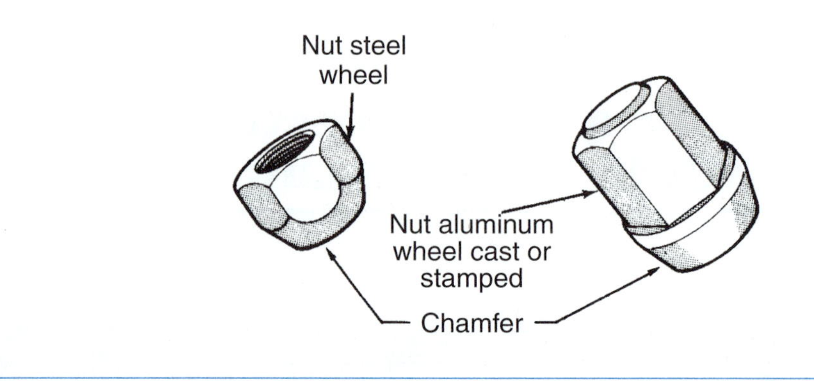

Figure 3-2 Types and parts of the lug nuts. (Courtesy of DaimlerChrysler Corporation)

fits over one of the lug nut studs. Use the key to remove the lock then pry off the cover. Be careful not to bend or distort the wheel cover or it will not fit back on properly.

The lug nuts hold the wheel onto studs. Passenger cars typically use either four or five lug nuts to hold on the wheel. These studs are attached to the brake rotor or brake drum. There are several designs of lug nuts (Figure 3-2). Lug nuts that hold on steel wheels are often different than those that hold on aluminum wheels. One side of the lug nuts are chamfered. The chamfered side always goes on toward the wheel rim. The chamfer fits in a matching chamfer on the wheel and centers the wheel over the studs as it is tightened.

The lug nuts are hex shaped or six sided to fit standard wrench sizes. The wrench sizes are different for different cars. Select a six-point impact socket (Figure 3-3) that fits the hex sides of the lug nut snugly.

SERVICE TIP: You should make match marks on the wheel rim and one of the studs with chalk or grease pencil. Always replace the wheel in the same position in case it has been balanced on the car.

WARNING: Using a loose fitting impact socket on a wheel lug nut can cause the corners of the hex to be rounded off and not fit a wrench properly.

CAUTION: Always use impact sockets with an impact wrench. Standard sockets can fracture and explode when used on an impact wrench. Always wear eye protection when using an impact wrench.

Figure 3-3 A six-point impact socket is used with an impact wrench.

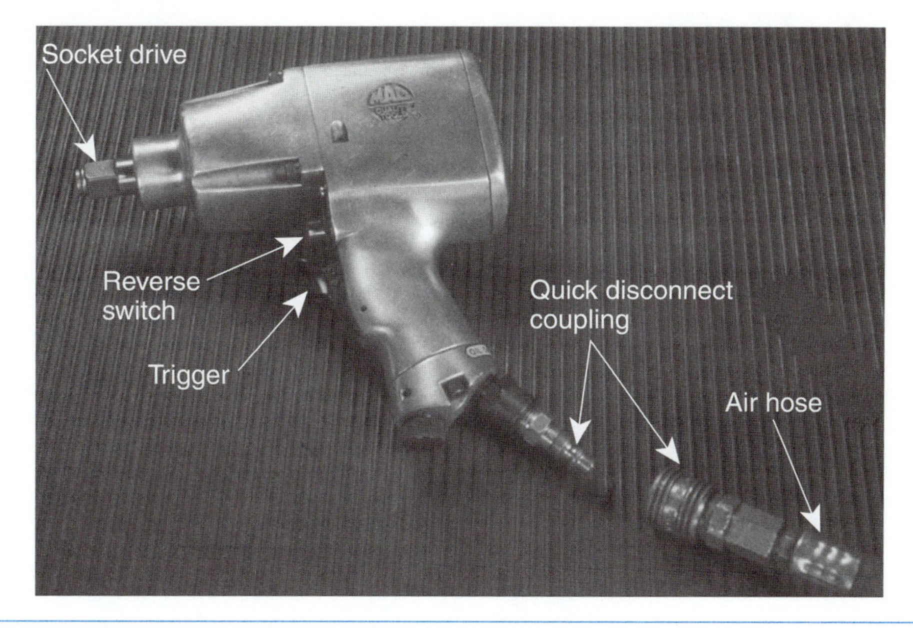

Figure 3-4 Parts and controls on a typical air impact wrench.

You are ready to set up and connect the air-operated impact wrench (Figure 3-4). The wrench has a socket drive on the end. Install the impact socket you plan to use on the wrench drive.

The wrench will need to be connected to the shop air source to get its power. The shop air hose is connected to the impact wrench with a part called a **quick disconnect coupling.** This coupling allows air lines to be attached to tools quickly without shutting off any air source. The coupling has two basic parts as shown (Figure 3-5). The part attached to the air hose is called the **female coupling.** The part attached to the wrench is called the **male coupling.** To attach the hose to the gun, pull back on the female coupling slide and insert it over the male coupling. Release the slide and it will lock the two parts together in an airtight fit. The coupling may be disconnected in the same way. Just pull the slide and the two parts of the coupling apart.

When you have the air line connected, you are ready to try the wrench. Depress the trigger. The wrench drive will spin the socket at a high rate of speed. Notice the direction the socket spins. In order to remove most lug nuts you must set the reverse switch on the wrench so the socket is driven counterclockwise. Try the reversing switch to see how it works, then set the wrench in the correct direction to remove the lug nuts.

A *quick disconnect coupling* is the part that allows the shop air hose to be easily attached to an air-operated tool like an air impact wrench.

A *female coupling* houses the locks that hold the male coupling in place after hookup. This coupling also has the release mechanism.

A *male coupling* slides into a mated female unit similar to the manner in which a ratchet's drive stud fits into a socket. The male is fitted with grooves and ridges for locking.

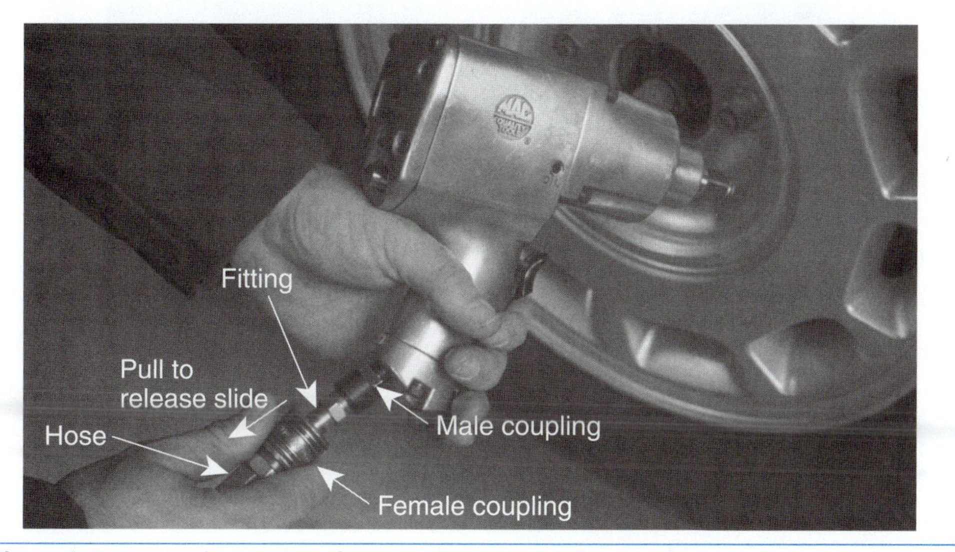

Figure 3-5 Parts and operation of a quick disconnect air line coupling.

Figure 3-6 Removing lug nuts with an air impact wrench.

WARNING: Some cars made in the 1940s and 1950s used left-handed threads on the lug nuts on the left side of the car. The designers did this so that if the nuts were loose, they would be automatically tightened by the rotation direction of the wheel. Look for this on old cars and do not mix up the lug nuts.

Insert the impact socket over one of the lug nuts. Pull the trigger and spin off the lug nut (Figure 3-6). Repeat this procedure on each of the other lug nuts. Support the wheel with your body as you remove the last lug nut to prevent the wheel from falling off the studs. Carefully lift the wheel off the studs. Use the proper lifting techniques described in the Classroom Manual.

To reinstall the wheel, first place it in position on the studs. You should always start the lug nuts by hand (Figure 3-7). Make sure they thread onto the stud several turns before using any wrench. Using a wrench to start the lug nuts could damage the threads.

You can use the air impact wrench to spin the nuts in contact with the wheel, but you cannot use a typical impact wrench to tighten the nuts. Reverse the direction of the impact wrench with the reverse switch. If the impact wrench has a pressure switch, set it to the lowest setting. Carefully spin each nut on the stud until it just begins to contact the wheel. Tightening must be done with a torque wrench, as explained in the next section.

Figure 3-7 Always start lug nuts by hand.

Figure 3-8 Wheel torque sockets can be used with an impact wrench to torque lug nuts.

If wheel torque sockets are available, select the correct one for the torque desired. Place the torque socket on the impact wrench in place of the regular impact socket used to remove the lug nuts (Figure 3-8). Use the impact wrench to torque the lug nuts.

Special Tools

Impact wrench

Wheel torque socket

Vehicle lift or jack

Adjusting and Using a Torque Wrench

Classroom Manual
page 37

Installing and tightening the lug nuts is a good way to learn how to use a torque wrench. Lug nuts should always be torqued with a torque wrench. Using the correct torque setting ensures that the lug nuts are tightened the right amount and evenly. If the lug nuts are tightened unevenly, the drum or rotor can be warped.

The first step is to look up the torque specifications for the lug nuts on the car you are servicing. An example of a wheel lug nut torque specification table is shown in Figure 3-9. This particular car has a recommended wheel nut torque of 140 N·m or 100 ft.-lbs. The 140 N·m is a metric system measurement for 140 Newton-meters. The conventional (English) system measurement is 100 foot-pounds. The measurement can be made with either a metric or conventional reading torque wrench.

Special Tools

Beam-type torque wrench

Dial readout-type wrench

Breakover hinge torque wrench

SPECIFICATIONS	
WHEEL NUT TORQUE	
All Wheels	140 N · m (100 ft.-lb.)
RUNOUT SPECIFICATIONS	
Maximum Radial Runout	
Steel Wheels	1.01 mm (0.040 inch)
Aluminum Wheels	0.76 mm (0.030 inch)
Maximum Lateral Runout	
Steel Wheels	1.14 mm (0.045 inch)
Aluminum Wheels	0.76 mm (0.030 inch)

Figure 3-9 A typical torque specification table. (Courtesy of Pontiac Motor Division)

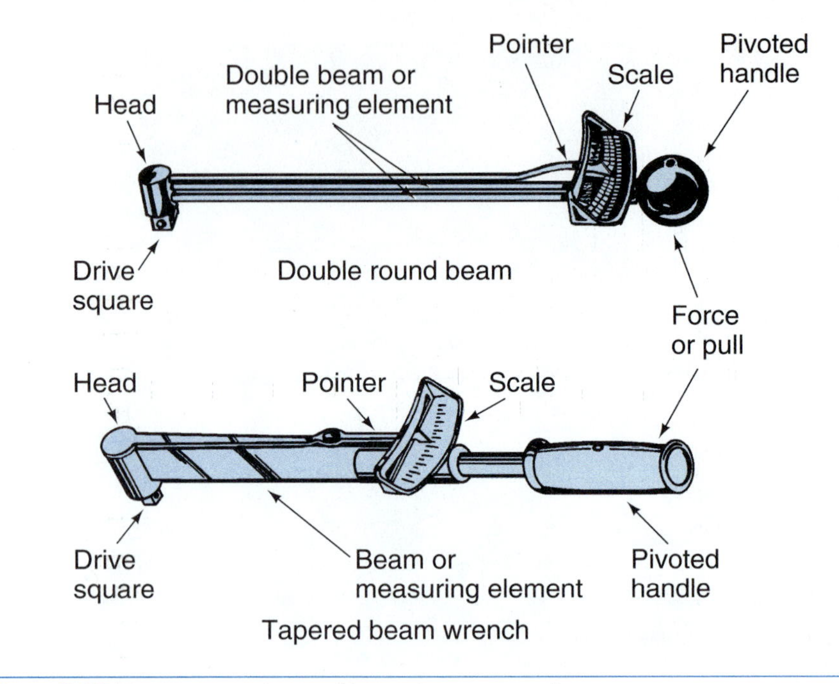

Figure 3-10 Two common types of beam torque wrenches.

There are two common beam-type torque wrenches (Figure 3-10): *a tapered beam* and a *double beam*. Both have a head with a square drive where the socket wrench is installed. As you pull on the pivoted handle, the beam bends in relation to the amount of force you apply. The amount of bend is shown by a pointer that measures the bending on a scale.

To tighten wheel lug nuts with a beam-type torque wrench, first select a wrench with the correct type of scale. In this case, it must measure in foot-pounds or Newton-meters. Then select the correct size socket that fits the lug nuts and install it on the torque wrench drive. Place the socket over the lug wrench and rotate the wrench by the handle while observing the scale (Figure 3-11). Accurate readings are only possible if you grip the handle in the center of the pivot as shown (Figure 3-12).

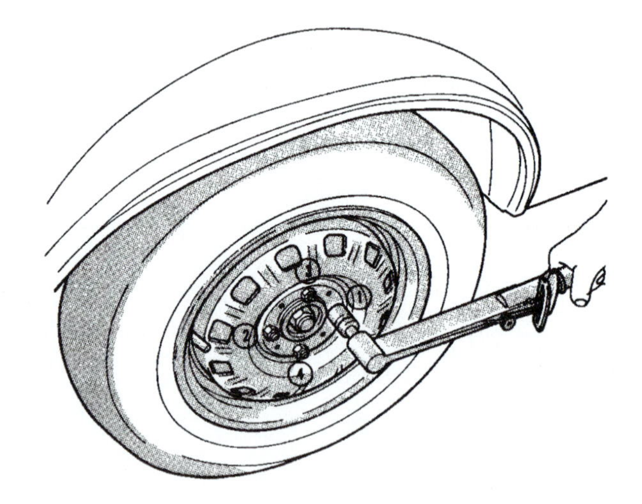

Figure 3-11 Tightening a lug nut while observing the pointer. (Courtesy of DaimlerChrysler Corporation)

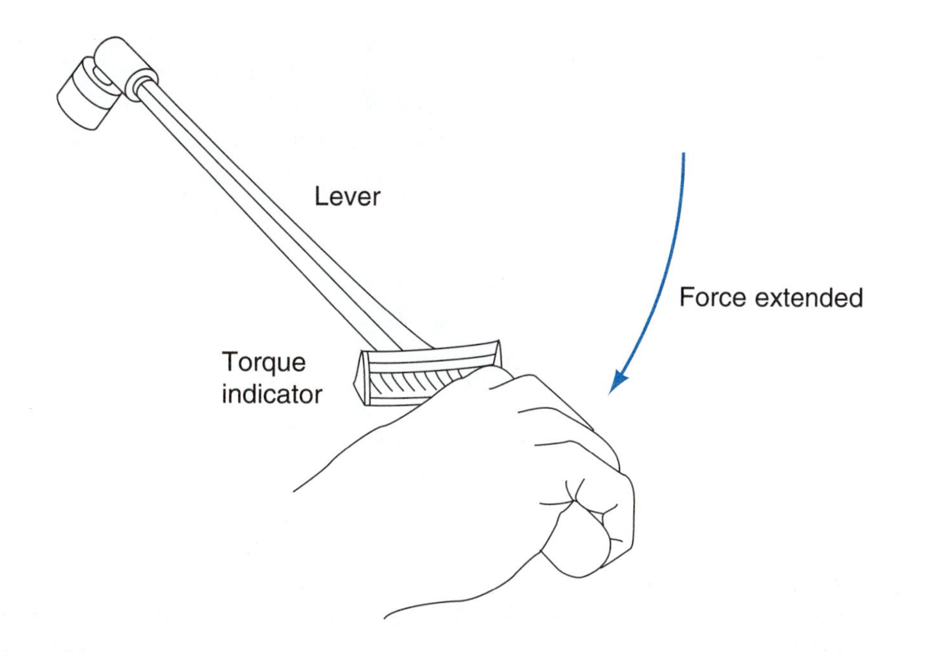

Figure 3-12 The right way to hold a torque wrench for accurate readings. (Courtesy of National Safety Council)

Stop turning the wrench when the pointer lines up with the 100 foot-pounds on the scale as shown in Figure 3-13. Most manufacturers recommend tightening each lug nut in stages of about 25 foot-pounds rather than tightening one all the way.

Another critical part of the job is the **tightening sequence**. The tightening sequence is the order in which each of the lug nuts is tightened. It was developed by the car manufacturer to prevent any distortion to the parts being tightened. Many automotive parts have a tightening sequence as well as a torque specification. A typical tightening sequence for a four and five lug nut wheel is shown in Figure 3-14.

A *tightening sequence* is the order in which the fasteners are tightened for a particular automotive part.

There are many styles of dial readout-type torque wrenches as shown in Figure 3-15. The wrench is operated just like the beam type. Take the reading by observing the needle inside

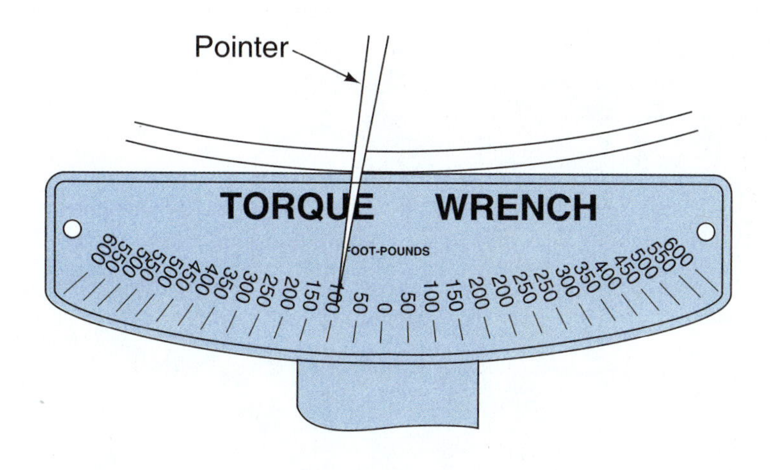

Figure 3-13 A pointer and scale reading of 100 ft.-lbs.

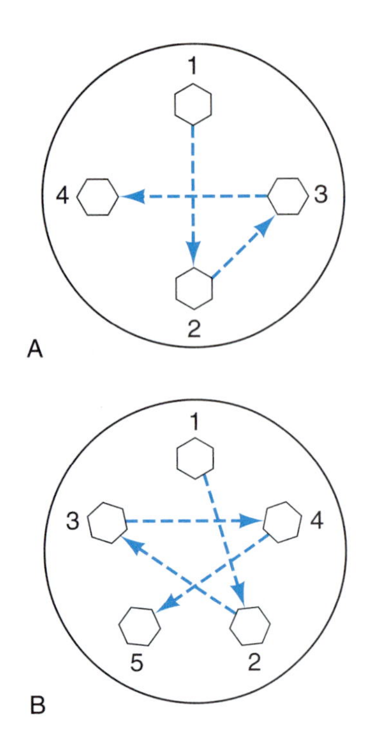

Figure 3-14 The tightening sequence for a four (A) and five (B) lug nut wheel. (Courtesy of National Safety Council)

Figure 3-15 Different systems of dial readout torque wrenches. (Courtesy of Central Tools, Inc.)

the dial. You should stop turning the wrench when the needle lines up with the specified torque as shown in Figure 3-16.

The breakover hinge-type torque wrench (Figure 3-17) has a hinge system near the head and a scale on the handle. The torque setting you wish to tighten to is set on the scale at the

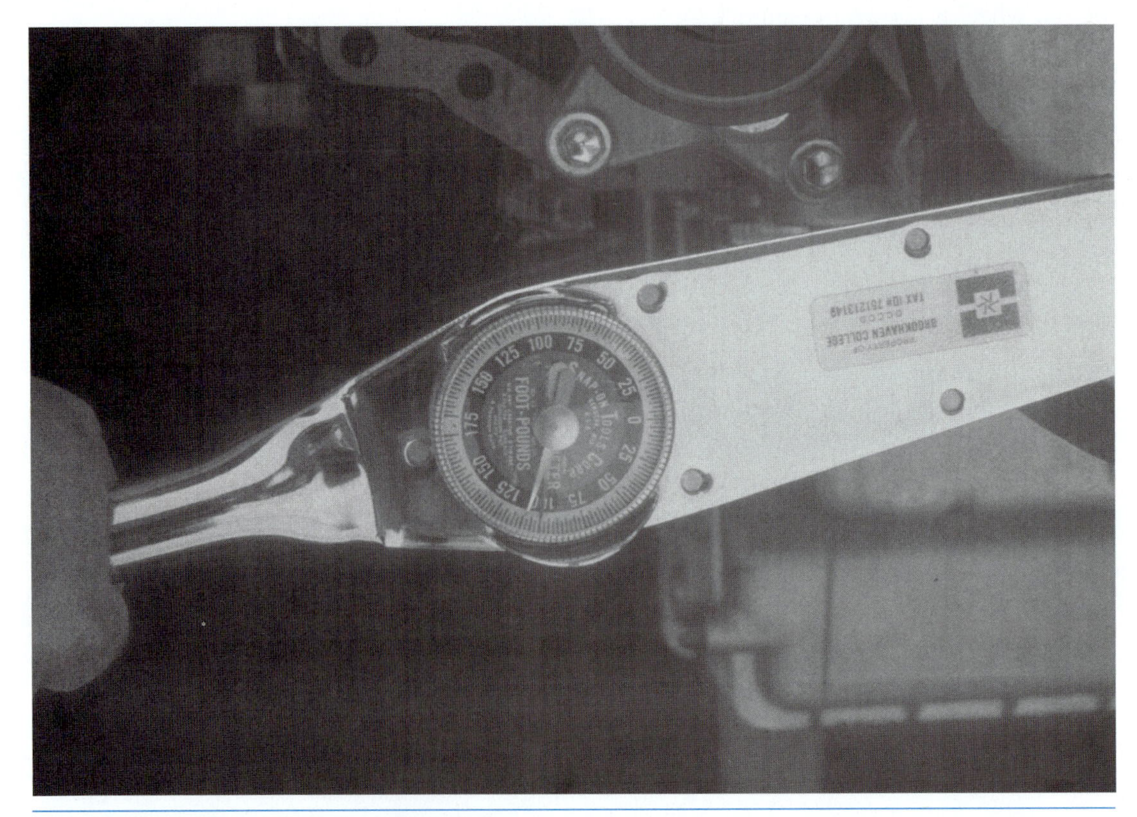

Figure 3-16 A dial reading of 100 ft.-lbs.

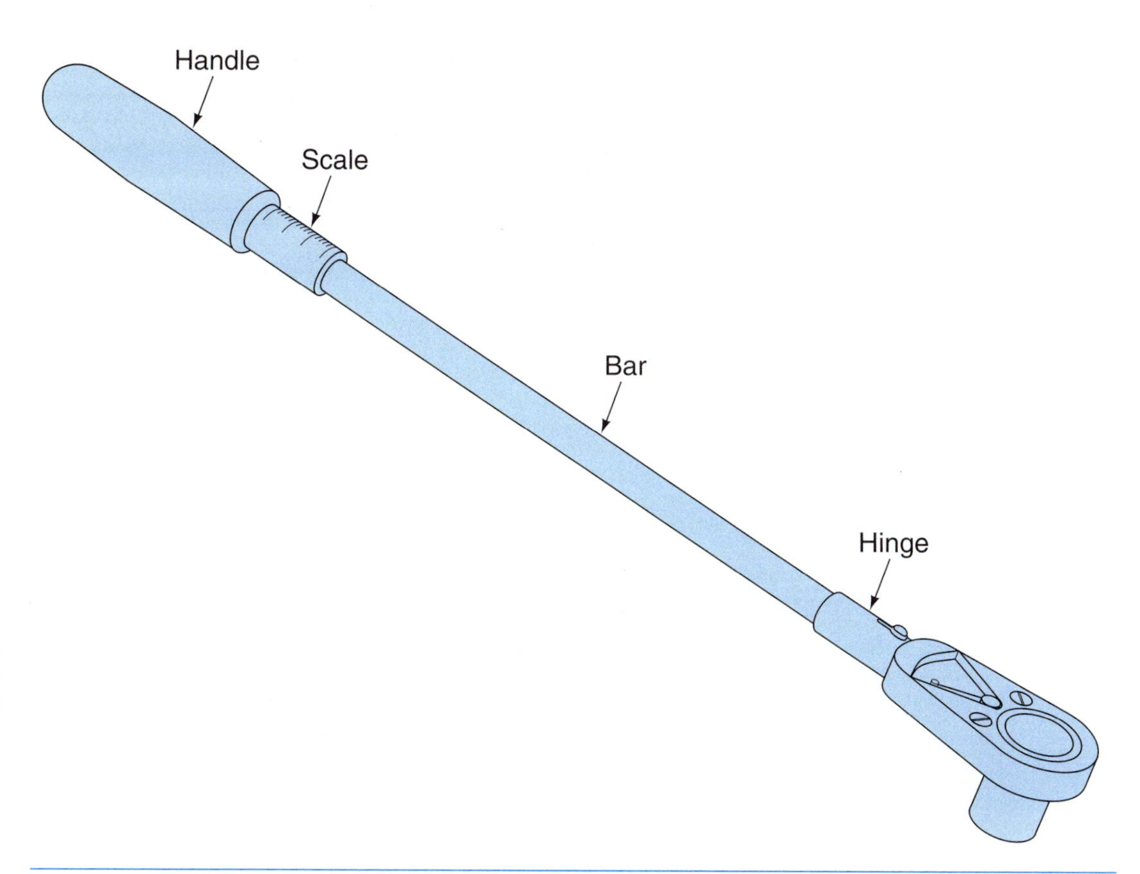

Figure 3-17 Parts of a breakover hinge-type torque wrench.

handle. When you tighten to that amount, the hinge is activated and the head swivels and makes a clicking noise. This lets you know you have reached the right torque.

You set the torque by turning the handle assembly. There is a scale on the bar of the wrench and index marks on the handle. Turn the handle until the 100 foot-pounds mark on the bar is indexed with the zero mark on the handle as shown in Figure 3-18. Then tighten the lug nuts until the head swivels and clicks.

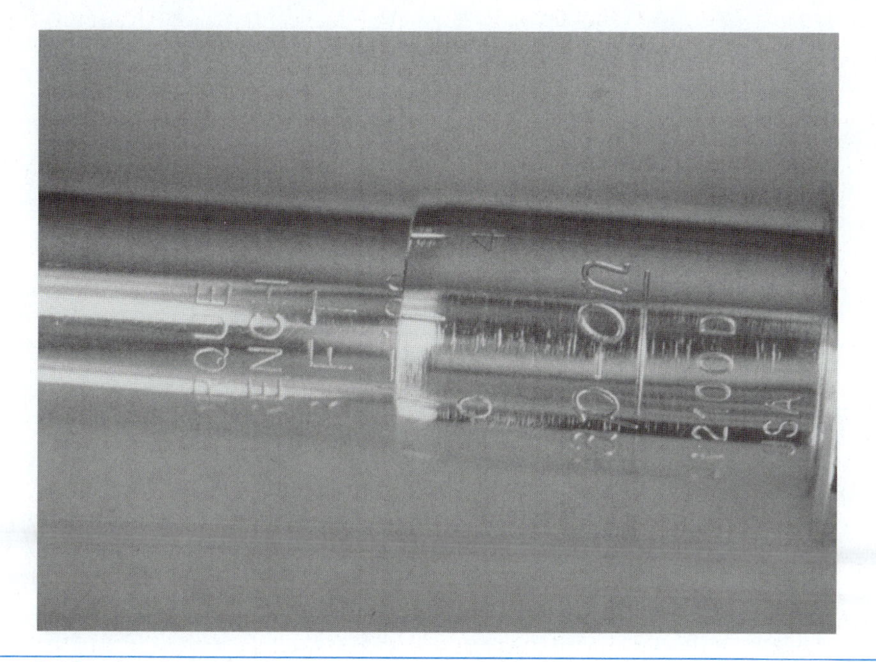

Figure 3-18 A breakover hinge handle setting of 100 ft.-lbs.

Classroom Manual
page 63

Special Tools

Selection of gear and bearing pullers

Pullers are used for many different types of service jobs. Technicians typically call them gear pullers.

Using Gear and Bearing Pullers

A good way to learn basic **puller** setup and use is to pull a part off a shaft. You may be able to find a scrap alternator around the shop to practice this job. The typical alternator (Figure 3-19) has a drive pulley that is installed tightly on a shaft. A puller must be used to remove this pulley.

The first step in the pulley removal is to remove any nut, set screw, or retainer holding it on the shaft. Then you are ready to select and use a puller. The correct type of puller is shown next to the alternator in Figure 3-20.

The puller has several different parts. The part that grips the part to be pulled is called the *jaw*. You must select a puller with jaws that fit snugly on the part to be pulled. Two- and three-jaw pullers are common. The jaws are attached to a *cross arm*. The cross arm swivels or

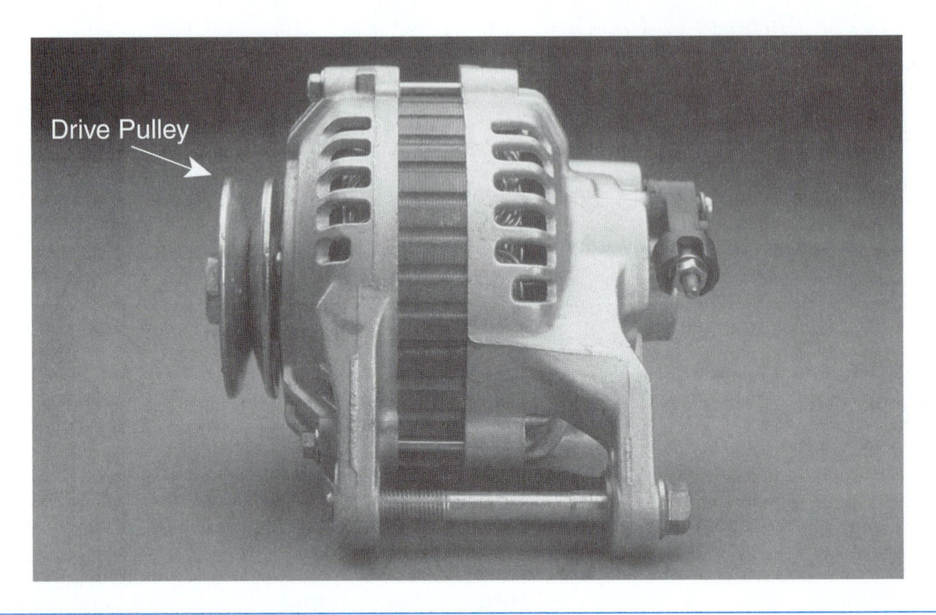

Figure 3-19 The alternator has a drive pulley attached to a shaft.

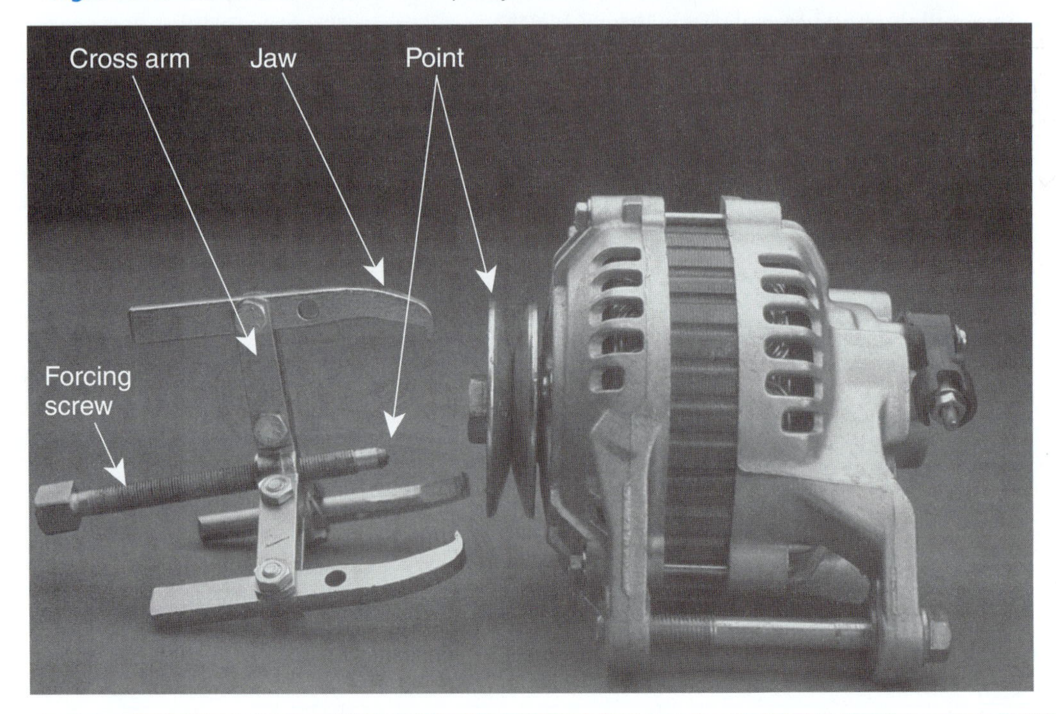

Figure 3-20 The correct puller to remove the alternator pulley.

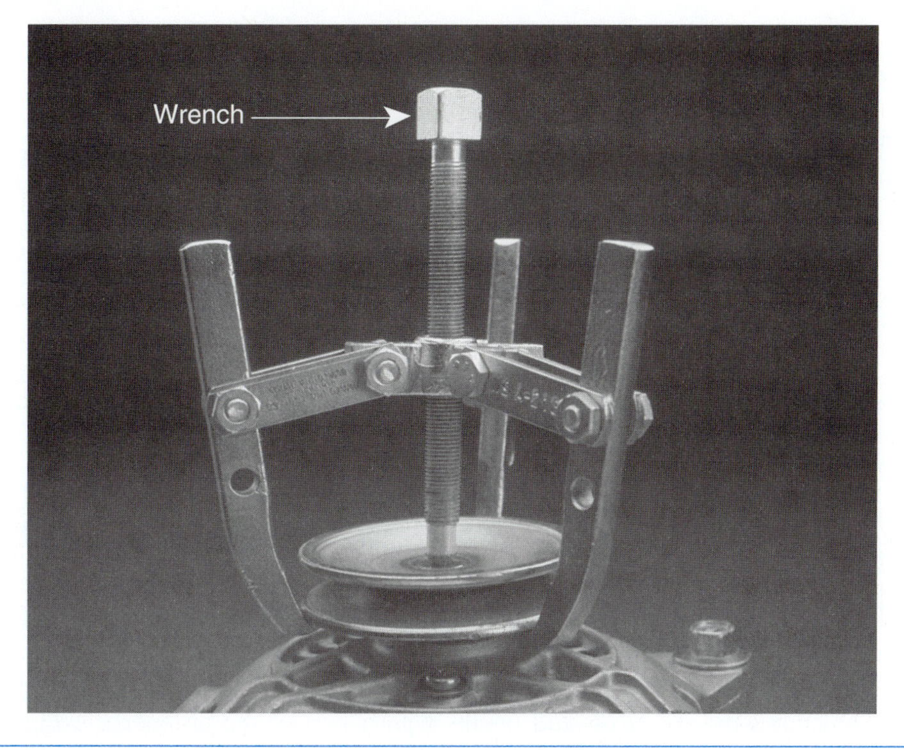

Figure 3-21 Using the puller to remove the alternator pulley.

slides so the jaws can be adjusted for different diameters of parts. The *forcing screw* is located in the middle of the puller. When the jaws are positioned on the pulley, this screw can be threaded down in contact with the shaft. Turning it with a wrench forces the pulley and shaft in different directions. This separates the two parts.

CAUTION: If the puller jaws do not fit the part, or the cross arm will not allow proper placement of the jaws, select another size puller. Using an incorrect size puller or not adjusting the puller correctly can allow the jaws to slip off during pulling and cause injury. Always wear eye protection when using a puller. Parts can break and cause eye injury.

Adjust the puller jaws and cross arms until the jaws fit a secure part of the pulley as shown in Figure 3-21. Choose the correct size wrench that fits the forcing screw and run the screw down in contact with the shaft. Make sure the point of the forcing screw is centered on the shaft. If not, readjust the puller cross arms or jaws. Slowly and cautiously tighten the forcing screw. Watch carefully for any evidence that the puller is trying to slip off or slip off-center. Make any necessary readjustments. If the puller is secure, tighten the puller forcing screw, and pull the pulley of its shaft.

Removing a Broken Stud or Screw

Classroom Manual
page 51

Regardless of how careful you are as a technician, sooner or later you will have the misfortune to break a screw or stud (Figure 3-22). The broken stud or screw will have to be removed before you can complete your service work. Broken screw removal is a very valuable skill. We suggest you try some of the techniques described here on some scrap parts before you do it on a customer's car.

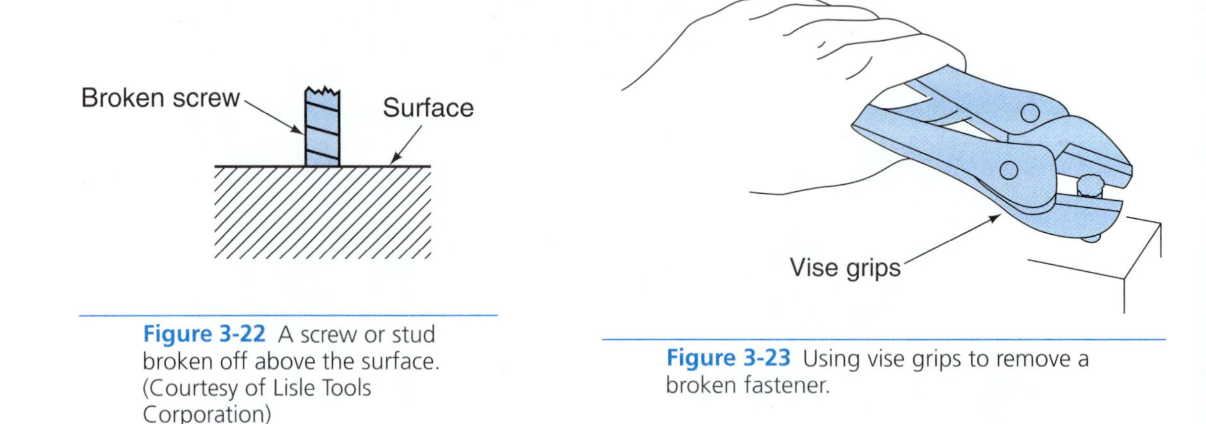

Figure 3-22 A screw or stud broken off above the surface. (Courtesy of Lisle Tools Corporation)

Figure 3-23 Using vise grips to remove a broken fastener.

Special Tools

- Vise grips
- Small cape chisel
- Set of screw extractors
- 3/8-inch portable drill

Screws fail for many reasons. Sometimes they are defective when manufactured. Often they are overtightened by a technician not using a torque wrench. Sometimes they are driven too deeply into a threaded hole and bottom out. Regardless of the cause, a broken screw has to be removed and a new fastener used in its place.

There are a couple of removal techniques that should be used before any special tools are tried. Sometimes the screw or stud breaks with a part of the fastener sticking above the surface. This often happens if a screw is bottomed out in the hole. First try a pair of vise grips. Adjust them to the proper size and grip the broken fastener as shown in Figure 3-23. Then turn the fastener in the correct direction for removal.

If the fastener is broken off even with or below the surface, the removal job will be more difficult. Sometimes the fastener breaks even though it is not extremely tight. A fast removal technique uses a small cape chisel. Place the chisel on the broken surface of the fastener. Use a hammer to tap it in a direction for removal as shown in Figure 3-24. If the fastener is loose enough, it may begin to rotate. Tap the fastener around until it sticks above the surface. Then grab and continue turning it with vise grips.

If these techniques fail to work, you will have to use a screw extractor. There are many different types of screw extractors. They often come in sets and are often supplied with the necessary size of drill as shown in Figure 3-25.

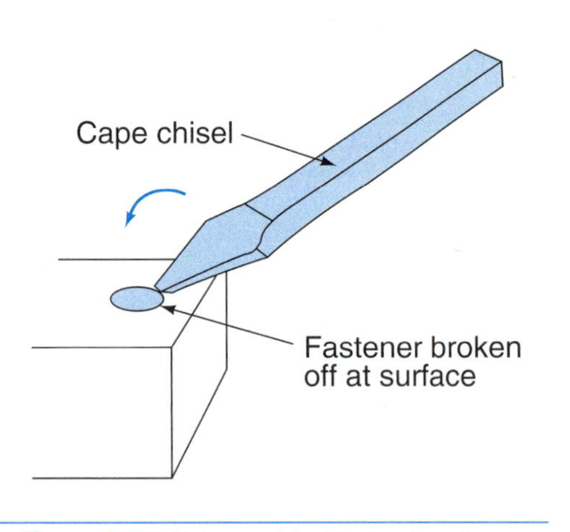

Figure 3-24 Using a chisel to back out a broken fastener.

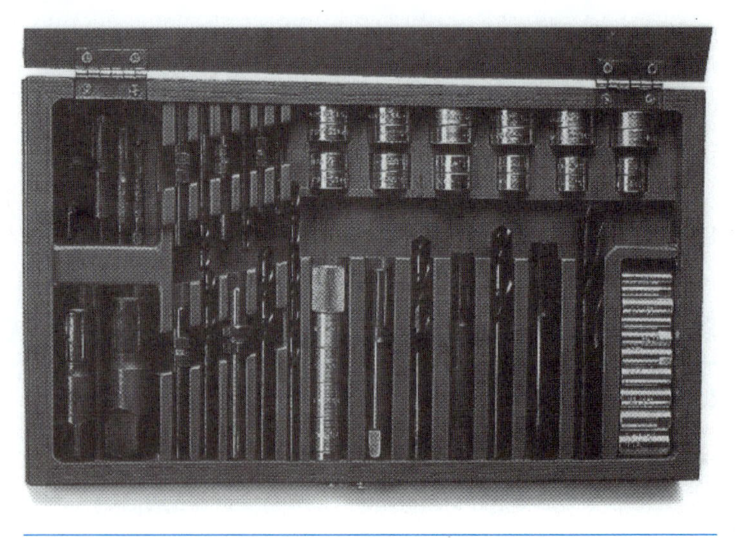

Figure 3-25 A screw extractor set with matching drills. (Courtesy of Snap-on Tools Company)

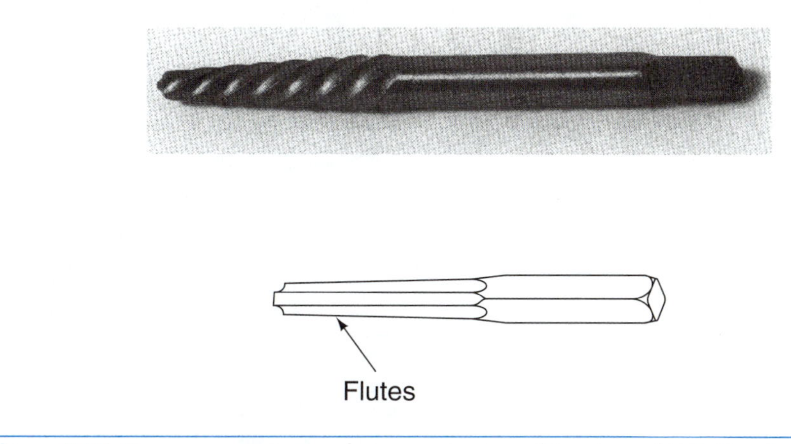

Figure 3-26 Two common types of screw extractors. (Courtesy of U.S. Navy)

Two common types of screw extractors are shown in Figure 3-26. One has *reverse threads* and the other has a set of *flutes*. In either case, the screw extractor is made to fit in a hole you drill in the fastener. Then you insert the screw extractor and turn it with a tap wrench. The reverse threads or flutes grip the fastener and allow it to be backed out of the threads.

When using a screw extractor, always follow the instructions supplied with the set. You must select the correct size drill and screw extractor for the job. The correct size to use is typically shown on the set instructions.

The first step is to center punch the exact center of the broken fastener as shown in Figure 3-27. Then install the correct size drill bit in a $^{3}/_{8}$-inch size portable drill motor. Drill through the center of the fastener to a depth recommended in the screw extractor instructions as shown in Figure 3-28. Be careful not to drill too deeply and damage the automotive part, or to drill off to an angle.

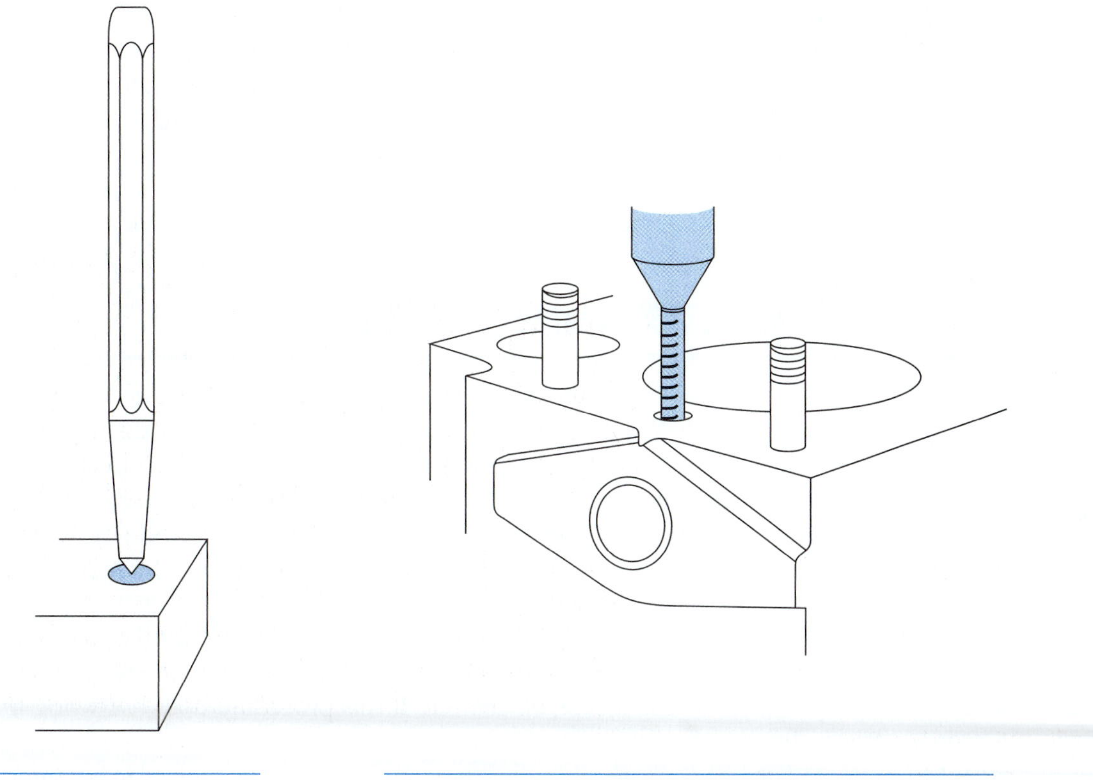

Figure 3-27 Center punching the center of the broken fastener.

Figure 3-28 Drilling a hole in the broken fastener.

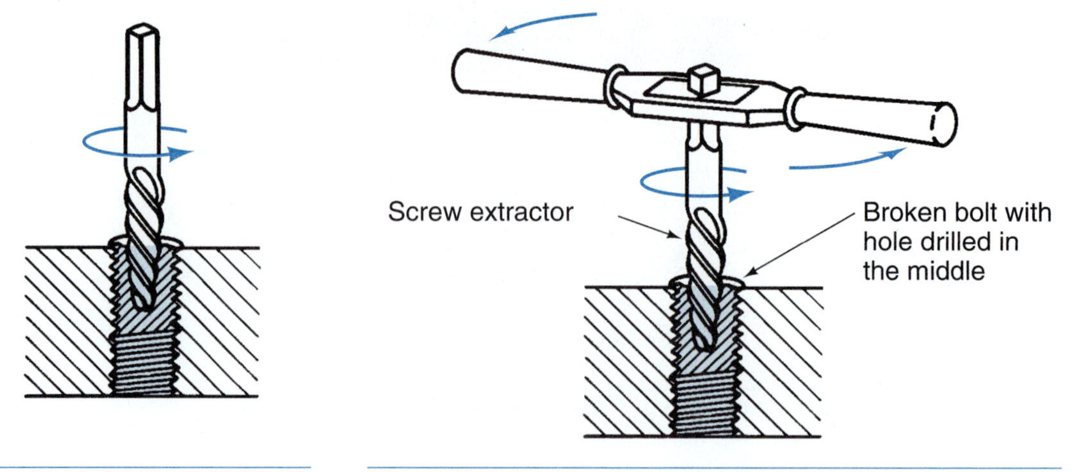

Figure 3-29 Installing the screw extractor in the drilled hole.

Figure 3-30 Using a tap wrench and extractor to remove the broken fastener.

When you have the hole drilled properly, you are ready to install the extractor. Be sure to choose the correct size extractor for the hole you drilled. Install the extractor as shown in Figure 3-29. Some extractors must be tapped into the hole with a hammer.

Use a tap wrench to turn the extractor in a direction to back the fastener out of the hole as shown in Figure 3-30. Some screw extractor sets have drivers that slide over the extractor and fit a standard automotive wrench.

CAUTION: Always wear eye protection when drilling and using a screw extractor. Parts could fracture and injure your eyes.

SERVICE TIP: Sometimes difficult-to-remove broken fasteners can be removed by carefully heating or cooling the area. Because the fastener and the threaded hole are made from different material, they will expand or contract at different rates. This can cause the fastener to loosen up. Heating can be done with an oxyacetylene torch. This technique can only be used where there is no fire danger or around components made of soft metals like aluminum. Heat the area around the fastener. Try to keep as much heat as possible from the fastener. This will speed the process of separating the fastener threads from the component.

Removing a Damaged Nut

Classroom Manual
pages 29–30

Special Tools

Cold chisel
Vise grips
Penetrating oil
Nut splitter
Antiseize compound

A good technician knows how to remove damaged nuts. Cars that have accumulated many years and high mileage and have been repaired many times often have damaged fasteners. For example, hex nuts may have rounded off corners due to the use of an incorrect wrench size (Figure 3-31). Or, a nut may have rusted to a bolt or stud like the one shown in Figure 3-32.

A nut that is rounded off at the corners often will not allow the correct size wrench to fit tightly enough to remove the nut. These usually result when a technician uses the wrong size wrench or when a nut is stuck or "seized" to the bolt or stud.

First try to use a file and remove the displaced metal. Then try the largest, adjustable, open-end wrench that will correctly fit on the nut. Be careful to push on the wrench with your hand positioned properly in case it slips off the nut. If this procedure does not work, try large vise grips. Adjust these to fit tightly on the nut. Try to turn the nut with the vise grips. Be careful to protect your hand in case the grips slip off the nut. If neither of these steps work, you will have to split the nut as described later.

Figure 3-31 An example of a nut with rounded corners.

Figure 3-32 An example of a nut that has rusted on a stud.

Another problem is nuts that are stuck on a bolt or stud. A nut may be stuck because it was overtightened and its threads are jammed or distorted into the threads of the stud or bolt. Corrosion and rust build up between the two fasteners and lock them together. This is called **condition seizing.** In either case, the nut may be impossible to loosen and pulling on it with too much force can cause the bolt or stud to break. This makes it even more difficult to remove the broken stud or bolt.

Condition seizing is a fastener that has seized because of conditions around and on the fastener. One condition is over tightening. A second condition may be exposure to water followed by rust.

When you find a stuck nut, first apply a **penetrating oil** as shown in Figure 3-33. Penetrating oil is made to seep into the threads between stuck fasteners and remove the corrosion. Follow the directions on the oil container. You will usually have to apply the material several times and wait a long period of time for it to work.

Penetrating oil is a fluid used to penetrate between stuck threads to allow the fasteners to be removed.

SERVICE TIP: Stuck fasteners can often be removed by heating them. You can use an oxyacetylene torch to warm the nut, which causes it to expand and break its bond with the stud or bolt. You can only use this technique on nuts where it is possible to heat without the danger of fire or around components that are not made of soft materials like aluminum.

When penetrating oil fails to remove the nut, or when the nut is too damaged to be removed, the last resort is to split the nut. Use a sharp flat chisel to make a complete cut between the nut and the stud or bolt as shown in Figure 3-34. Try not to contact the threads of

Figure 3-33 Penetrating oil can free a rusted nut.

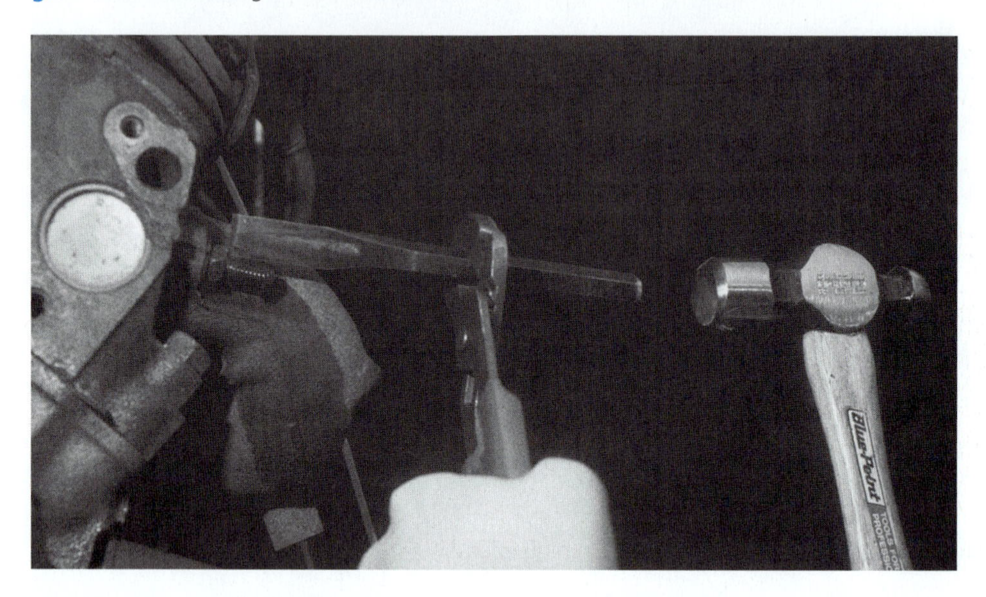

Figure 3-34 Splitting a nut with a flat chisel.

the bolt or stud. Usually one cut will free the nut. Sometimes you will have to cut it on opposite sides to get it off.

A *nut splitter* is a tool used with a wrench to split a stuck nut for removal.

A tool called a **nut splitter** is available to split a nut (Figure 3-35). It is positioned over the stuck nut. A forcing screw driven by a wrench causes a cutter to penetrate and split the nut.

Fasteners used in areas of the car subject to heat and corrosion often seize over time. An example is the fasteners used under the car on the exhaust system. Seizing is also a problem when steel fasteners are used in aluminum parts. A good prevention technique is to coat these fasteners with an antiseize compound when they are assembled. This will make your job much easier if you have to remove the parts later. An example of an antiseize compound is shown in Figure 3-36.

CUSTOMER CARE: When a little extra time is taken to do a professional job, like using a torque wrench or coating fasteners with antiseize compound, the repair job will last longer. A tecnnician's reputation as a craftperson will pay off in the long run when customers want her or him to work on their cars.

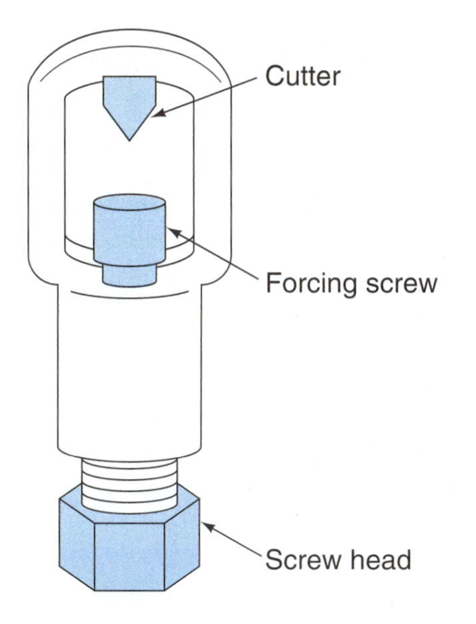

Figure 3-35 A nut splitter.

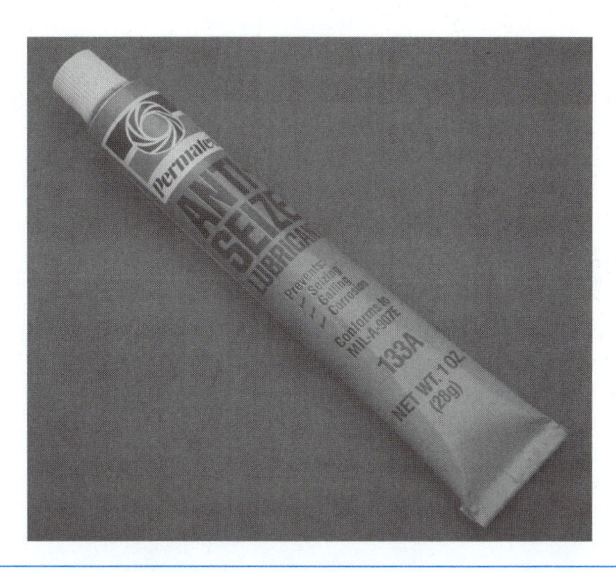

Figure 3-36 Antiseize compound prevents nuts from sticking to bolts and studs.

Setting Up and Using an Oxyacetylene Torch for Heating

Classroom Manual
page 58

Special Tools

Welding goggles

Welding gloves

Oxyacetylene welding and cutting outfit

Striker

CAUTION: Oxygen and acetylene cylinders must be secured in an upright position. Never allow a cylinder to fall over. If a cylinder were to fall over and break off its valving, it could become a rocket. When cylinders are being stored, they must have a steel protecting cap over the valve area.

The oxyacetylene welding and cutting outfit is used in the shop for welding, cutting, or heating. Photo Sequence 3 shows a typical procedure for setting up and lighting an oxyacetylene torch for heating. Welding and cutting are skills many technicians develop in order to do many advanced repair procedures. A good first step to mastering these skills is to learn how to set up the system for heating. Heating parts to aid in their removal is an important use of the acetylene torch.

WARNING: The installation of welding regulators and hoses on the oxygen and acetylene cylinders should be done by a technician with special training. We describe the set-up of the system from the torch handle only.

Before beginning any set-up of the oxyacetylene outfit, make sure the gas flow from both the oxygen and acetylene cylinder is off. Look at the hose pressure and cylinder pressure gauge on top of each cylinder (Figure 3-37). Both gauges should be at zero. If not, turn the cylinder valve on top of the cylinder clockwise to close. Repeat this procedure for the other cylinder.

If the valve was open and there is a pressure reading on either gauge, bleed the gas out of the system before going any further. Make sure the cylinder valve is closed. Open the oxygen valve at the torch handle. Turn the oxygen regulator handle clockwise (in). This will allow oxygen in the line to flow through the hose and out the torch handle. Both gauges will drop to zero. Turn the regulator handle counterclockwise (out). Repeat this same procedure with the acetylene cylinder.

Now you are ready to install the torch handle (Figure 3-38) onto the hoses. The handle used for heating is the same one used for welding. It has two connections: one is for the oxygen hose and the other is for the acetylene hose. The torch handle connections are marked "AC" for acetylene and "OX" for oxygen. To ensure that the correct hose is connected to the

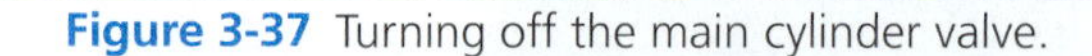

Figure 3-37 Turning off the main cylinder valve.

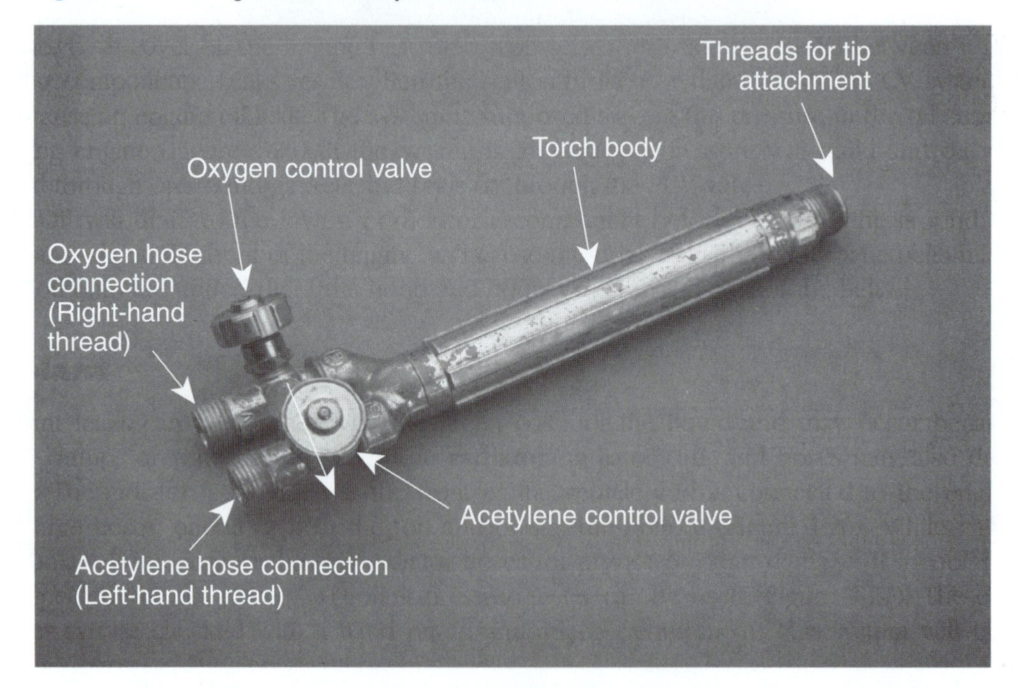

Figure 3-38 Parts of oxyacetylene torch handle.

Photo Sequence 3
Typical Procedure for Setting Up and Lighting an Oxyacetylene Torch for Heating

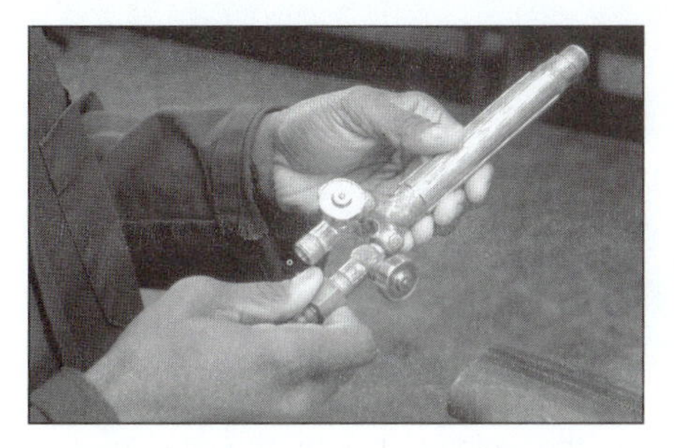

P3-1 Connect the green or black oxygen hose to the torch handle connection marked "OX" by turning it clockwise (right-hand threads).

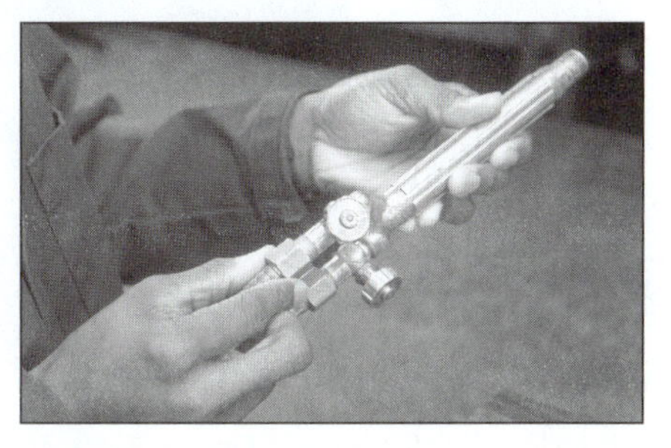

P3-2 Connect the red acetylene hose to the torch handle connection marked "AC" by turning it counterclockwise (left-hand threads).

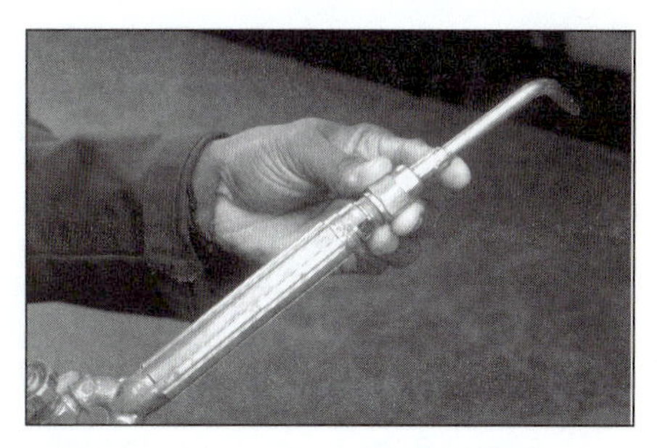

P3-3 Connect the correct size torch tip to the end of the torch handle.

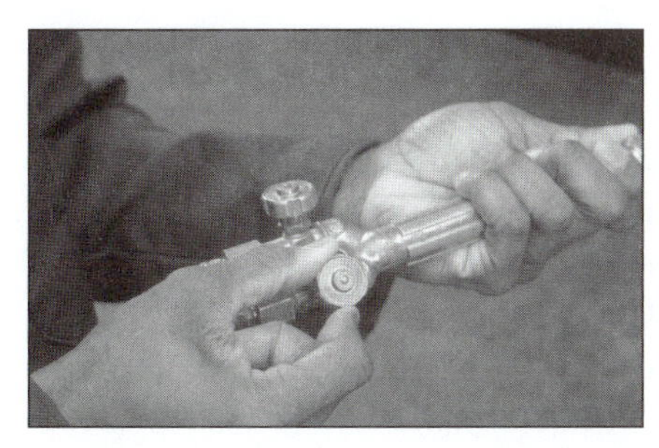

P3-4 Close both torch handle valves by turning them clockwise.

P3-5 Turn the main cylinder valve on the oxygen cylinder valve on by turning it counterclockwise.

P3-6 Turn the main cylinder valve on the aceytlene cylinder valve on by turning it counterclockwise.

P3-7 Turn the oxygen regulator valve clockwise until the working gauge reads 10 psi.

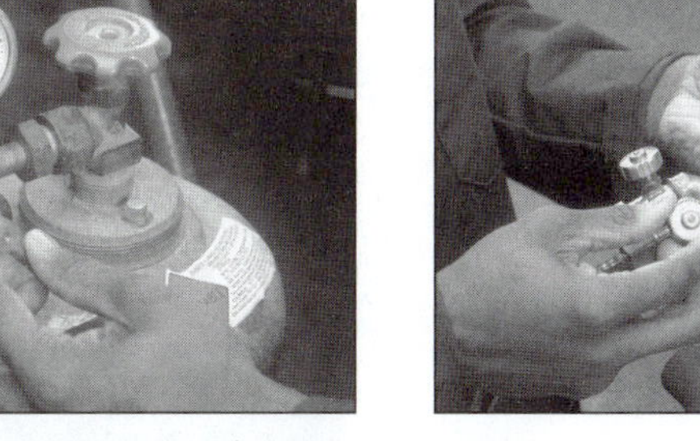

P3-8 Turn the acetylene regulator valve clockwise until the working pressure gauge reads 5 psi.

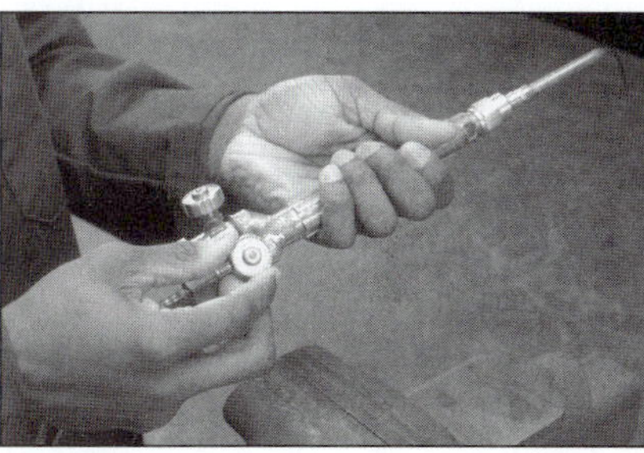

P3-9 Turn the torch handle acetylene valve on slightly.

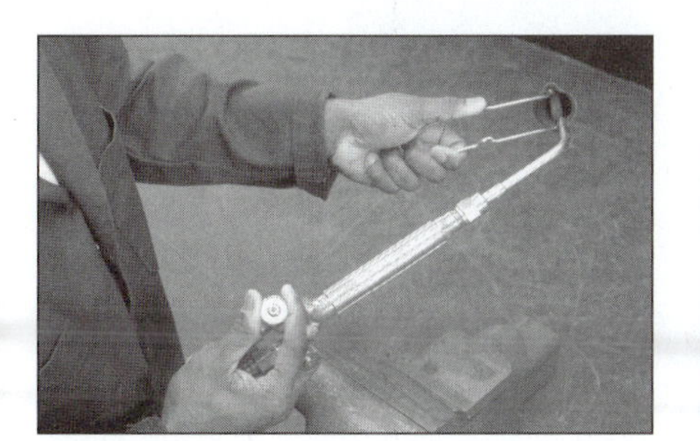

P3-10 Use a striker to ignite the acetylene gas at the tip of the torch.

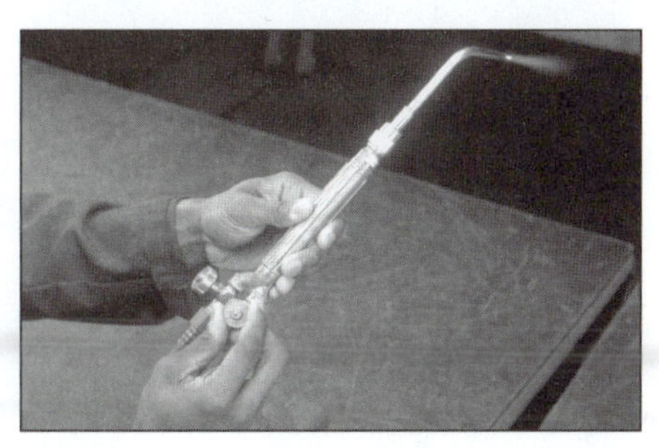

P3-11 Open the oxygen valve at the torch handle and add oxygen to achieve a netural flame.

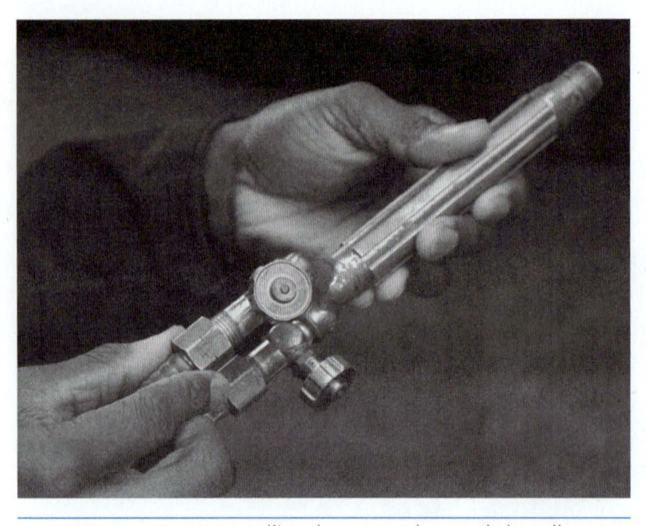

Figure 3-39 Installing hoses to the torch handle.

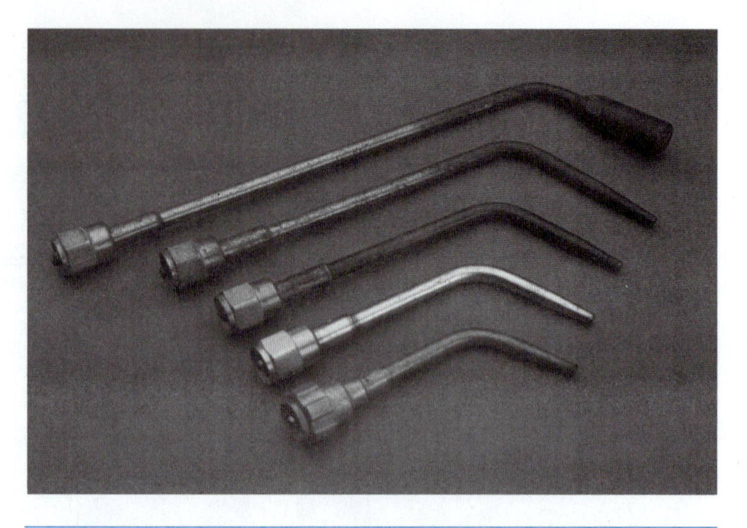

Figure 3-40 The larger the welding tip size, the more heat it can develop.

proper connection, the hoses are color coded and have left- and right-handed threads. The oxygen hose is green or black and its connection has standard right-hand threads. The acetylene hose is red and its connection has left-hand threads.

Connect the red hose to the AC connection on the torch handle, observing the left-hand threads. Connect the green or black hose to the "OX" connection on the torch handle. Use a wrench to gently tighten the hose connections as shown in Figure 3-39.

A *welding tip* is the part attached to an oxyacetylene torch handle where the gases are mixed to create the flame.

The torch handle can be equipped with a number of different size **welding tips.** Welding tips are stamped with a number. The larger the number, the larger the tip and the more heat that can be developed. Common sizes are 1 to 4 as shown in Figure 3-40. Most heating is done with a small tip such as a number 1. Select a size 1 welding tip and screw it onto the end of the torch handle as shown in Figure 3-41.

Now adjust the gas pressure for heating. There are two valves on the torch handle. The one next to the oxygen hose controls the flow of oxygen to the tip. The one next to the acetylene hose controls the flow of acetylene to the tip. Close both these valves by turning them clockwise.

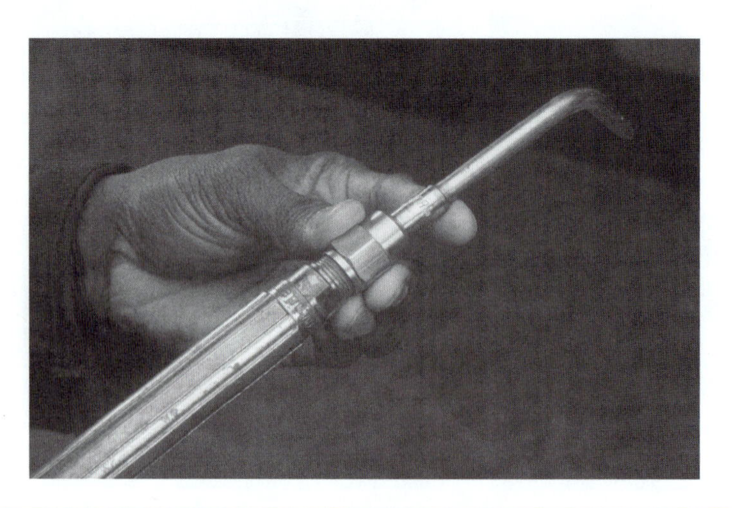

Figure 3-41 Installing the welding tip on the torch handle.

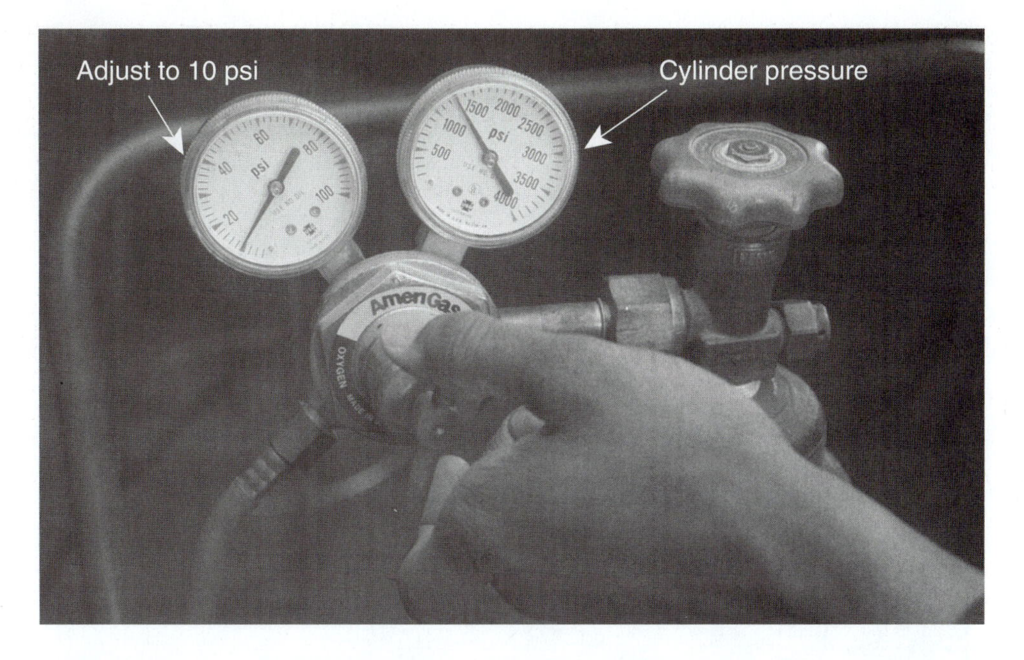

Figure 3-42 Adjusting the working pressure with the regulator handle.

Turn the main valve on the top of each cylinder counterclockwise to open the valve. The needle on the cylinder pressure gauge will rise to show the pressure in the cylinder. Turn the handle on the oxygen regulator clockwise (in) until the needle registers 10 psi (Figure 3-42). Turn the handle on the acetylene regulator the same way until the needle points to 5 psi. This will be the working pressure for heating.

CAUTION: The oxyacetylene torch develops dangerous ultraviolet light rays. Always wear shaded welding goggles when using this equipment. The flame at the end of the torch is very hot. Wear gloves and be careful not to direct the flame at another person or anything that could ignite.

You are now ready to ignite the torch. Put on your welding goggles and welding gloves. The torch is ignited with a tool called a **striker**. The striker (Figure 3-43) has a *striker bar* and *flint* that cause a spark when rubbed together. The spark is used to ignite the torch flame. There is a cup on the striker that protects the operator from burns during startup.

The *striker* has a flint that creates a spark to ignite an oxyacetylene flame.

CAUTION: Always use a striker to ignite a torch. Never use matches or a cigarette lighter because these would put your hand next to the flame. Never allow the flame to point at any part of the welding outfit. When you heat something, always mark it "hot" with chalk so a co-worker will not get burned picking it up.

Hold the torch handle in your hand with the tip pointing downward away from your body and away from the welding cylinders. Hold the striker by the handle and position it near the tip. Turn the acetylene valve on the torch handle just slightly open. Rub the striker flint over the striker bar to create a spark (Figure 3-44). The spark will ignite the acetylene gas coming out of the tip. Open the acetylene valve slowly until you see the sooty smoke start to disappear. Then slowly open the oxygen valve on the torch handle.

As you open the oxygen valve, you will see the color of the flame change at the tip. The yellow flame of the acetylene will change to a blue as you add oxygen. Continue to slowly

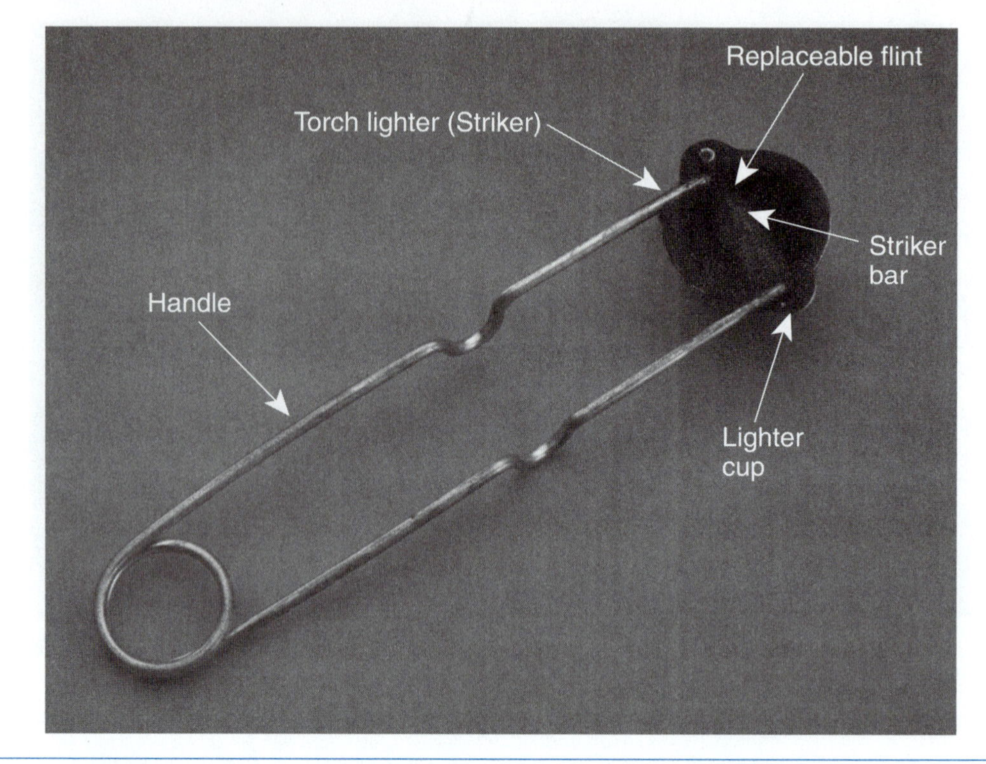

Figure 3-43 The striker is used to create a spark to ignite the torch flame.

A *neutral flame* is an oxyacetylene flame with a small, sharp inner cone used for heating and welding.

add oxygen until you can see a small, sharp blue inner cone in the middle of the flame envelope as shown in Figure 3-45. This is called a **neutral flame** and is the type of flame wanted for heating.

After heating, it is important to know how to shut the equipment down. First, turn off the acetylene valve on the torch handle. This will extinguish the flame. Then turn off the oxygen valve on the torch handle. Turn the main cylinder valve clockwise on top of both cylinders. Open the valves on the torch handle to bleed the system. Turn the oxygen and acetylene regulator handles counterclockwise until they become loose. Close both valves on the torch handle. Move the cylinders to their storage area.

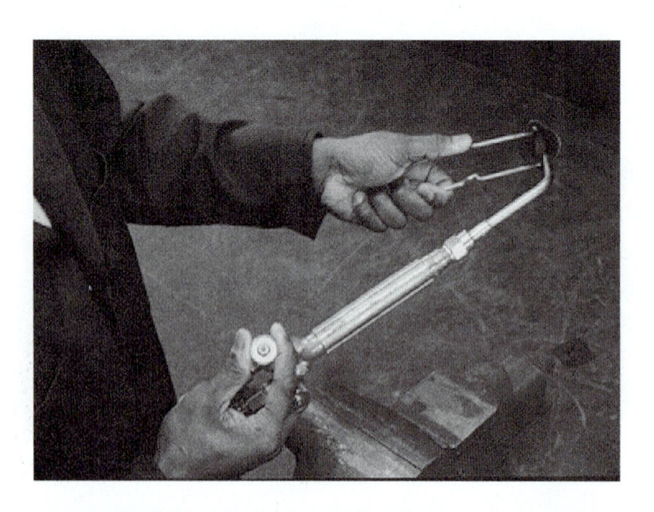

Figure 3-44 Igniting the torch flame with a striker.

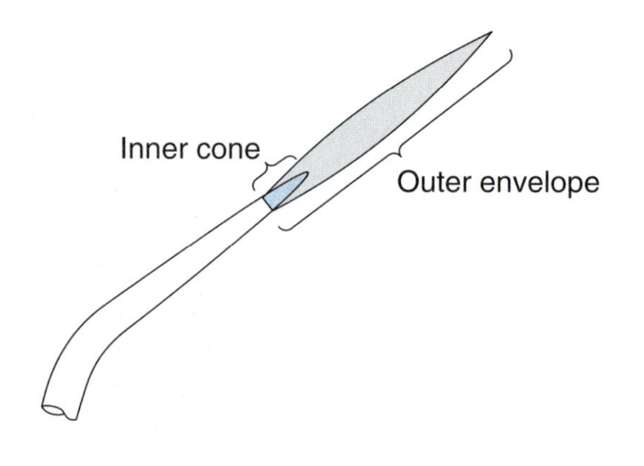

Figure 3-45 A neutral flame has a small, sharp inner cone.

Lifting a Vehicle

To work safety and quickly, it is sometimes required to raise the vehicle to a good working height. If brakes or tires are being done, a good working height has the wheels at about chest level. Easier engine and transmission oil changes can be done with the vehicle high enough for the technician to walk under it.

Classroom Manual
pages 60–63

Special Tools

Vehicle double-post above-the-ground lift

Service manual showing lift points

Double-Post Lifts

Double- and single-post lifts work similarly with the primary difference being the position of the lift cylinder. Procedures discussed here can be used to some extent with single-post lifts. Before lifting any vehicle with any type of lift or jack, consult the driver's manual or service manual for the location of the vehicle's lift or jack points (Figure 3-46).

WARNING: Never guess at the vehicle's lifting points. What appears to be solid metal may bend or be crushed when the vehicle's weight is transferred to it. Damage to the vehicle and possible injury could occur.

Move the vehicle so it is centered between the posts. The front doors should be roughly even with the posts. Most lifts of this type have shorter lift arms at the front. This places most of the vehicle's weight closer to the lifting cylinder and provides better balance. Most double-post lifts have a **spring-loaded switch** button that operates the motor and pump. Pressing and holding the button down will power the motor. Any time the switch is released for any reason, the spring will automatically push the switch to OFF.

This type of *spring-loaded switch* is sometimes referred to as a "dead-man switch."

After identifying the lift points, rotate the lift arms under the vehicle and position the pads directly under the lift points. On longer cars and trucks, it may be better to raise the lift arms until the pads are just below the vehicle. This makes it easier to properly place the pads.

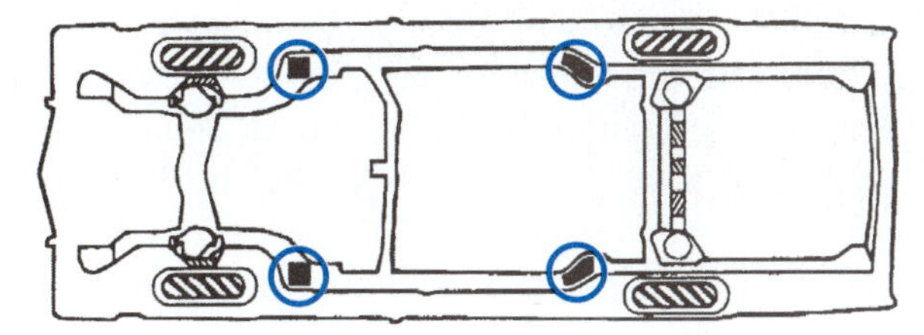

UNIBODY

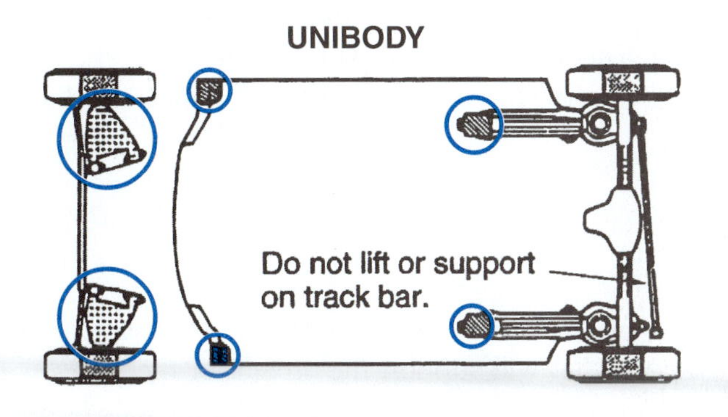

Figure 3-46 The highlighted points are typical lift and jack points.

With the pads aligned, lift the arms until there is contact between the pad and vehicle. Check the contact points to ensure there is good proper alignment between the pads and lift points. Make adjustments as necessary.

As the vehicle is being to raised, note that both ends of the vehicle are moving upward evenly. Pickup trucks are hard to keep even because of their length and the position of the lift points. If the vehicle is not even from end to end and side to side, lower it back to the ground and move the vehicle or arms until both ends rise at the same time. Continue to raise the vehicle until all wheels are off the floor. Attempt to rock the vehicle by hand. If positioned properly on the lift, it should not rock at all. In fact, it will probably have a more solid side-to-side movement then when sitting on its wheels.

When the vehicle is at the desired work height, stop the upward movement. Lower the vehicle until the safety locks on the lift engage. The vehicle can now be repaired.

To lower the vehicle, first raise it high enough to clear the locks. There is usually a lock on each post. Move both to the unlock position. Some lifts require the technician to hold an unlocking lever in place while the lowering valve is operated. A small lever attached to another spring-loaded switch usually operates the lowering valve. Lower the vehicle until the pads break contact. Continue lowering the lift arms until they are at the bottom of their travel. Swing the arms from under the vehicle.

Floor Jacks

Special Tools

Floor jack

Wheel blocks

Service showing jack point

Jack stands

CAUTION: Never perform any work under a vehicle supported only by a jack. Serious injury could result if the vehicle slips from the jack or the jack moves.

CAUTION: Never perform any work other than a tire change using a tire-changing jack. This jack is designed to raise one corner of the vehicle to change a tire. It was not made for other repairs. Serious injury could result when the jack and vehicle move.

Some work can be done underneath the vehicle using a floor jack and jack stands. It may be more difficult, but for some jobs it may be quicker than using a lift.

Before jacking the vehicle, decide where the jack's lift pad will be placed and how much of the vehicle is being lifted. Shift the transmission to PARK (forward gear on manual transmissions) and set the parking brake. Position the wheel blocks behind and in front of at least one wheel that is being left on the floor. The surface should be level and enough for the jack wheels to roll.

Unlock the jack handle and align the jack's lift pad under the jacking point of the vehicle. Rotate the valve knob on the end of the handle clockwise to close the valve. Move the handle up and down to raise the jack pad until it makes contact with the vehicle. Check the contact and if it is correct, continue to jack up the vehicle. The jack should roll further under the vehicle as it is raised. This is to keep the pad directly under the jacking point. If the jack cannot roll, it will either pull the vehicle toward the jack handle or slide the pad from its contact point.

CAUTION: Never work under a vehicle supported only by a jack of any type. Injuries could occur if the jack slips.

When the vehicle is at the desired height, slide the jack stands under the vehicle. They should be placed at a point that will support the weight of the vehicle. This is usually under the frame, under the lower end of the suspension, or under the axle housing. The stands should be opposite each other at the same height to keep the vehicle level. With the stands in place, slowly rotate the valve knob counterclockwise. The knob will probably stick a little because of

Photo Sequence 4
Typical Procedure for Jacking and Supporting a Car

P4-1 Drive the car on to a solid, level surface.

P4-2 Set the parking brake.

P4-3 Put the transmission in PARK.

P4-4 Block the front and rear of one of the wheels.

P4-5 Position the jack pad on a major strength point under the car.

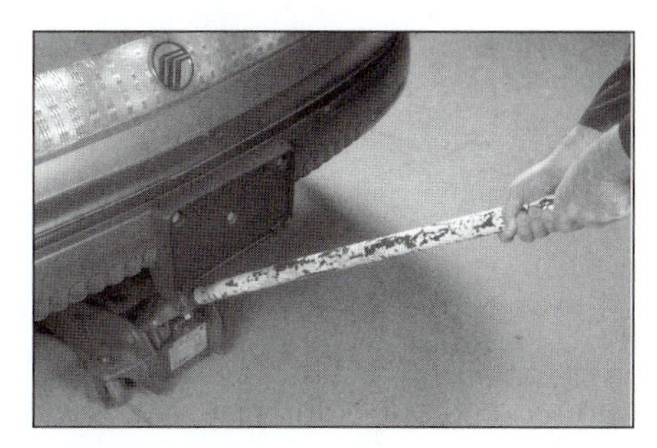

P4-6 Turn the jack handle valve to the lifting position.

P4-7 Raise the car carefully.

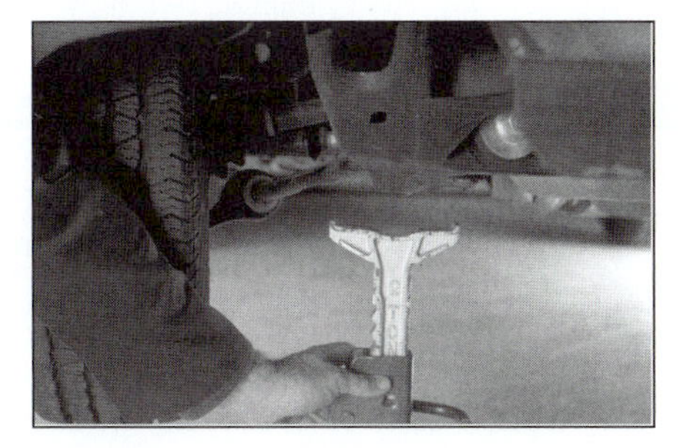

P4-8 Adjust and position jack safety stands to properly support the car.

P4-9 Slowly turn the jack handle to the lowering position and lower the car down on the jack safety stands.

P4-10 Shake the car to test its stability on the jack safety stands.

the hydraulic pressure against the valve. As the knob begins to turn, the jack will begin to lower. Proceed slowly until the vehicle is resting securely on the jack stands. Remove the jack, stand the handle straight up, and lock it in place. Do not leave the jack in a walking area and *never* leave the handle lying down.

To remove the stands and lower the vehicle, carefully place the jack in position to make the same contact as when raising the vehicle. Unlock the handle and close the valve. As before, once the pad makes contact with the vehicle, check it before raising the vehicle. Raise the vehicle just enough to remove the stands. With the stands removed, open the jack valve slowly and lower the vehicle to the ground. Store the jack properly. Ensure that the parking brake is set before removing and storing the wheel blocks.

Classroom Manual
page 65

Collecting Service Data

Most service data is easily obtained using paper or computerized service manuals. However, there are times when the data is not located in an expected area of the manual. Torque specifications for a certain component like cylinder head bolts may be listed in the specification table in one manual while another manufacturer places it in the section concerning engine assembly or in the cylinder head installation section.

The best method is to determine the correct term for the part and the system in which it is located. Most manufacturers use the same terms for the majority of automotive parts and components. They may, however, list a term as part of a subsection. For instance, data on the cooling system may be in a separate section on cooling or in a subsection under engine performance or diagnosis. Experience in the use of manufacturer service manuals is the best method to determine how each is set up.

Data banks are basically a computerized version of a paper manual. The two largest automotive repair data systems, Mitchell and AllData, use manufacturer data for the information, but both have some differences in the way their data bank is set up.

CASE STUDY

Older technicians have been heard to say "I have an educated arm, I do not need a torque wrench." They think they can feel the correct tightness for the fasteners they are tightening and do not need a torque wrench.

A group of technicians got a chance to prove this at a recent automotive tools trade show. A torque wrench manufacturer had an engine on a stand with the cylinder head bolts ready for tightening. Technicians were offered a chance to tighten a head bolt to the specified 85 ft.-lbs. without a torque wrench. If they could get to within 10 ft.-lbs. of the specification they would win a prize. Technician after technician took a turn. After each tightening, the torque on the head bolt was measured. The trade show lasted three days and not one prize was given. So much for the educated arm theory!

Terms to Know

Condition seizing
Double beam
Female coupling
Lug nuts
Male coupling
Neural flame
Nut splitter
Penetrating oil
Quick disconnect coupling
Striker
Taper beam
Tightening sequence
Welding tip

ASE Style Review Questions

1. The use of an air impact wrench is being discussed. *Technician A* says the air impact wrench works well for tightening wheel lug nuts.
   Technician B says the air impact wrench should be used primarily for disassembly work. Who is correct?
   A. A only
   B. B only
   C. Both A and B
   D. Neither A nor B

2. The use of an air impact wrench is being discussed. *Technician A* says standard twelve-point sockets can be used on the impact wrench.
   Technician B says six-point impact sockets must be used on an impact wrench. Who is correct?
   A. A only
   B. B only
   C. Both A and B
   D. Neither A nor B

3. The use of an air impact wrench is being discussed. *Technician A* says to check the direction with the reverse switch before using the wrench.
   Technician B says eye protection must be worn when using an air impact wrench. Who is correct?
   A. A only
   B. B only
   C. Both A and B
   D. Neither A nor B

4. The use of a torque wrench to tighten lug nuts is being discussed. *Technician A* says torque specifications are found in the shop service manual.
   Technician B says the rule is to get the lug nuts as tight as possible. Who is correct?
   A. A only
   B. B only
   C. Both A and B
   D. Neither A nor B

5. The use of a torque wrench is being discussed. *Technician A* says wheel lug nut specifications are in newton-meters.
   Technician B says the specifications are in foot-pounds. Who is correct?
   A. A only
   B. B only
   C. Both A and B
   D. Neither A nor B

6. The use of a torque wrench is being discussed. *Technician A* says fasteners must be tightened to the correct torque.
   Technician B says fasteners must be tightened in the correct sequence. Who is correct?
   A. A only
   B. B only
   C. Both A and B
   D. Neither A nor B

7. The use of a gear puller is being discussed. *Technician A* says gear pullers can be used to pull parts of a shaft.
   Technician B says gear pullers can be used to pull parts out of a hole. Who is correct?
   A. A only
   B. B only
   C. Both A and B
   D. Neither A nor B

8. The removal of a broken stud is being discussed. *Technician A* says first try to remove the broken stud with a screw extractor.
   Technician B says first try to remove the broken stud with a chisel. Who is correct?
   A. A only
   B. B only
   C. Both A and B
   D. Neither A nor B

9. The removal of a rounded off nut is being discussed. *Technician A* says it can be removed with a flat chisel.
Technician B says it can be removed with a nut splitter. Who is correct?
A. A only
B. B only
C. Both A and B
D. Neither A nor B

10. A nut is rusted onto a stud. *Technician A* says soak it in antiseize compound.
Technician B says soak it with penetrating oil. Who is correct?
A. A only
B. B only
C. Both A and B
D. Neither A nor B

Table 3-1 Guidelines for Service

Connect and use an air impact wrench to remove the lug nuts on a wheel.

Steps		Classroom Manual	Shop Manual
	1. Raise car on a hoist or floor jack and support on safety stands.	60	63
	2. Remove wheel cover and mark lug to wheel position.	—	39
	3. Connect air impact wrench to air source.	50	43
	4. Set impact wrench to reverse and install correct size socket.	50	43
	5. Remove lug nuts.	50	43
	6. Reinstall wheel and start lug nuts by hand.	50	44
	7. Reverse switch on impact wrench and spin lug nuts into position to be torqued.	50	44

SAFETY

Use lift locks.
Wear eye protection.
Use impact socket.

Table 3-2 Guidelines for Service

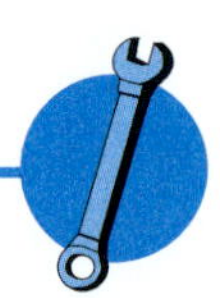

Look up the torque specifications for wheel lug nuts and torque wheel lug nuts to the correct torque.

Steps		Classroom Manual	Shop Manual
	1. Look up torque setting for lug nuts and tightening sequence in service or owner's manual.	64	45
	2. Connect correct size socket to torque wrench.	32	46
	3. Tighten the lug nuts in stages to correct torque.	37	47
	4. Tighten lug nuts in the correct sequence.	37	47
	5. Replace wheel covers.	—	44
	6. Lower car from jack or hoist.	60	65

SAFETY

Wear eye protection.

Table 3-3 Guidelines for Service

Select, install, and use a gear puller to correctly remove a pulley from a shaft.

Steps		Classroom Manual	Shop Manual
	1. Remove the retaining nut from the pulley.	63	50
	2. Select the correct size and type of gear puller.	63	50
	3. Adjust the cross arm and jaws to install the puller on the pulley.	64	51
	4. Drive the forcing screw in to force the pulley off the shaft.	64	51
SAFETY	Wear eye protection.		

Table 3-4 Guidelines for Service

Use a screw extractor to remove a broken stud or screw.

Steps		Classroom Manual	Shop Manual
	1. Select correct size screw extractor and drill for broken fastener.	57	51
	2. Drill correct size hole in center of broken fastener.	52	53
	3. Install screw extractor in drilled hole.	58	54
	4. Install correct wrench or driver on screw extractor and rotate it to remove broken fastener.	58	54
SAFETY	Wear eye protection.		

Table 3-5 Guidelines for Service

Use a flat chisel to split and remove a stuck nut from a stud or bolt.

Steps		Classroom Manual	Shop Manual
	1. Soak stuck nut with penetrating fluid.	—	55
	2. Cut the nut completely along one side with a cold chisel.	52	55
	3. Remove the nut with a wrench or pliers.	—	56
SAFETY	Wear eye protection.		

Table 3-6 Guidelines for Service

Set up and use an oxyacetylene torch for heating.

Steps	Classroom Manual	Shop Manual
1. Install torch handle on acetylene and oxygen hoses.	58	58
2. Install correct size welding tip on torch handle.	58	60
3. Turn on main cylinder valves.	58	68
4. Use regulator handles to adjust correct working pressure.	58	61
5. Use striker to ignite acetylene flame.	—	61
6. Adjust acetylene and oxygen for neutral flame.	—	62
7. Follow correct procedure to shut down equipment.	—	62

SAFETY

Wear eye protection.
Wear welding gloves and apron or coat.

Table 3-7 Guidelines for Service

Raise a vehicle with a lift.

Steps	Classroom Manual	Shop Manual
1. Position the vehicle between the posts.	61	63
2. Position the lift arms and pads under the vehicle.	61	63
3. Raise the vehicle until its wheels clear the floor. Check for stability.	61	64
4. Lift the vehicle to work height and set the locks.	61	64

Table 3-8 Guidelines for Service

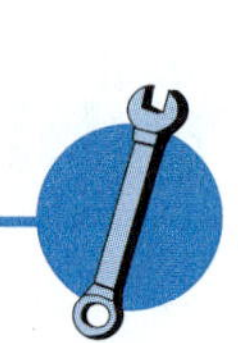

Lift a vehicle with a jack.

Steps	Classroom Manual	Shop Manual
1. Position the vehicle, place in park, and block wheels.	59	64
2. Place jack lift pad under the vehicle's lift point.	59	64
3. Unlock the handle and knob.	59	64
4. Operate the handle to raise the jack.	59	64
5. Place jack stands under the vehicle.	59	64
6. Lower the vehicle to the jack stand.	59	64

Job Sheet 5

5

Name ______________________________ Date ______________

Raising a Vehicle

Upon completion and review of this job sheet, you should be able to raise a vehicle using a double-post lift.

Tools Needed

None

Procedures

1. **A.** Identify the vehicle's lift points.

   Make___________ Model___________ Year___________

   B. Indicate lift points on the illustrations below.

   Full Frame

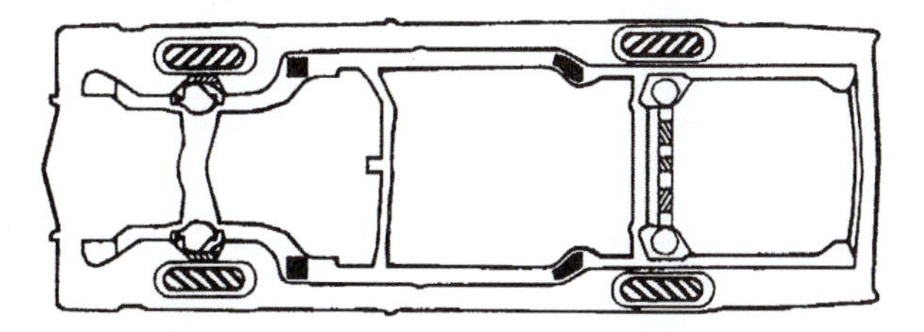

   Unibody

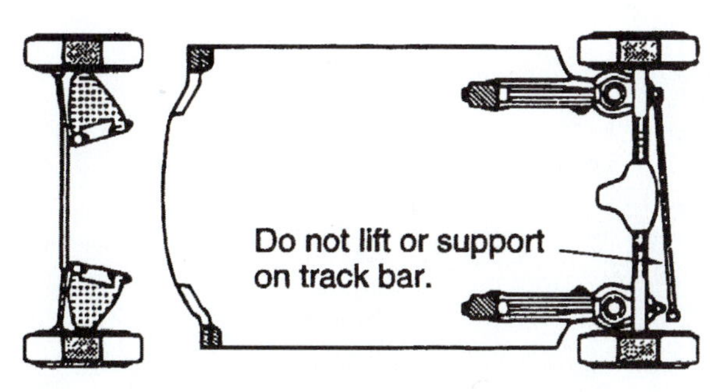

2. Center the vehicle between the two posts. The front door handle should be about even with the posts.
3. Rotate and extend lift arms under the vehicle.
4. Align lift pads with lift points.
5. Raise lift arms until pad contacts points by holding the switch in the ON position.
6. Check contact between pads and lift points.

7. Raise the vehicle until all wheels clear the floor. Check the vehicle by shaking to ensure it is securely on the lift.
8. Continue to raise the vehicle until the center of the wheel assembly is at eye level.
9. Lower the vehicle until the locks engage the lift arms.
10. Raise the vehicle to remove weight from the locks.
11. Place the locks in the unlocked position.
12. Lower the vehicle to the floor and the arms to their lowest travel.
13. Swing the arms from under the vehicle and move the vehicle from the bay.

____

____

Job Sheet 6

6

Name ______________________ Date ______________

Using an Impact Wrench and a Torque Wrench

Upon completion and review of this job sheet, you should be able to remove and install a tire assembly using an impact wrench and a torque wrench.

Tools Needed

1/2-inch-drive impact wrench
Set of 1/2-inch-drive impact sockets
1/2-inch-drive torque wrench (50–250 ft.-lbs.)
Hub cap hammer or long, flat screwdriver
Lift or jack and jack stands

Procedures

1. Locate the following vehicle service information.

   Lug nut wrench size ______________

   Lug nut torque ______________________

2. Lift the vehicle to a good working height. Use Job Sheet 5 for instructions.
3. Remove the hubcap by using a hubcap hammer or a flat tip screwdriver.
4. Select the correct impact socket and install on impact wrench.
5. Connect the impact wrench to the air supply.
6. Set the impact wrench to reverse and set the socket over the first lug nut.
7. Use one hand to support the wrench and squeeze the impact wrenches' trigger.
8. Release the trigger as soon as the nut screws from the stud.
9. Repeat steps 6 through 8 to remove other lug nuts. If necessary, use one hand to keep the tire assembly on the hub while the last nut is removed.
10. Lay the wrench aside and grasp the tire assembly at the 4 o'clock and 8 o'clock positions. (12 is straight up)
11. Keeping the back straight and bending the knees, remove the tire assembly from the hub and place on floor.

Installation of the tire assembly is the reverse:

12. Grasp the tire at the 4 o'clock and 8 o'clock positions.
13. Keep the back straight and use the knees to lift the tire assembly to the hub.
14. Align the wheel with the studs and slide the wheel assembly onto the studs.

15. Use one hand to keep the assembly on the studs while the first lug nut is started.
16. Start each lug nut by hand.
17. Inspect the impact wrenches' controls. If there is an adjustment, set the wrench to its lowest forward setting. If there is no adjustment, screw the nuts on as far as they go by hand.
18. Use the impact wrench to screw the nuts into contact with the wheel.
19. Lower the vehicle until the wheels just touch the floor. A second option is to have someone apply the brakes while the nuts are being torqued.
20. Set the torque wrench to the specified torque and transfer the socket from the impact wrench to the torque wrench.
21. With the wheels held, use the torque wrench to tighten the first lug nut. Once the torque is achieved, move to the next nut.
22. The second nut should be across the wheel from the first one.
23. Torque each lug nut following a cross pattern.
24. Install the hubcap.
25. Lower the vehicle and clean the area.
26. Complete the repair order.

CHAPTER 4

Diagnosing by Theories and Inspection

Upon completion and review of this chapter, you should be able to:

- ❑ Explain how understanding theories of operation can help a technician diagnose a fault.
- ❑ Discuss how the air/fuel mixture can affect thermodynamics and engine performance.
- ❑ Locate and inspect components from under the hood.
- ❑ Inspect drive belts.
- ❑ Inspect the top of the engine for leaks.
- ❑ Locate components under the vehicle.
- ❑ Inspect under the vehicle for fluid leaks.
- ❑ Inspect the exhaust system for leaks.
- ❑ Inspect vehicle lighting and wipers for operation

Introduction

One important step in diagnosing a problem is to look at the simple things first. Understanding the theories of operation will assist the technician in determining the fault. In addition, an inspection of the vehicle will give the technician insight into how the vehicle was operated and maintained. The two should give the technician an idea of what to expect as the diagnosis and repairs are being performed.

Diagnosing by Theory

Classroom Manual
page 73

When a system fails it is because some part of it did not fulfill the requirements of a theory. Often, a lamp does not work because the electricity did not flow from the power source, through the lamp, and back to the power source. A lamp may be "burned out," but the electricity failed to return to the power source, hence, no light. The same diagnostic procedure can be applied to any system on the vehicle. Every technician will encounter a fault that fails to respond to typical diagnostic tests. By thinking about the theory or theories that make the system work, the technician can track through the system and find the point where the theory breaks. In this area the fault will be found. It may be caused by the lack of electricity, fuel, or air. It may be a broken part or any other item that is needed to make the system work.

Engine Operations

In the Classroom Manual we discussed the need for fuel, oxygen, and spark to be present in the combustion chamber to produce power. Two of those items, fuel and oxygen, require pressure and vacuum in order to be drawn into the chamber. The air and fuel must also be mixed in the correct **ratio**.

A *ratio* is the proportional amount of one element to another. An ideal air/fuel ratio is 14.7 parts of air to one part of fuel (14.7 to 1).

The laws of motion also apply to the movement of air and fuel. The engine is designed to make maximum use of physics to move, mix, fire the fuel, and transfer the produced power to the vehicle's driveline.

Understanding the theory of operation of vehicle systems will give the technician a head start on diagnosing. Let us discuss a possible situation. The technician does not have any high-tech diagnostic tools available, but has a vehicle that will not start. As discussed, the engine must receive air and fuel in the correct mixture. The mixture is then fired with a spark. The easiest item to check is the airflow. Remove the air filter housing and try to start the vehicle. It is not often that an air intake is so plugged that no air can pass. But a dirty air filter could reduce the amount of air and cause a rich mixture or too much fuel for the air available. This could cause a no-start situation in the engine, but more likely will allow it to start but run poorly.

A broken or missing vacuum line will cause the opposite condition or a lean mixture, which is too much air for the fuel available. Both conditions reduce the power produced and cause more harmful emissions. Most vacuum line connections are on the intake under the air filter housing or ducting.

A *tachometer* is a meter used to measure the speed of a device, i.e., engine speed. It is measured in revolutions per minute (rpm).

Let us consider a second situation concerning a poorly operating engine. The fact that it is running indicates that fuel, air, and spark are present and being used. Apparently, something is wrong with one of the three. Conduct a quick test to see if the problem is the same at idle and higher engine speeds. After allowing the engine to warm, speed up the engine until it reaches 1,200 to 1,500 rpms. If a meter **(tachometer)** is not available, hold the accelerator at about one-quarter throttle.

If the engine smoothes out and runs evenly, then the problem exists only at idle. The engine intakes little air and fuel at idle. The air valve, sometimes called a throttle valve, is close and air enters through an air control valve (Figure 4-1). Carbon buildup can partially block the valve and reduce the oxygen available to the engine. The air/fuel ratio is rich and combustion cannot be supported. Since the valve has no effect when the engine is off idle, the engine will run smoothly at speed.

A second problem that affects idle speed is a vacuum (air) leak around the intake manifold. As mentioned above, a vacuum leak will cause the air/fuel ratio to be lean, but unless the leak is very large, the engine will run fairly smoothly off idle. A fuel or spark problem will probably be present at idle and may increase at higher engine speeds. In this example, we have found that a small valve or a leak can upset the application of thermodynamic theory by changing the air/fuel ratio so combustion is not completed.

CAUTION: Never look into a carburetor when the engine is running. The fuel in a cylinder could ignite early and cause a flame to run back though the intake system and cause injury to the eyes and face.

A *carburetor* is old air/fuel mixture device. It is found on older cars, light trucks, and most small equipment engines.

The fuel delivery cannot be checked easily on a fuel injection system. It is best left to technicians with the proper test equipment. On a **carburetor** system, operating the throttle once or twice without the engine running will check the fuel delivery. Observe the inside of

THROTTLE VALVE

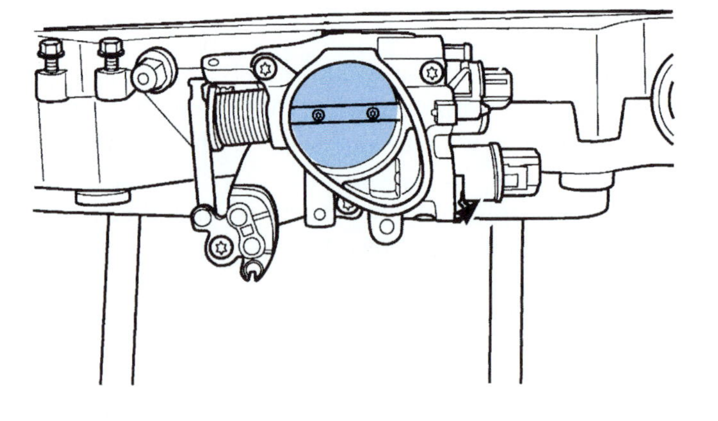

Figure 4-1 The throttle valve is operated by the driver, and it controls the amount of air entering the engine. (Courtesy of DaimlerChrysler Corporation)

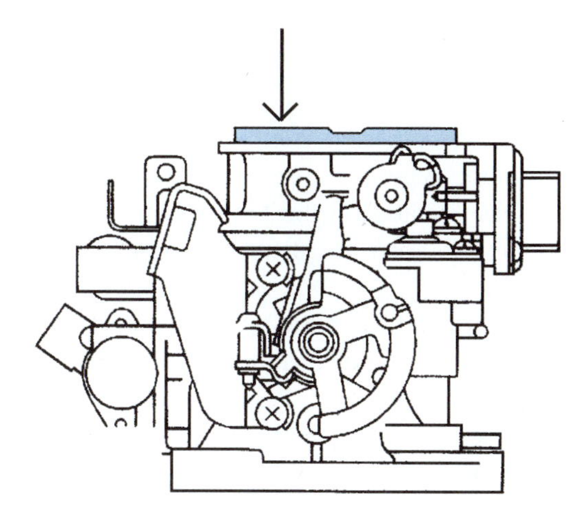

Figure 4-2 Peer downward into the opening without the engine running to check for fuel as the throttle is operated. (Courtesy of Nissan North America)

the carburetor while moving the throttle (Figure 4-2). There should be a squirt of fuel sprayed into the carburetor each time the throttle is moved. Once or twice is sufficient.

Remember that in order for the engine to work properly, the laws of physics must be followed. If one of the above checks fails, then the faulty system has been found. From this point, the technician can inspect and test that system until the root cause is found.

This was a simplistic diagnostic test using a theory of operation. Similar diagnosing occurs everyday, except we now rely on electronic devices to do the initial testing for us. An electronic sensor "stuck on lean" indicates that there is not enough fuel or that there is too much air in the mixture. Fuel injection relies on constant pressure and volume. If a sensor indicates the mixture is lean, then the fuel is insufficient or the air volume may be too large. On most **engines** using a fuel injection system, the fuel delivery is the most common cause of a lean condition. The fuel may not be present in either pressure, volume, or both. A little common sense tells us that if there is no volume, then there can be no pressure, so in this instance we can assume there is insufficient fuel available. The most common cause of this problem is a partially clogged fuel filter. A twenty dollar part could restore engine performance and ensure that thermodynamics is accomplished.

When the first electronically-controlled *engines* began to show up for repairs, many good sensors were replaced due to a lack of understanding about system and test equipment operation.

A point to remember as technicians advance in their careers is to avoid replacing a sensor until it is determined that the sensor is bad. In most cases, the sensor is doing exactly what it was designed to do. In the case above, the oxygen sensor is telling the technician that the fuel flow is insufficient for the amount of air available.

Component Location and Inspection

Classroom Manual
page 76

CAUTION: Do not attempt to perform an inspection or repairs on hot engines. Serious burns may result or injury could occur if a temperature-sensitive cooling fan switches on.

The various components of the vehicle's systems are located in many different areas of the body and frame. Belt driven components are mounted on and driven by the engine. Steering, brake, and suspension systems have components mounted on the engine, body, and frame. It is also sometimes difficult to locate and view a component.

Transverse means the engine is mounted crossway to the length of the vehicle.

Location and Identification

With front-wheel-drive and **transverse** engines there is not much room under the hood. Most engines have the alternator, power-steering pumps, water pump, and air conditioning compressor at the front of the engine where the drive belts are located (Figure 4-3). On a front-wheel-drive vehicle, this may be located on the left or right side of the vehicle. When dealing with a vehicle's left or right, always look at it from the vehicle's point of view. The left side is always the driver's side in most countries. One or all components may require the vehicle to be lifted, the engine mounts removed, or both for access and repair. In addition, the battery may be located within the wheel well or elsewhere in the vehicle and requires the removal of other components for access and service. In most cases, the battery has remote terminals mounted in the engine compartment for jump starting and electrical tests (Figure 4-4).

Rear-wheel-drive vehicles usually have sufficient room under the hood and elsewhere to repair or replace components (Figure 4-5). In addition, the major components are easy to identify. For that reason, we will discuss component location and inspection using a typical front-wheel-drive vehicle. Left and right directions apply to any vehicle.

Figure 4-3 Most front-wheel-drive vehicles have the engine set crossway (transverse) in the vehicle.

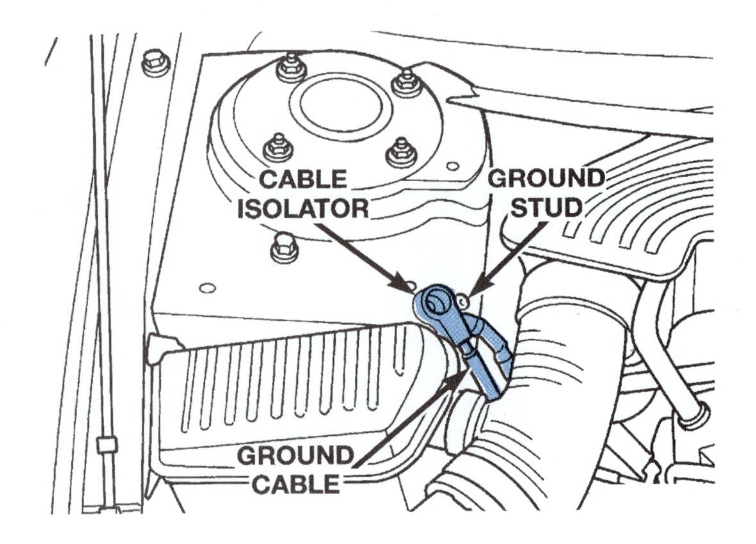

Figure 4-4 This is the negative cable remote for the battery. A red one is mounted nearby for the positive side. (Courtesy of DaimlerChrysler Corporation)

Figure 4-5 The engine compartment of a rear-wheel-drive vehicle leaves room to access the components.

Before starting the inspection, study the engine compartment and locate the alternator (Figure 4-6). It is usually at the top of the engine, but may be mounted low on the engine. It will usually have a single groove pulley with a fan behind it (Figure 4-7). The alternator is short, round, and will appear to be made of aluminum. It will have slots running across the rear cover next to the area where the wires are positioned. The end opposite the pulley will have a two-wire electrical connection and usually a heavy, red wire attached to another terminal. In most cases, the alternator will be the easiest belt-driven device to access.

While checking for the alternator, locate any other belt-driven components. The **water pump** may not be located on the outside of the engine, but there should be at least five pulleys, counting the crankshaft pulley at the bottom center of the engine. There may be up to eight pulleys depending on idler and tension pulleys used and the number of belt-driven components available.

Some *water pumps* are driven by a belt positioned behind a cover on the front of the engine.

Figure 4-6 This alternator is mounted at the top of the engine. Note the drive belt and pulley.

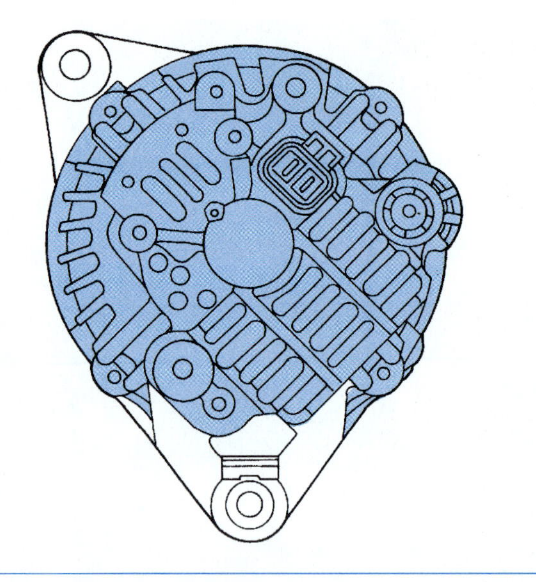

Figure 4-7 The rear of the alternator is slotted to allow air to be forced into the exhaust by the fan. (Courtesy of DaimlerChrysler Corporation)

The power-steering pump and reservoir are oddly shaped (Figure 4-8). The easiest way to identify these devices is to locate a drive pulley that has a small cap behind the metal bracket between the pulley and pump (Figure 4-9). The cap is used for checking the fluid level. However, some vehicles have a remote-mounted reservoir with two hoses, one of which routes down to the pump. In either case, the filler cap will be marked in some way to indicate that this is the power-steering reservoir. Usually, the marking will indicate that only a certain type of power-steering fluid should be used in that system.

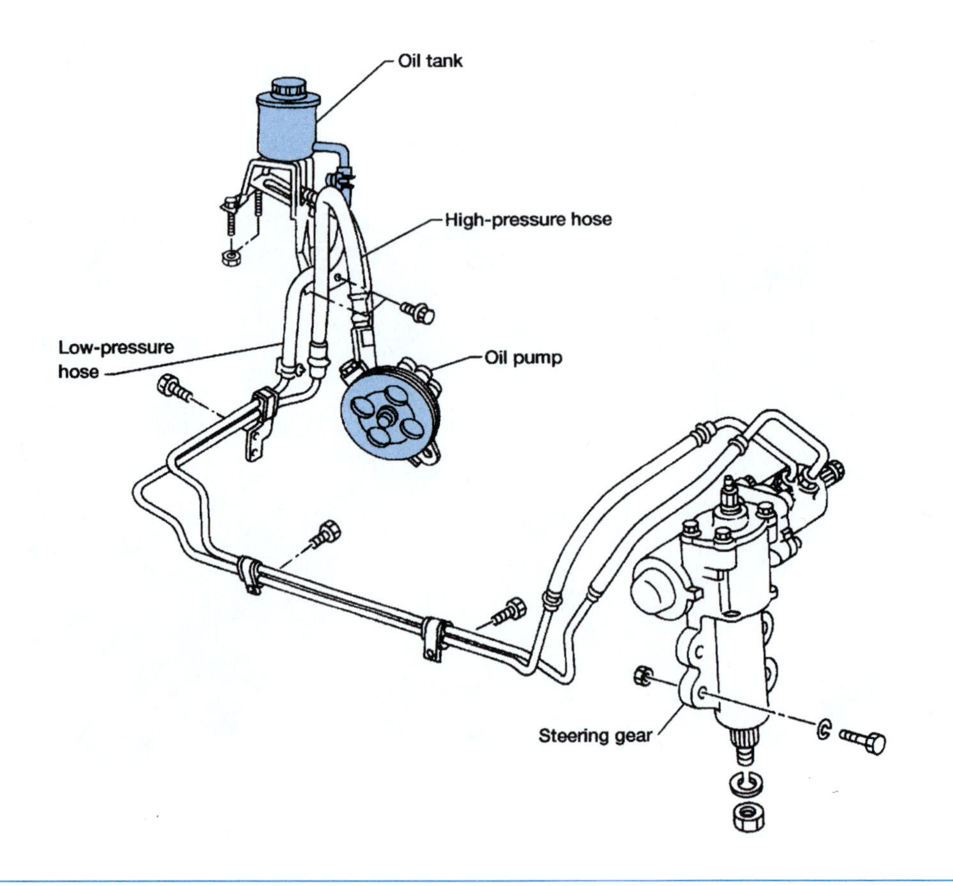

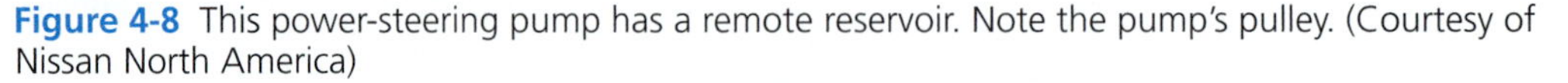

Figure 4-8 This power-steering pump has a remote reservoir. Note the pump's pulley. (Courtesy of Nissan North America)

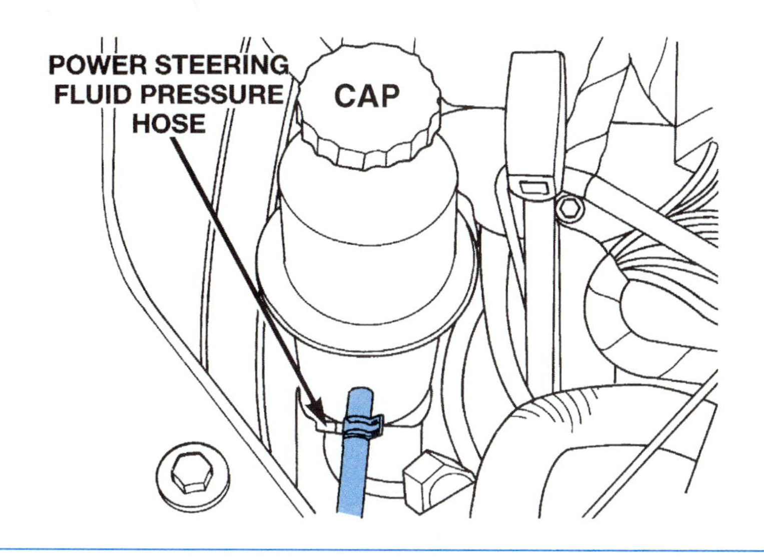

Figure 4-9 If the power-steering reservoir is remote mounted, the technician can follow the hose to locate the pump. (Courtesy of DaimlerChrysler Corporation)

Many times, the power-steering pump is located to the lower left side of the engine in rear-wheel-drive vehicles and the upper right side of the engine in front-wheel drive. Again, remember that the directions are based on the vehicle or engine, *not* the position the technician is facing. The front of an engine is almost always the end with the drive belts, and the left and right of an engine may not be the left and right of the vehicle (Figure 4-10).

Inspection

An inspection of the vehicle, particularly the engine compartment, can give a technician an idea of how the vehicle was maintained and used. An inspection may also help in the technician's diagnoses.

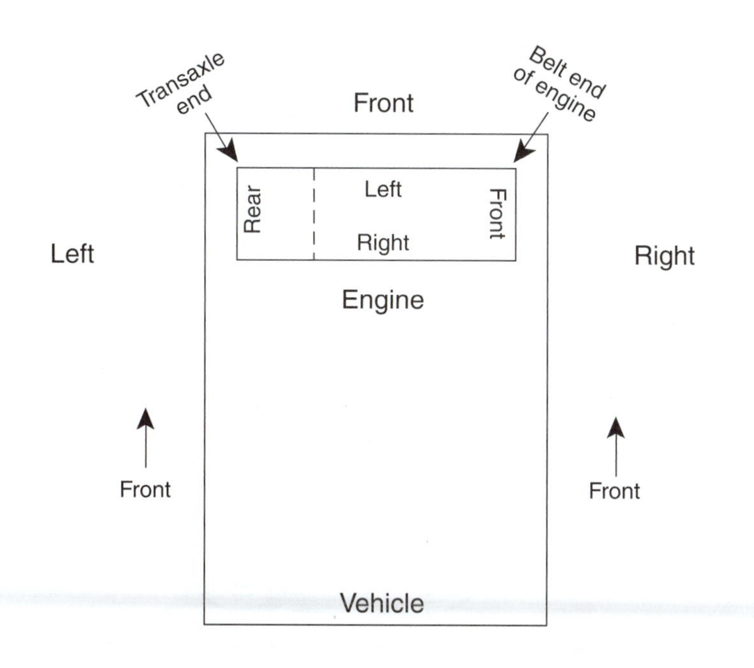

Figure 4-10 Directions are based on the front of each major component.

The *V-belt* is so named because of its shape. It is used to drive accessory components mounted at the front of the engine.

The *serpentine belt* is named because of its winding route among the various pulleys at the front of the engine. It may replace between two to four V-belts used on older systems.

A *span* is the space between two pulleys where the belt has no support.

Drive Belts

The first and most obvious check is the drive belts. There are several different types of belts in use today. The most common is the **V-belt** which was used on almost all cars and light trucks in some manner until about 1990 (Figure 4-11). Some V-belts are ribbed along the backside. A flat, inside-ribbed belt was installed over a period of several years to replace some V-belts. The newest belt is also a flat, ribbed belt called a **serpentine belt** (Figure 4-12). In many cases, it replaced all of the previous V-belt applications with one continuous belt (Figure 4-6). Some vehicles still use a V-belt in conjunction with a serpentine belt, but this is becoming uncommon with each new model.

Look for any unraveling at the edges of the drive belt. Unraveling edges may indicate that the pulleys are out of alignment. Loose mounting bolts on the driven component or its mounts will allow the pulley to shift to the side and rub the edge of the belt. Observe the tension of the drive belt. A typical method of checking drive belt tension is to press down on the belt across it longest **span** (Figure 4-13). If the belt depresses between $^3/_8$ inch to $^1/_2$ inch, the tension is correct. If there is doubt about the tension, use a belt tension gauge like the one in Figure 4-14. If the belts are a dual or matched belt system, both must be checked individually and have the same tension. Matched belt sets must be replaced as a set even if only one belt is bad.

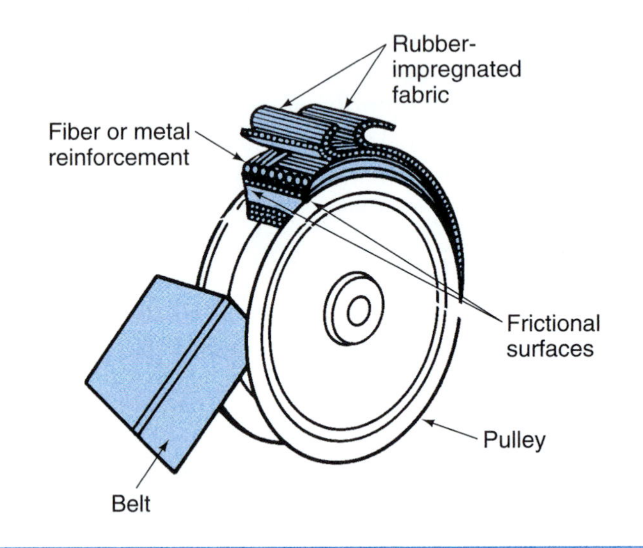

Figure 4-11 A V-belt is made of several parts and drives many accessories. (Courtesy of DaimlerChrysler Corporation)

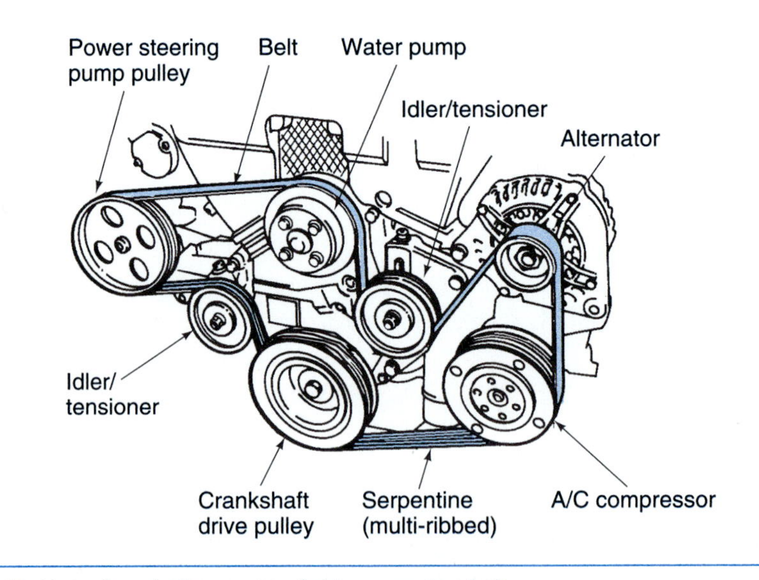

Figure 4-12 Note the winding route of this serpentine belt.

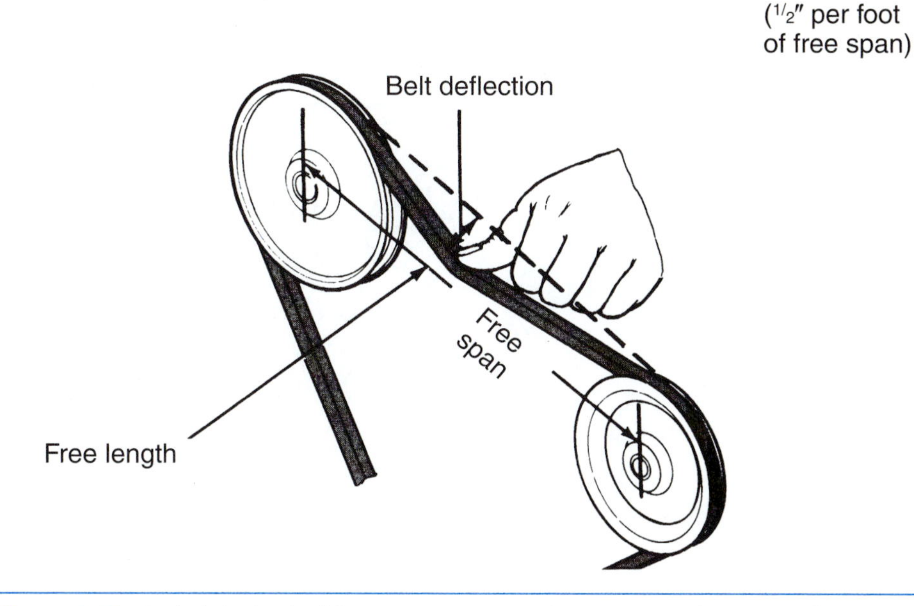

Figure 4-13 The belt is checked for tension across its longest span. (Courtesy of DaimlerChrysler Corporation)

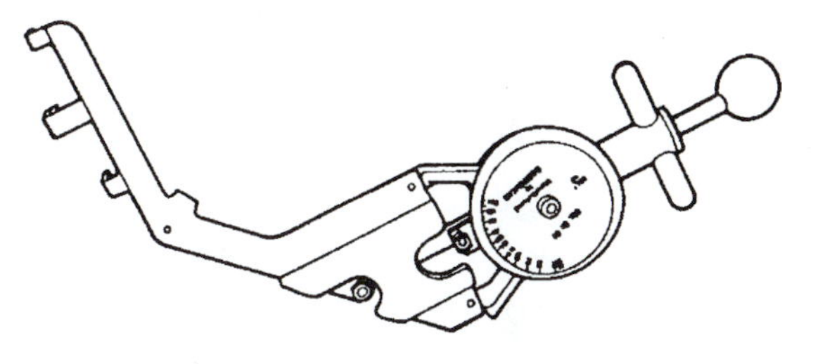

Figure 4-14 Belt tension can be checked using a gauge like this one. (Courtesy of DaimlerChrysler Corporation)

CAUTION: Use care when moving an automatic belt tensioner. The wrench can slip off and catch fingers under the belt being removed.

Most serpentine belts use an *automatic tensioner* (Figure 4-15). The tensioner is bolted to the engine block. A large, heavy tension spring is used to keep the belt at the proper tension. A serpentine belt runs at higher tension than a V-belt, and the tensioner may be difficult to access. The belt tension can be released by using a wrench or special tool to hold the tensioner back as the belt is moved off one of the pulleys. The wrench must be secured on the tensioner before any attempt to move it is made. The serpentine belt runs in a pattern that allows it to go around each pulley. The routing is not always obvious. Before removing a serpentine belt, study its routing or find a chart showing the routing. Manufacturers usually place a placard on the radiator or hood with this information, but the placard may be missing or unreadable. A service manual may be required to install the new belt properly.

While checking tension, also check the underside of the belt. A V-belt should be smooth but not slick. Also, there should be no missing pieces (Figure 4-16). Note the condition at the bottom of the pulley groove. The wear pattern should not extend to the bottom of the groove, and the bottom, the narrowest part of the belt's V, should not show wear. If it does, the belt is too narrow and is running at the bottom of the groove. A belt should only contact the pulley along the sides of the V. The belt will slip and cause a squealing noise even when it is new and running on the bottom. The ribs on the inside of the serpentine belt will probably be

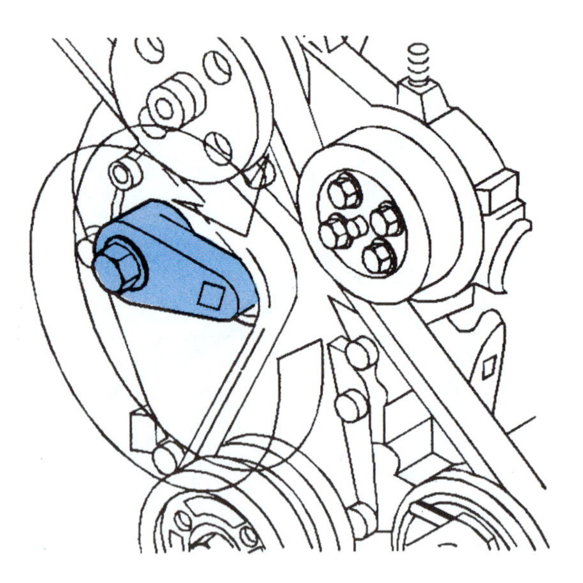

Figure 4-15 This type of tensioner keeps the belt correctly tightened during engine operations. (Courtesy of Chevrolet Motor Division, General Motors Corporation)

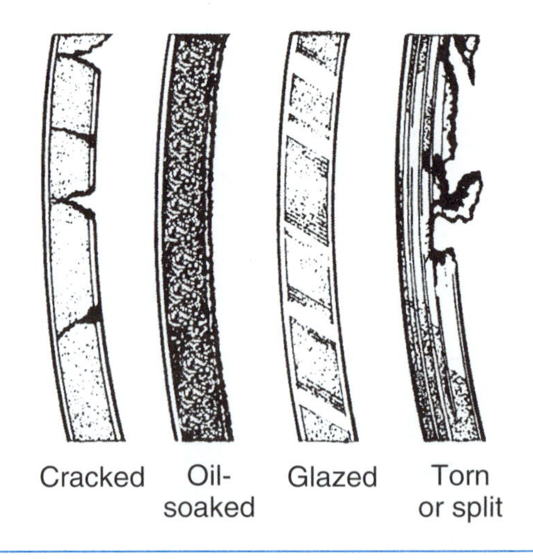

Figure 4-16 The oil-soaked condition will lead to the other conditions. (Courtesy of DaimlerChrysler Corporation)

cracked. This is because of the way the belt is wound over the pulleys. The best point to remember is to consider the total view of each belt. Key points that should require belt replacement are:

1. The belt has unraveling at its edges.
2. The belt has missing or torn parts.
3. The V-belt is slick to the point of slipping under the load.
4. Excessive cracks appear in the serpentine belt ribs.
5. The belt appears old and dry.
6. The belt is wet with engine oil or antifreeze.
7. Cracks appear on the outside of the belt.

Here are a few other points to remember concerning drive belts. One method to prevent a comeback on a repair job is to replace all belts at the same time. Sometimes that does not apply if it is oblivious that one or more belts are fairly new. Ask the owner before replacing a belt that appears to be new. A new belt, however, will not last long if there is oil or antifreeze leaking onto it. Both chemicals have acids in them from the interior of the engine and will erode the belts quickly. The leak should be fixed before new belts are installed. Also remember that matched sets should be replaced as a set and all must be replaced with belts of matching length, width, and type.

Wiring

WARNING: Before performing any electrical repairs, disconnect the negative cable from the battery. Damage to electronic components or injury to the technician may result.

There are two things that quickly kill electrical wiring and batteries: lack of maintenance and **corrosion**. If preventive maintenance is performed, corrosion will not be a problem. But along with antifreeze changes, battery preventive maintenance does not appear to be a great concern of some drivers until the vehicle quits working.

Corrosion can be caused by oxidation (oxygen acting on metal) or by acid or acid fumes. The battery cables, terminals, and tray are exposed to oxygen, acid, and fumes from the battery.

Inspect the wiring at the back of the alternator for tightness and routing (Figure 4-17). The battery end of the cable is the most probable place for massive corrosion. This is partially due to the heavy current needed during cranking. Sometimes the corrosion can be removed, but if it extends under the cable's insulation it is best to replace the cable (Figure 4-18). The cable's terminals can also become a point of corrosion. Terminals can be replaced without changing the complete cable. There are temporary terminals that can be clamped to the bare cable strands, but they tend to increase resistance and lose voltage between the clamp and cable. The best method is to use the type of terminal that has to be soldered to the cable (Figure 4-19).

Other starting and charging system inspections include security of the battery and corrosion on the battery tray/box. The tray should be cleaned and the battery held in place by proper holddowns. **Baking soda** and water work well to remove corrosion and electrolyte leakage. Use plenty of water to wash all of the soda from the vehicle. A final check includes the routing of all starting and charging system wires and harnesses. Ensure that there are no contacts between the wire and engine hot spots.

Baking soda is a common household product. It is classified as a chemical base that neutralizes acids. Some chemical bases can cause corrosion that resembles acid corrosion.

Figure 4-17 Ensure that connections on the alternator and all other connections are not damaged and are properly connected. (Courtesy of Chevrolet Motor Division, General Motors Corporation)

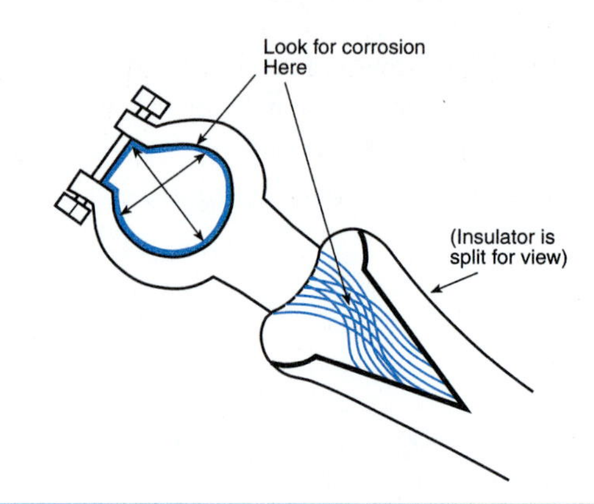

Figure 4-18 Once corrosion gets under the cable's insulation, the entire cable should be replaced.

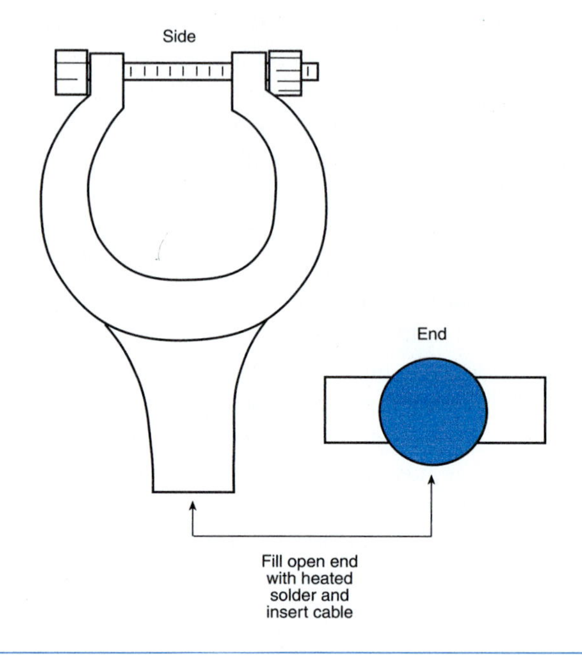

Figure 4-19 The permanent type of battery terminals are best suited for preventing corrosion.

Classroom Manual
page 76

Checking for Leaks in the Engine Compartment

There are many places for leaks to occur under the hood. Among them are fuel clamps and hoses, air intake hoses, cover gaskets, and engine seals. The origin of some leaks may not be easy to locate, but they are evident based on the amount of liquid on the outside of the engine and engine mounted components. Smell and color can identify leaking fluids and their possible source.

1. Pink or red—Power steering fluid; automatic transmission fluid; some manual drivelines.
2. Green—Engine coolant, antifreeze; has distinct odor and very greasy to touch; newer fluids may be pinkish.
3. Black or very dark—Engine oil; manual drivelines; driveline fluid has distinct odor.
4. Clear—Brake fluid; some power-steering fluid; brake fluid is very slick compared to other fluids; fluid may look dirty.
5. Clear with distinct odor—Gasoline; will feel cold even in warm weather.
6. Dirty clear with distinct odor—Diesel fuel; oily to the touch.

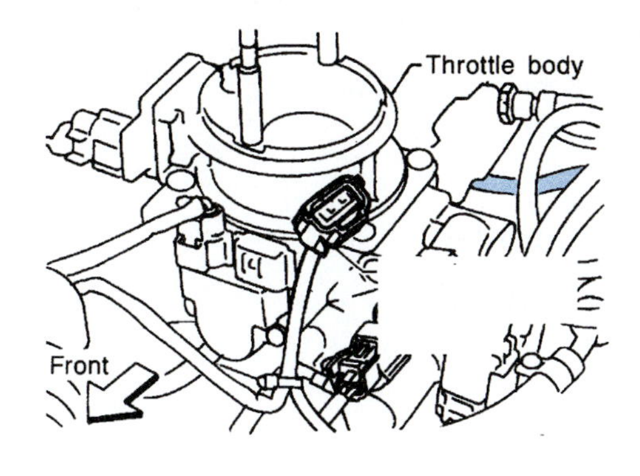

Figure 4-20 A throttle body like this one has two fuel lines on the rear. (Courtesy of Nissan North America)

The oil will be dirty and the color not readily apparent sometimes, but a little investigation by the technician can pinpoint the type and system.

A *throttle body* is used to mount one or two injectors and regulate the fuel pressure.

Fuel System

The fuel system ends at the intake manifold. It may be just one fuel line connected to the carburetor or multiple lines to a throttle body.

Electronic fuel injection (EFI) systems use a pressure line and a return line. On a **throttle body** system, both lines are close to each other (Figure 4-20). EFI systems using individual injectors and **fuel rails** may have both lines close to each other or the lines may be at opposite ends of the fuel rail (Figure 4-21). The checks are the same on either of the systems. Visually check each connection in the lines and each connection to the fuel rail, throttle body, or carburetor. Feel under each connection with a finger.

Fuel rails provide a common channel to maintain constant fuel pressure and volume at each injector in a multi-injector system.

WARNING: Use a flashlight for illumination if a fuel leak is suspected. An electric light may create an electrical spark when it is switched on and off, thereby causing a fire.

There must be no leaks or any type of **seepage** in any fuel system. In other words, if there is dampness at a fuel connection or anywhere on the line, find the problem before releasing the vehicle. This is especially true if fuel lines were disconnected during repair work. While checking the lines, also inspect the fuel rail connection to the injectors, the fuel injectors, and the carburetor or throttle body to the extent possible. Some of the fittings, hoses, lines, and components may not be visible. Older carburetors tend to seep at their various seals and gaskets. The carburetor should be rebuilt or replaced.

Seepage is seen as a damp spot with no visual of fluid drops. Seepage is considered to be normal. Many seals will not function correctly without some seepage.

Air Intake

The air system is fairly easy to check. Air usually enters the intake system behind one of the headlights and travels through some flexible or rigid tubing to the air cleaner housing (Figure 4-22). Many times there is a **mass airflow sensor (MAF)** located in the air tubing (Figure 4-23). Any air entering the engine that does not flow through the sensor will result in poor engine performance. The sensor measures the amount of air entering the engine and the **powertrain control module (PCM)** selects the amount of fuel based on that measurement. Inspect air ducts for cracks or improper mounting and fastening.

Vacuum leaks like the ones mentioned earlier also allow air to enter the system without going through the sensor. The end result is a lean mixture because of the extra air plus a wrong air volume measurement being sent to the PCM.

The *MAF sensor* measures the amount of air entering the engine's intake systems.

The *PCM* is an on-board computer that controls the electronic components of the engine and sometimes the transaxle.

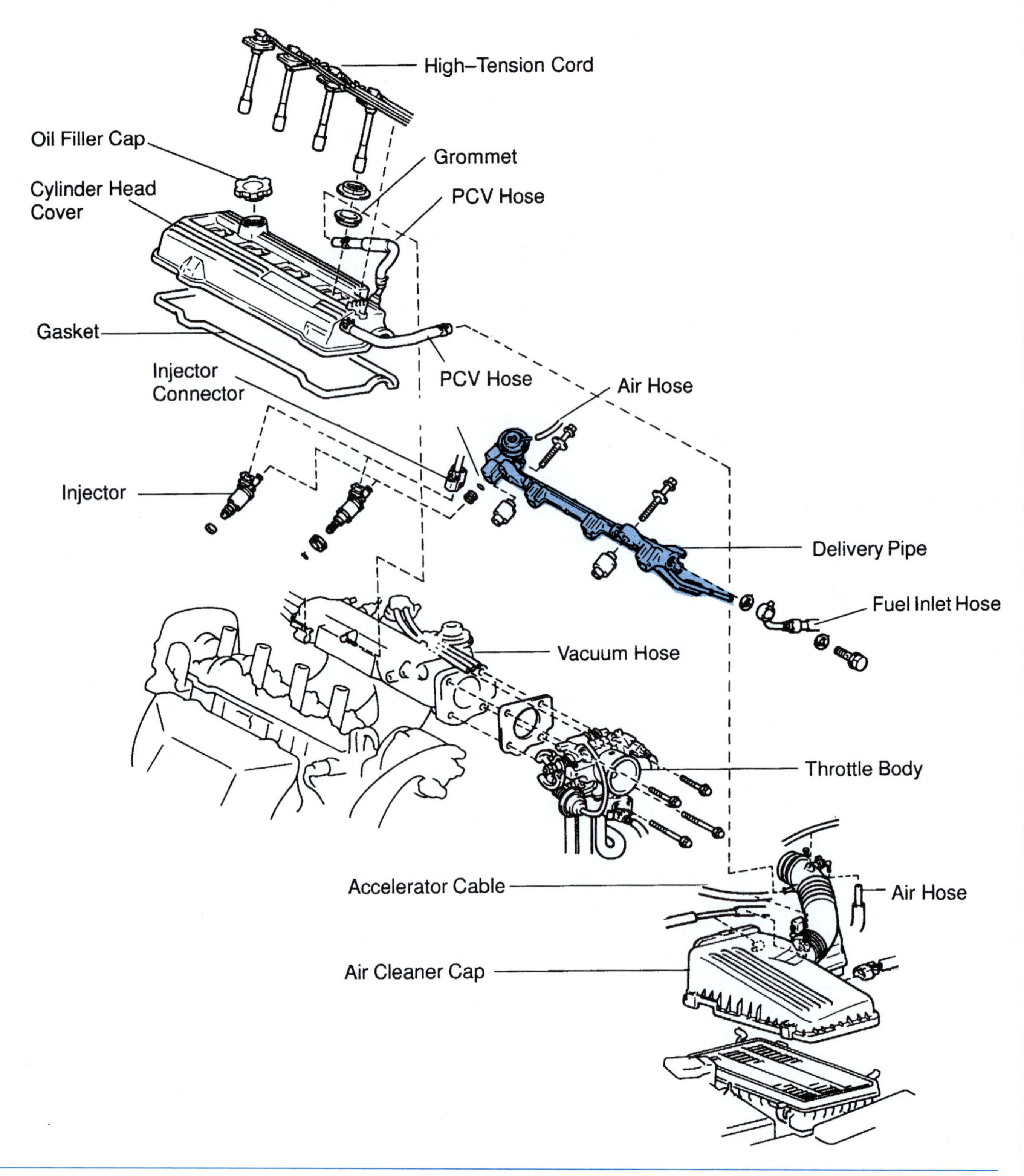

Figure 4-21 The fuel rail or delivery pipe provides constant volume to each fuel injection concurrently. (Reprinted with permission)

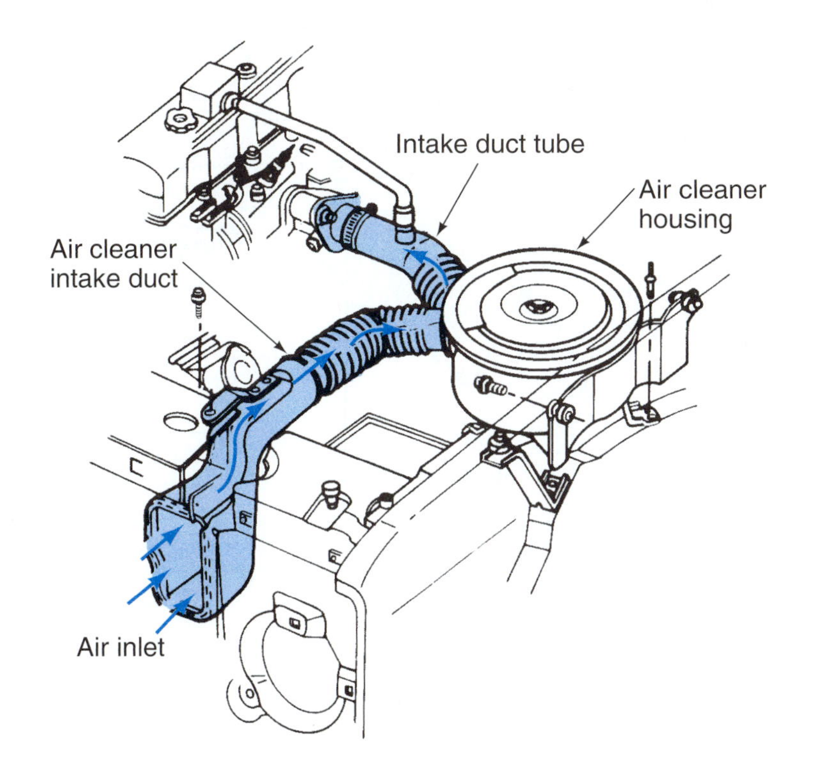

Figure 4-22 The intake air ducting can become cracked because of engine servicing or age.

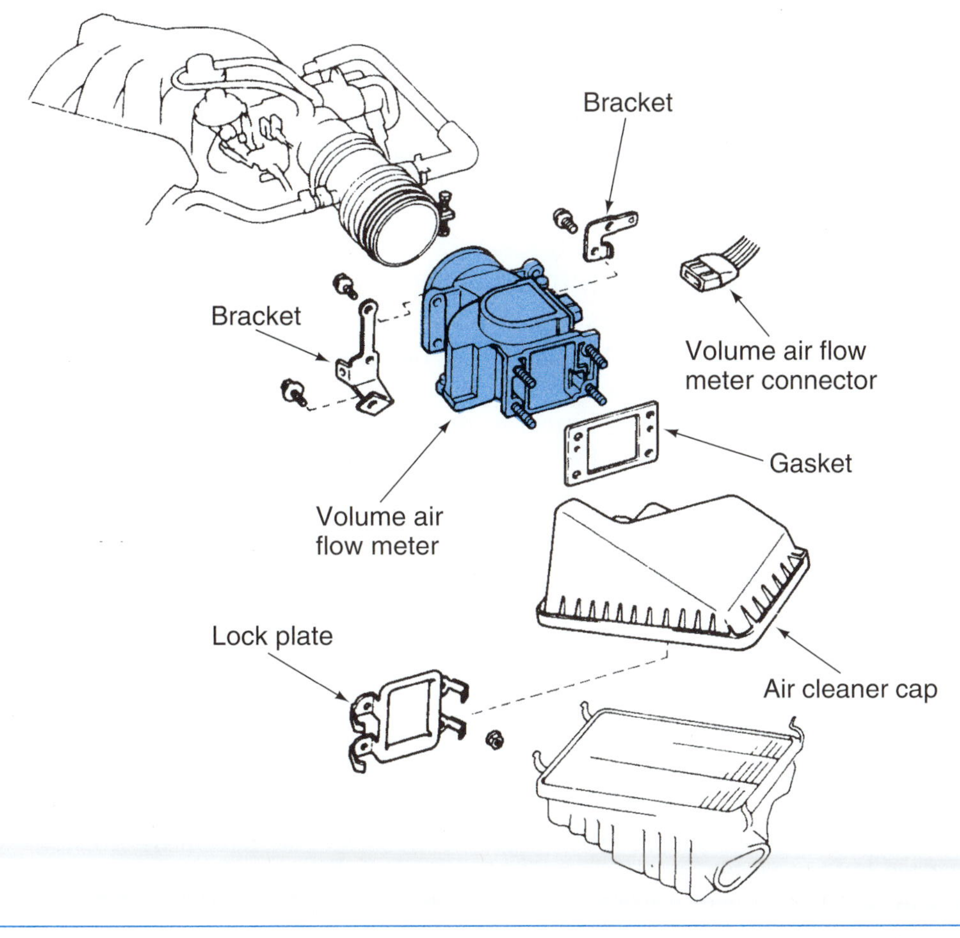

Figure 4-23 The mass airflow sensor measures the amount of air entering the engine. (Reprinted with permission)

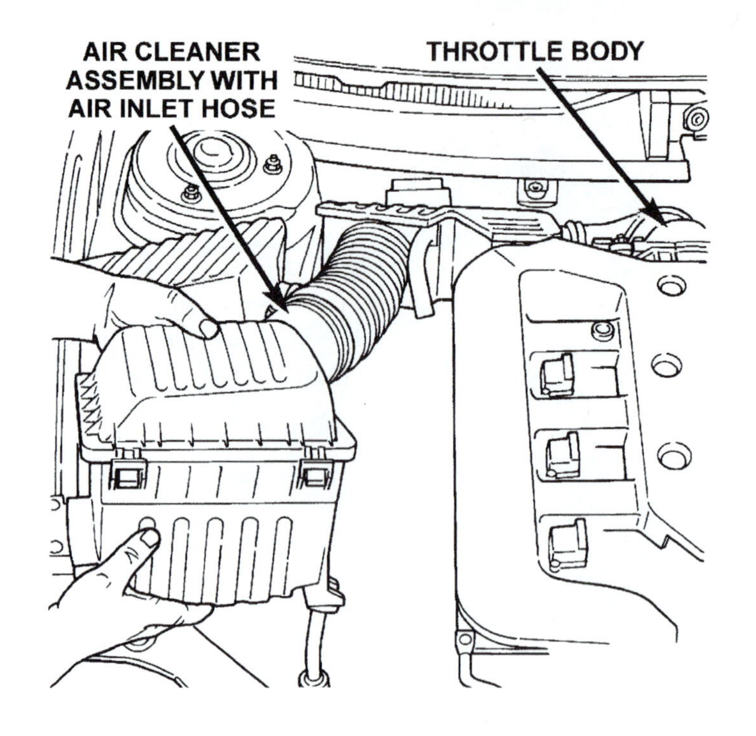

Figure 4-24 A dirty air cleaner affects electronic fuel injection systems much more than the old carburetor. (Courtesy of DaimlerChrysler Corporation)

The *PCV system* is used to vent the pressure created inside the engine by heat and pressure. Failure to vent the engine will cause oil leaks.

The air cleaner in the housing cleans the air before it enters the intake manifold (Figure 4-24). A dirty air filter restricts airflow and results in poor engine performance by causing a rich mixture. While checking the air filter, look for any oil or oil residue in the filter housing. The presence of oil there indicates a problem with the **positive crankcase ventilation (PCV)** system, which is covered in Chapter 8, "Engine Maintenance."

Oil

Oil leaks are noticeable by black, oily dirt around the engine covers and on the sides of the engine. Each cover on the engine has some type of seal or sealing material between it and its mating components. Leaks result when the seal is broken or damaged. The PCV system may also cause an engine oil leak. The system is supposed to vent the pressure in the oil sump of a running engine (Figure 4-25). If the system is inoperative, pressure will build and oil will be forced through or around gaskets and seals or through the PCV valve.

Oil will drain to the lowest portion of a component before dropping the ground. Leaks around seals on the front of the engine will cause the drive belts to be oily. The belts may also sling oil onto the underside of the hood and body panels in the area of the belts.

Coolant

A *radiator* is connected to a vehicle's engine by two hoses. Damage or rot to these hoses can cause coolant leaks.

Coolant leaks can be identified by the green color of the liquid and may occur because of loose clamps or water pump seals. The **radiator**, radiator cap, and hoses may also develop leaks. The radiator is mounted at the front of the vehicle and is connected to the engine by two large hoses, one at or near the top and one at the bottom (Figure 4-26). While checking for coolant leaks, inspect the two radiator hoses for dry rot or other damage. If a coolant leak is suspected but not visible, a coolant pressure tester can be used (Figure 4-27). The cooling system can be checked with a hand pump and adapter equipment. This system will be covered in Chapter 8, "Engine Maintenance."

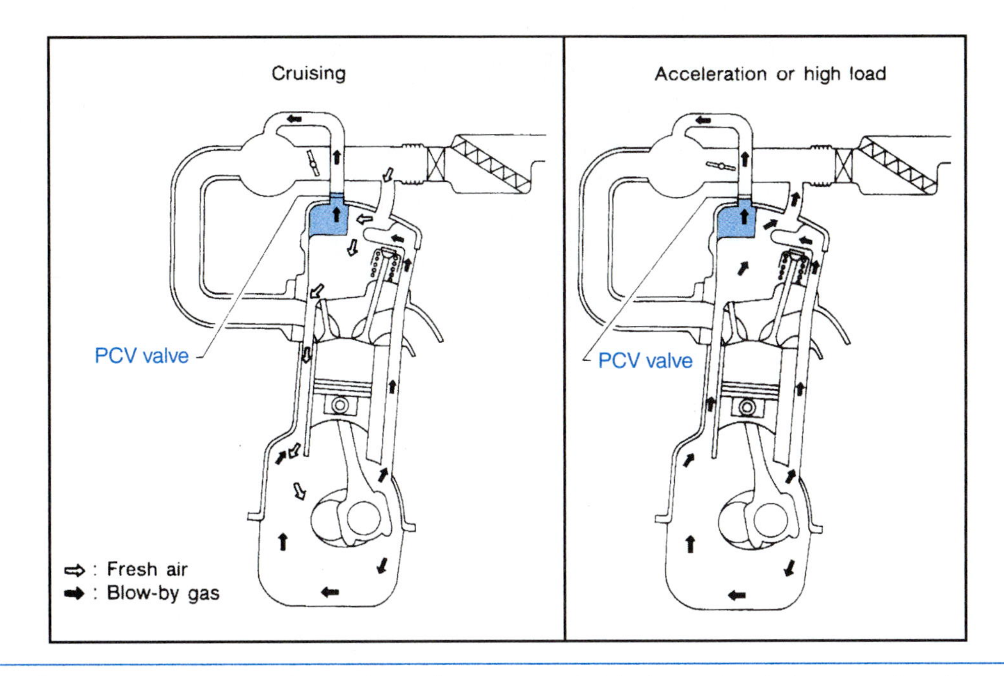

Figure 4-25 The PCV is designed to reduce pressure inside the engine during operation. (Courtesy of Nissan North America)

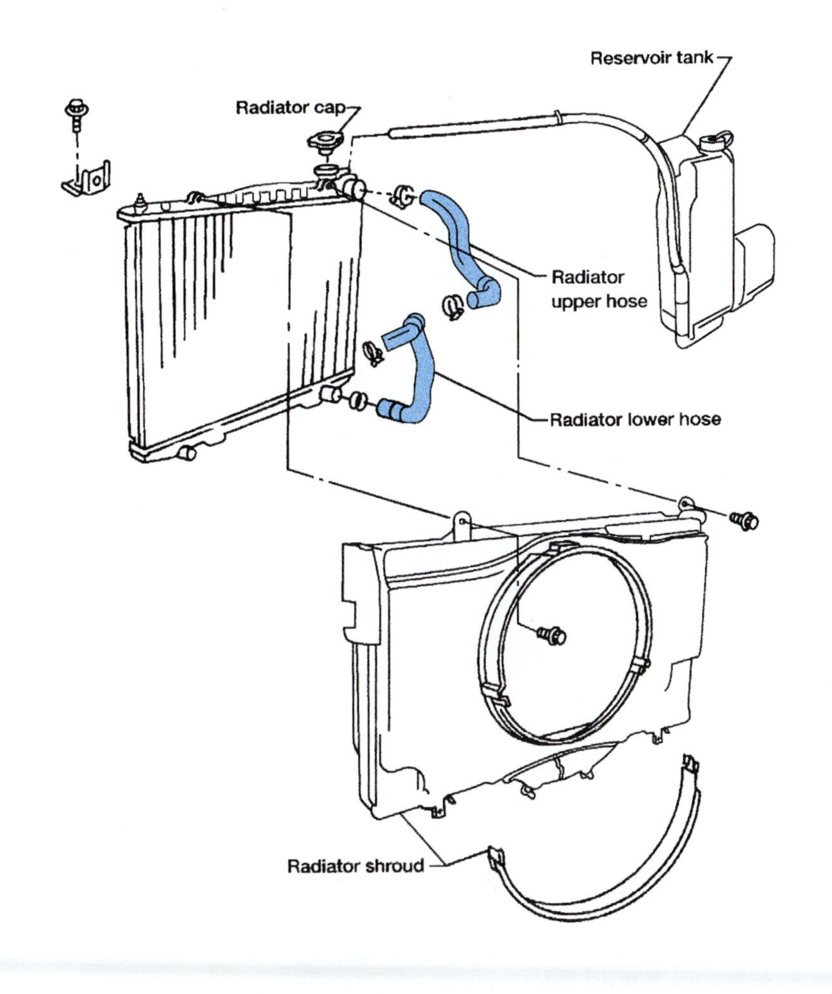

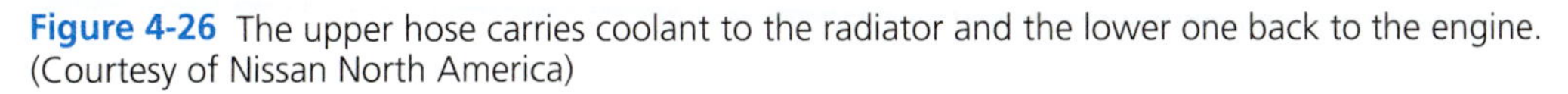

Figure 4-26 The upper hose carries coolant to the radiator and the lower one back to the engine. (Courtesy of Nissan North America)

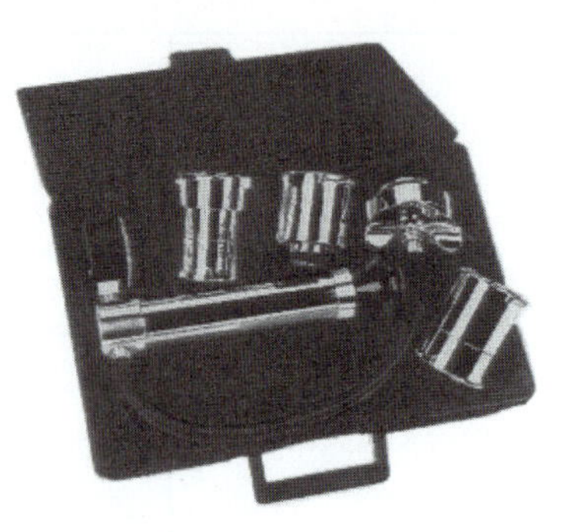

Figure 4-27 This is a common coolant system pressure tester. (Courtesy of Snap-on Tools Company)

Exhaust

CAUTION: The exhaust system gets hot very quickly after the engine is started. Extreme care must be taken when checking or working around the exhaust system. Serious burn injury could result.

Sound is the best and safest method to check the exhaust system for leaks. The sound of a leaking exhaust is different from almost any other sounds from the vehicle. If the sound is coming from the engine area, it usually means the leak is at the connection between the exhaust manifold and engine or between the manifold and exhaust pipe (Figure 4-28).

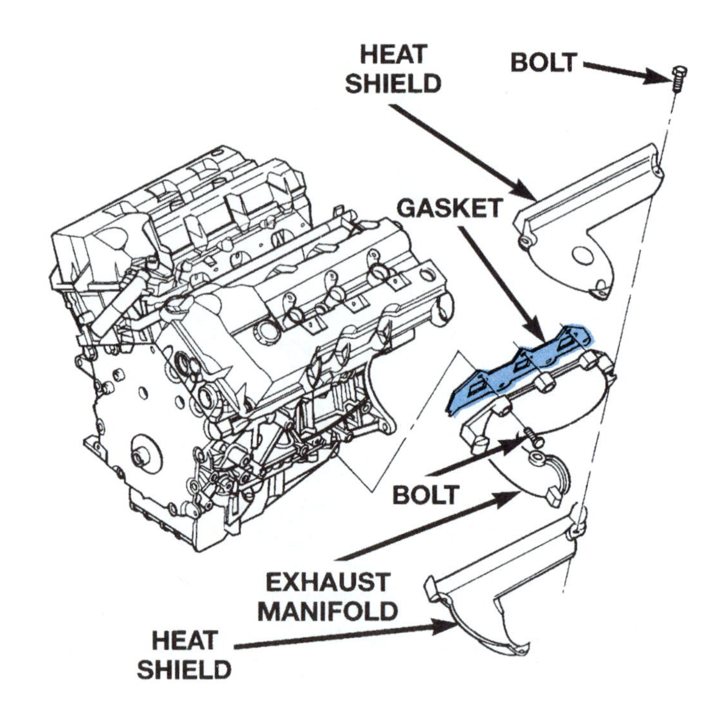

Figure 4-28 Some exhaust leaks occur when the manifold gasket burns through. (Courtesy of DaimlerChrysler Corporation)

Inspection and Leak Checking under the Vehicle

Classroom Manual page 76

WARNING: Before lifting a vehicle, ensure that the vehicle's lift points are identified and the lift pads are properly placed. Failure to do so may result in damage to the vehicle and the creation of a work hazard. If necessary, review Chapter 3, "Automotive Tools," on using a lift.

CAUTION: Never work under a vehicle that is not supported by jack stands or the locks of a lift. The vehicle could slip off a jack and cause serious injury.

Under-vehicle inspection and work can be done using a floor jack and jack stands. However, for this discussion, it is assumed that the vehicle will be raised on a double-post, above-ground lift.

Position the vehicle between the posts and align the lift's pads with the vehicle's lift points. Raise the vehicle slightly above the desired work height and then lower it until the lift's locks engage and support the vehicle's weight. Start at the front of the vehicle and work rearward observing for components and leaks not visible from the top of the engine. First, inspect the front of the engine for components not found or not readily visible from the top of the engine.

Probably the most visible major component at this point is the air conditioning (A/C) **compressor**, which is a type of pump (Figure 4-29). A/C compressors come in many different sizes but modern ones usually have a round body, which is slightly less in diameter than the length of the compressor. At the rear or on one side of the compressor will be two large, aluminum lines. This is the only component that will have this type of line. One end of the compressor will have the pulley and drive belt. Just behind the compressor pulley is a magnetic locking device with a small electrical plug. Make sure the plug is connected securely to the compressor and there are no broken or damaged wires visible.

A *compressor* is used to pressurize a vapor. In an A/C system, a hot vapor is pressurized and forced through a cooling device to change the vapor to a cooler liquid state.

On some vehicles it may be easier to see the power-steering pump from this viewpoint. This is not necessarily the position from which to remove the pump, but the technician may be able to find a hidden fastener that is invisible from the top. It may also be easier to see the bottom of the alternator from this view, but its removal will probably be from the top.

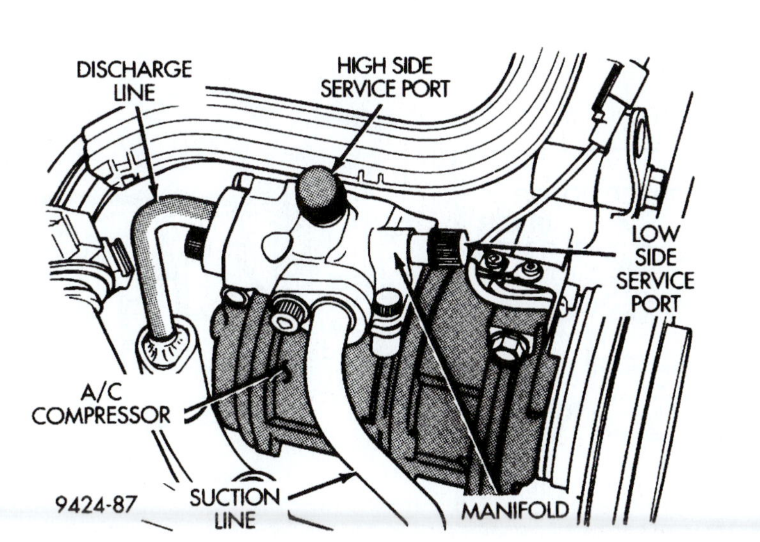

Figure 4-29 The A/C compressor is capable of producing up to 500 or 600 psi in the A/C system. (Courtesy of DaimlerChrysler Corporation)

Other components visible or at least partially visible include those of the engine, transmission, steering, suspension, and brake systems. Each system is discussed in later chapters. Directly under the lower portion of the engine is the oil pan. The oil filter is usually mounted low on the engine near the top edge of the pan. The filter may be found on either side or near either end of the engine. A plug will be located at the lowest part of the oil pan for easy removal of the engine oil (Figure 4-30).

Check around the oil filter, oil pan, and the engine in general for evidence of oil leaks. Engine oil will be black. Since liquids will flow to the lowest point before dropping off, do not assume that oil on the pan means the leak originated there. The oil could be coming from the top of the engine. Sometimes it is necessary to completely clean the entire engine compartment, test drive the vehicle, and then attempt to find the source of the leak.

A rear-wheel drive vehicle uses a transmission instead of the *transaxle*. They basically work in the same manner.

Behind the engine on the opposite end from the drive belts is the **transaxle** (Figure 4-31). The transaxle allows the driver to select gears for movement of the vehicle. There is a pan under this unit as well. It is usually flat and parallel to the floor, but it may be angled because of the space available for the transaxle. Usually, there is no place to drain the oil. The pan has to be removed for service. Most transaxles have the filter inside the pan. Inspect the transaxle at its pan and anywhere else that appears to be a connection between the two parts. Most of the external parts of a transaxle have some type of seal between its mating parts. Transaxle oil or fluid is usually red but may be very dark because of dirt and overheating. One fairly common leak area is the connection between the engine and the transaxle. Fluid found there might be engine oil or transaxle fluid, and both could be dark in color and difficult to identify. However, burned transaxle fluid will have a very burned smell.

Rear-wheel drive vehicles usually do not use a *sub-frame*.

A vehicle with four- or all-wheel drive may have one or two *control arms* at each wheel.

The engine and transaxle are mounted on a **sub-frame** which reaches from the passenger floorboard to the front bumper (Figure 4-32). It also extends from side to side under the vehicle. Extending from the sub-frame to an area near the bottom of each wheel assembly is a lower **control arm** (Figure 4-33). This provides a pivot and support point for the wheel.

Looking upward past the arm and into the fender area, the lower part of the upper suspension components is visible. Also in this area on both sides will be a long, slender rod extending from the center of the vehicle to an attachment near the wheel. These are the outer parts of the steering system (Figure 4-34). The braking devices for each wheel may also be visible. If this is a late model Chrysler car, notice the fasteners that hold the inner panel for the left fender. The vehicle's battery is located behind that panel.

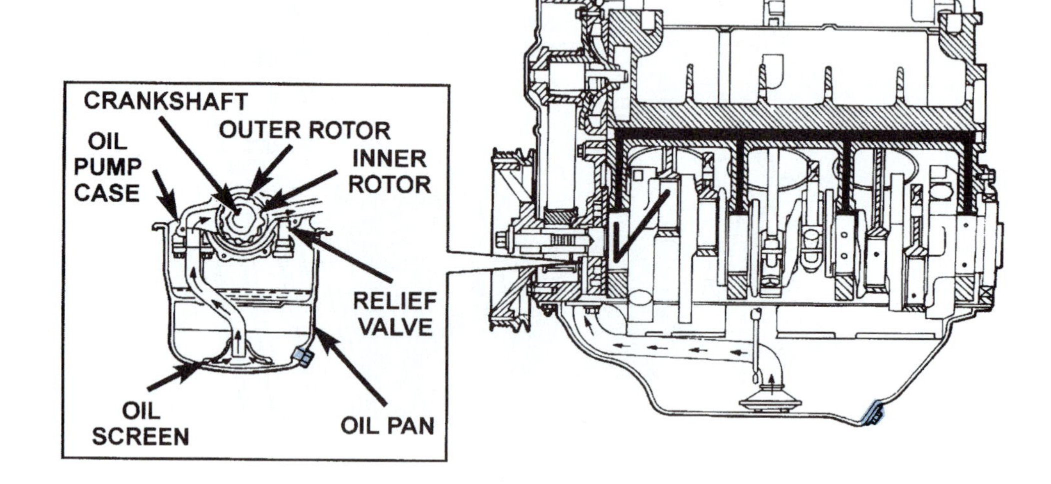

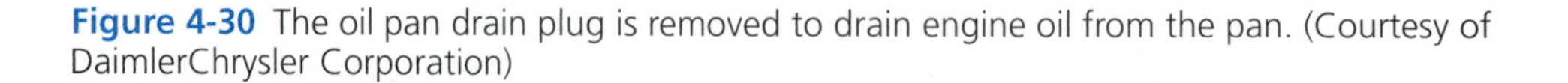
Figure 4-30 The oil pan drain plug is removed to drain engine oil from the pan. (Courtesy of DaimlerChrysler Corporation)

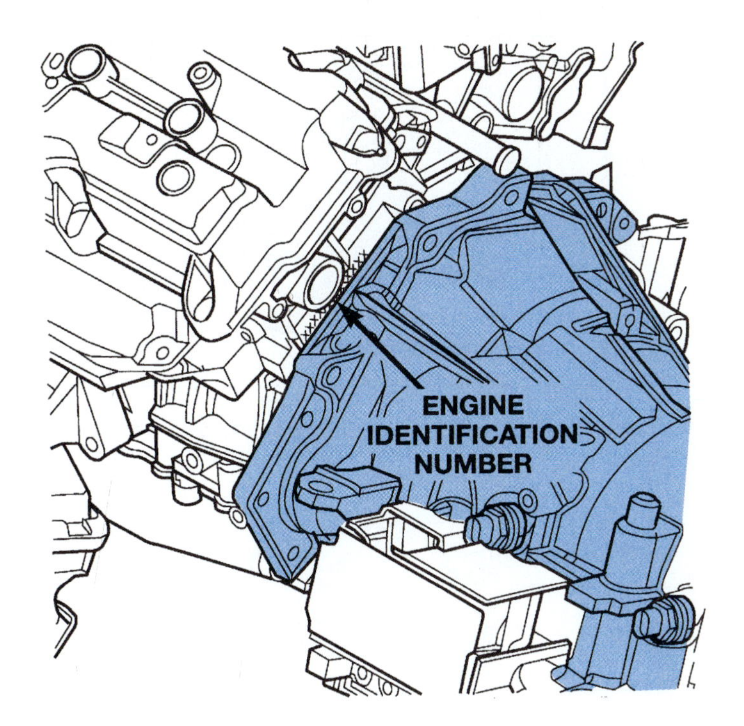

Figure 4-31 The transaxle, like the transmission, is mounted to the rear of the engine. (Courtesy of DaimlerChrysler Corporation)

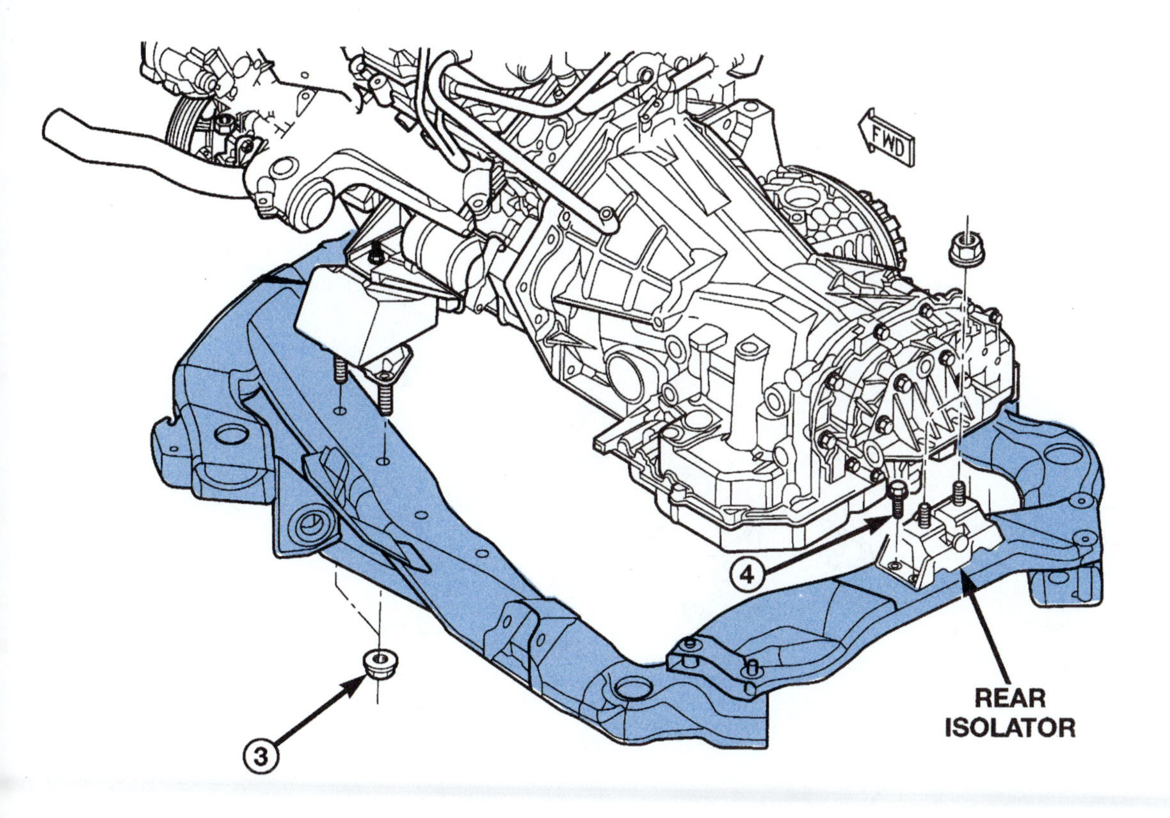

Figure 4-32 The sub-frame supports the engine and transaxle in a front-wheel-drive vehicle. (Courtesy of DaimlerChrysler Corporation)

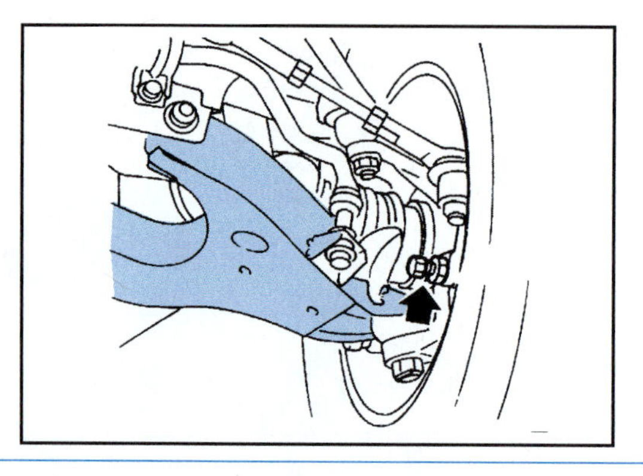

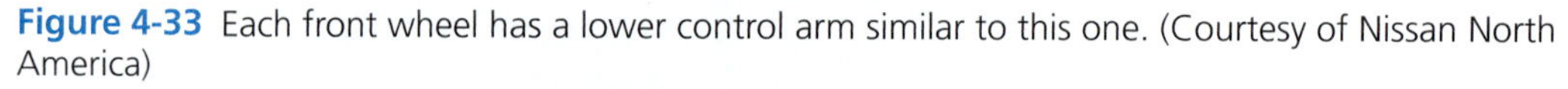

Figure 4-33 Each front wheel has a lower control arm similar to this one. (Courtesy of Nissan North America)

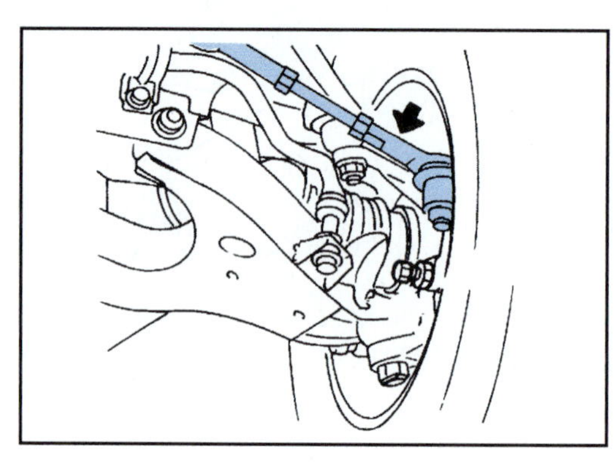

Figure 4-34 This is the outer end of the steering linkage. (Courtesy of Nissan North America)

Rear-wheel *drive axles* are usually in enclosed in a large axle housing.

Extending from each side of the transaxle out to the wheels is large, round bar with bellowed rubber boots at each end (Figure 4-35). These are the **drive axles**. They transfer power from the transaxle to the wheels.

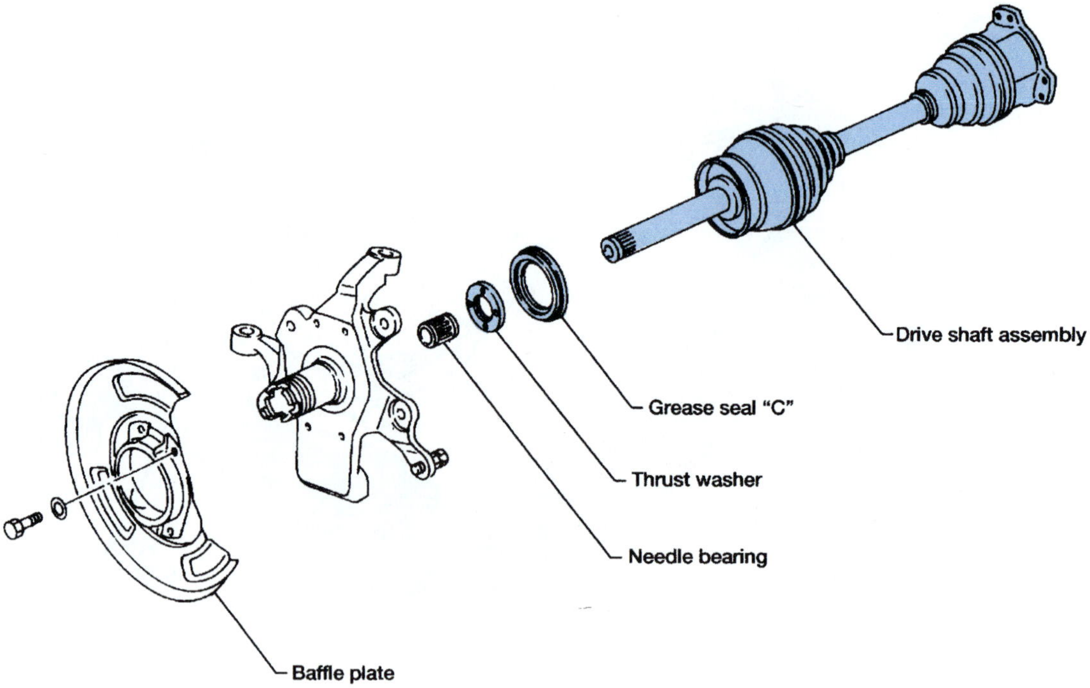

Figure 4-35 This drive axle represents the type used by most front-wheel-drive vehicles. (Courtesy of Nissan North America)

Usually, no fluid leaks are found in these areas unless the brake system is leaking, but grease may be slung around the inside of the fender and any components mounted there. This is a result of a torn rubber boot. The boot is designed to hold special-purpose grease. A tear in the boot, combined with the high-speed rotation of the drive axle, slings the grease out in a vertical, circular pattern.

Moving rearward from the engine, the technician will find the floor pan or floorboard for the passenger compartment. Along one or both sides of the vehicle are the brake and fuel lines and the electrical conductors. On some vehicles, the lines and conductors are installed inside a tubular section of the frame and are not visible from this view. There should be no leaks or seepage along any of the lines.

At the rear of the vehicle is the most obvious component: the fuel tank. The suspension for each wheel is also visible. There can be several different components visible depending on the design of the suspension system. Almost everything visible between the body and the wheel is some part of either the suspension or the brake system (Figure 4-36). Each of the systems mentioned in this section is covered in detail in later chapters.

The inside of the wheel assemblies and the mounting components should be clean and free of any oil. Fluid or oil at this point usually indicates a leaking brake system. Naturally, there should be no leakage or seepage from the fuel tank or any of the lines and hoses under the vehicle. On the side of the vehicle where the fuel cap is located, check the area between the fuel tank and the fender or body for a large and small hose (Figure 4-37). The larger one is the fuel inlet hose. The second and smaller hose allows air to escape the tank during filling. Each is held in place with clamps. Like all fuel hoses, there should be no leaks or seepage.

CAUTION: Do not handle or touch any exhaust component until it is completely cool. Very serious burns could result.

Inspect the exhaust system that runs under the vehicle (Figure 4-38). There should be no soot at the connections. The larger parts of the exhaust are silencers called *mufflers* or *resonators*, and they may have small holes near one of the welded seams. Ideally, there should be no soot around the holes either, but they should be examined closely and corrective action taken as needed.

A large unit in the exhaust system near the engine is part of the emission control system. It operates at a temperature near or exceeding 1,600 degrees Fahrenheit. It can be uncomfortable working near this system, so it should be allowed to cool before repairs are made. There should be no kinks, holes, or cracks anywhere else in the exhaust system. If such defects exist, the sections in which they appear must be repaired or replaced.

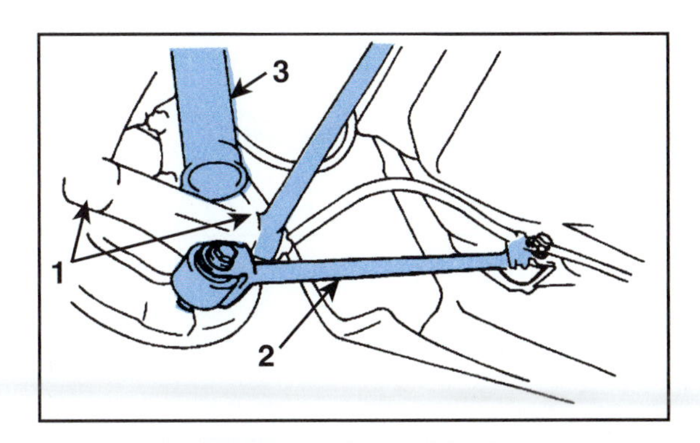

Figure 4-36 Items 1 and 2 keep the wheel assembly in line with the body. Item 3 is the lower end of a suspension part. (Courtesy of Nissan North America)

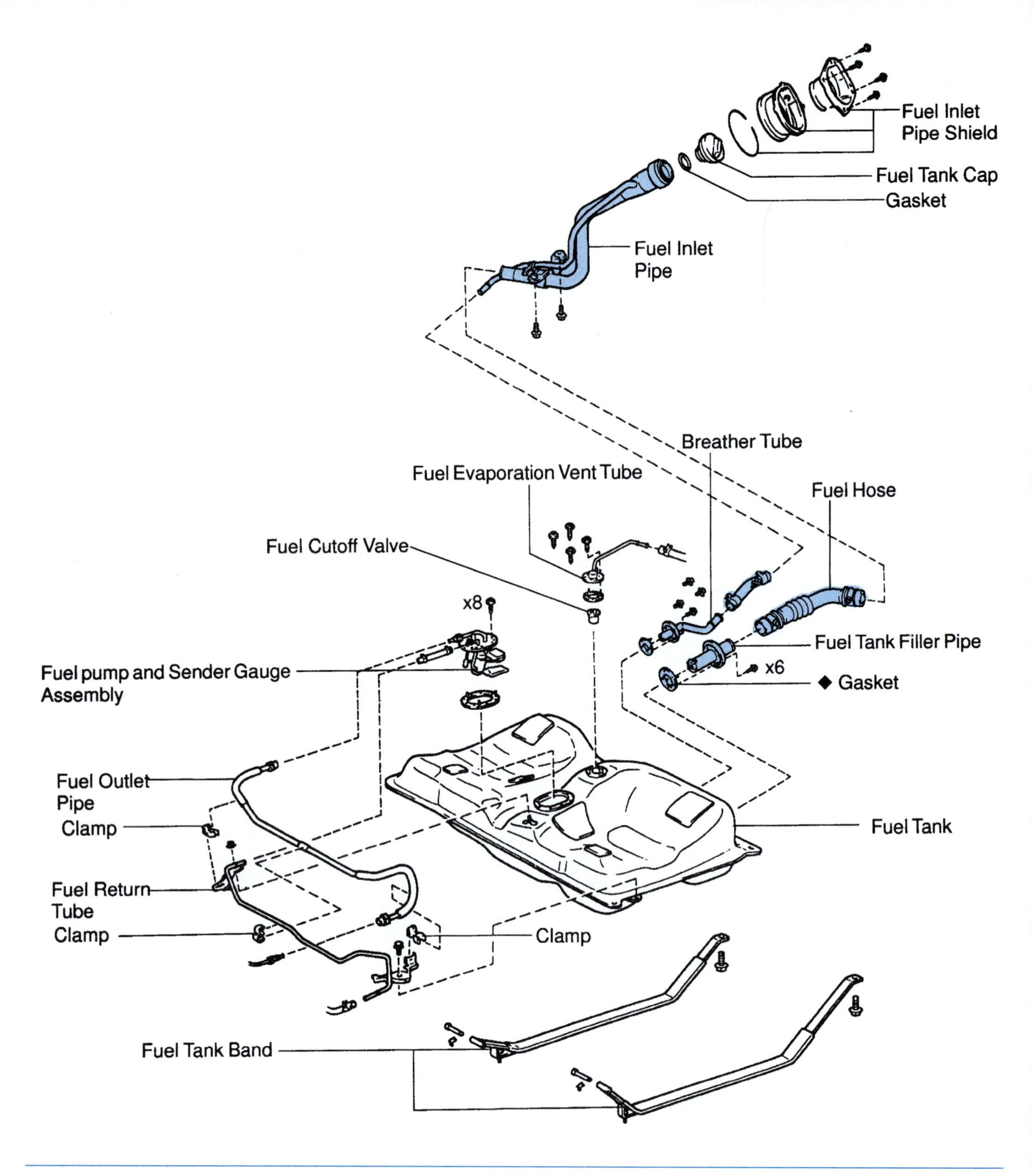

Figure 4-37 The filler and breather pipes (hoses) have several clamps to hold them in place. (Reprinted with permission)

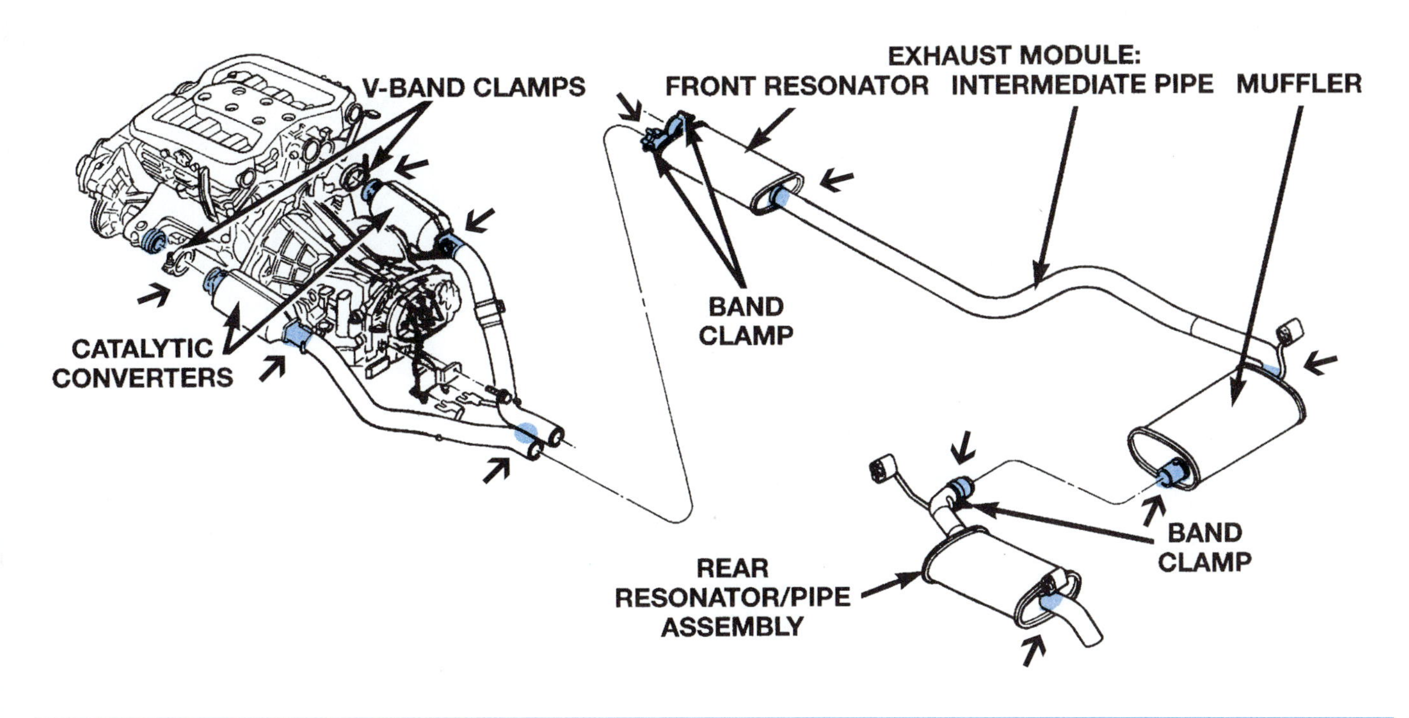

Figure 4-38 Exhaust systems most probably leak at the points noted, but may rust through at any point. (Courtesy of DaimlerChrysler Corporation)

Inspecting Body-Mounted Components

The entry-level technician performs only basic repairs to body-mounted components. He or she can, however, inspect and test many body-mounted components.

Classroom Manual
page 76

Lights

Inspecting the **lighting system** is a relatively simple process, but it should be done every time the vehicle is brought in for servicing or repairs. An inoperative lamp may not be apparent to the operator until too late. Repairs to the lighting system are covered in Chapter 12, "Auxiliary System Service."

Two *lighting systems* are considered emergency lighting. They are the brake lights and the hazard warning lights. They should operate in almost any condition as long as electrical power is available.

Start your inspection with the lights that must be connected directly to the battery. In other words, the ignition switch does not need to be in the RUN position.

Turn on the parking lights and walk around the vehicle (Figure 4-39). All parking lights should be on. Switch on the headlights. Some imported vehicles require the ignition switch to be in the RUN position before the headlights will work.

Even though light may be visible on the wall in front of the vehicle, do not assume everything is correct and that the lights are in working order. Move to the front of the vehicle and make sure both headlamps are operating. Check each headlight in its high- and low-beam position and remember to check the high-beam (bright) indicator to be sure it is working in the dash.

Check the hazard warning lights next. The headlights and parking lights may be left on or off, but ensure that the ignition switch is off. The hazard lights are on a circuit by themselves and must operate regardless of the position of any other switches. The hazard switch will be on the right side of the steering column near the ignition key slot, on the top center of the steering column just forward of the steering wheel, or on the instrument panel (Figure 4-40). It will be marked with a red marking in most cases.

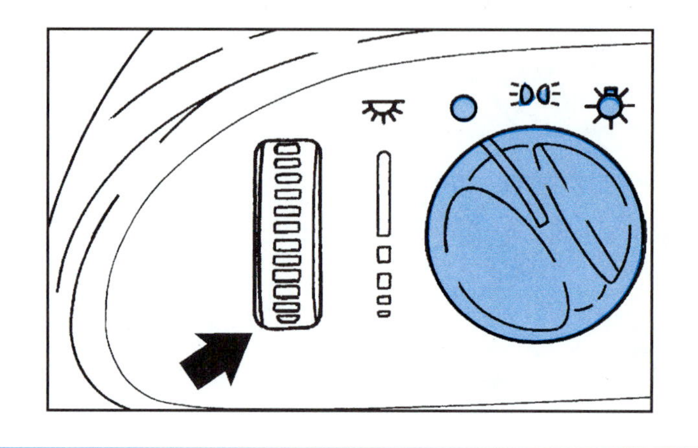

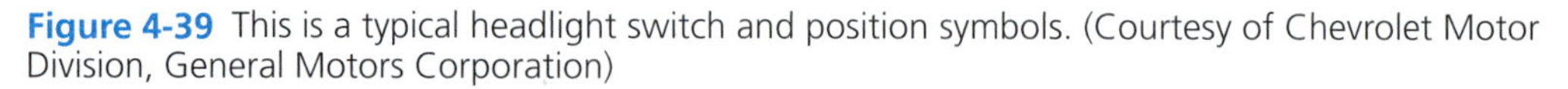

Figure 4-39 This is a typical headlight switch and position symbols. (Courtesy of Chevrolet Motor Division, General Motors Corporation)

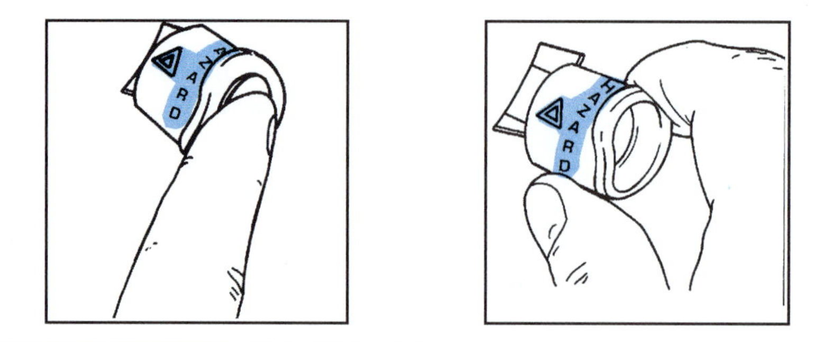

Figure 4-40 A typical hazard light switch. (Courtesy of Chevrolet Motor Division, General Motors Corporation)

Move the hazard switch to the ON position and observe the direction or turn signal indicators in the dash. Both should be flashing. The two front parking lights should be flashing, as should the brake lights in the rear. This system will work even if three lamps are burned out. The fourth light will still flash.

With the hazards working, apply the brakes. The front lights should still flash, but the rear lights should shine solidly without flashing. Releasing the brakes will allow the rear lights to flash again. Switch off the hazard lights and retest the brakes.

The ignition switch must be in the RUN position to test most turn or signal lights. Switch on the ignition and select the left turn signal. The turn signal indicator should blink. Select the right turn signal and observe the right indicator. With the right turn signal on, check the rear and front lights on the right side of the vehicle. Reselect the left turn signal and check the left side.

Leave the left turn signal on and apply the brakes. Depending on the design of the lighting circuits, one of several things should happen. If this is a typical two-lamp system, the rear left turn signal will flash and the right brake light will glow solidly. On a three-lamp system, the brake and turn signal are on separate lamps, so there will be brake lights on both sides in the rear and a flashing light for the left turn. After checking the brake and left turn signal, do the same test for the right turn signal.

With the ignition still in the RUN position, operate the headlights and hazard lights again in the same manner noted earlier. Headlights that require the ignition switch to be in the RUN position can be checked now. One other headlight system that requires checking is the daytime running lights. In this system, the headlights are automatically switch on when the ignition switch is on. The lights will be dimmer then normal because of the low voltage and current supplied. If in doubt, turn on the ignition switch and observe the headlight illumination. Switch on the headlight switch and the illumination should brighten considerably.

Check the lights in the dash while the ignition switch is on. There will be several red lights visible. The system at this time is operating in a test mode of the warning lights. The warning lights and gauges will be covered in Chapter 12, "Auxiliary System Service."

Figure 4-41 Some interiors lights have switches mounted on the light frame. (Courtesy of Chevrolet Motor Division, General Motors Corporation)

There should be one or more interior lights that work when a door is opened. In addition, the interior lights usually have a switch that can be used for control. Observe the interior lights as each door is opened and closed. Remember to close the other doors as one is checked. Close all doors and locate the switch for the interior lights. Check the lights using this switch. Some interior lights will only work with the doors while others may have independent switches for each light (Figure 4-41). Check all of them.

Windshield Wipers

Windshield wipers tend to be left out of the maintenance process. A badly damaged wiper blade can scratch the windshield glass and will not sufficiently clear it of rain or debris. Repairs on the wiper system are covered in Chapter 12, "Auxiliary System Service."

WARNING: Do not allow the wiper arm to snap against the glass. The glass may chip or crack, resulting in a high cost of replacement.

WARNING: Do not operate the windshield wipers on dry glass. Using the windshield wipers on dry glass could scratch the glass and/or damage the new wipers.

Inspect the rubber insert portion of the wiper that lays on the windshield (Figure 4-42). The rubber should not be torn or ripped. If so, replace the insert before proceeding. With

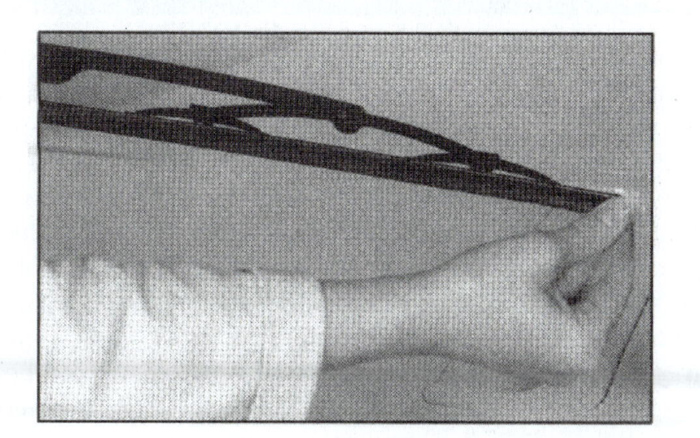

Figure 4-42 The rubber insert should be the only portion of the wiper to touch the glass. (Courtesy of Chevrolet Motor Division, General Motors Corporation)

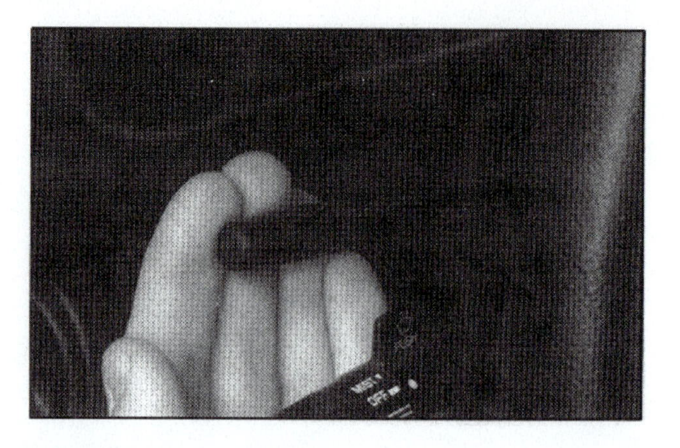

Figure 4-43 The little lever on this wiper switch operates the windshield washer. (Courtesy of Chevrolet Motor Division, General Motors Corporation)

good inserts on each side, turn the ignition to RUN. The wiper control is on a lever on the left or right side of the steering column or on the dash. Most vehicles have interval wipers, meaning that a position on the switch allows the wiper to sweep, pause for a set amount of time, and then make another sweep. Most wiper systems have two speed positions in addition to the interval. The system includes a washer position on the wiper lever or a separate switch for the washer. Before operating the wipers, use the washer to wet the glass.

Operate the wiper in each switch position and ensure that the glass is wet by using the washer. In the interval position adjust, the time by operating the interval switch (Figure 4-43). Some vehicles, however, do not have the option of adjusting the time. Also check the rear windshield. Some vehicles have a wiper and washer for this glass and it should be checked in the same manner as the front.

Doors

While checking the interior lights with the door, observe the movement of the door. It should be smooth and quiet. A grinding or scraping noise indicates a lack of lubricant. The door should shut solidly and swing in an even arc through it entire range. A sagging door will not close quietly and may have to be slam shut to latch properly. This usually indicates that the door hinges are worn and allowing the door to drop out of alignment. Repairs to the door should be performed by an experienced technician and in some cases by a body shop technician.

CASE STUDY

A customer complained of adding coolant every week or so. He stated that he could not find any coolant on the garage floor or inside the engine compartment. The technician performing the diagnosis noticed that the overflow reservoir was completely full, but the engine was only warm. Further testing with a pressure tester revealed no leaks, including the radiator cap. The technician located the end of the overflow reservoir drip hose. There was an indication of coolant in the end of the hose. It was decided that the cap was allowing coolant out of the system as designed, but not drawing it back into the system as it was supposed to do. As a result, the overflow would fill to capacity as the car was running and the excess would drain into the air stream flowing under the car. Since the driver checked the coolant regularly, the vehicle never overheated. But the driver did not check under the vehicle immediately after stopping. A replacement radiator cap solved the problem.

Terms to Know

Baking soda
Compressor
Control arm
Corrosion
Drive axle
Fuel rail
Mass air flow sensor (MAF)
Powertrain Control Module (PCM)
Positive crankcase ventilation (PCV)
Radiator
Ratio
Seepage
Serpentine belt
Span
Sub-frame
Tachometer
Throttle body
Transaxle
Transversal
V-belt

ASE Style Review Questions

1. *Technician A* says pressure and vacuum can be directly influenced by a leak.
   Technician B says pressure and vacuum are used to move the air and fuel into the engine. Who is correct?
   A. A only
   B. B only
   C. Both A and B
   D. Neither A nor B

2. Theories of operation are being discussed. *Technician A* says Newton's Laws affect the movement of the air and fuel.
   Technician B says the laws of motion can affect thermodynamics. Who is correct?
   A. A only
   B. B only
   C. Both A and B
   D. Neither A nor B

3. *Technician A* says a dirty air filter can affect thermodynamic theory.
   Technician B says thermodynamics will be affected by fuel flow through the fuel filter. Who is correct?
   A. A only
   B. B only
   C. Both A and B
   D. Neither A nor B

4. Leaks are being discussed. *Technician A* says oil on the bottom of the oil pan means the oil pan is leaking.
   Technician B says the source of an oil leak may be determined by the color of the fluid. Who is correct?
   A. A only
   B. B only
   C. Both A and B
   D. Neither A nor B

5. The air/fuel mixture is being discussed. *Technician A* says a leak in a vacuum line could cause the PCM to set the amount of fuel injected incorrectly.
   Technician B says a crack in the air intake ducting will result in a rich mixture. Who is correct?
   A. A only
   B. B only
   C. Both A and B
   D. Neither A nor B

6. *Technician A* says the alternator may be located near the top of the engine.
   Technician B says the compressor is used for power steering. Who is correct?
   A. A only
   B. B only
   C. Both A and B
   D. Neither A nor B

7. Light operation is being discussed. *Technician A* says all headlamps should operate with the ignition switch on or off.
   Technician B says the rear turn signal should flash with the brakes applied on a two-lamp system. Who is correct?
   A. A only
   B. B only
   C. Both A and B
   D. Neither A nor B

8. Drive belts are being discussed. *Technician A* says oil and coolant could shorten the life of the belts.
   Technician B says all water pumps use an exposed external drive belt for operation. Who is correct?
   A. A only
   B. B only
   C. Both A and B
   D. Neither A nor B

9. *Technician A* says a serpentine belt should have no cracks on the outside.
Technician B says belts must be replaced with new belts of like size in length and width. Who is correct?

A. A only
B. B only
C. Both A and B
D. Neither A nor B

10. *Technician A* says fluid seepage on a rear wheel is acceptable.
Technician B says power-steering fluid may be the same color as transaxle fluid. Who is correct?

A. A only
B. B only
C. Both A and B
D. Neither A nor B

Table 4-1 ASE Task

Inspect the engine assembly for fuel, oil, coolant, and other leaks; determine necessary action.

Problem Area	Symptoms	Possible Causes	Classroom Manual	Shop Manual
Leaks	Black oil. Spots on.	**1.** Oil pan gasket leaking.	—	85, 90, 94
		2. Valve cover gaskets leaking.	—	85, 90
		3. Crankshaft seals leaking.	—	—
		4. Positive crankcase ventilation plugged or inoperative.	—	90, 91, 94
		5. Engine oil overfull.	—	94
	Fuel smell.	**1.** Loose fuel line fittings.	—	87, 97
		2. Carburetor or fuel injections leaking.	—	87
		3. Ruptured fuel lines.	—	87, 97
		4. Overfilled fuel tank.	—	97

SAFETY

Wear safety glasses.
Allow the engine to cool.
Move a fire extinguisher near the vehicle.
Warn all others about possible fuel leakage.

Table 4-2 ASE Task

Diagnose power-steering gear noises, vibration, looseness, hard steering, and fluid leakage problems; determine needed repairs.

Problem Area	Symptoms	Possible Causes	Classroom Manual	Shop Manual
Leaks	Red, pink oil spots on floor.	**1.** Transaxle oil pan leaking.	—	94
		2. Transaxle lines fitting loose.	—	94
		3. Seals leaking.	—	94
		4. Cracked or split lines or hoses.	—	94
		5. Transaxle oil overfull.	—	94

SAFETY

Wear safety glasses.
Allow the engine and transaxle to cool.

Table 4-3 ASE Task

Verify the driver's complaint and road test the vehicle. Determine necessary action.

Problem Area	Symptoms	Possible Causes	Classroom Manual	Shop Manual
Performance	Driver's complaint.		73	75

SAFETY

Road test only as necessary.
Secure driver and supervisor permission for road test.
Use short trips if possible, obey traffic regulations.

Table 4-4 ASE Task

Determine if no crank, no-start, or hard starting condition is an ignition system, cranking system, fuel system, or engine mechanical problem.

Problem Area	Symptoms	Possible Causes	Classroom Manual	Shop Manual
Performance	Engine hard to start, poor idle.	**1.** Rich air/fuel mixture.	76	75
		2. Lean air/fuel mixture.	76	75
		3. Dirty air filter.	75	87–90
		4. Dirty fuel filter.	76	87
		5. Vacuum leak.	77	76
		6. Air bypassing sensors.	77	88
		7. Worn ignition system.	80	85
		8. Worn engine, poor cranking vacuum.	77	76

SAFETY

Wear safety glasses.
Allow the engine to cool.
Move a fire extinguisher near the vehicle before servicing the fuel system.

Table 4-5 ASE Task

Inspect brake lines and fittings for leaks, dents, kinks, rust, cracks, or wear; tighten loose fittings and supports.

Inspect flexible brake hoses for leaks, kinks, cracks, bulging, or wear; tighten loose fittings and supports.

Problem Area	Symptoms	Possible Causes	Classroom Manual	Shop Manual
Brakes	Poor braking.	**1.** Brake lines cracked.	78	94
		2. Brake hoses cracked, split.	96	94
		3. Low brake fluid.	96	86
		4. Loose brake line/hose fitting.	96	94

SAFETY

Wear safety glasses.
Allow the engine to cool.
Move a fire extinguisher near the vehicle before servicing the fuel system.

Table 4-6 ASE Task

Inspect, clean, fill, or replace the battery.

Problem Area	Symptoms	Possible Causes	Classroom Manual	Shop Manual
Electrical	Poor starting. Poor charging. Dim or no lights.	**1.** Corrosion on cables, terminals.	82	85
		2. Broken or loose wires, connection.	82	85
		3. Weak, damaged, old battery.	—	85
		4. Loose alternator drive belt.	—	82
		5. Bad alternator.	—	79

SAFETY

Wear safety glasses.
Allow the engine to cool.
Disconnect the battery's negative cable first before servicing the battery or alternator.

Table 4-7 ASE Task

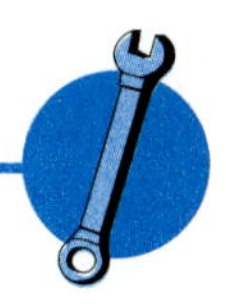

Inspect, replace, and aim headlights and bulbs.

Inspect, test, adjust, repair, and replace all switches, bulbs, sockets, connectors, and wires for parking lights and taillight circuits.

Inspect, test, and repair or replace switches, flasher units, bulbs, sockets, connectors, and wires for turn signal and hazard light circuits.

Inspect, test, adjust, and repair or replace switches, flasher units, bulbs, sockets, connectors, and wires for stoplight (brake light) circuits.

Problem Area	Symptoms	Possible Causes	Classroom Manual	Shop Manual
Lights	No brake light. No turn signal(s). No tail (park) light(s).	**1.** Blown bulb.	91	99
		2. Blown circuit protection device.	91	99
		3. Broken, loose conductor and connectors.	91	99

SAFETY

Wear safety glasses.

Table 4-8 ASE Task

Inspect, service, and replace the exhaust manifold, exhaust pipes, mufflers, resonators, catalytic converters, tail pipes, and heat shield.

Problem Area	Symptoms	Possible Causes	Classroom Manual	Shop Manual
Exhaust	Noise.	**1.** Loose, missing exhaust fasteners.	75	92, 97
		2. Missing, damaged, loose exhaust seals.	75	92, 97
		3. Holes in exhaust system components.	75	92, 97

SAFETY

Wear safety glasses.
Allow the vehicle and exhaust to cool before repairing.

Job Sheet 7

7

Name ____________________ Date ____________

Locating the Source of a Leak

Upon completion of this job sheet, you should be able to locate the source of a leak.

Tools Needed

Basic tool set
Flashlight
Mechanic mirror
Service manual

Procedures

1. Identity the type of liquid and system leaking.

   Color and smell of liquid ____________________

   Suspected system ____________________

2. Raise the hood and inspect any visible components of the suspected system. Use the flashlight and mirror as needed.

   Remarks ____________________

   ____

3. Inspect the sides and covers of the engine and other components on or near the suspected system components.

   Remarks ____________________

   ____

4. Is the leak source visible from this view? If not, prepare the vehicle for lifting.

   Remarks ____________________

   ____

5. For the purpose of this job sheet, assume that the vehicle has to be lifted. Lift the vehicle to a good working height.

6. Observe the point at which the liquid drops from the vehicle's components.

7. Use the light and mirror to trace the liquid up or over the components until a dry area is found. A dry area is defined as any component that does not show signs of the leaking liquid.

■ **NOTE:** It may take some effort and time to locate the highest point of the leak. Practice with the flashlight and mirror to become accustomed to the different angles of sight with the mirror.

Remarks ____

____

8. Identify the source of the leak. Students may want the assistance of the instructor or an experienced technician for the proper terminology at this point in their training.

Remarks ____

____

9. Lower the vehicle and complete your comments. A recommendation for repair or service can be made if desired. Use a service manual for information as needed.

Remarks ____

____

____

____

____

____

____

____

____

____

____

Fasteners

Upon completion and review of this chapter, you should be able to:

- ❑ Use a pitch gauge to identify English (U.S) and metric thread sizes.
- ❑ Use a tap to repair damaged internal threads in an automotive component.
- ❑ Remove and replace a stud with a stud remover.
- ❑ Remove and replace a stud with a jam and drive nut.
- ❑ Strip wire and install a solderless terminal and solderless butt connector.
- ❑ Use a soldering gun or soldering iron and rosin core solder to solder an electrical terminal to a wire.
- ❑ Use a soldering gun or soldering iron and rosin core solder to splice two wires together.

Introduction

Basic technician's tool set

Eye protection

Almost every job a technician does involves the use of fasteners. In the Classroom Manual we described the types and uses for fasteners. This chapter describes several tasks involving fasteners. It describes how to use a pitch gauge to identify threads, how to use a tap to repair internal threads, how to remove and replace a damaged stud, how to install solderless electrical connectors, and how to solder wires and electrical connectors.

Using a Pitch Gauge to Identify Threads

Classroom Manual
page 117

Special Tools

English (U.S.) pitch gauge

Metric pitch gauge

Selection of English (U.S.) and metric fasteners

Replacing damaged or lost fasteners is a common problem when reassembling an automotive component. Whenever a fastener must be replaced, be sure to get a replacement that is the exact quality and thread size as the original. One of the easiest ways to select the correct fastener is to use an original bolt or nut to determine the correct replacement size. For example, if a replacement bolt is needed, take the nut and find the bolt that fits. Make sure the replacement bolt threads into the nut.

There are times when matching bolts and nuts is impossible. In these cases you may need to measure the thread on a fastener to determine its size. The tool used to do this is the pitch gauge. A pitch gauge (Figure 5-1) has a series of blades, each with teeth on one side. Each blade is marked with a thread size. The teeth on the blades are used to match up with threads on a fastener. If they match up, determine the thread size. There are pitch gauges for both English (U.S.) and metric threads.

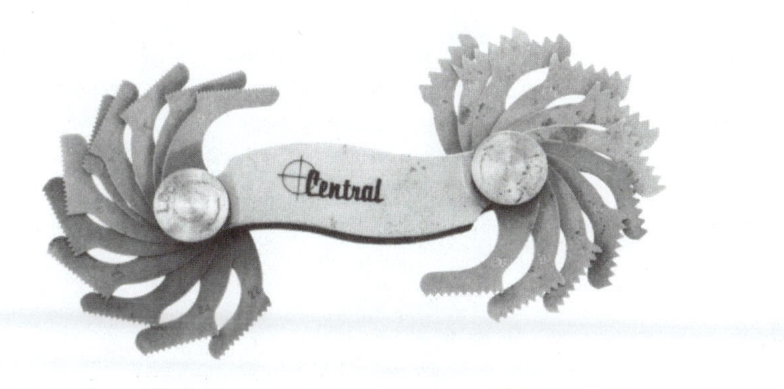

Figure 5-1 A pitch gauge is used to find the thread size of a fastener. (Courtesy of Central Tools, Inc.)

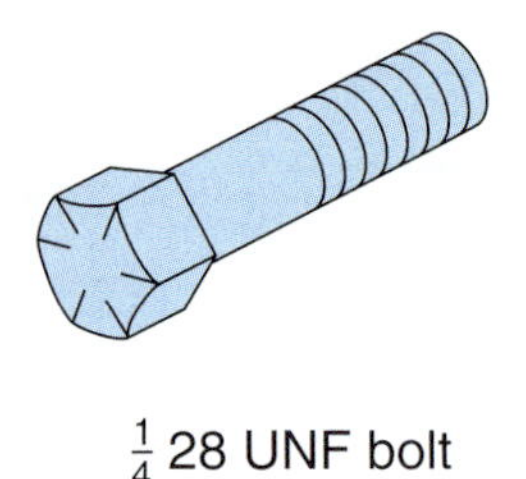

Figure 5-2 A bolt with threads that are 1/4 28 UNF.

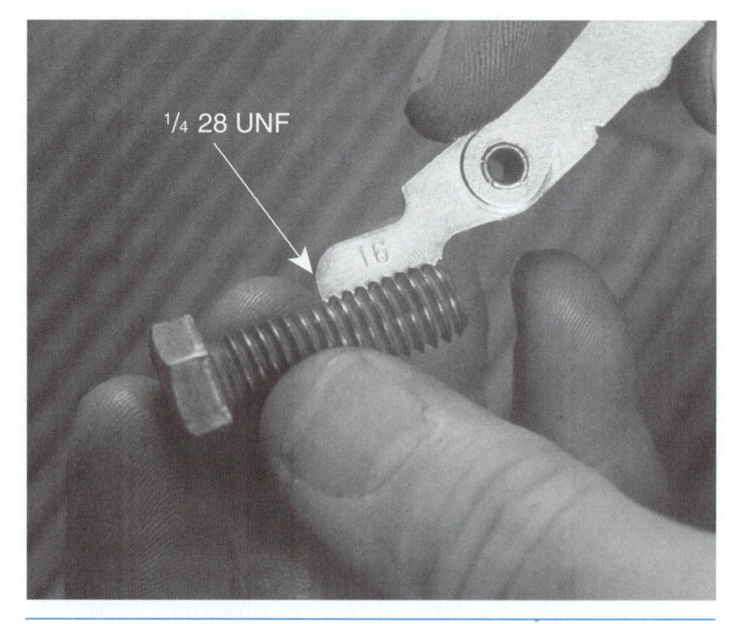

Figure 5-3 The pitch gauge blade that matches the threads reads 16 threads per inch.

The best way to learn how to use a pitch gauge is to practice with a bolt or hex head cap screw with a known size. This way you can get to see how the blades of the pitch gauge should match the threads on a fastener.

Begin with a common English (U.S.) thread size like the 1/4 28 UNF bolt or hex head cap screw shown in Figure 5-2. Pull each blade out of the pitch gauge tool and set the teeth on the fastener as shown in Figure 5-3. Keep trying the blades until a set of teeth fits perfectly into the threads. When a match is found, thread pitch can be identified. The next step is to simply read the thread pitch off the pitch gauge blade. In our example, the pitch gauge blade reads 16 threads per inch.

Now the technician is ready to try a fastener with an unknown thread size. Choose a metric fastener with an unknown thread size like the one shown in Figure 5-4. This time, use a metric pitch gauge. Pull out each blade and set it on the threads as done before. When a perfect match is found, determine the thread pitch. The thread pitch of the fastener is 1.5, as shown on the pitch gauge blade in Figure 5-5.

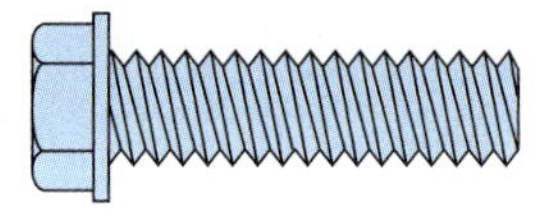

Figure 5-4 A metric bolt with an unknown thread size.

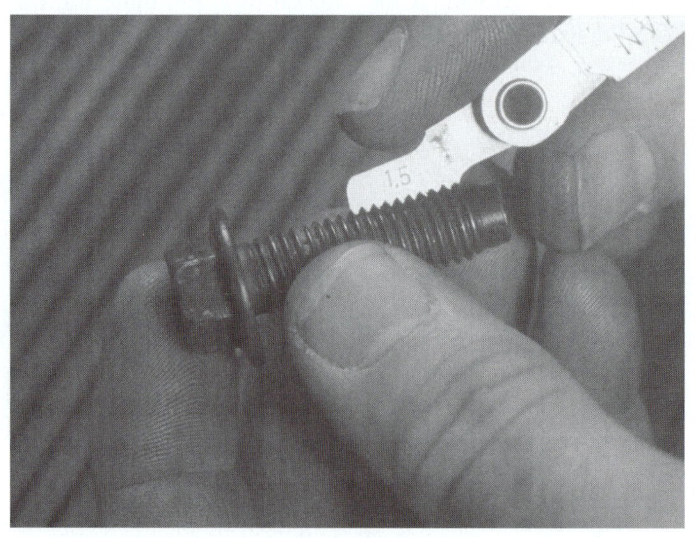

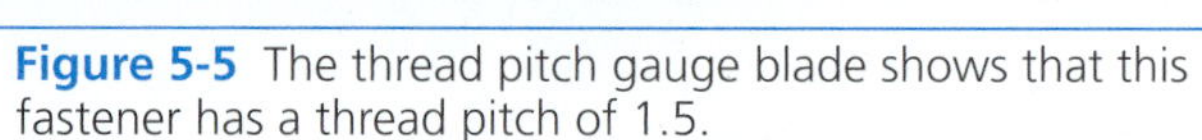

Figure 5-5 The thread pitch gauge blade shows that this fastener has a thread pitch of 1.5.

Using a Tap to Repair Internal Threads

Classroom Manual
page 117

Special Tools

English (U.S.) or metric tap and die set

Machinist square

Stripped threads have been damaged by cross threading.

Technicians often find fasteners that have been damaged. Photo Sequence 5 shows a typical procedure for repairing damaged threads with a tap. A common problem is fasteners that have not been started properly by hand. When a fastener is started into a threaded hole crooked, and then a wrench is used to force the fastener into the threaded hole, the mismatched threads are often damaged. This damage (Figure 5-6) is often called cross threading or **stripped threads.**

The damaged internal threads must be repaired before the automotive component can be reassembled. The tools made to repair internal threads come in a set called a tap and die set. There are many different types and sizes of tap and die sets as shown in Figure 5-7. There are sets for both metric and English (U.S.) thread sizes. Most shops have a set that contains the most common sizes used for the work done in the shop.

The tool used to repair or make new internal threads is called a *tap*. The tap (Figure 5-8, page 115) is a hardened cutting tool. It has a square drive end that is used to rotate it for thread cutting. The shank is the round part of the tap and is where the size of the tap is found. The actual thread cutting is done with sharp cutting threads spaced between flutes where the cut metal can collect. The end of the tap is chamfered or tapered so that it can fit easily into the hole to be tapped.

Before using a tap to repair the threads, make sure to select the correct size and type. This often means the technician will have to determine the thread size of the screw that fits into the threaded hole you are repairing. This can be done with the pitch gauge as explained earlier.

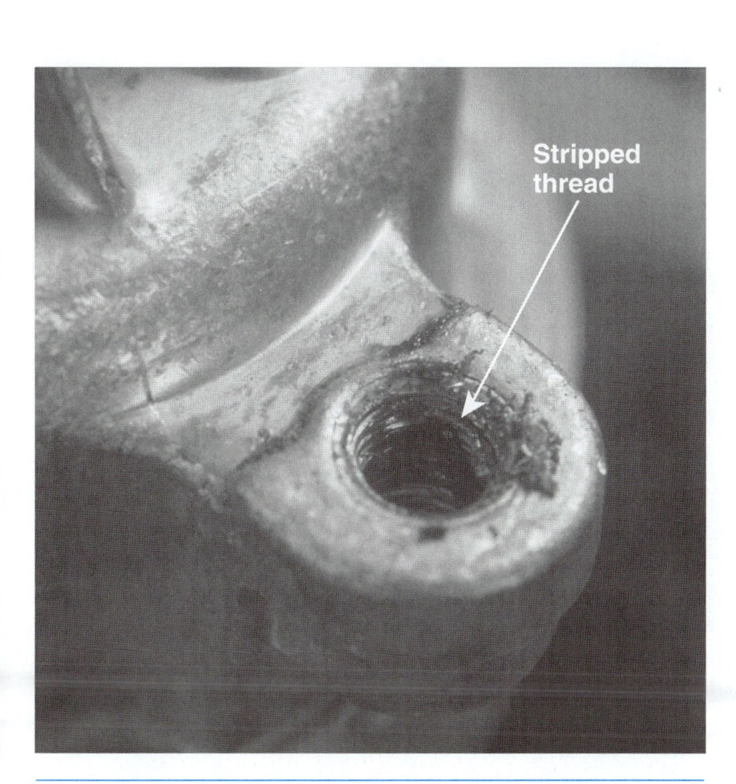

Figure 5-6 A damaged or stripped internal thread.

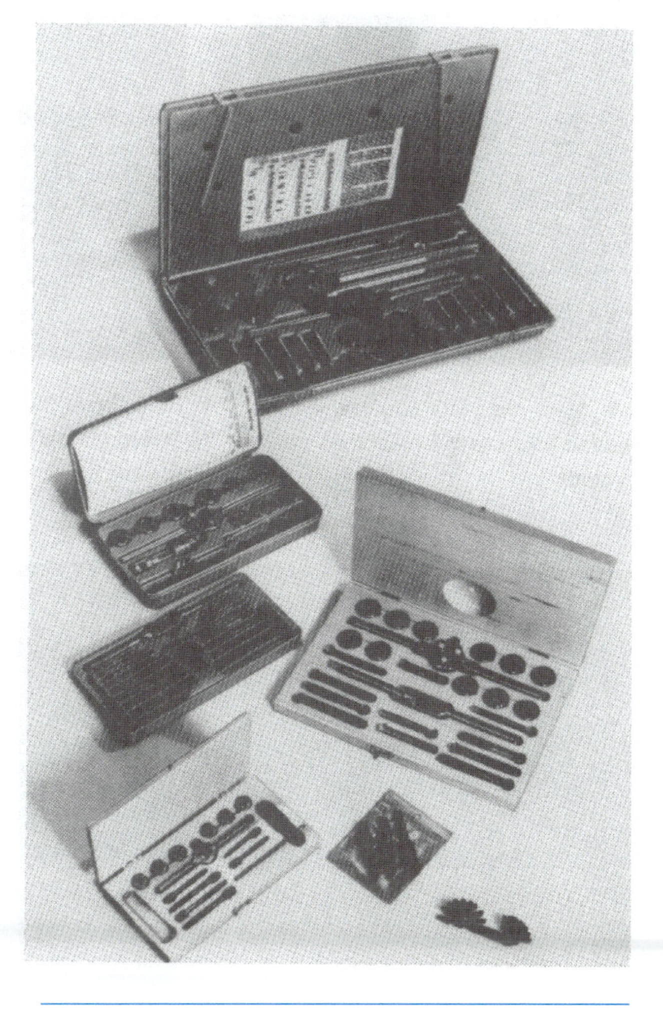

Figure 5-7 Tap and die sets. (Courtesy of Snap-on Tools Corporation)

Photo Sequence 5

Typical Procedure for Repairing Damaged Threads with a Tap

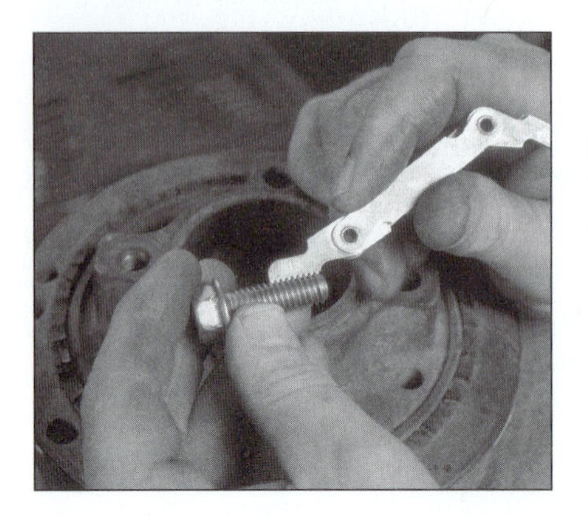

P5-1 Determine thread size of fastener that fits into damaged internal threads by matching fastener to pitch gauge.

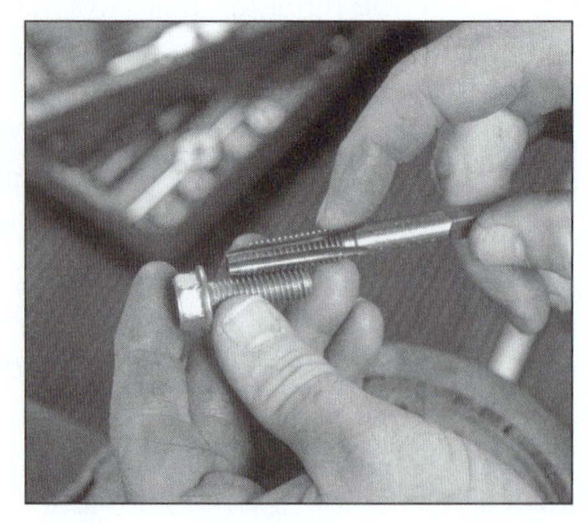

P5-2 Select the correct size and type of tap for the threads to be repaired.

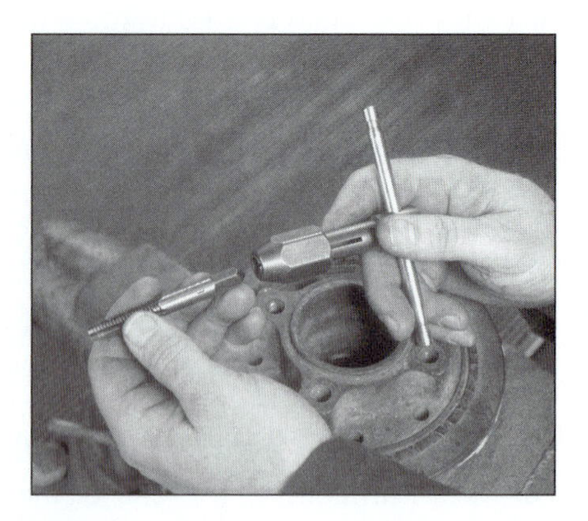

P5-3 Install the tap into a tap wrench.

P5-4 Start the tap squarely in the threaded hole using a machinist square as a guide.

P5-5 Drive the tap clockwise down the threaded hole the complete length of the threads.

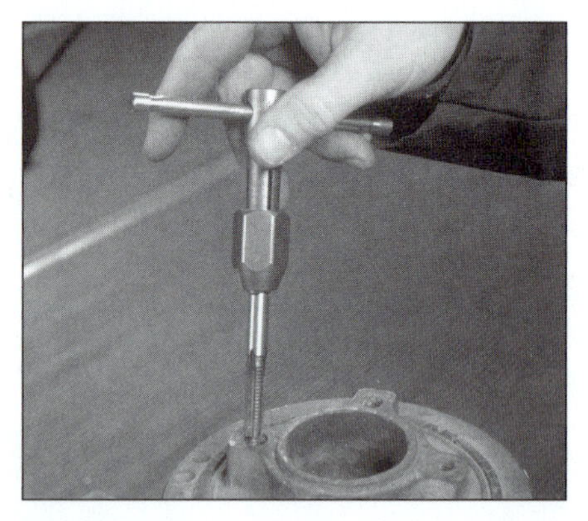

P5-6 Drive the tap back out of the hole by turning it counterclockwise.

P5-7 Clean the metal chips left by the tap out of the hole.

P5-8 Inspect the threads left by the tap to be sure they are acceptable.

P5-9 Test the threads by threading the correct size fastener into the threaded hole.

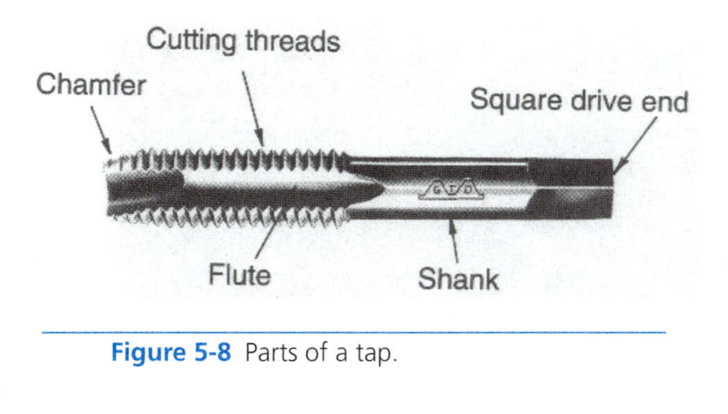

Figure 5-8 Parts of a tap.

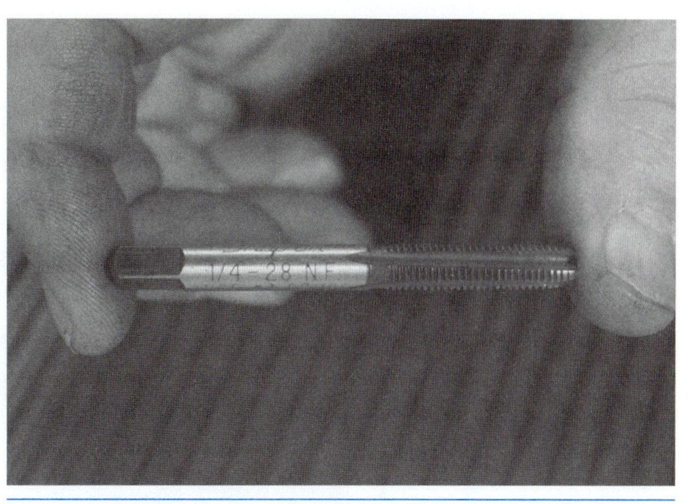

Figure 5-9 A tap with the size 1/4 28 UNF on the shank.

If a technician were to measure the thread on the screw that fits into a damaged internal thread as 1/4 28 UNF, then a tap with the size 1/4 28 UNF on the shank would be selected as shown in Figure 5-9.

A technician must also determine if she or he is repairing a thread that goes all the way through a part, called a **through hole.** The other type of hole does not go all the way through and is called a **blind hole.** Refer to the illustration (Figure 5-10) of a blind and a through hole.

There are different types of taps used for blind and through holes. A **taper tap** is used to repair threads in a through hole. The taper tap has several incomplete threads at the end to help it start in the hole. A **bottoming tap** is used to repair threads in a blind hole. The bottoming tap has only two incomplete threads at the bottom so it can tap threads all the way to the bottom of a hole. A taper and a bottoming tap are shown in Figure 5-11.

A *through hole* goes all the way through a part.

A *blind hole* does not go all the way through a part.

A *taper tap* is used to repair threads in a through hole.

A *bottoming tap* is used to repair threads in a blind hole.

WARNING: Practice using a tap on a scrap automotive component before attempting this job on a good part. Using a tap incorrectly can cause further damage to threads.

After selecting the correct type and size of tap, attach the tap to a **tap wrench**. There are two common types of wrenches used to drive a tap. One is called a T-handle and the other is a *hand-held tap wrench* (Figure 5-12). The T-handle has an adjustable chuck on the end that is

A *tap wrench* is used to drive the square shank of a tap.

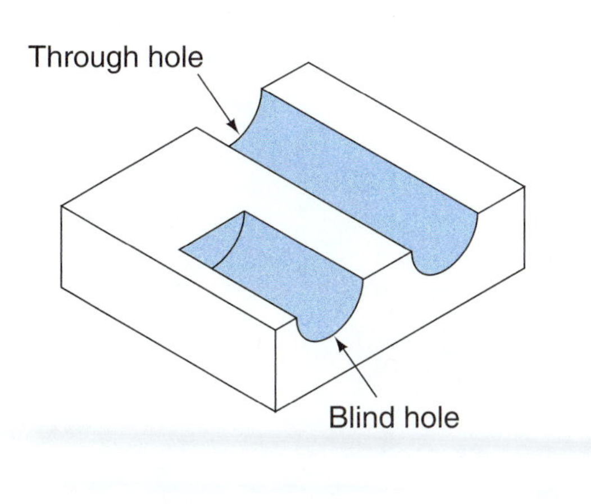

Figure 5-10 A through hole and a blind hole.

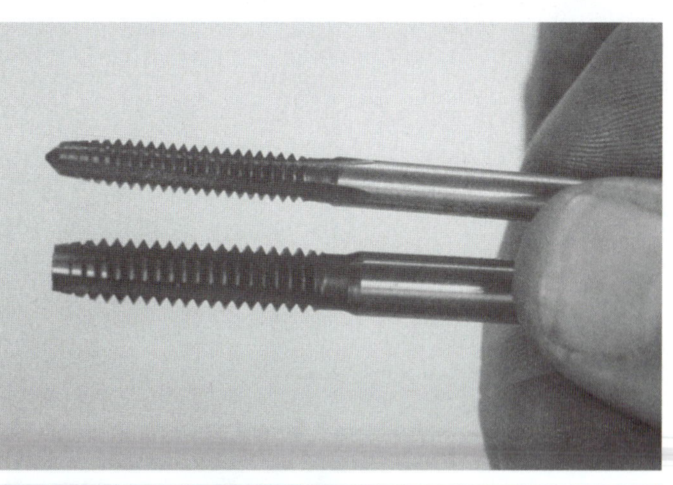

Figure 5-11 A taper tap is used for through holes and a bottoming tap is used for blind holes.

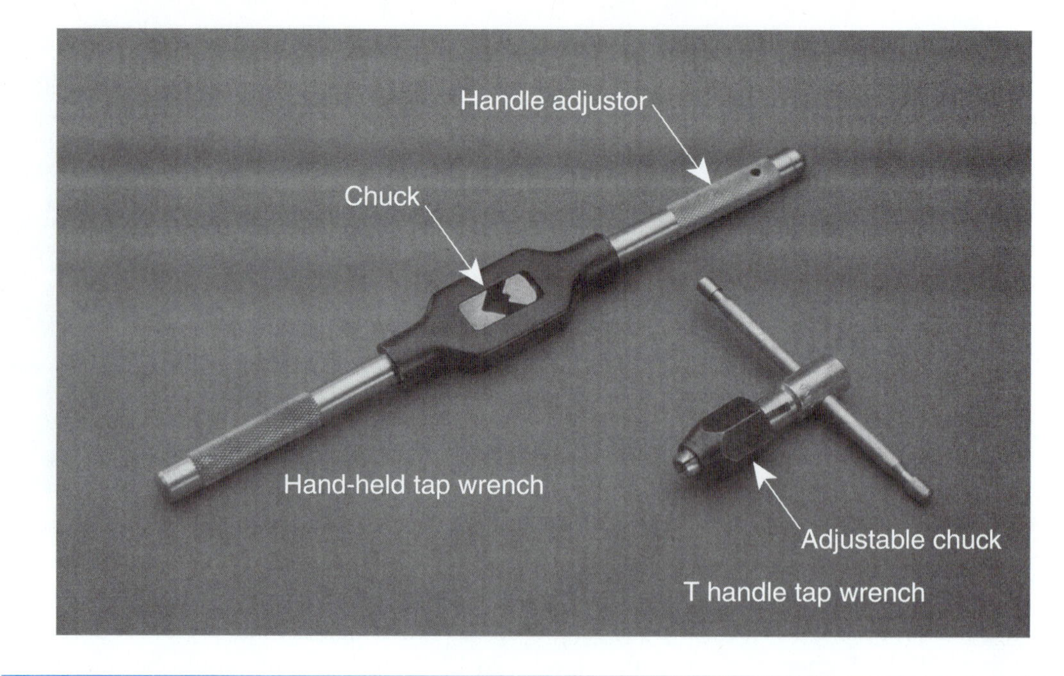

Figure 5-12 Two common types of tap wrenches.

opened and tightened over the square drive of the tap. The hand-held tap wrench also has an adjustable chuck that is opened and closed by rotating one of the handles. The T-handle is usually the best choice for repair work, especially on smaller threads like the 1/4 28 UNF.

CAUTION: Always wear eye protection when using a tap. Taps are hardened tools and may shatter if they break.

Slide the square drive end of the tap into the chuck of the T-handle tap wrench. Tighten the chuck to hold the tap securely in the wrench. Set the tap in the hole to be repaired and rotate the tap carefully one or two turns. Make sure the tap enters the hole square and not on an angle. A good way to check this is to place a machinist square next to the tap as shown in Figure 5-13.

Figure 5-13 The tap must be started square in the hole.

Carefully and slowly rotate the tap through the threaded hole. When the bottom is reached, slowly rotate the tap counterclockwise up and out of the thread. Clean up all metal filings that result from cutting out **damaged threads.** Test the repaired thread with the correct size screw.

Using a tap or die to repair *damaged threads* is often called "chasing" the threads.

SERVICE TIP: Sometimes threads are too damaged to repair. Badly damaged threads can be repaired with the installation of a part called a "helicoil." A special helicoil tap is used to form a special thread in the hole. The helicoil is a metal coil that is threaded into the hole using a special tool. The coil forms new threads that are the same size as the damaged ones.

Removing and Replacing a Damaged Stud

Classroom Manual
page 117

Special Tools

Six-inch scale

Stud remover

Penetrating fluid

Thread sealant

Antiseize compound

A technician will often find damaged external threads on screws, bolts, and studs. These can be repaired with a die much like we used a tap to repair internal threads. However, because the screws, bolts, and studs are easily replaced with new fasteners, we rarely perform this job. All technicians need to master the replacement of studs.

Studs often get damaged from overtightening or from cross threading a nut on the threads (Figure 5-14). A damaged stud is replaced by unscrewing it from the part and installing a new one of exactly the same size.

Stud replacement can be difficult because the stud has often been in place for a long period of time. There has been constant heating and cooling, corrosion, and rust buildup between the stud threads and the internal threads of the part. These factors combine to make some studs very difficult to remove.

The first step in stud removal is to use penetrating fluid to remove the corrosion to free the stud from its mating threads. Soak the area of the threads with penetrating fluid as shown in Figure 5-15. Allowing the fluid to soak into the threads overnight will make it easier to remove the stud.

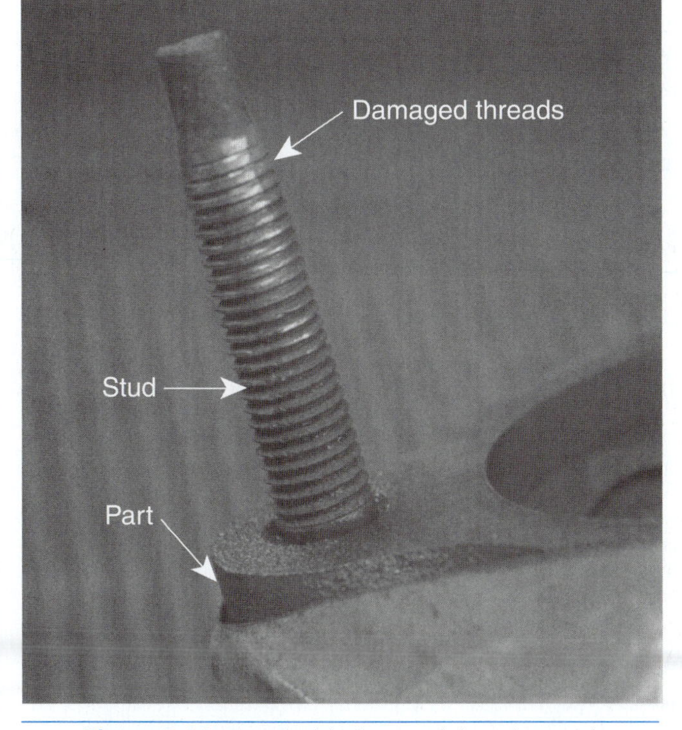

Figure 5-14 A stud with damaged threads must be replaced.

Figure 5-15 Use penetrating fluid around the stud threads before removing them.

Figure 5-16 A stud remover is used to unscrew a stud.

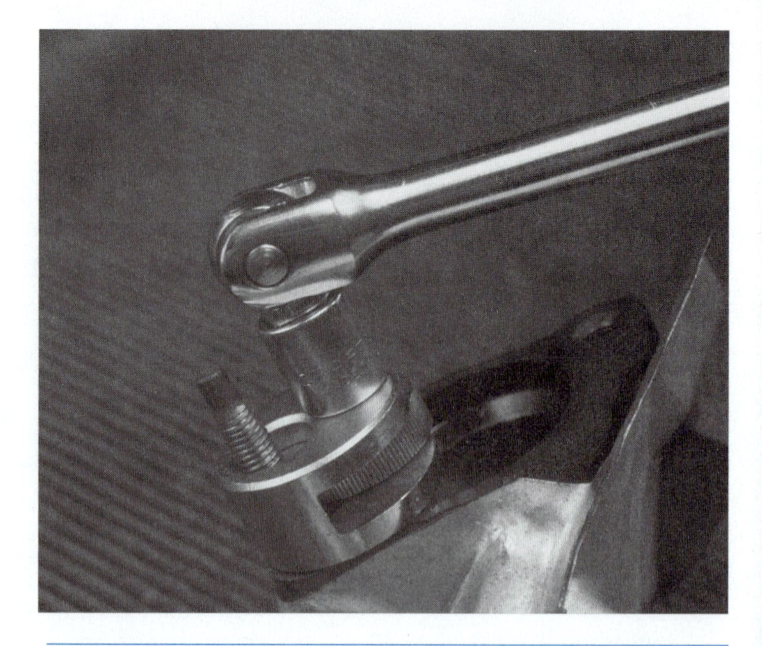

Figure 5-17 Using a stud remover to remove a stud.

Before removing the old stud, measure the distance it sticks up from the surface. This measurement will be needed later when installing the new stud. Use a 6-inch scale to measure from the part surface to the top of the stud. Write the measurement down so it can be referred to later.

A *stud remover* grips a stud and can be driven by a wrench to unscrew a stud.

A **stud remover** is used to remove studs (Figure 5-16). It is installed over the stud. The jaws on the stud remover grip the outside of the stud. A wrench fits on the stud remover and allows a technician to rotate the stud in a counterclockwise direction to remove the stud as shown in Figure 5-17.

If a stud remover is not available, a stud can be removed with two nuts. Locate two nuts that are the correct thread size to thread onto the stud. Start one nut and thread it all the way down to the bottom of the stud. This nut will be the drive nut. Start another nut and thread it down until it contacts the first nut. This is called the *jam nut.*

Put a wrench on the bottom drive nut and hold it in place. Put another wrench on the jam nut and tighten, or "jam," it against the drive nut. The jam nut will now hold the drive nut in position on the stud (Figure 5-18).

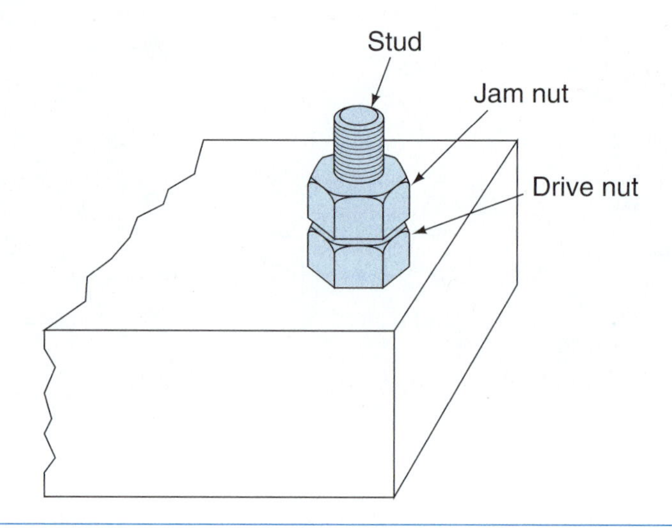

Figure 5-18 A drive nut and jam nut installed on a stud.

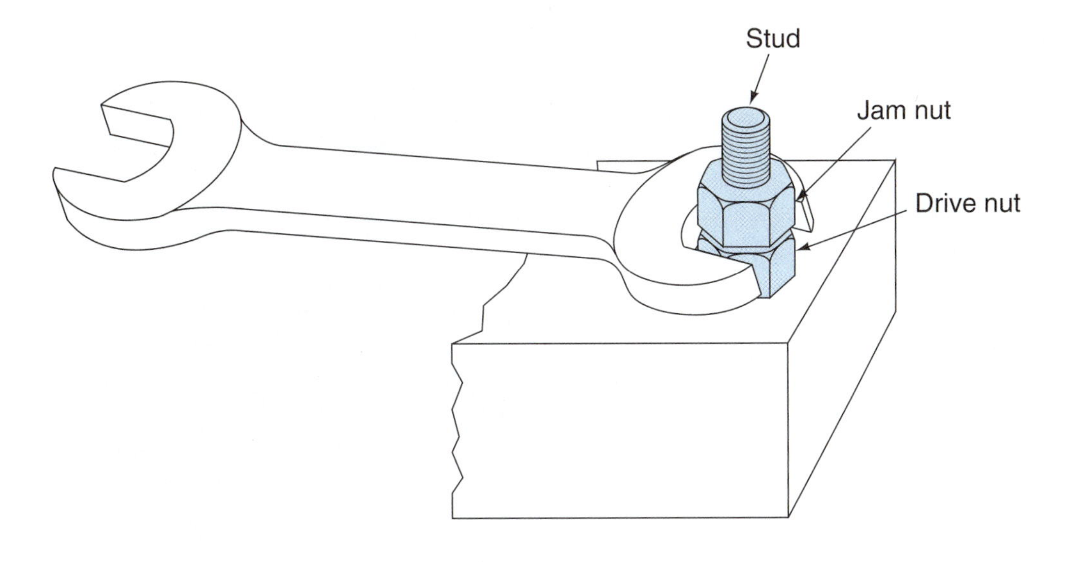

Figure 5-19 Use a wrench on the drive nut to unscrew the stud.

Now put an open-end wrench on the bottom drive nut. Turn the nut in a counterclockwise direction. Turning the nut in this direction causes it to want to unscrew up the stud (Figure 5-19). The jam nut, however, prevents the drive nut from moving up the stud. Instead the forces cause the stud to unscrew.

When the old stud is out, inspect the internal thread. If it appears rusty or damaged, clean up the thread by running the correct size tap through the threads as previously explained. Compare the new stud with the old one (Figure 5-20). The studs should be exactly the same thread size and the same length.

Check the shop service manual for the car being worked on to determine if the threads of the new stud should be coated. If the stud should be locked in place and not easily removed, the technician may be instructed to use a **threadlocking compound** or **thread sealing compound** (Figure 5-21). Threadlocking compounds are used on studs and other fasteners when vibration might cause them to unscrew. Thread sealants are used when a stud extends into an area where liquids, such as oil or coolant, could get on the fastener.

Threadlocking compound is used on threads to prevent them from unscrewing during vibration.

Thread sealing compound is a material used on threads to prevent liquids such as coolant or oil from damaging the threads.

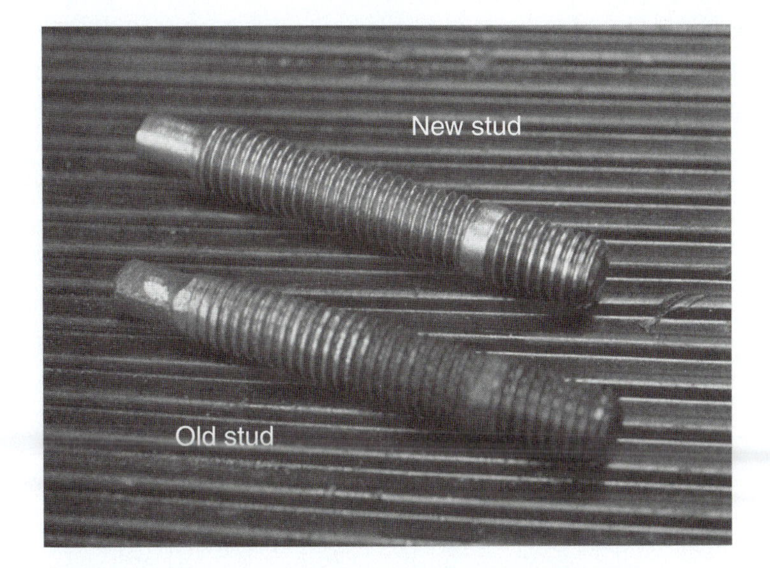

Figure 5-20 The replacement stud must be the same size as the old stud.

Figure 5-21 The stud may need to be coated with threadlocker or sealant before installation. (Courtesy of Saverio Bono)

Figure 5-22 Studs that may need to be removed are coated with antiseize compound. (Courtesy of Saverio Bono)

Antiseize is used on stud threads to make removal easier and to prevent a reaction between the stud and the metal of the internal threads.

Antiseize compound (Figure 5-22) is used on the stud threads to prevent the stud from reacting with the metal of the internal threads. If this happens, the stud could stick or seize. Antiseize compound prevents this reaction and makes the stud easier to remove next time.

After the new stud is properly coated, it can be installed. Start the stud by hand, making sure it enters the threads squarely. Turn the stud in as far as possible by hand before using any tools. Then use two nuts as described earlier to drive the stud into the part. Use the depth measurement made on the old stud to be sure it is driven in the correct depth (Figure 5-23).

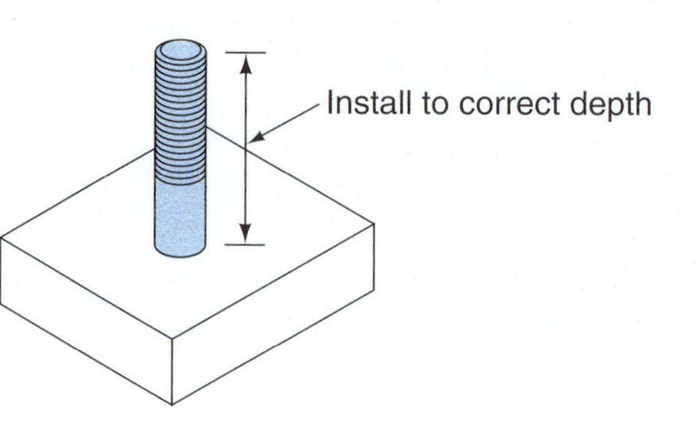

Figure 5-23 The new stud must be installed to the correct depth.

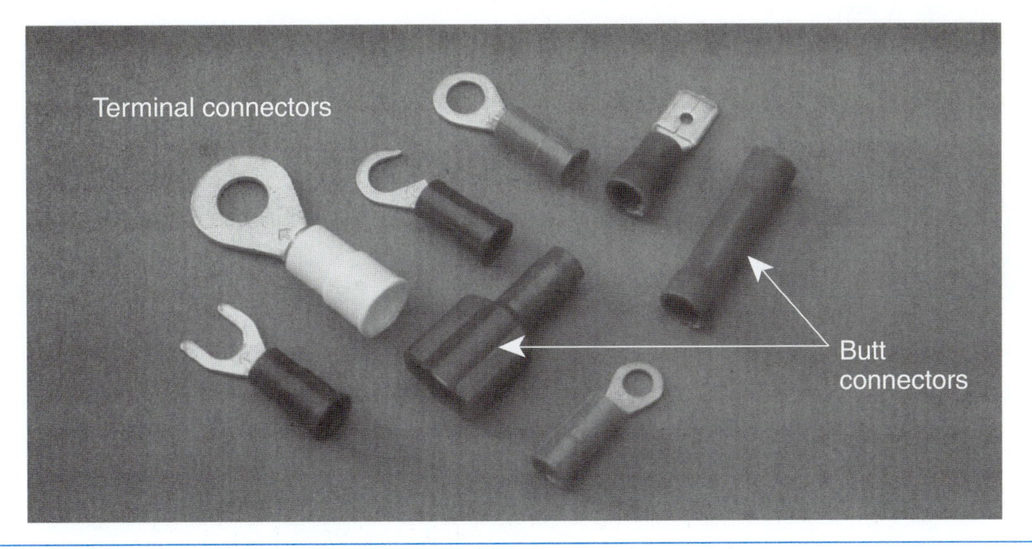

Figure 5-24 Solderless connectors are available in many sizes and shapes.

Installing Solderless Electrical Connectors

Replacing burned or damaged wires is a common electrical service procedure. When installing new wires in an electrical system, a technician will need to be able to install the connecting hardware on the wires. There are two general types of wire connectors: *solderless* and *soldered*. This section describes how to use the solderless type. The next section shows how to solder on wire connectors.

Classroom Manual
page 125

Special Tools

Wire stripping and crimping pliers

Solderless connectors

WARNING: Some manufacturers do not recommend the use of solderless butt connectors for making wiring repairs. They recommend soldering the two wires together or complete wire or wiring harness replacement.

There are many types and styles of solderless connectors (Figure 5-24). The two basic groups are *butt connectors* used to connect two wires together and *terminal connectors* used to connect a wire to a terminal. Both types have the same basic parts shown in Figure 5-25. The connector is made in two parts. The part that connects to the wire or terminal is made from metal that is a good electrical conductor. A plastic insulator fits over the parts that must not be allowed to touch another conductor. There are **crimping** tabs on the end of a terminal connector and on each end of a butt connector. The tabs are used to attach the wire. The crimping tabs are formed in a circular shape so that wire can be inserted.

Crimping is a procedure used to join a wire to a solderless connector.

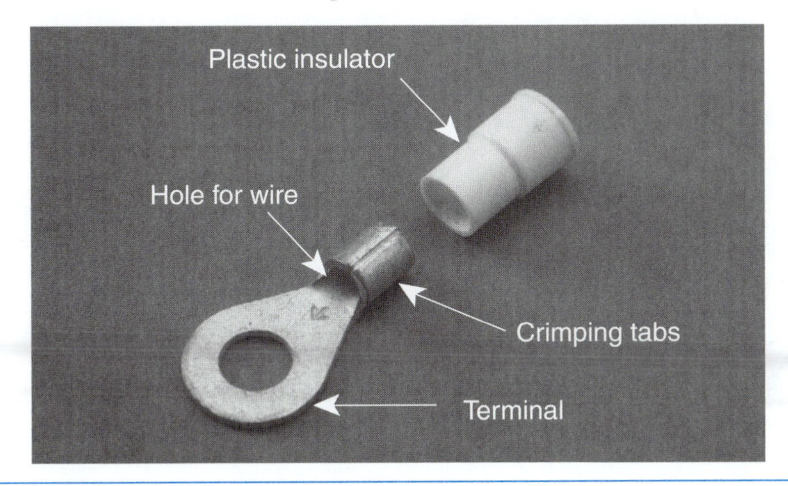

Figure 5-25 Parts of a solderless connector.

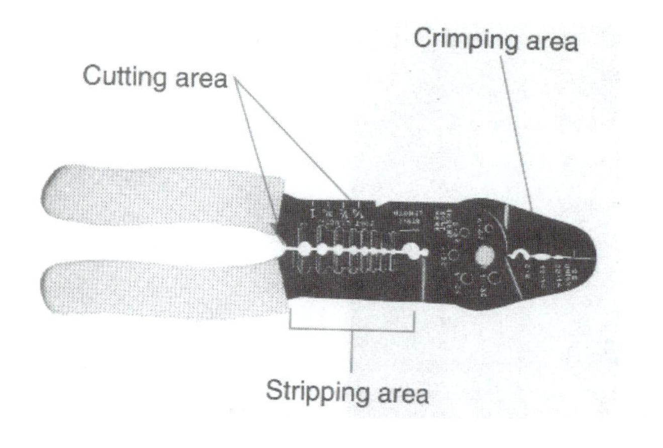

Figure 5-26 Parts of stripping and crimping pliers. (Courtesy of Klein Tools)

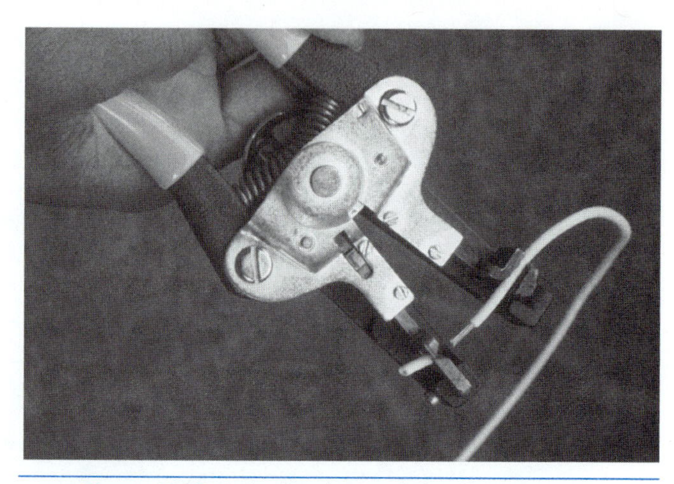
Figure 5-27 Stripping the insulation from a wire.

Connectors are classified according to the size wire they fit. A 16-gauge (1.0 mm²) connector has a crimping tab hole that will fit a 16-gauge (1.0 mm²) wire. The wire size for solderless connectors is printed on the package.

WARNING: Always use the correct size connector for the wire you are installing. Wire that is too small or large for the connector will not be retained properly in the connector.

Stripping and crimping pliers are used to strip insulation from wire and crimp a solderless terminal on wire.

Stripping is the procedure used to remove insulation from an electrical wire.

The basic tool used to install solderless connectors is the **stripping and crimping pliers** shown in Figure 5-26. The jaw area of the tool is used to crimp or squeeze the crimping tabs of the connector onto the wire. The area behind the jaws has a set of cutters used to strip away the insulation on the wire.

The first step in installing the connector is to strip the insulation off the end of the wire. The wire stripper has several sets of cutters in the stripping area. Each is labeled with a wire gauge size. When **stripping** a 16-gauge (1.0 mm²) wire, insert approximately 1/4 inch of the end of wire between the stripper cutters. Squeeze down on the tool handles. This action cuts the insulation but not the wire. Pull the wire out of the stripping area while holding the handles together. The insulation should be removed from the end of the wire as shown in Figure 5-27.

Insert the stripped end of the wire through the terminal crimping tabs as shown in Figure 5-28. When installing a butt connector, install a wire in both ends as shown in Figure 5-29. Be sure the stripped wire fits completely into the crimping area.

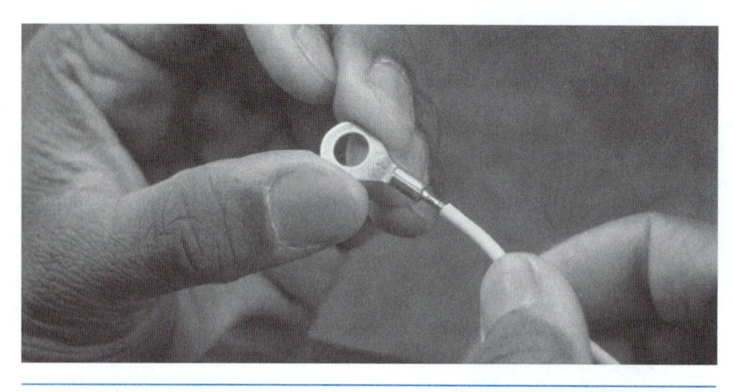
Figure 5-28 The stripped end of the wire is inserted into the crimping taps.

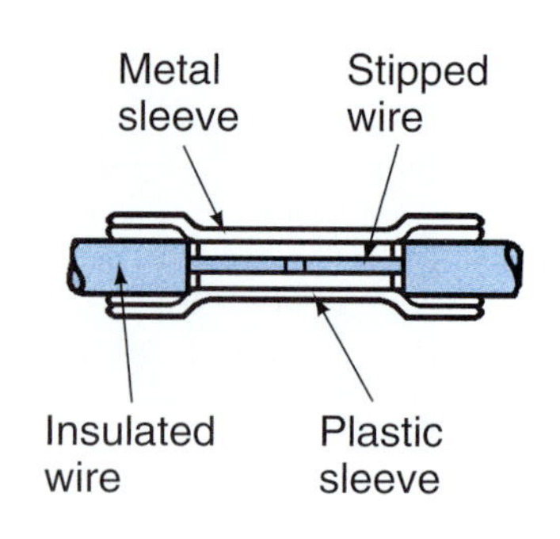

Figure 5-29 How two wires fit in a butt connector.

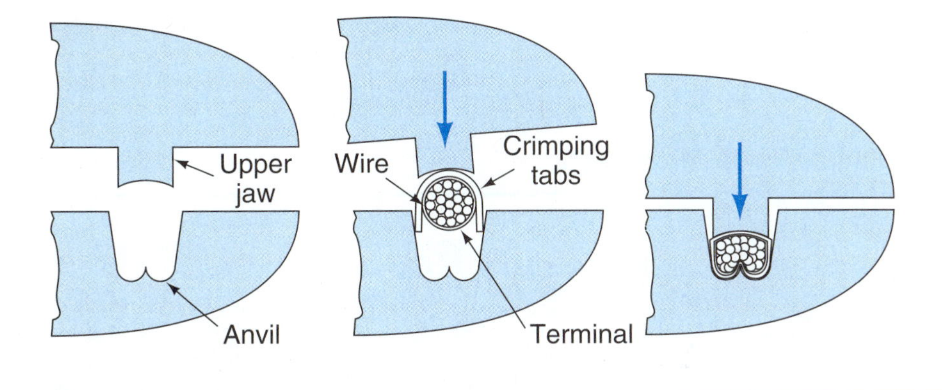

Figure 5-30 Crimping a solderless connector in crimping jaws.

The crimping jaws on the end of the pliers are used to squeeze the crimping tabs down on the wire. Most stripping and crimping pliers have several sets of jaws that are marked with different wire gauge sizes. Use the one that is correct for your solderless terminal size.

Place the terminal and wire in the tool jaws as shown in Figure 5-30. Squeeze on the tool handles to crimp the tabs against the wire. The two parts of the tool jaw will flatten part of the terminal and turn in the crimping tabs on the wire as shown in Figure 5-31.

Test the connection by gripping the wire in one hand and the terminal in the other. Gently try to pull the two apart. A properly installed connector will stay on the wire.

Soldering Wires and Electrical Connectors

Classroom Manual
page 125

SERVICE TIP: A good way to practice the wire stripping and soldering techniques presented in this chapter is to make a set of jumper leads for circuit testing. Cut a 24-inch length of 16-gauge (1.0 mm^2) wire. Strip both ends of the wire. Buy two alligator clips from the local auto parts or electronics store. Solder these on each end of the wire.

Special Tools

- Soldering gun or soldering iron
- Rosin core solder
- Stripping and crimping pliers

Another way to join electrical wires or terminals to wires is by **soldering**. Photo Sequence 6 shows soldering wire and electrical connectors. This is a method of joining wires and connectors using a metal called solder. Solder is a metal combination, or alloy, of tin and lead that melts at a low temperature. It can be melted and then allowed to cool back to a solid to join electrical components. Solder has excellent electrical conductivity.

Solder is a low-melting-point metal made from tin and lead used to join electrical wires and terminals.

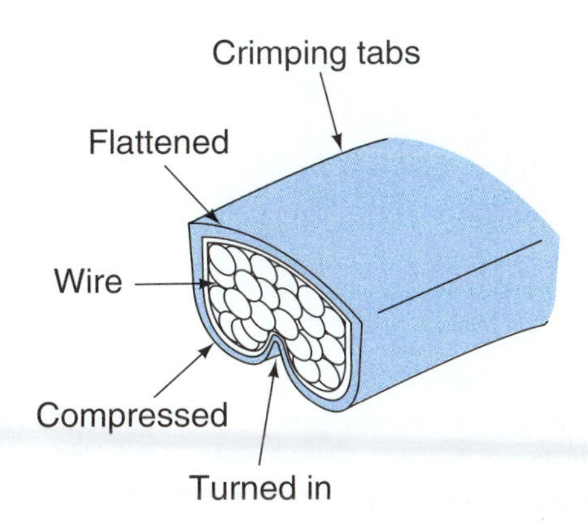

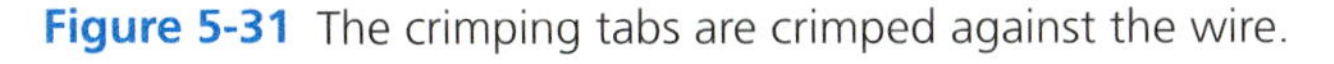

Figure 5-31 The crimping tabs are crimped against the wire.

Photo Sequence 6
Typical Procedure for Soldering Electrical Terminals and Joining Wires

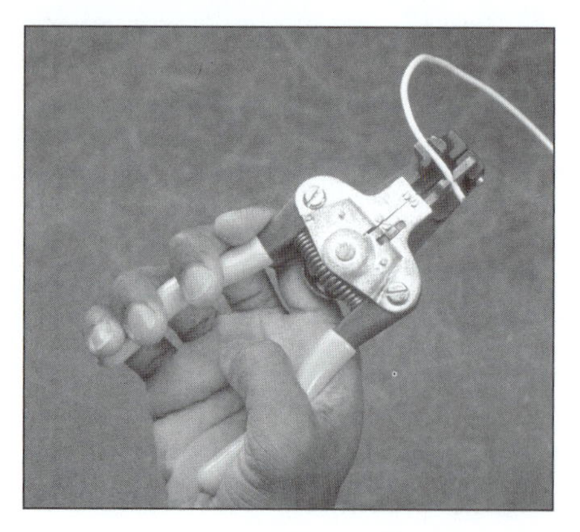

P6-1 To solder a terminal to a wire, first place the end of the wire in the correct size cutter on the stripping and crimping pliers.

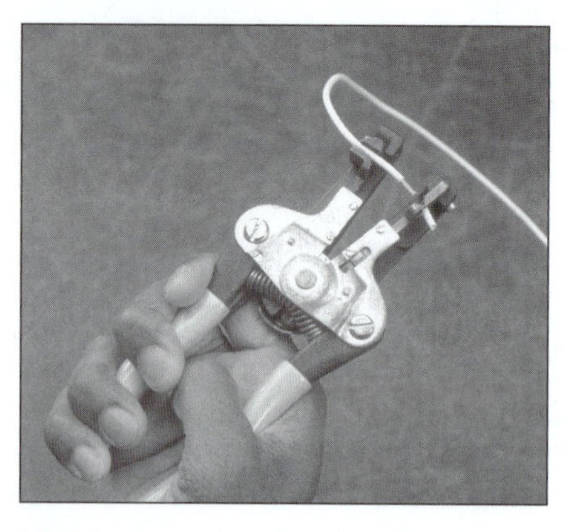

P6-2 Squeeze the pliers' handles and pull the insulation off the end of the wire.

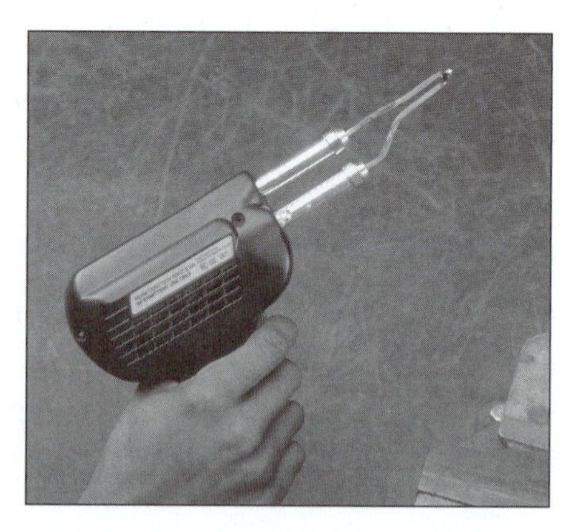

P6-3 Pull the trigger on the soldering gun to heat the end of the gun.

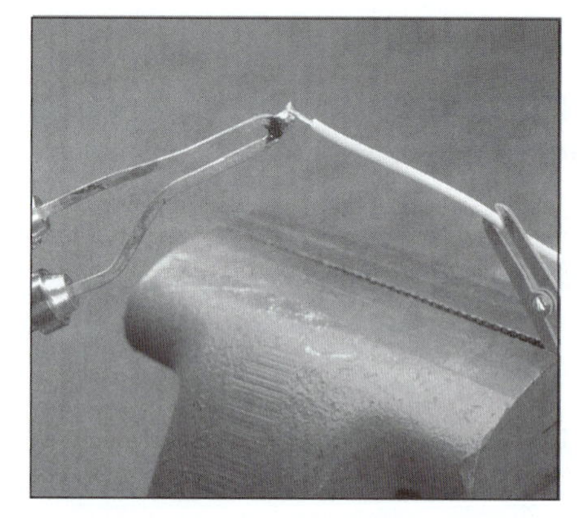

P6-4 Touch the heated end of the soldering gun to the stripped end of the wire to heat the wire.

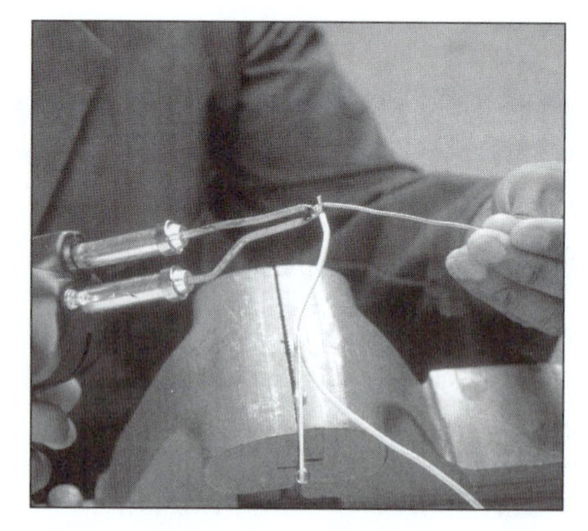

P6-5 Melt a small amount of rosin core solder over the surface of the wire to "tin" the wire.

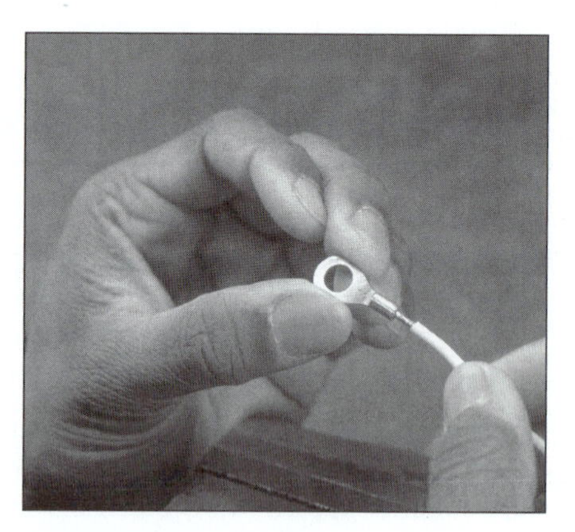

P6-6 Install the end of the wire into the correct size electrical terminal.

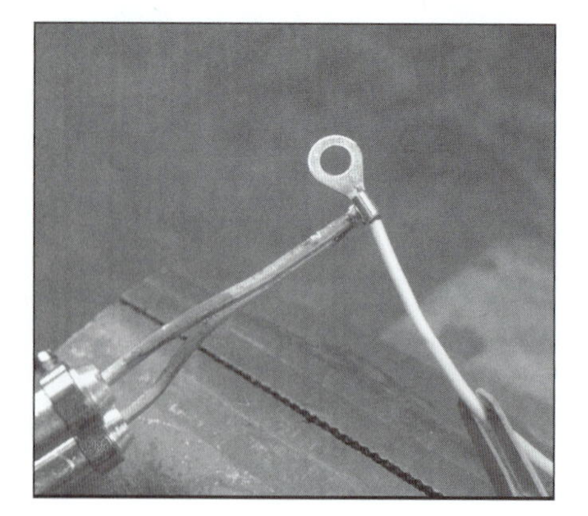

P6-7 Heat the electrical connector and wire with the soldering gun.

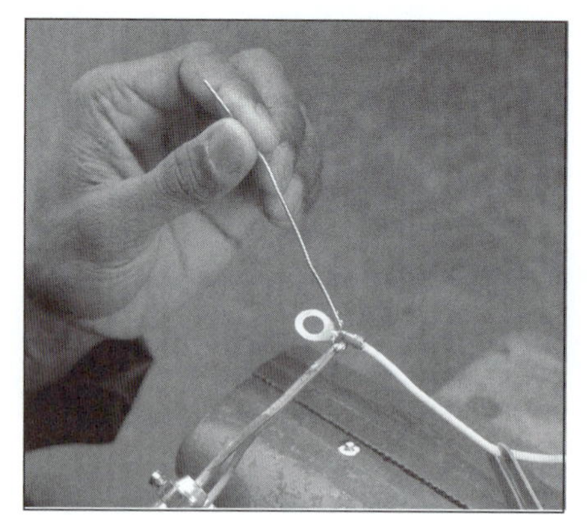

P6-8 Melt solder into the joint between the wire and connector and allow to cool.

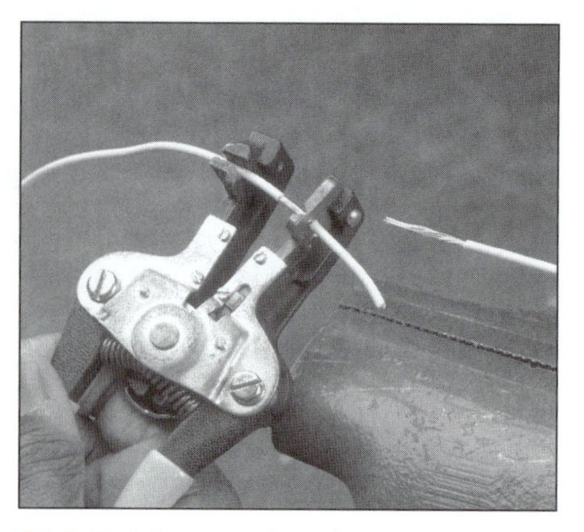

P6-9 To join two wires, first use the stripping and crimping pliers to strip the ends of both wires.

Typical Procedure for Soldering Electrical Terminals and Joining Wires (continued)

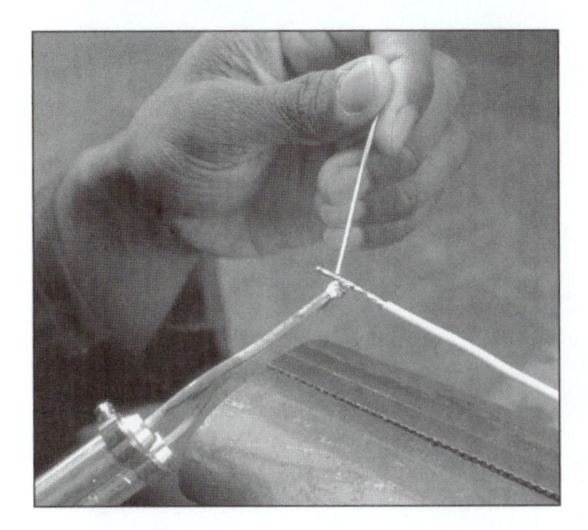

P6-10 Use the soldering gun to heat each wire end and melt solder on the wire to tin the wire.

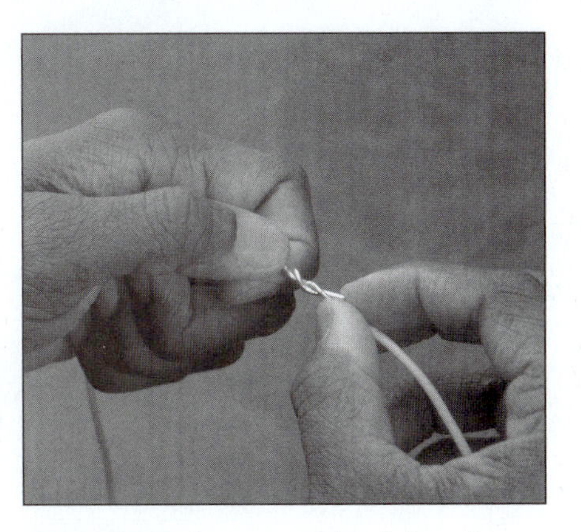

P6-11 Wrap the two ends of the wire together to form a tight connection.

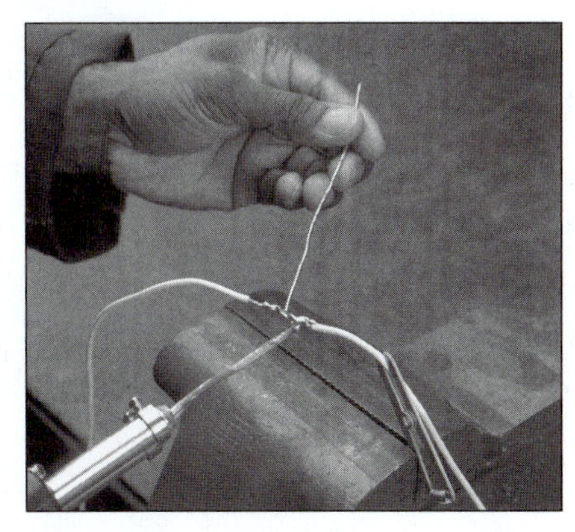

P6-12 Melt solder between the two wires to form a connection.

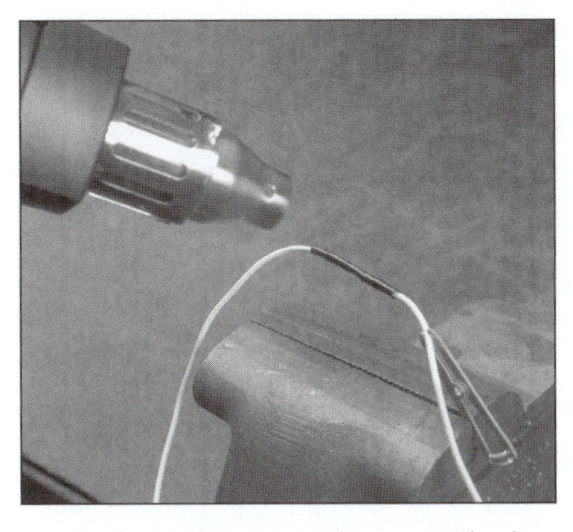

P6-13 Cover the exposed wires with electrical tape.

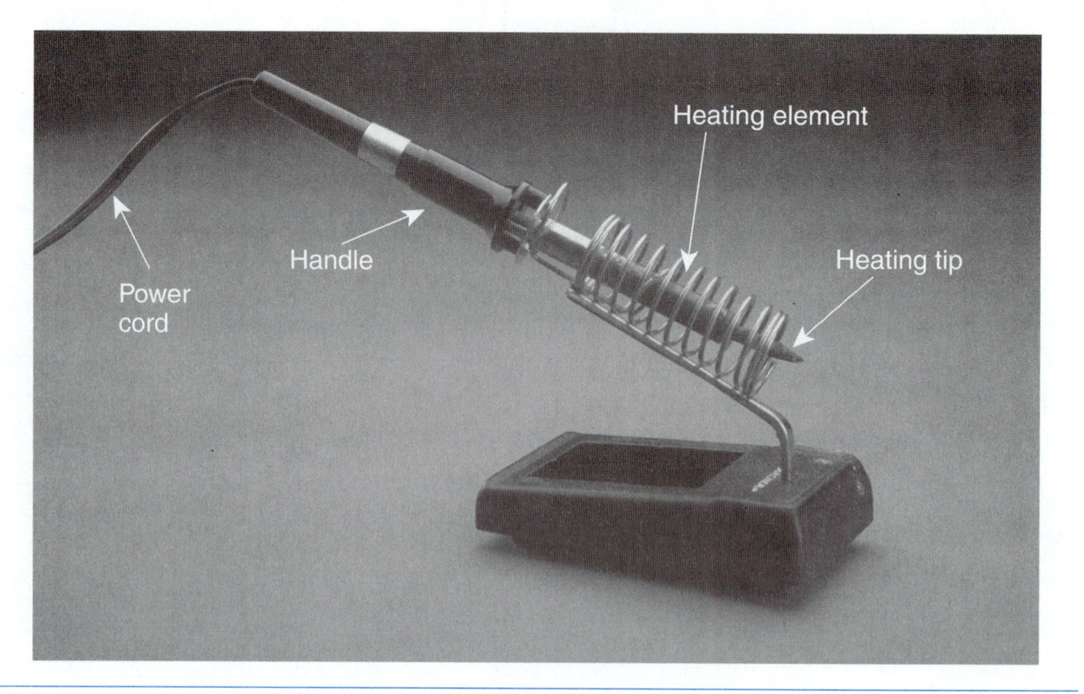

Figure 5-32 A typical electrical soldering iron.

Soldering irons and guns have electric heating elements used to melt solder.

In order to solder parts together, the solder must be heated to its melting point. This is done with either a **soldering gun** or a **soldering iron.** An electric soldering iron is shown in Figure 5-32. The unit has a power cord that is plugged into an electrical outlet. Electricity flows through a heating element inside the unit and heats the tip hot enough to melt solder. An insulated handle allows you to hold the handle safely. Inexpensive soldering irons do not have on-off switches. Once plugged in, they take a short time to heat up to get the tip hot enough to melt solder.

The soldering gun gets its name from the fact that it looks a little like a gun and has a trigger as shown in Figure 5-33. The soldering gun power cord is plugged into an electrical outlet. The tip is heated instantly by a heating element inside the gun whenever the trigger is pulled. When the trigger is released, the tip cools down. Some units have a light that illuminates when the tip is hot enough to melt solder. Both the soldering gun and soldering iron are used the same way to melt solder.

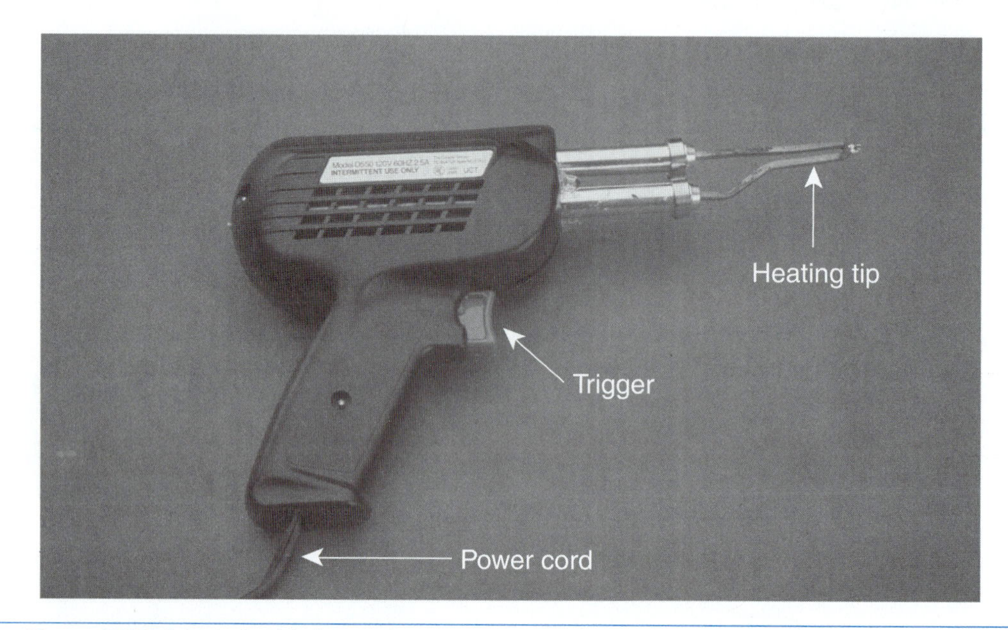

Figure 5-33 A typical soldering gun.

Figure 5-34 Spool of 40-60 rosin core solder.

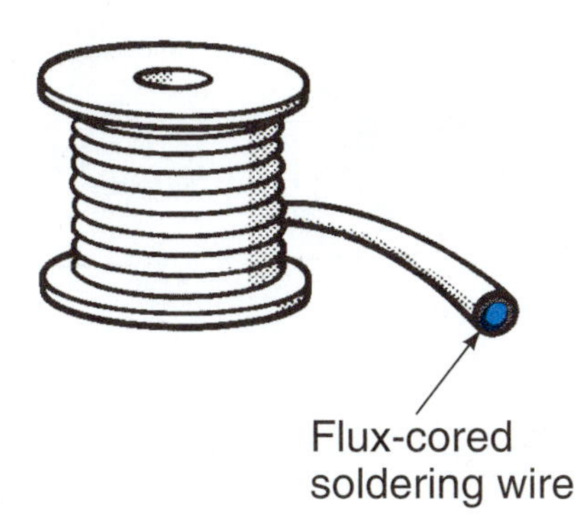

Figure 5-35 The center of the solder has a core of rosin flux.

CAUTION: The heating tip of the soldering gun and soldering arm gets to a temperature in excess of 300°F. Be careful not to touch the tip or it will cause burns.

There are many types of solder to do many different joining jobs. Solder used for electrical work is sold in spools (Figure 5-34). It is shaped in the form of a wire. Many different diameters are available. A solder wire diameter of 0.062 inch is commonly used for general electrical work. In addition, many different alloys of tin and lead are available. Most electrical work is done with 40/60 alloy. The diameter and alloy are marked on the solder spool.

Soldering must be done with a material called soldering **flux.** Soldering flux is used to remove oxides from the wires and terminals. Oxides are formed when metal is exposed to air. The flux cleans the wires and terminals. It also helps the solder flow more easily. Although flux is available separately, solder for electrical work typically has the flux inside the core of the solder wire as shown in Figure 5-35.

Flux is made from acid or rosin and is used when soldering to clean and remove oxides from the solder joint.

There are two general types of flux: *acid* and *rosin.* Acid-type flux is used for joining sheet metal. It is too corrosive for electrical work. Electrical soldering is done with a noncorrosive flux called rosin. The solder spool will indicate the type of flux in the solder core. Choose solder that is marked "rosin core."

WARNING: Never use acid core solder on electrical components. It can cause corrosion that can interrupt electrical flow.

With the correct solder and the soldering gun or iron heated, try soldering a terminal to a wire. Just as in solderless connectors, choose wire and terminals that are the same gauge size.

CAUTION: Both the flux and the solder cause fumes that are dangerous to breathe for a long period of time. Always solder in a well-ventilated area. Always wear eye protection when soldering.

Use the stripping and crimping pliers to strip the insulation off approximately 1/4 inch of the end of the wire as shown in Figure 5-36. A layer of solder must be melted over the wire

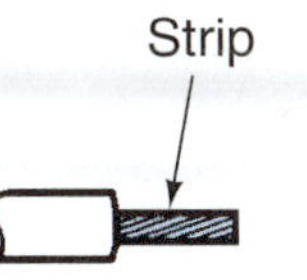

Figure 5-36 Stripping the insulation off the wire.

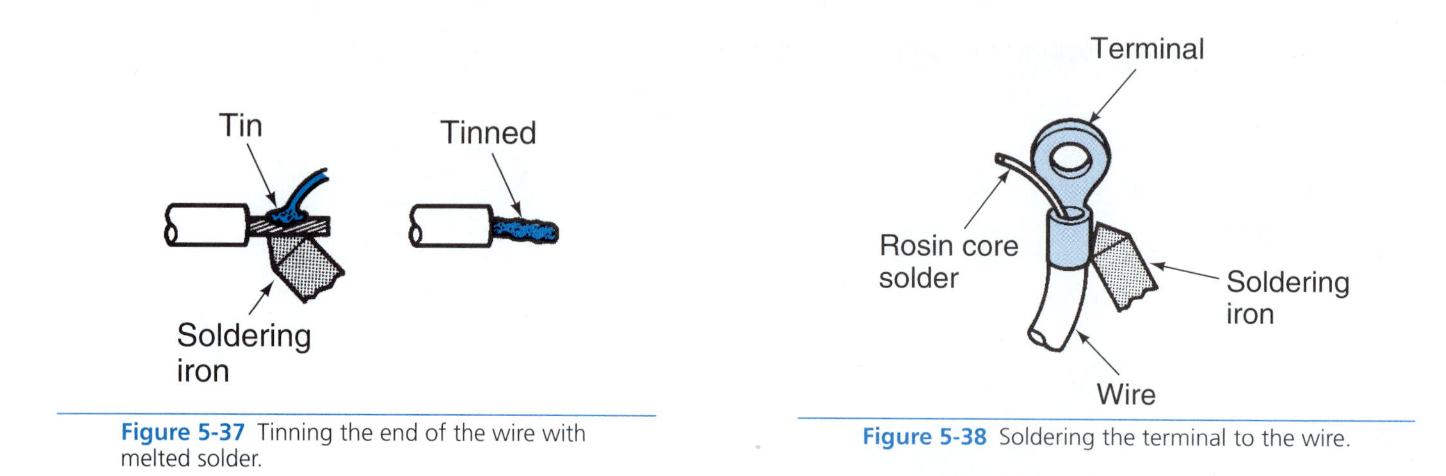

Figure 5-37 Tinning the end of the wire with melted solder.

Figure 5-38 Soldering the terminal to the wire.

Tinning is a procedure of coating solder over the wires of the conductor prior to soldering on a terminal or other conductor. It gives a much better connection with no added resistance.

Splicing is a term used to describe the joining of two or more electrical wires.

before it is joined to the terminal. This procedure is called **tinning.** Tinning improves the final joining of the terminal and wire. Touch the tip of the soldering gun or iron to the wire and allow it to heat a moment. Then touch the solder to the heated wire and allow the solder to flow completely over the wire. Tinning the wire is shown in Figure 5-37.

Place the tinned wire into the terminal. Heat the terminal with the soldering iron tip. Give it a moment to heat up, then apply solder into the terminal as shown in Figure 5-38. Allow the solder to solidify and cool. Then test the connection by holding onto the terminal and pulling on the wire. A defective solder joint will break under pressure.

Soldering can also be used to join, or **splice,** two or more wires together. The common methods of joining wires are shown in Figure 5-39. In all the methods, the wire ends are stripped and tinned, then they are soldered as shown in Figure 5-40. The exposed area of the wires and solder will then have to be covered with the correct type of electrical tape or *heat shrink.*

SERVICE TIP: Replacement wires can be covered with a product called "heat shrink." Heat shrink is a thin-wall plastic tubing. The tubing is available in many sizes to cover a single wire or a number of wires to form a loom. Heat the tubing after the wires are installed through the tubing. The tubing shrinks into a tight fit around the wires.

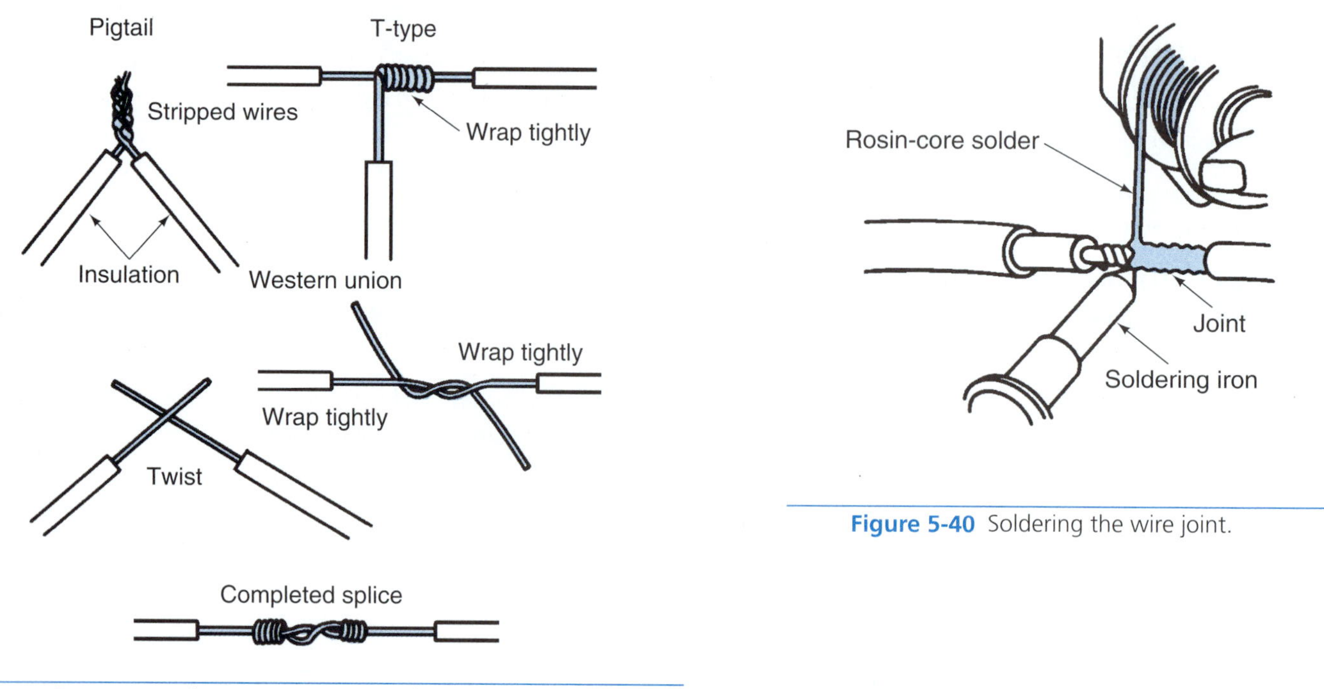

Figure 5-40 Soldering the wire joint.

Figure 5-39 Three common ways of splicing wire.

CASE STUDY

Using the wrong type of soldering flux can have disastrous results. A friend related a problem several years ago. He was doing a ground-up restoration on an old 356 Porsche. All the old wiring had been pulled out of the car in order to put in a complete new wiring harness. After pricing commercially available wiring harnesses, he decided to make his own. He bought and cut hundreds of feet of new wire. He wanted to do a good job, so he soldered each terminal and connector in his new loom.

The newly restored car had been on the road a couple of years when he started having electrical problems. Light switches, turn signals, and accessories stopped working for no apparent reason. He began to notice that a corrosion was building up on his electrical connections. Further investigation showed that the corrosion was caused by his use of an acid core solder. He was quite unhappy when he found out that all of the hundreds of connections had to be redone with the new terminals and the correct type of rosin core solder.

Terms to Know

Antiseize
Blind hole
Bottoming tap
Crimping
Flux
Soldering gun
Soldering iron
Splice
Stripped threads
Stripping
Stripping and crimping pliers
Stud remover
Tap wrench
Taper tap
Threadlocking compound
Thread sealing compound
Through hole
Tinning

ASE Style Review Questions

1. The use of a pitch gauge is being discussed. *Technician A* says the pitch gauge can be used to identify English (U.S.) threads. *Technician B* says the pitch gauge can be used to identify metric threads. Who is correct?
   A. A only
   B. B only
   C. Both A and B
   D. Neither A nor B
2. The use of a tap to repair internal threads is being discussed. *Technician A* says the tap must have the correct thread size on the shank. *Technician B* says one tap fits many different thread sizes. Who is correct?
   A. A only
   B. B only
   C. Both A and B
   D. Neither A nor B
3. The use of a tap to repair internal threads is being discussed. *Technician A* says a taper tap is used for a through hole. *Technician B* says a taper tap is used for a blind hole. Who is correct?
   A. A only
   B. B only
   C. Both A and B
   D. Neither A nor B
4. The use of a tap to repair internal threads is being discussed. *Technician A* says a taper tap has numerous incomplete threads at its bottom. *Technician B* says a bottoming tap has numerous incomplete threads at its bottom. Who is correct?
   A. A only
   B. B only
   C. Both A and B
   D. Neither A nor B

5. The removing and replacing of a stud is being discussed. *Technician A* says a stud remover can be used to remove the stud.
   Technician B says two nuts can be used to remove the stud. Who is correct?
   A. A only
   B. B only
   C. Both A and B
   D. Neither A nor B

6. The removing and replacing of a stud where vibration is a problem is being discussed.
   Technician A says to coat the stud with antiseize compound.
   Technician B says to coat the stud with threadlocking compound. Who is correct?
   A. A only
   B. B only
   C. Both A and B
   D. Neither A nor B

7. The removing and replacing of a stud that goes into an area with engine coolant is being discussed.
   Technician A says to coat the stud with antiseize compound.
   Technician B says to coat the stud with sealing compound. Who is correct?
   A. A only
   B. B only
   C. Both A and B
   D. Neither A nor B

8. The installation of a solderless electrical connector is being discussed. *Technician A* says the connector and wire must have the same wire gauge size.
   Technician B says the wire must be one wire gauge smaller than the connector. Who is correct?
   A. A only
   B. B only
   C. Both A and B
   D. Neither A nor B

9. The soldering of a terminal on a wire is being discussed. *Technician A* says to use acid core solder.
   Technician B says to use rosin core solder. Who is correct?
   A. A only
   B. B only
   C. Both A and B
   D. Neither A nor B

10. The soldering of a terminal on a wire is being discussed. *Technician A* says to tin the stripped wire before installing it in the connector.
    Technician B says to tin the wire after installing it in the connector. Who is correct?
    A. A only
    B. B only
    C. Both A and B
    D. Neither A nor B

Table 5-1 ASE Task

Inspect and repair damaged threads where allowed.

Problem Area	Symptoms	Possible Causes	Classroom Manual	Shop Manual
Poor operation, broken parts	Noise, parts inoperative.	**1.** Loose fasteners.	117	119
		2. Missing, damaged fastener threads.	112	113
		3. Crankshaft seals leaking.	108	113

SAFETY

Wear safety glasses.

Table 5-2 ASE Task

Inspect, test, adjust, and repair or replace switch, bulbs, sockets, connectors, and wires of parking lights and taillights circuits.

Inspect, test, and repair or replace switches, flasher units, bulbs, sockets, connectors, and wires of turn signal and hazard lights circuits.

Inspect, test, adjust, and repair or replace switch, flasher units, bulbs, sockets, connectors, and wires of stoplight (brake light) circuit.

Inspect, test, adjust, and repair or replace switch, bulbs, sockets, connectors, and wires of headlight circuits.

Inspect, test, and replace wiper motor, resistors, switches, relays, controllers, connections, and wires of wiper circuit.

Inspect, test, and repair or replace washer motor, pump assembly, switches, relays, connections, and wires of washer circuit.

Problem Area	Symptoms	Possible Causes	Classroom Manual	Shop Manual
Lights	No brake light. No turn signal(s). No tail (park) light(s). No wipers. No windshield washer. No headlight. No parking lights.	**1.** Loose, damaged, corroded connections. **2.** Broken, loose conductors.	124 125	121 121

SAFETY

Wear safety glasses.
Disconnect the negative cable of the battery.
Disable the air bag system—see the service manual for procedures.

8

Job Sheet 8

Name ______________________________ Date ______________

Restore Damaged External Threads

Upon completion of this job sheet, you should be able to restore damaged external threads on a fastener.

Tools Needed

Thread restoring file
Thread restoring tool
Vise
Basic tool set
Gloves

Procedures

1. Determine the type of restoring device desired.
   Restoring file—Go to step 2.
   Restoring tool—Go to step 10.
2. Match the fastener threads to a set of teeth on the file.
3. Locate a nut that fits the fastener.
4. Mount the fastener firmly in a holding device. A table-mounted vise is usually sufficient.
5. Don the gloves.
6. Aligned the selected file teeth with the fastener threads. Allow the teeth to overlap onto undamaged threads if possible. This will act as a guide.
7. Work the file over the damaged area. The fastener may have to be repositioned within the holding device for easy access to all damage.
8. Use the nut to make periodic checks of the repairs during the filing.
9. When the nut screws easily onto the fastener, the task is complete.
10. When using a restoring tool, select the tool insert that fits the damaged threads.
11. Select a nut that fits the fastener.
12. Install the insert into the tool handle with teeth facing the clamping section of the tool handle.
13. Mount the fastener into a holding device. A table-mounted vise is usually sufficient.
14. Clamp the tool around an undamaged thread area.
15. Turn the tool so the insert runs over and through the damaged threads.
16. Move the tool back and forth over the damaged area until it turns smoothly.
17. Tighten the clamp slightly and move it over the damaged area several more times.
18. Remove the tool and test the repair with the nut.
19. Repeat steps 14 through 18 until all repairs are complete.

☑ **Instructor's Check** ______________________________

Making and Reading Measurements

Upon completion and review of this chapter, you should be able to:

- ❑ Use a feeler gauge for measuring spark plug gap.
- ❑ Use an outside micrometer to measure shaft diameter.
- ❑ Use an outside micrometer and telescopic gauges to measure a bore.
- ❑ Measure a brake disc for thickness with an outside micrometer.
- ❑ Measure bores with inside micrometer.
- ❑ Use a dial indicator to measure runout and gear backlash.
- ❑ Measure pressure.
- ❑ Measure vacuum.

Introduction

This chapter will guide technicians through methods of making measurements using tools studied in the Classroom Manual. The tasks discussed will be typical of the types of measurement jobs performed in a shop.

Measuring Spark Plug Gap

Classroom Manual
pages 135–139

The *gap* is the distance between the two electrodes on the spark plug. The electrical spark that jumps across this gap ignites the air/fuel mixture. "Gapped" is the term used to identify the procedure to adjust the gap.

An *electrode* is the last end of an electrical circuit. It is not intended as a terminal, but it is used to aim or control current from the circuit to another object, usually another electrode.

Even with the best electronics, all gasoline-fueled engines use spark plugs to fire the air/fuel mixture. While many vehicles only require spark plug replacement every 100,000 miles, the new plugs will have to be correctly gapped (Figure 6-1). If the **gap** is too wide, the spark may be weak in strength and short in duration because of the high voltage needed to bridge the gap. With a too-narrow gap, the spark will have a low voltage and not much heat. In both cases, the spark will have difficulty igniting the air/fuel mixture, resulting in a poorly-operating engine.

There are or will be scrapped spark plugs in almost every shop or technical school. Select five or six of them and a round, wire feeler gauge (Figure 6-2). If necessary, clean the plugs before gapping. Assume that the correct gap for the first one is 0.035 inch.

Locate the 0.035-inch wire from within the gauge set. Attempt to slide the wire between the plug's two **electrodes** (Figure 6-3). If the gap is too small or too wide, use the gauge set's small bending tool to move the outer electrode. Check the gap again and correct any further space by using the bending tool on the gauge set. With one plug correct, select different gaps for the other plugs and set them. Most of the current vehicles use spark plug gaps between 0.035 inch and 0.046 inch. Some older General Motors vehicles use 0.080-inch gaps.

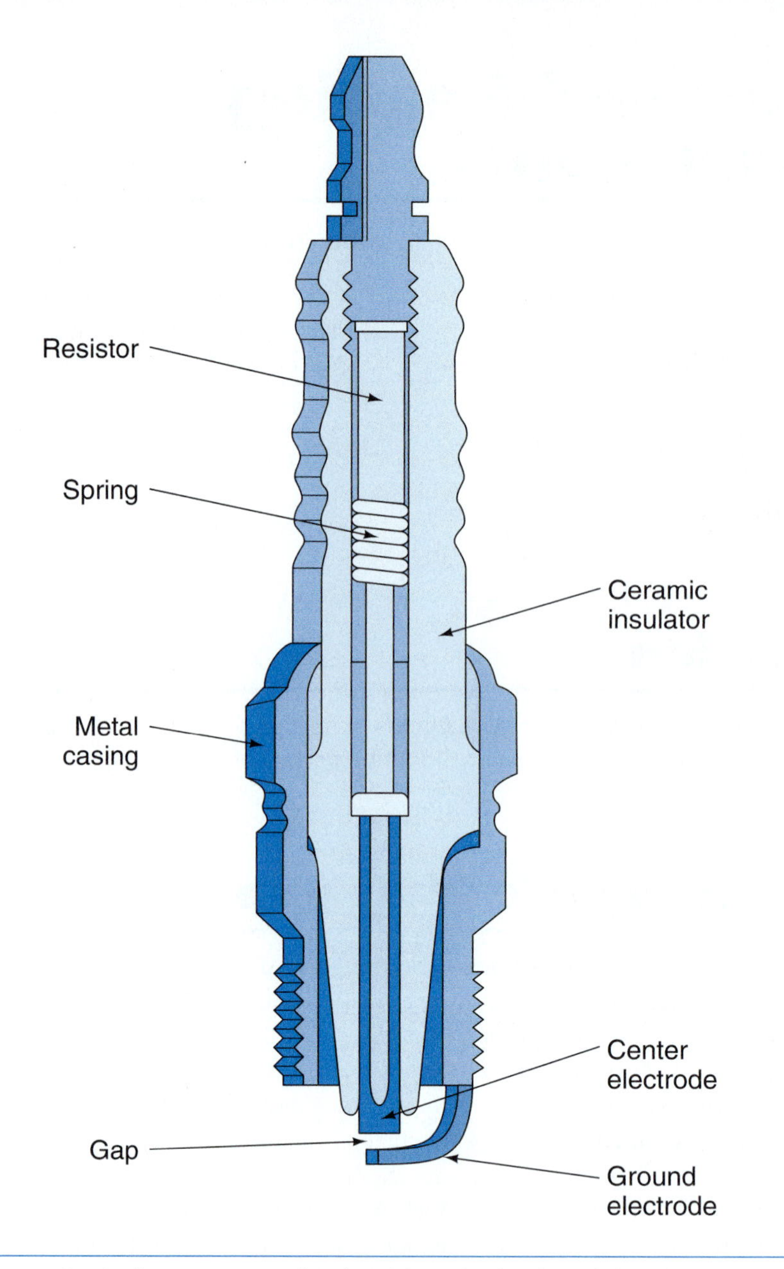

Figure 6-1 Spark plug gap causes the electricity to bridge (jump) the space and create a spark. (Courtesy of Cooper Automotive/Champion Spark Plug)

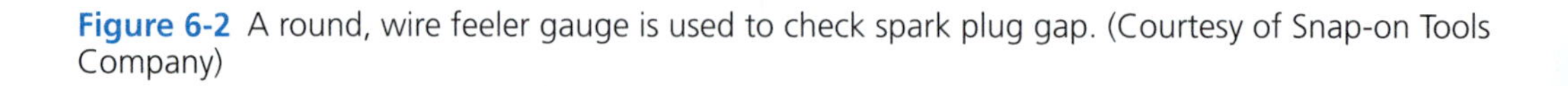

Figure 6-2 A round, wire feeler gauge is used to check spark plug gap. (Courtesy of Snap-on Tools Company)

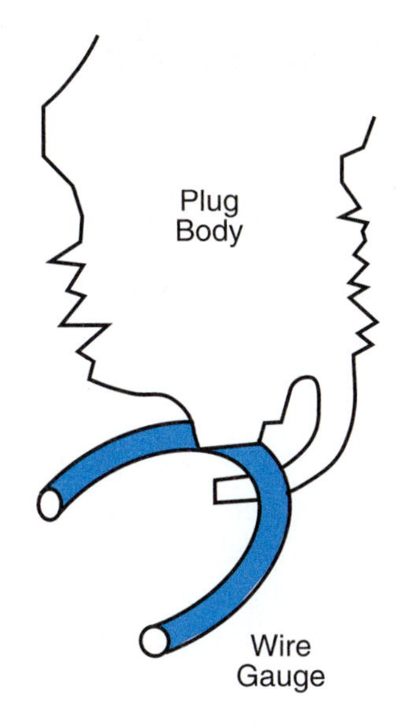

Figure 6-3 The gauge wire should have a slight drag as it is moved between the two electrodes.

Measuring Shaft Diameter and Wear

Classroom Manual
page 140

There will be times when a technician is required to measure various shafts to determine if they are serviceable or need machining or replacement. Many shops do not rebuild engines in-house but still have to determine if a replacement is needed. The same applies to transmissions and transaxles. Sometimes the noise made by a bad engine does not mean that the engine is nothing more than scrap metal. The amount, loudness, or lack of noise does not accurately indicate the amount of damage.

Select a shaft that has one or more precision-machined areas on it (Figure 6-4). An output shaft from a scrapped transmission, a crankshaft, or a camshaft works well for this exercise. Select a set of USC and metric micrometers that will measure the various areas of the shaft.

Figure 6-4 The shiny, highlighted areas are the crankshaft bearing journals.

Special Tools

Outside USC micrometers

Shafts

Using a USC Outside Micrometer to Measure a Bearing Journal

Before starting the measurement, inspect the shaft and journals. It is a waste of time to measure a journal if it is damaged beyond repair unless it is being done for practice. We will review the classroom information on reading measurements as we proceed through the rest of this chapter. Select the correct USC micrometer for the first area to be measured. The area selected should be machined (Figure 6-5). Measuring the rough or unmachined area is not needed in most instances. If possible, locate some data for the shaft from the instructor or service manual, specifically, data on the correct diameter and the amount of wear that may exist before the shaft has to be replaced is needed.

Move the spindle away from the anvil far enough for the micrometer to be slipped over the shaft (Figure 6-6). Rotating the thimble counterclockwise opens the gap between the two measuring faces. With the two measuring faces, anvil, and spindle 180 degrees apart and placed over the shaft, rotate the thimble clockwise until it contacts the shaft. Keep the micrometer in place as the ratchet is turned clockwise until it slips. Lock the spindle and check the alignment of the faces on the shaft. If they are positioned correctly, remove the micrometer so the measurement can be easily read. Unlock the micrometer and reposition it if the faces are not in place correctly.

Begin reading the measurement by noting the size of the total width of the micrometer (Figure 6-7). For instance, if this is a one-inch to two-inch micrometer, the first number will be 1 and will be to the left of the decimal point.

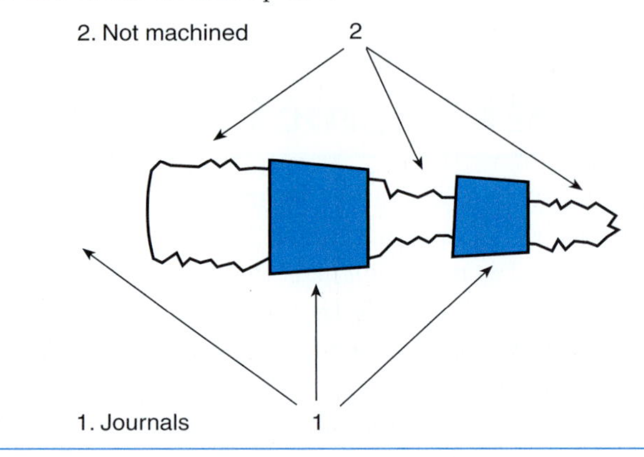

Figure 6-5 Most shafts has several bearing journals to support the shaft evenly along its length.

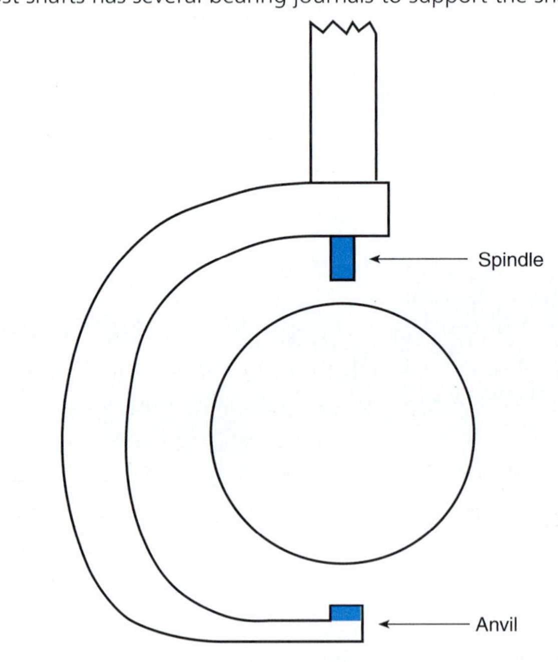

Figure 6-6 Open the micrometer and slide it over the component being measured.

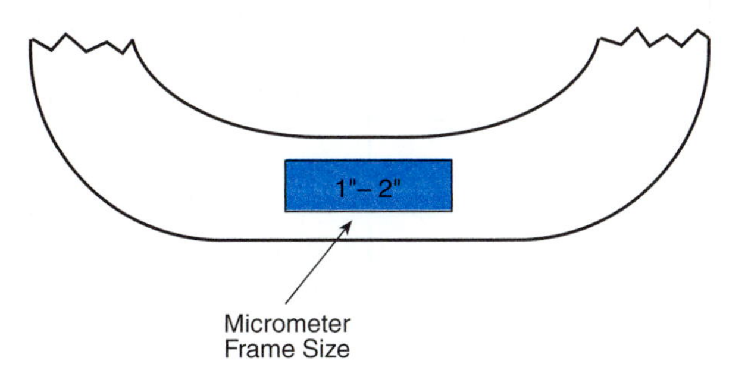

Figure 6-7 The micrometer frame size is the digit to the left of the decimal point.

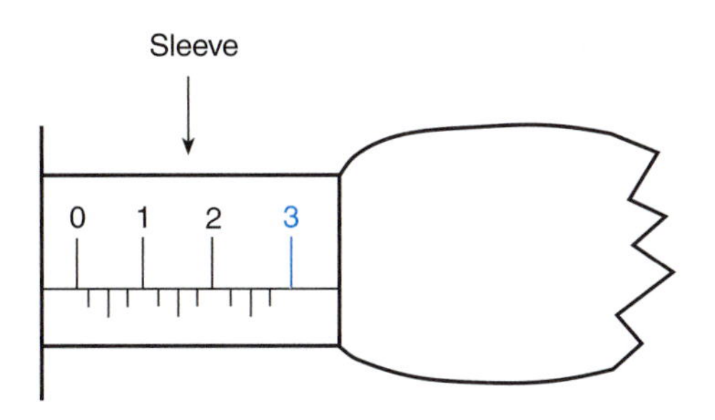

Figure 6-8 The highest exposed sleeve number is the first digit to the right of the decimal point.

The second number and the first to the right of the decimal point will be read from the index line on the sleeve (Figure 6-8). This will be the highest visible number along the index line. We will assume for this exercise that the number is 3.

The second number to the right of the decimal comes from the graduations under the index line on the sleeve (Figure 6-9). Each mark is worth 0.025 inch. If the second graduation mark is the last one showing, then the number will be 0.050 inch. Using the values from the last three paragraphs, we can say the shaft is at least 1.350 inches thick. The numbers are found as follows:

1 inch from the micrometer frame plus
0.3 inch from the index line plus
0.050 inch from below the index line for a total of 1.350 inches

The last number comes from nearest graduation mark on the thimble that most nearly aligns with the index line on the sleeve. There may be some room for interpreting the reading at this point. If the index line is near the center between two thimble marks, two technicians may come up with two different readings (Figure 6-10). This 0.001-inch difference between the two is not of concern on this type of shaft. If the two readings are at the limits of the wear tolerance, the shaft should probably be replaced anyway. The last number derived from the thimble is added to the previous readings. Assume that the number is 0.004 inch and it is added to the readings above, then the shaft is 1.354-inches thick at this **journal.**

A *journal* is a finely machined portion of a shaft that is fitted through a bearing or housing. This machined surface reduces the friction and heat as the shaft operates.

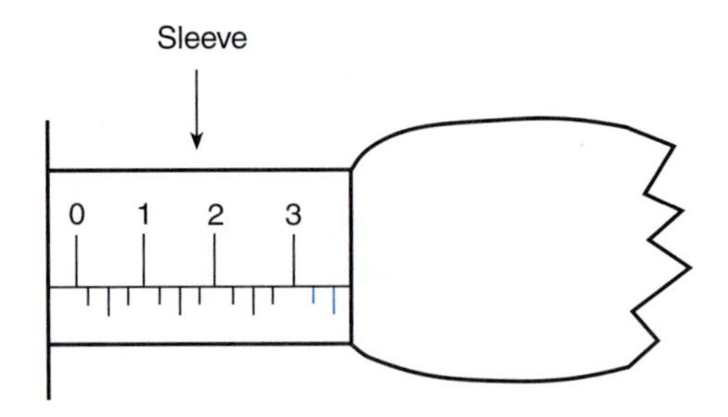

Figure 6-9 The number of exposed lines past the top number is counted and multiplied by 0.025.

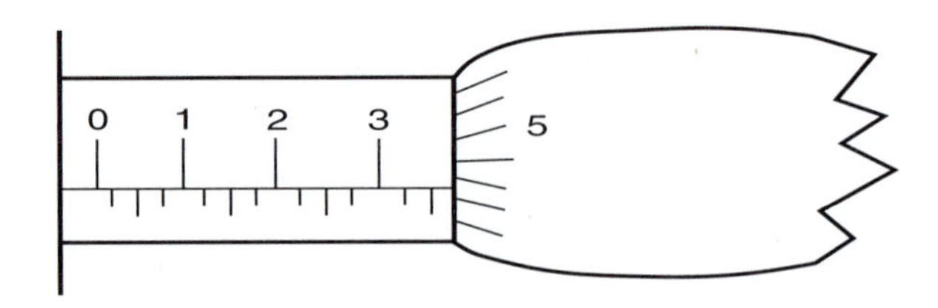

Figure 6-10 The closest thimble graduation to the index line is always selected as the last number of the measurement.

The term *out-of-round* only applies to a journal or shaft that is supposed to be perfectly round.

A second measurement should be made to determine if the journal and shaft is **out-of-round**. If the journal is not a perfect or nearly perfect circle, the journal and bearing will make contact. The lives of the journal, shaft, and bearing will be severely shortened. Other damage may also result from the failure of this bearing.

Move the micrometer to a position 90 degrees to the one where the last measurement was made (Figure 6-11). Place the measuring faces in the same position relative to each other. Operate the thimble, ratchet, and lock to move the faces to the journal surface. Remove the micrometer and record the measurement. Compare the two measurements to each other and the specifications for the journal.

If one of the two measurements is smaller then the other, the journal is out-of-round and has a slight oval shape. A difference of about 0.0005 inch to 0.001 inch may be allowed on some shafts. Consult the service data for the allowable tolerance (Figure 6-12). The journals on some shafts can be machined and undersize bearings can be used.

While checking for out-of-round data, also check for the specified minimum diameter of the journal. Even a perfectly round journal that is too thin may result in a failure.

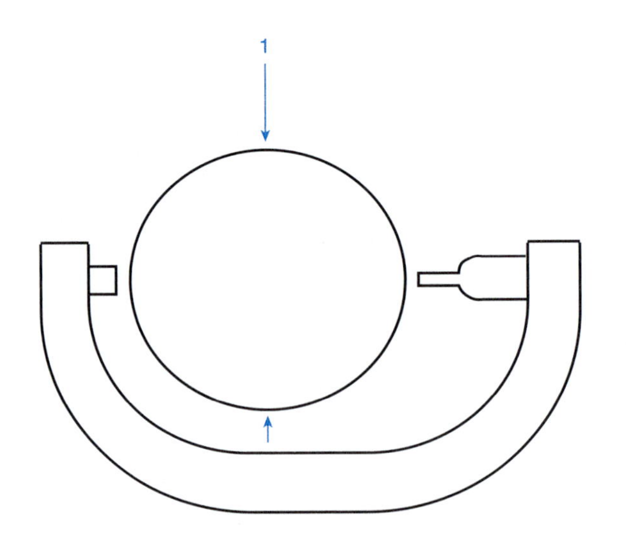

Figure 6-11 The second shaft measurement is at a right angle to the first (arrows).

Crankshaft main journal diameter:

Mark "0"	54.998–55.003 mm (2.1653–2.1655 in.)
Mark "1"	54.993–54.998 mm (2.1651–2.1653 in.)
Mark "2"	54.998–54.993 mm (2.1649–2.1651 in.)

Figure 6-12 There is only 0.005 mm (0.0002 inch) of wear allowed on a crankshaft Mark "1" main journal (54.998–54.993 = 0.005).

Measuring with a metric micrometer is done the same way except for the actual compilation of the measurement. Once the micrometer has been positioned and locked, remove it for easy reading. The numbers above the index line are graduated one millimeter apart. The last and largest exposed number on the index line will be the digit to the left of the decimal point, just the opposite of the USC micrometer (Figure 6-13). If the journal is more than 25 mm, this first number must be added to the frame size.

Digits to the right of the decimal are a combination of the graduations below the index line and the nearest line on the thimble. See if the .5 mm line below the index line is exposed (Figure 6-14). Determine which line of the thimble most closely aligns with the index line and

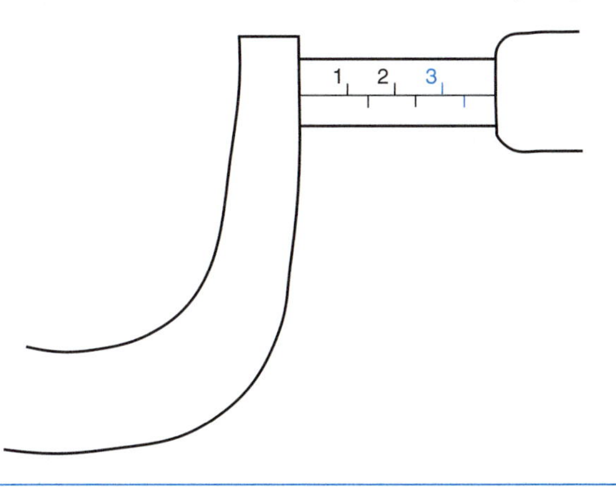

Figure 6-13 On a metric micrometer, the last exposed number on the index line is the digit to the left of the decimal point.

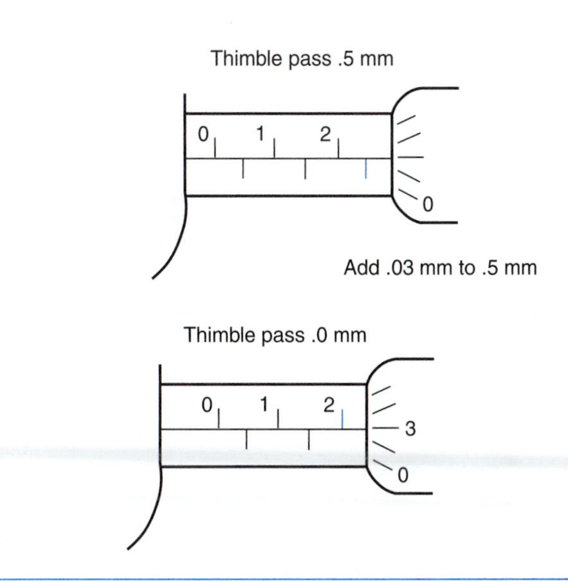

Figure 6-14 Ensure that the below-the-line reading is correct. A 0.5 mm error could be made.

add that number to the 0 mm or .5 mm line found one step back in Figure 6-14. The digits to the left of the decimal point come from above the index line plus the frame size. Numbers below the line and the thimble all go to the right of the decimal.

Classroom Manual
page 147

Special Tools

Outside UCS and metric micrometers

Telescopic gauges

A component with machined bores

A *bore* is usually defined as an internally machined hole as opposed to a hole that is not finished to a smooth surface.

Thrust wear is caused by expanding gases in the combustion chamber that push away from the spark plug and apply an angular force against one side of the piston. In turn, that side of the piston is pushed toward the cylinder wall.

Ring travel area goes from near Top Dead Center (TDC) to near Bottom Dead Center (BDC). It is the area where the piston's sealing rings move up and down in the cylinder.

Measuring Bore with an Outside Micrometer and Telescoping Gauge

The diameters of machined **bores** in vehicle components have to be just as perfect as the various bearing journals. In fact, some of the bores may actually act as bearing journals and all are a very important part of vehicle operation. For instance, the bores or cylinders in an engine tend to wear more on one side. This is **thrust wear** and results in an out-of-round cylinder (Figure 6-15). The bore walls also wear throughout the piston **ring travel area** (Figure 6-16). This is normal wear for an engine, but the bore should be corrected when the engine is being rebuilt. Bores that used to house bearings may also wear and allow the bearings to move, thereby causing damage.

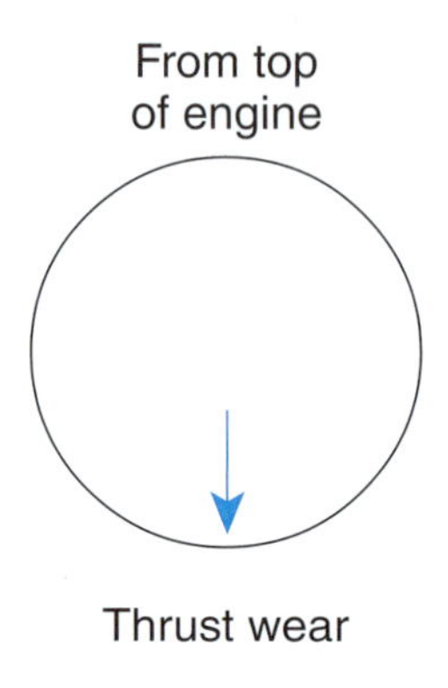

Figure 6-15 Thrust wear will cause a bore to elongate or become somewhat egg-shaped.

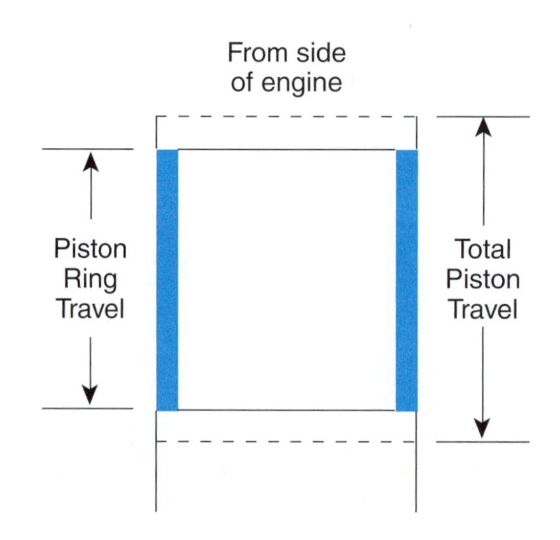

Figure 6-16 The piston's rings are mounted just below the top of the piston, making a smaller wear pattern than the amount of piston travel would seem to indicate.

As with the journal above, inspect the bore for any damage. The component operating in the bore requires a smooth finish to reduce friction, heat, and wear.

With the inspection complete, select a telescopic gauge that fits into the bore and will extend far enough to reach the walls. Select an outside micrometer to match the gauge. Insert the gauge about a quarter of the way down the bore and release the spindles. Move the gauge until it is square in the bore. This means the gauge is at right angles (90 degrees) to the vertical centerline of the bore (Figure 6-17). It must also touch the opposite walls at the widest point. If an engine cylinder bore is being measured, the spindles should be parallel to the length of the block or at a right angle to the length of the block. When the gauge is properly positioned, lock the spindles by turning the lock thimble clockwise.

Remove the gauge from the bore and place it between the measuring faces of the micrometer (Figure 6-18). Rotate the thimble and ratchet until the faces are set against the ends of the spindles. Lock the micrometer, remove the telescopic gauge, and read and record the measurement. A second measurement must be made to determine if the bore is out-of-round.

The same setup will be used for the second measurement except that the gauge is turned 90 degrees from the first measurement position within the bore (Figure 6-19). The measurement should be at the same height as the first measurement. Position the telescopic gauge, lock it, and remove it from the bore. Measure the distance between the spindles' ends with the micrometer and record the measurement.

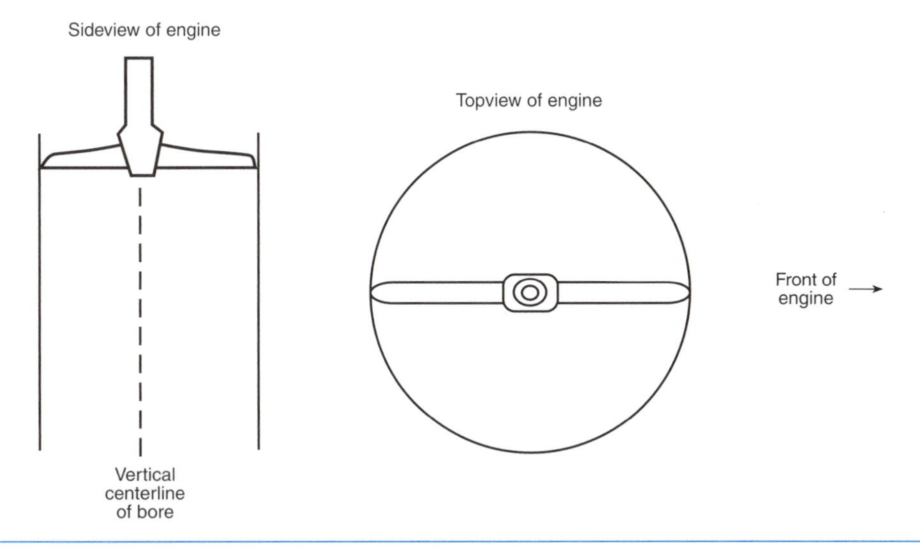

Figure 6-17 The gauge must be square in the bore for an accurate measurement.

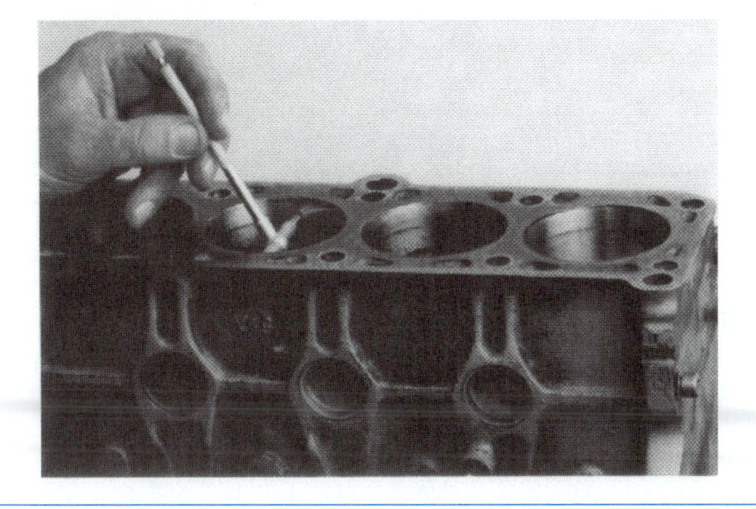

Figure 6-18 Use an outside micrometer to measure the extended telescoping gauge. Note the placement of the gauge within the bore in the left side of the figure.

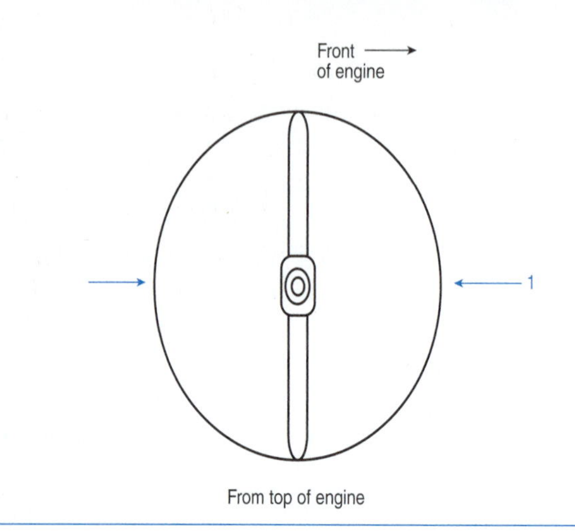

Figure 6-19 Like the shaft, the second measurement of the bore is at a right angle to the first.

If one of the two measurements is smaller than the other, the bore, like the journal, is out-of-round and has a slight oval shape. Most bores have a little more out-of-round tolerance than some shafts. Consult the service data for the allowable tolerance (Figure 6-20). Most bores can be machined and oversize components can be used.

While checking for out-of-round data, also check the specified maximum diameter of the bore. A bore that is too large will have thin walls. High compression or stress could cause the walls to crack and may cause serious damage.

Any shaft or bore can be measured with an outside micrometer and telescopic gauges if space and the correct measuring procedures are used. The two primary factors ensuring a true measure are the placement of the device's measuring faces and the proper reading of the scales. Costly mistakes can be made if either of the two is not done correctly.

Classroom Manual
page 148

Special Tools

Inside micrometer set

A component with machined bores

Measuring Bore with an Inside Micrometer

Measuring a bore with an inside micrometer is a little easier and can be more accurate than the procedures mentioned above. This is because the micrometer itself is placed inside the bore and makes a direct measurement.

If the bore is more than one inch in diameter, select an extension spindle to fit the micrometer. If necessary, review page 148 of the Classroom Manual on how to select an extension. Fit the extension to the micrometer and rotate the thimble to bring the micrometer to zero.

Standard diameter:

Mark "1"	87.000 – 87.010 mm (3.4252 – 3.4256 in.)
Mark "2"	87.010 – 87.020 mm (3.4256 – 3.4262 in.)
Mark "3"	87.020 – 87.030 mm (3.4260 – 3.4264 in.)

Maximum diameter:

STD	87.23 mm (3.4342 in.)
O/S 0.50	87.73 mm (3.4350 in.)

Figure 6-20 This engine only allows a maximum wear tolerance of 0.23 mm for the Mark "1" (87.23–87.000). (Reprinted with permission)

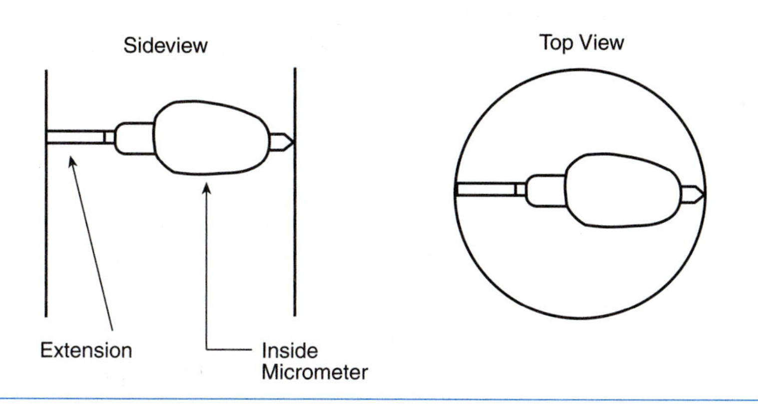

Figure 6-21 An inside micrometer is placed in a position similar to the placement of a telescoping gauge.

Place the micrometer inside the bore in a position similar to the placement of a telescopic gauge (Figure 6-21). Rotate the thimble until the end of the extension and the end of the micrometer are square in the bore. Once positioned and extended to the widest diameter of the bore, lock the micrometer and remove it from the bore.

Read the measurement on the micrometer in the same manner as an outside micrometer. Remember to add the size of the extension to the left side of the decimal (Figure 6-22). For instance, if the micrometer reading is 0.037 inch and the extension is 2 inches long, then the bore is 2.037 inches in diameter.

Place the micrometer back in the bore at a right angle (90 degrees) to the first position (Figure 6-23). Position and extend the micrometer and extension to the widest point of the bore. Lock, remove, and read the measurement of the second measurement.

Like previous measurements, compare the two readings and determine if the cylinder is out-of-round. Also make a decision on any service that may be needed on the bore. Data on wear limits and guidance on repair or replacement can be found in the service manual.

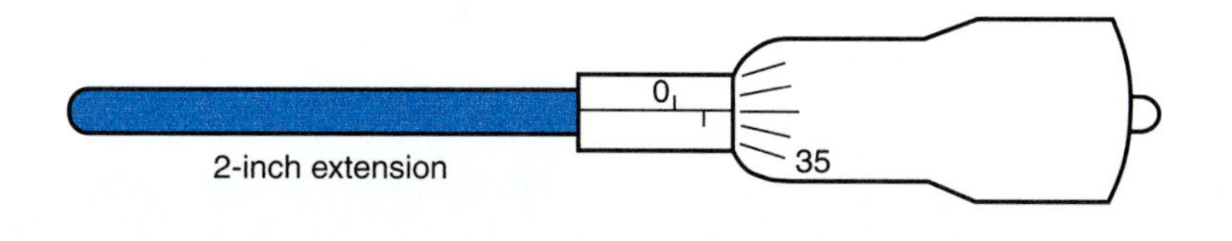

Figure 6-22 Remember to add the length of the extension spindle to the reading on the micrometer.

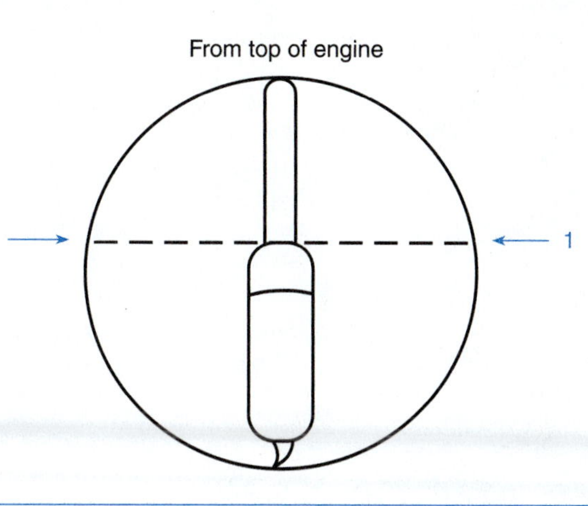

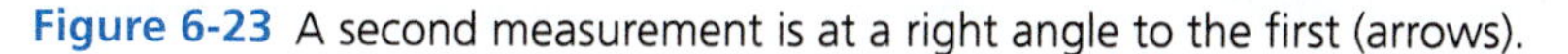

Figure 6-23 A second measurement is at a right angle to the first (arrows).

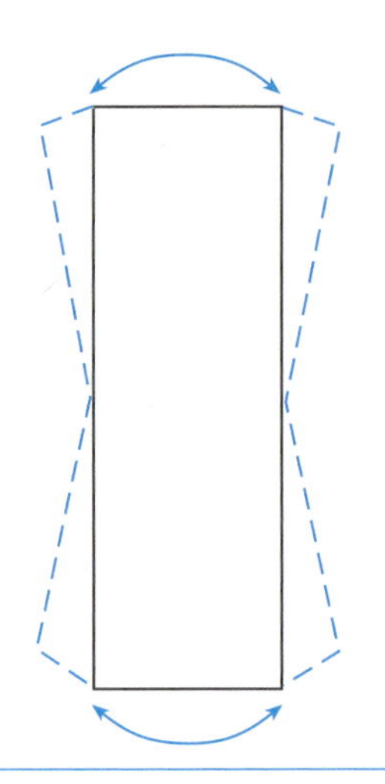

Figure 6-24 The side-to-side movement of the outer edge of a disc is called "runout."

Classroom Manual
page 153

Special Tools

Dial indicator set

A vehicle with the disc brake and wheel assembly removed

Runout is the amount of side-to-side wobbling in a wheel or disc.

Measuring Runout on a Disc

Runout is the term used to determine if there is a wobble in a wheel or disc (Figure 6-24). Generally, a disc or wheel should spin without any side to side movement. Movements of this type cause a vibration or shake and if serious enough could cause damage. It can be frustrating to passengers, when the vibration is transmitted throughout the vehicle. In most cases, technicians do not measure runout on a brake disc, but it is a good training device.

For this job we will use a brake system disc mounted on the vehicle which allows an easy mounting area for the disc and the dial indicator. The dial indicator requires a mounting bracket that can be attached to a solid portion of the engine block. It may be a *clamp type* or a *magnetic type* (Figure 6-25). For this setup, a clamp should be used. Sometimes the shape and position of the mounting area may be such that only one type of mount can be used. Regardless of the mount type, ensure that it is securely fastened so the dial indicator can be moved and positioned properly against the component being measured. The dial indicator should be the balanced type, but a continuous type can be used.

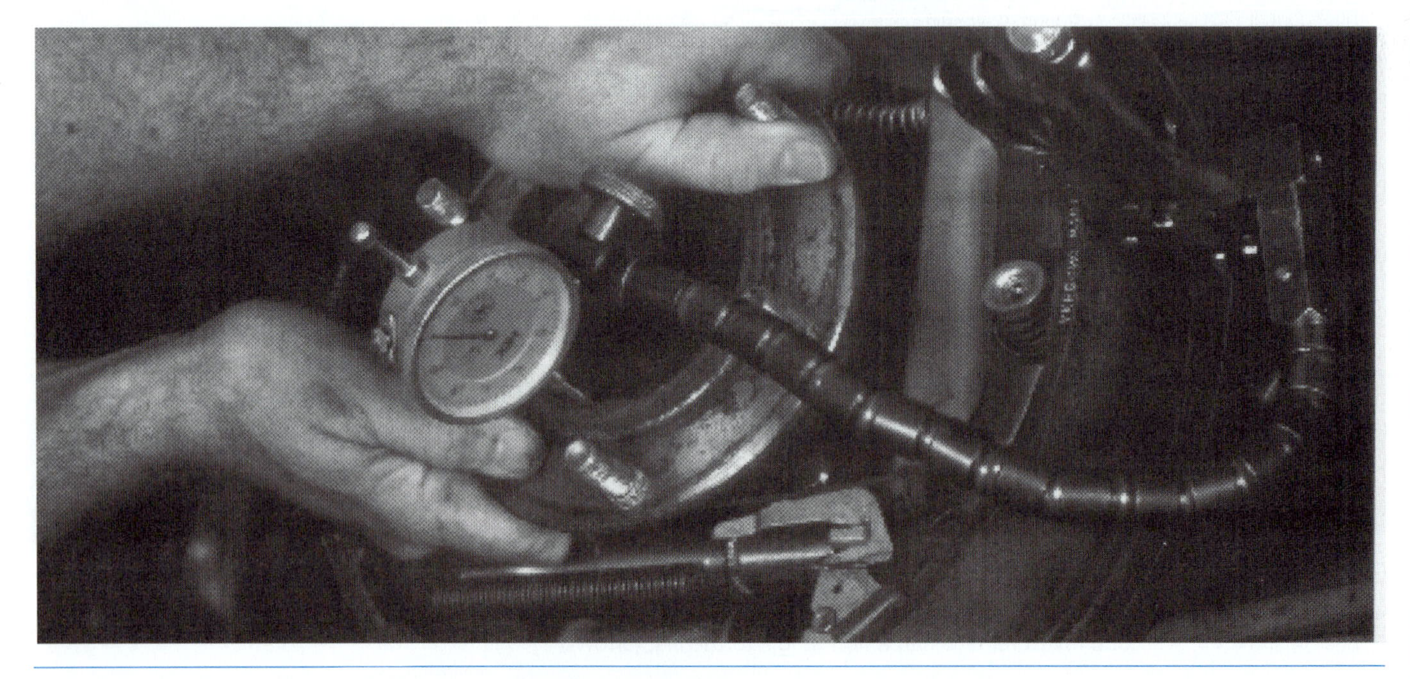

Figure 6-25 This dial indicator is mounted with a clamping mount called a "vise-grip mount."

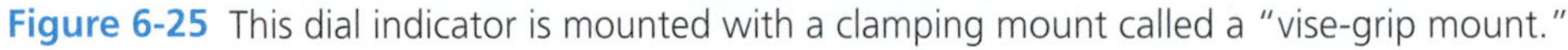

Set the disc on the hub and install at least half of the lug nuts. The nuts only need to be tight enough to hold the disc in place. Once the disc is positioned, mount the dial indicator bracket to the brake caliper mount. Assemble the various swivels and rods so the links reach over the disc (Figure 6-26). Before mounting the dial indicator to the mount, release the indicator's plunger completely. Install the indicator on the mounting rod and shift the linkage until the plunger is resting against the disc about midway from the machined area (Figure 6-27).

Move the dial indicator toward the disc until about half of the plunger is retracted into the indicator. At this point, tighten all of the mount's swivel or links to hold the dial indicator in place. Hold the dial indicator with one hand and adjust the scale until the zero mark is directly under the needle (Figure 6-28). It may take a few tries to get it correct.

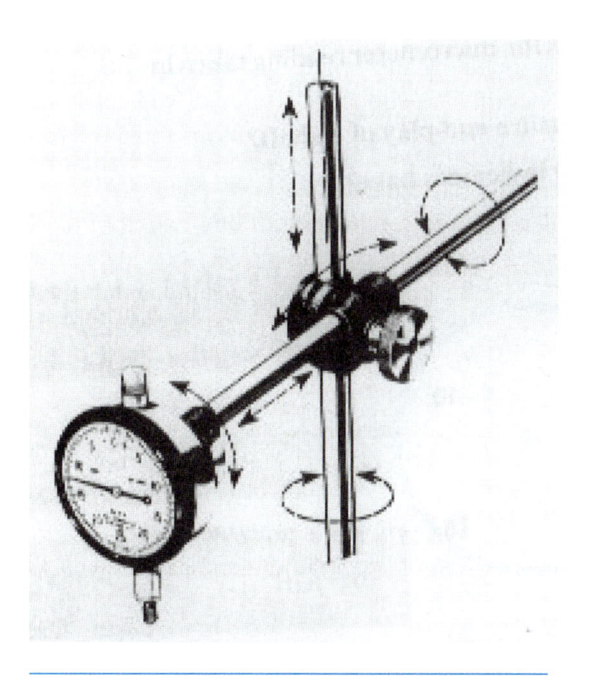

Figure 6-26 Most dial indicator mounts can be adjusted to fit almost any mounting situation. (Courtesy of C. Thomas Olivo Associates, *Fundamentals of Machine Technology*)

Figure 6-27 This dial indicator is mounted using a magnetic mount against a suspension component. Note the position of the dial plunger and the needle. (Reprinted with permission)

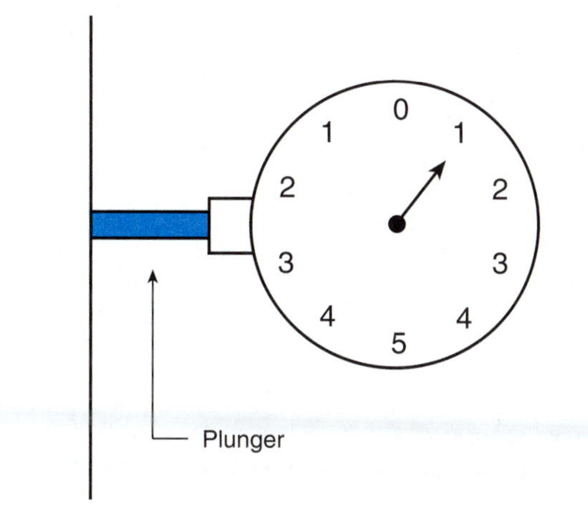

Figure 6-28 With the plunger retracted about halfway, the needle will probably not be at zero.

When the dial indicator is firmly mounted, the plunger is partially retracted, and the scale set to zero, the measurement can be taken. Using a screwdriver between the wheel lugs or by hand, rotate the disc in one direction slowly and evenly. Sharp, sudden movements should be avoided. Instead, try to make a smooth, continuous rotation to prevent moving the dial indicator. As the disc is rotated, read the dial indicator's scale for movement of the needle (Figure 6-29).

The needle will probably move to the left and right of zero (Figure 6-30). Generally, any movement exceeding 0.002 inch indicates the disc needs to be machined. On a balanced scale, the reading can be read from the numbers directly under the needle. If a continuous dial indicator is used and the disc's rotation caused the needle to move counterclockwise, the marks will have to be counted (Figure 6-31). Start at zero and count counterclockwise until the mark directly under the needle is reached. Check the size or calibration of the scale. It will be marked on the face as either $^1/_{1000}$ inch or $^1/_{10,000}$ inch. Each mark on the scale represents one of the calibration units. Usually, it is best to make the measurement two or three times to ensure an accurate reading.

There are several flat, disc-type components including the brake disc and flywheel that should be measured for runout. A warped component that is supposed to be flat will cause its mated component to vibrate and shake, thereby resulting in driver discomfort and possible damage.

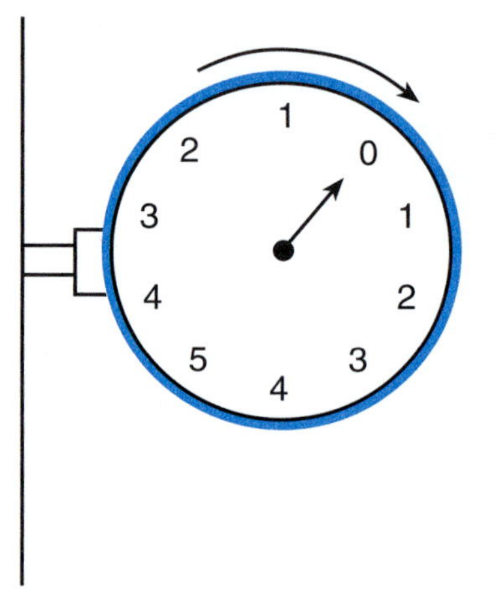

Figure 6-29 Rotate the knurled ring to turn the scale so that zero is under the needle. The indicator will have to be held in place.

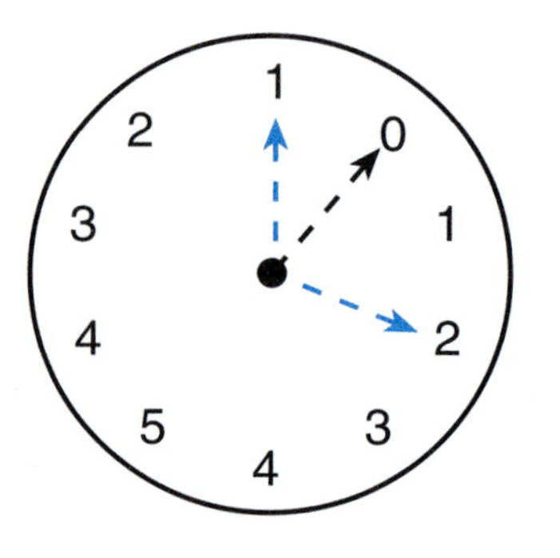

Figure 6-30 The needle may move to the left and right of zero as the disc is rotated.

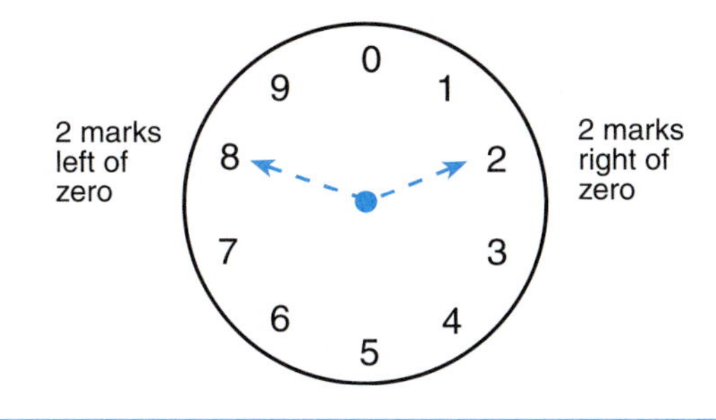

Figure 6-31 A continuous scale dial indicator requires the technician to count the number of graduations the needle moved to the left of zero.

Using Brake Calipers to Measure the Thickness of Brake Discs

Classroom Manual
page 140

Special Tools

Brake Disc

Brake caliper (micrometer)

The brake disc is also known as the brake *rotor*.

During brake repair, the disc (or **rotor**) should be checked for excessive wear and warpage. The procedure discussed above will give the technician an idea of the flatness of one side of the rotor, but a more common measurement is the thickness and parallelism of the two opposing friction areas. Thickness can be measured with a regular outside micrometer or a special brake disc micrometer (Figure 6-32). The primary difference between micrometers is quick thimble action and larger numbers on the brake micrometer. This makes it easier and quicker to use. Warpage or parallelism between the two opposing friction areas can also be checked with a micrometer or a disc parallel gauge.

Brake disc thickness can be checked on the car, a table, or a brake lathe. Find the minimum thickness specification for the disc. Many times it is stamped on the inside or outside of the hub portion of the disc and most are listed as metric measurements (Figure 6-33). If the brake disc micrometer is USC, convert the specification to USC.

Figure 6-32 A special micrometer for the quick, easy measurement of a brake disc. (Courtesy of MATCO Tool Company)

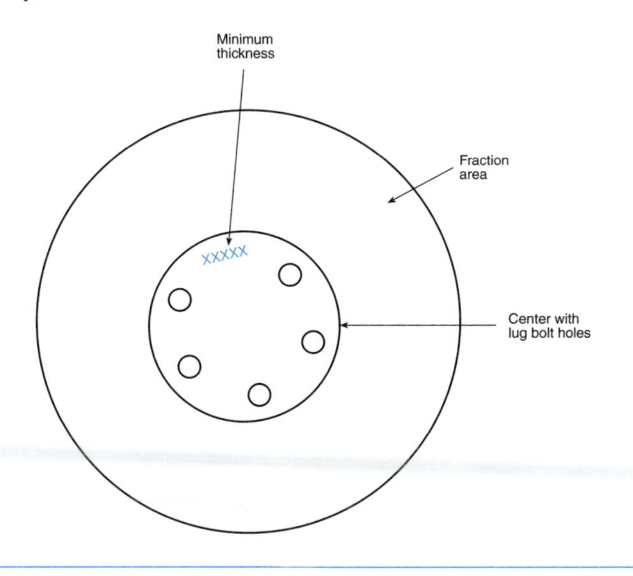

Figure 6-33 The minimum thickness is usually on the disc, but it can also be found in the service manual.

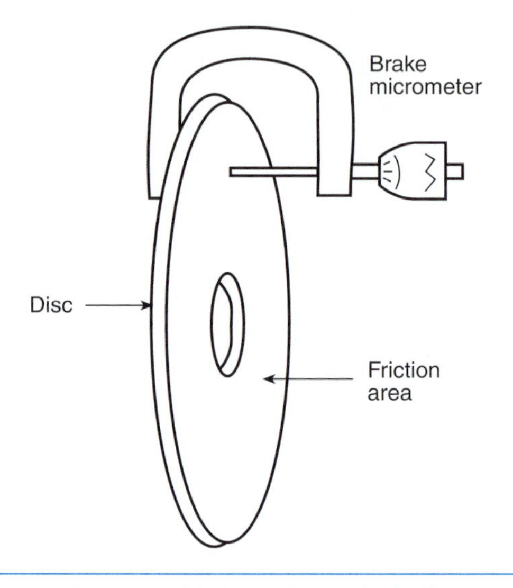

Figure 6-34 The measuring faces should always be at the halfway point of the friction area. The outer edges generally wear more than the area closer to the hub.

Place the micrometer's measuring faces on each side of the disc midway to the wear area (Figure 6-34). Rotate the thimble until the spindle and anvil faces are against the disc. Read and record the measurement. In almost all cases, the measurement will be less than one inch. Repeat the measurement at a minimum of twelve different points around the disc (Figure 6-35). Remember to measure at the mid-point of the friction areas at each of the measuring points. Compare your readings to the specified minimum thickness. Any measurement that is less than specified means the disc is not to be reused and should be discarded. Differences over 0.002 inch between any of the twelve means the disc's friction areas are not parallel. If the disc is not too thin, it can be machined and reused.

Measuring Pressure

Classroom Manual
page 156

Special Tools

Tire pressure gauge

Vehicle

Air conditioning systems use *pressure* to force the refrigerant to change its physical state from liquid to vapor and then back to liquid again.

Pressures on an automobile is normally measured on the fuel, oil, air conditioning, power steering, brake, and automatic transmission systems. With antilock brake systems being used on most vehicles, special gauges are used to check brake pressure. Pressure tests on these systems are used to detect malfunctions in fluid delivery. Tapping into the fluid flow in some way is the method to perform any pressure test (Figure 6-36). Gauges and taps must be selected depending on the system being tested. Gauges may be either a pressure/vacuum type, measuring only up to 10 psi, or a pressure type that measures up to 200 psi or more. Normally, the pressure/vacuum gauges

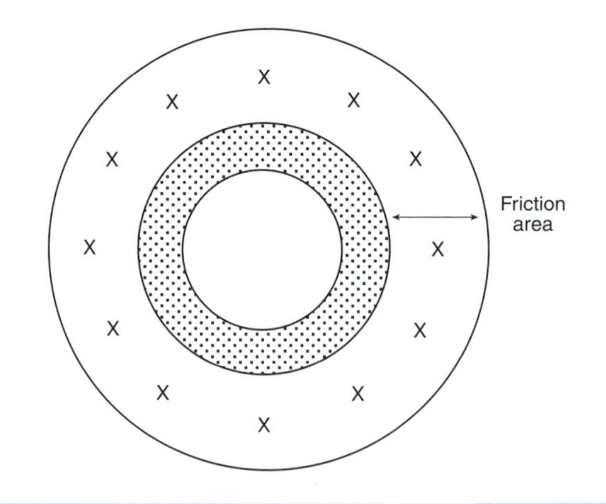

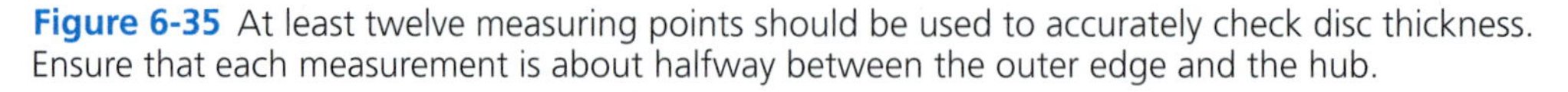

Figure 6-35 At least twelve measuring points should be used to accurately check disc thickness. Ensure that each measurement is about halfway between the outer edge and the hub.

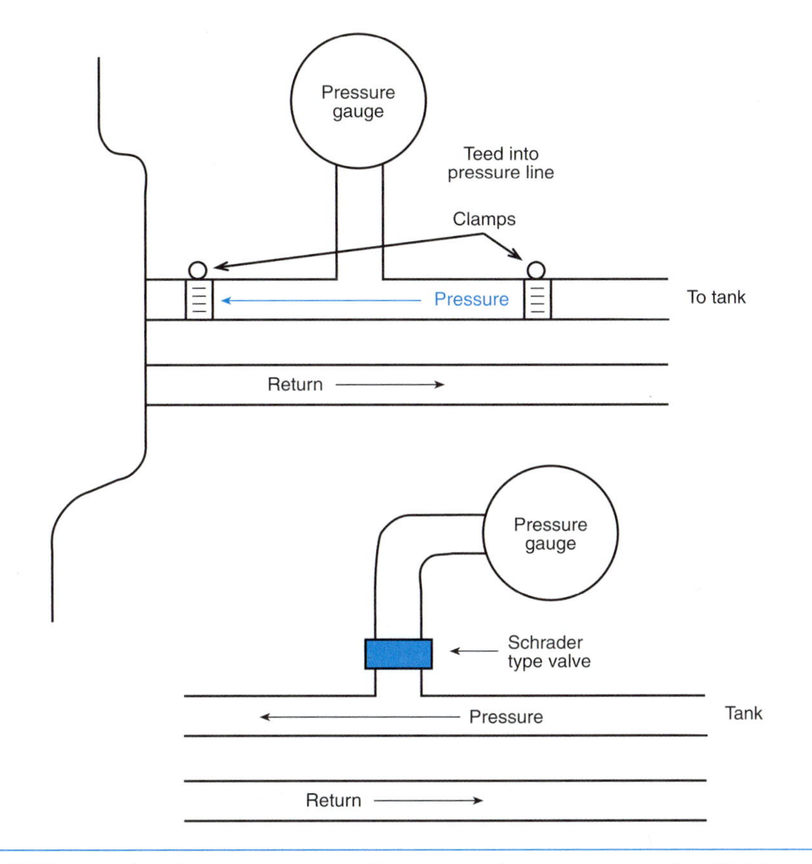

Figure 6-36 Note that the pressure gauge is connected to the pressure line. This is true in almost every measurement taken on any hydraulic system.

are only used on carburetor fuel systems. Almost all other systems require gauges that can read high pressures. Only an experienced technician should take pressure measurements, especially on the high-pressure systems. For our discussion, we will measure the pressure of a tire.

CAUTION: Never add air to a large truck or equipment tire that is low on air. The design of some large wheel rims can allow the tire or locking equipment to become dislodged and can erupt in a dangerous explosion. Very serious injury or death can occur. The wheel assembly must be removed and the tire broken loose from the rim before inflating.

To obtain a correct **tire pressure** measurement, the tire must not have been driven for more than two or three miles. In other words, it must be "cold." Driving heats the air in the tire, thereby increasing pressure. With the vehicle on the ground or floor, remove the cap from the tire's valve stem.

Tire pressure directly affects the fuel economy, handling characteristics, and riding comfort of the vehicle.

CAUTION: Never exceed the tire's maximum pressure rating. An explosion could occur if the tire splits open and injury or death could occur. The pressure rating can be found on the side of the tire.

CAUTION: Never stand directly to the side of the wheel assembly. A tire, new or old, may have a weak spot that will burst with a high explosive effect, thereby causing injury or death. Technicians should beside the fenders and attempt to check or adjust pressure with only his or her hands near the tires.

Place the connection end of the gauge firmly over and down the end of the valve stem (Figure 6-37). There should be no noise of escaping air around the gauge. Read the pressure with the gauge held in place. A typical passenger car should have between 32 psi to 35 psi. If the reading is high, remove the gauge and release some air from the tire. If too low, add air through the valve stem. Do not exceed the tire's maximum allowable pressure.

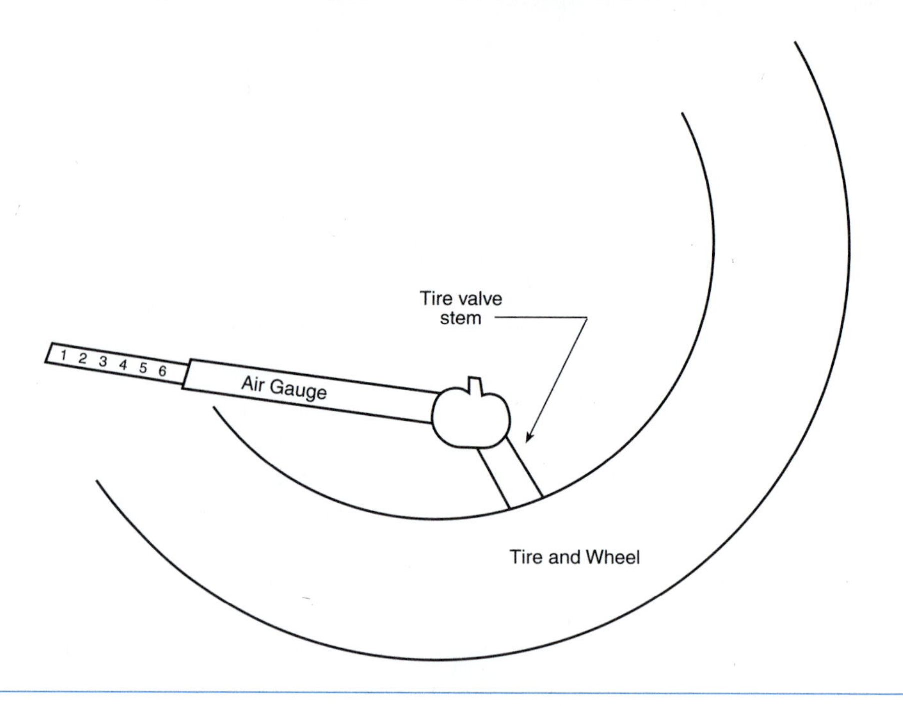

Figure 6-37 Press the gauge firmly against the valve stem.

Classroom Manual
page 156

Measuring Engine Vacuum

Special Tools

Vacuum gauge or

Pressure/vacuum gauge

Vehicle

Measuring vacuum on an automobile engine can tell the technician a great deal about the internal condition of that engine. The downward movement of the pistons within their cylinders creates engine vacuum. Anything that causes an engine's vacuum to drop will result in lower performance. The causes and results of improper engine vacuum will be covered in Chapter 8, "Engine Maintenance."

WARNING: An experienced technician or instructor should guide the student or new technician through this section the first time. Disruption to the engine's performance or injury to the technician could occur if skill, attention to detail, and safety procedures are not used.

A vacuum test is performed using a port that is tapped directly into the engine's intake manifold (Figure 6-38). When possible, a multiple connection, sometimes known as a "vacuum tree,"

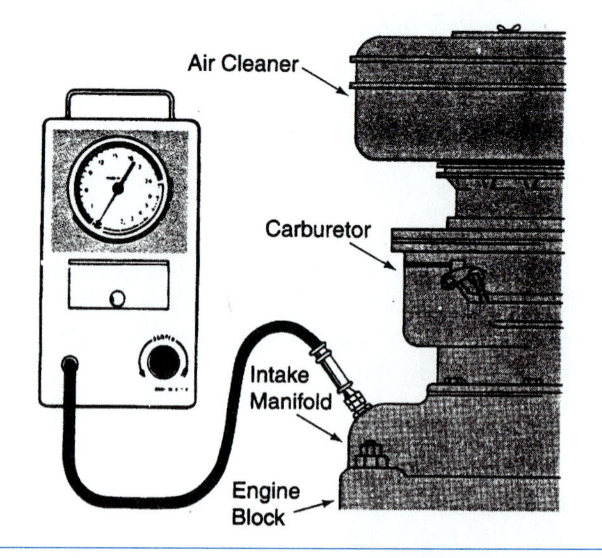

Figure 6-38 This gauge hookup will give direct engine vacuum under all engine speeds.

should be used for the gauge's connection (Figure 6-39). If an empty port is not available, the gauge will have to be teed into one of the vacuum lines from the intake manifold. If a vacuum line is disconnected and a gauge is put in its place, the device operated by the vacuum line may cause a malfunction in engine operations and cause errors in the measurements (Figure 6-40). This can lead an inattentive technician in diagnostic circles.

With the vacuum gauge connected, start the engine and allow it to warm. Observe the vacuum measurements at idle and high idle engine speeds (Figure 6-41). Readings of 17 to 20 in. Hg are normal at both speeds. Note that there will be a drastic drop in vacuum as the engine is accelerated. A steady, below-specified vacuum or a fluctuating needle normally denotes an engine needing repairs.

CASE STUDY

An engine was checked for a poor idle condition. One of the tests was a measurement of the engine's vacuum. A vacuum gauge was connected to an intake port for the adjustment. As soon as the vacuum line was disconnected and the gauge connected, the engine begin to run smoothly. A closer inspection found that a hose connected to a small vacuum motor in the heating system was disconnected. This allowed a substantial leak and leaned the air/fuel mixture. However, it was not enough of a leak to cause a noticeable difference at cruising speeds. The disconnected hose was under the dash, and road noise and the radio muffled the sound of the leak.

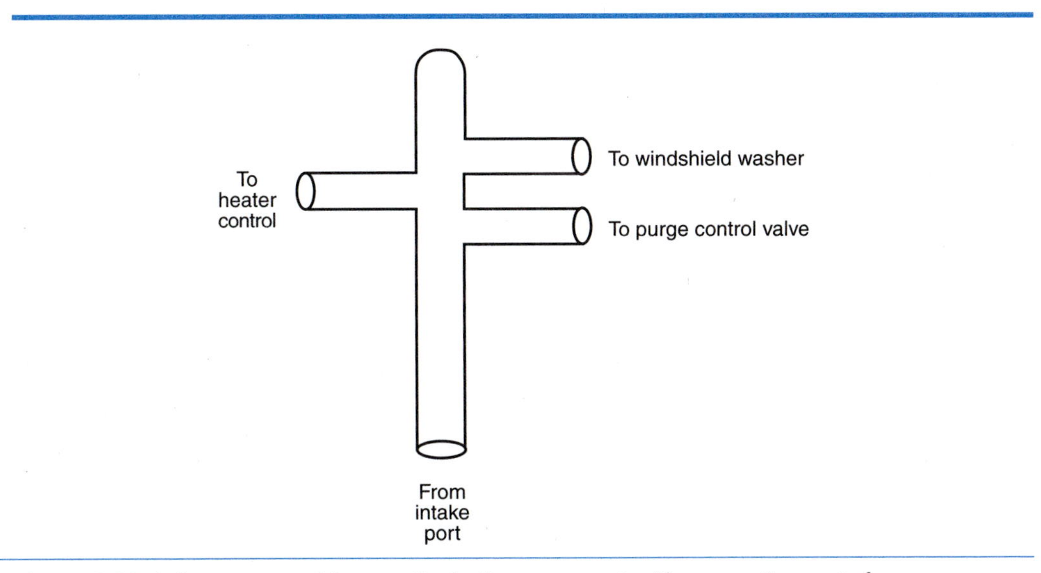

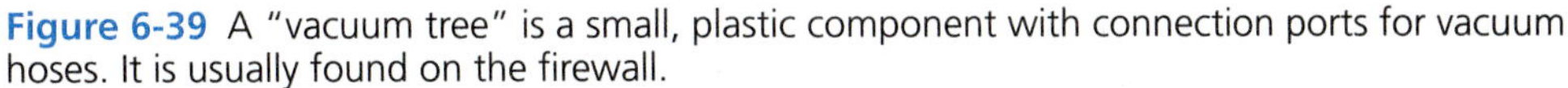

Figure 6-39 A "vacuum tree" is a small, plastic component with connection ports for vacuum hoses. It is usually found on the firewall.

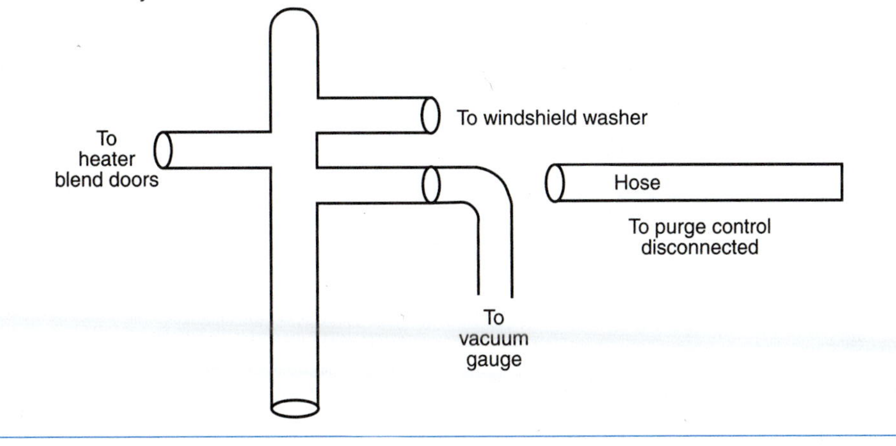

Figure 6-40 Disconnecting a system hose while the engine is running could cause the engine to perform poorly during the test.

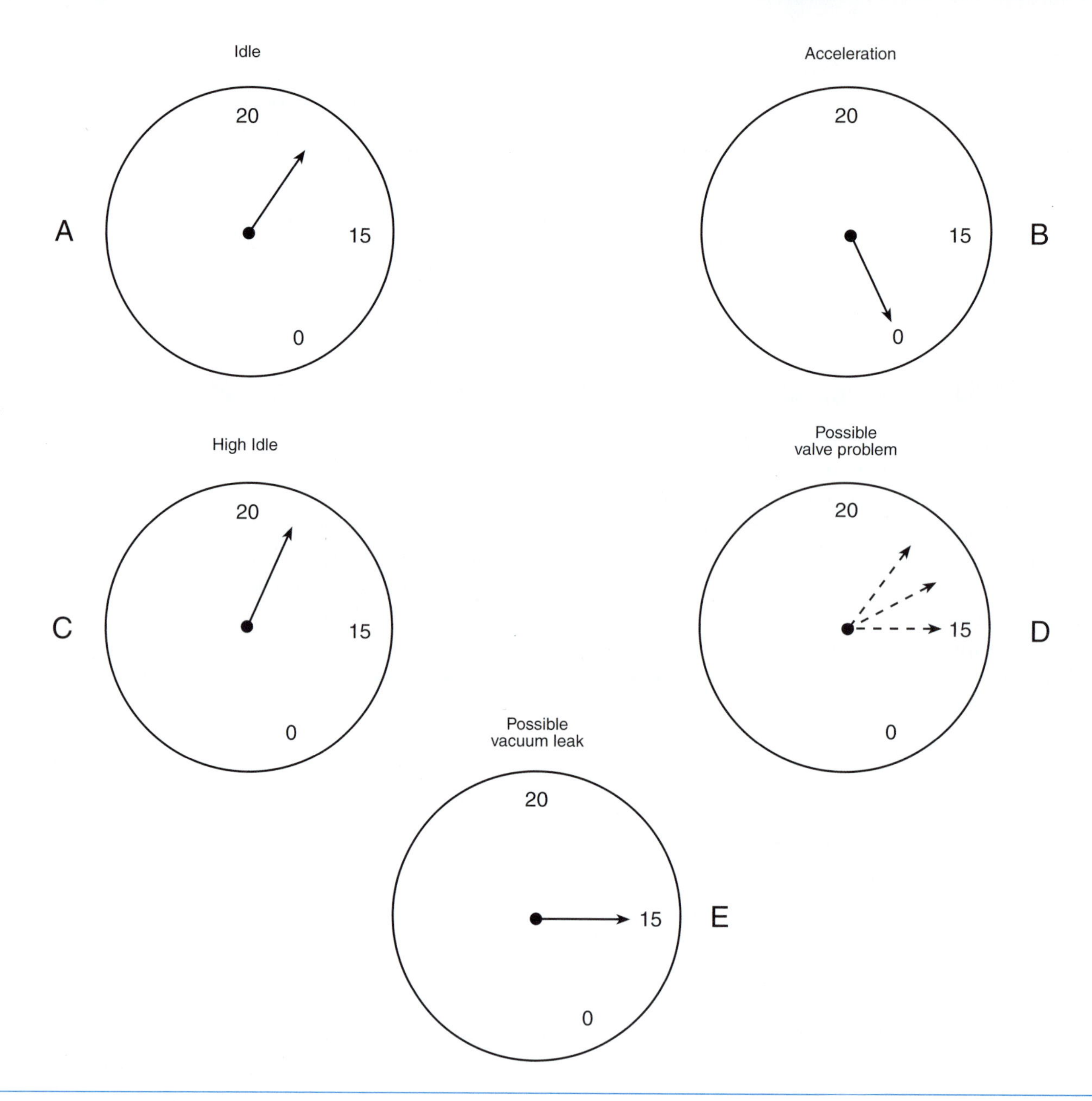

Figure 6-41 Dials A, B, and C indicate a good engine while D and E could indicate a mechanical problem with the engine.

Terms to Know

Bore
Electrodes
Friction area
Gap
Journal
Out-of-round
Ring travel area
Rotor
Thrust wear

ASE Style Review Questions

1. Pressure and vacuum measurement is being discussed. *Technician A* says a pressure gauge is used to check some brake systems.
   Technician B says an engine vacuum test may indicate the problems with a poorly performing engine. Who is correct?
   A. A only
   B. B only
   C. Both A and B
   D. Neither A nor B

2. *Technician A* says pressure may be used within the steering system.
   Technician B says tire pressure must be checked when the tire is cold to obtain an accurate measurement. Who is correct?
   A. A only
   B. B only
   C. Both A and B
   D. Neither A nor B

3. *Technician A* says any out-of-round wear in a cylinder requires the replacement of the component with the cylinder.
   Technician B says an out-of-round measurement can be made with just a telescopic gauge. Who is correct?
   A. A only
   B. B only
   C. Both A and B
   D. Neither A nor B

4. Measurement systems are being discussed.
   Technician A says a metric micrometer may be used to measure an USC crankshaft journal.
   Technician B says a conversion factor is used to convert measurements between measuring systems. Who is correct?
   A. A only
   B. B only
   C. Both A and B
   D. Neither A nor B

5. Cylinder measurements are being discussed.
   Technician A says a larger-than-specified measurement may indicate an out-of-round cylinder.
   Technician B says only one measurement is made to determine if a cylinder has an out-of-round condition. Who is correct?
   A. A only
   B. B only
   C. Both A and B
   D. Neither A nor B

6. Brake disc measurements are being discussed.
   Technician A says a dial indicator is used to check disc thickness.
   Technician B says an outside micrometer is normally used to determine if the fiction areas are parallel. Who is correct?
   A. A only
   B. B only
   C. Both A and B
   D. Neither A nor B

7. *Technician A* says a bearing journal is measured twice.
   Technician B says the journal measurements are performed 180 degrees apart. Who is correct?
   A. A only
   B. B only
   C. Both A and B
   D. Neither A nor B

8. Shaft measurements are being discussed. *Technician A* says a measurement smaller than specified requires the shaft to be machined or replaced.
   Technician B says diameter measurements larger than those specified usually indicate wrong data or a poor measurement. Who is correct?
   A. A only
   B. B only
   C. Both A and B
   D. Neither A nor B

9. Dial indicators are being discussed. *Technician A* says a dial indicator may have a scale that is numbered from zero and goes up in two different directions.
   Technician B says a brake disc caliper is a modified dial indicator. Who is correct?
   A. A only
   B. B only
   C. Both A and B
   D. Neither A nor B

10. Micrometers and dial indicators are being discussed. *Technician A* says a dial indicator can be used without a mount.
    Technician B says the use of a micrometer requires a clamping mount. Who is correct?
    A. A only
    B. B only
    C. Both A and B
    D. Neither A nor B

Table 6-1 ASE Task

Clean, inspect, and measure the disc with a dial indicator and a micrometer; follow manufacturers' recommendations in determining what needs to machined or replaced.

Problem Area	Symptoms	Possible Causes	Classroom Manual	Shop Manual
Worn rotors	Rough braking.	**1.** Excessive hard braking.	140	144, 147
		2. Wrong pads.	—	144, 147
		3. Assembled wrong.	—	144, 147
		4. Caliper or piston sticking.	—	144, 147

SAFETY

Wear eye protection.
Use a device to catch brake dust during the removal of pads.

Table 6-2 ASE Task

Measure and adjust tire pressure.

Problem Area	Symptoms	Possible Causes	Classroom Manual	Shop Manual
Short tire life	Pulls to one side, and wears on the edges.	**1.** Low air pressure.	156	156
		2. Tire worn out.	—	156
		3. Hole in the tire.	—	156

SAFETY

Wear eye protection.
Determine recommended tire pressure.

Table 6-3 ASE Task

Inspect and measure cylinder walls; remove cylinder wall ridges; hone and clean cylinder walls; and determine need for further actions.

Problem Area	Symptoms	Possible Causes	Classroom Manual	Shop Manual
Engine performance and oil consumption	Blue exhaust.	**1.** Worn piston rings.	148	140–143
		2. Worn cylinders.	148	140–143

SAFETY

Wear eye protection.

Table 6-4 ASE Task

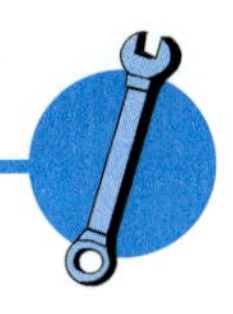

Visually inspect the crankshaft for surface cracks and journal damage; check the oil passage condition; measure journal wear; check the crankshaft sensor reluctor ring (where applicable); determine necessary action.

Problem Area	Symptoms	Possible Causes	Classroom Manual	Shop Manual
Engine operations	Noise and poor engine performance.	**1.** Worn bearings.	142	135
		2. Damaged/worn journals.	142	136–140
		3. Damaged/worn piston rod.	—	136–140

SAFETY

Wear eye protection.

Table 6-5 ASE Task

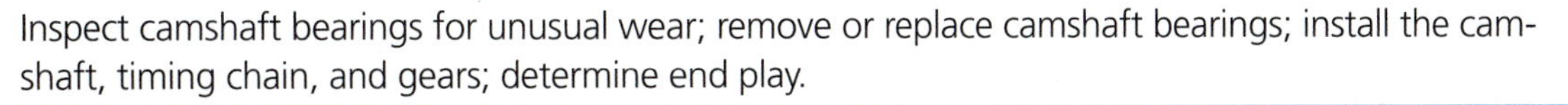

Inspect camshaft bearings for unusual wear; remove or replace camshaft bearings; install the camshaft, timing chain, and gears; determine end play.

Problem Area	Symptoms	Possible Causes	Classroom Manual	Shop Manual
Worn valve components	Noise and poor engine performance.	**1.** Worn valve components.	—	135
		2. Worn camshaft or bearings.	142	136–140

SAFETY

Wear eye protection.

Job Sheet 9

9

Name ______________________________ Date ______________

Measuring Groove Depth with a Depth Micrometer

Upon completion of this job sheet, you should be able to measure the depth of a groove cut into a block.

Instructor's Note: This task is written on the assumption that some type of grooved component is available. The author uses two blocks of scrap metal cut by the machine tool technology class. Each block has two grooves with three different depths within each groove. Grooved and planed blocks of wood or an engine with cylinder heads removed will work just as well with minor changes in the procedures.

Tools Needed

Component or material with grooves
OR
Engine with head removed and pistons installed
Depth micrometer

Procedures

1. Secure the material to be measured so it will not move as the measurement is made.
2. Select an inside micrometer and, if necessary, extension spindles.
3. If there is specification data available, record it below. Space is also available to record the actual measurement.

First Groove Specification	Actual
______________	______________
______________	______________
______________	______________

Second Groove Specification	Actual
______________	______________
______________	______________
______________	______________

Third Groove Specification	Actual
______________	______________
______________	______________
______________	______________

4. Fit the extension to the micrometer as needed.

5. Retract the micrometer until it will not reach the bottom of the groove when it is positioned.
6. Ensure that the surface where the micrometer is to be set down is flat and smooth.
7. Place the micrometer over the hole with the spindle pointing into the hole.
8. Position the base of the micrometer so it extends over two edges of the groove.
9. Holding the micrometer firmly in place by hand, rotate the thimble until the spindle contacts the bottom of the groove.
10. Lock the micrometer and remove it from the material for easy reading.
11. Record the measurement.
12. Perform any other measurements needed to complete this exercise and record all readings next to the specifications.
13. If specifications were available and recorded, compare them to the actual measurements and determine any needed repairs.

____

____

Job Sheet 10

10

Name ________________________ Date ______________

Measuring with an Outside Micrometer

Upon completion of this job sheet, you should be able to measure a shaft with an outside micrometer.

Instructor's Note: For this task the author uses four different shafts from transmissions. Each shaft has three to five machined areas that can be measured. Each machined area is labeled so the answers can be tracked and graded.

Tools Needed

Shafts to be measured
Outside micrometer (at least one should have metric measurement)

Procedures

1. Select the correct size micrometer for the areas to be measured.
2. Open the micrometer enough to slide over the area to be measured.
3. Close the measuring faces until they make contact with the shaft.
4. Position the measuring faces 180 degrees apart at a right angle to the shaft.
5. Rotate the thimble with the ratchet until it slips.
6. Slide the micrometer back and forth over shaft to ensure the micrometer is closed and squared correctly on the shaft.
7. Flip the thumb lock on and remove the micrometer from the shaft.
8. Read the measurement and:
   - **A.** Record the size of the micrometer, i.e., 0 inch, 1 inch, 2 inches. ______________
   - **B.** Record the last number on the sleeve uncovered by the thimble. ______________
   - **C.** Record the number of thousandths of an inch (millimeters) between the uncovered sleeve number and the edge of the thimble. ______________
   - **D.** Read and add the thimble graduation aligned with the index line to the number arrived at in item C above. ______________
   - **E.** Record the total measurement in the appropriate blank at the end of this job sheet. Note that the number of blanks may not match the number of measurements. ______
   - **F.** Repeat the steps above for the other measurements.

Measurements

1. Shaft 1

   Area 1 ________ + ________ + ________ = ________ inches/millimeters
   (Size) (Sleeve) (Thimble)

   Area 2 ____________ Area 3____________ Area 4____________

2. Shaft 2

Area 1 ______________ Area 2______________ Area 3______________

3. Shaft 3

Area 1 ______________ Area 2______________ Area 3______________

Use the blanks below for additional measurements.

____ ________ ____________ ________ ____________

____ ________ ____________ ________ ____________

____ ________ ____________ ________ ____________

____ ________ ____________ ________ ____________

Servicing Bearings and Installing Sealants

Upon completion and review of this chapter, you should be able to:

- ❑ Remove and install a tapered roller bearing and wheel seal.
- ❑ Inspect bearings.
- ❑ Repack wheel bearings.
- ❑ Remove and install an outer bearing race in a hub.
- ❑ Install chemical sealants.
- ❑ Install a lip seal.

Introduction

Replacing and servicing bearings are typical tasks for automotive technicians. Reassembling automotive components requires the installation of different types of gaskets, seals, and chemical sealants. This chapter will give you a short explanation of these tasks and some items that need to be checked as the repairs or servicing are completed.

Removing a Wheel Bearing and Seal

We will discuss the removal of a wheel bearing and seal here and their installation later. A typical **hub** with two bearings and one seal will be used for this purpose. The procedures can be followed on other application except that some bearings may need to be pressed out and in. This is best done by an experienced technician.

Lift the vehicle and remove the wheel. Use a screwdriver or the special tool to remove the dust cover (Figure 7-1). Under the dust cover, a nut with a **cotter key** will be visible. Remove the key and nut and place to the side. A flat, keyed washer is removed next. Many times the outer bearing will slide out with the washer. Shaking the hub will dislodge the bearing and washer for an easier grip.

With the outer bearing, washer, nut, and cotter key removed, the hub can be removed. However, read the next short section on how the inner bearing and hub can be removed.

Thread the nut back onto the spindle about two or three threads. Grasp the edges of the hub and pull outward and downward (Figure 7-2). The inner bearing will hang on the nut and force the seal out. Force is not needed to perform this task. If the nut will not fit through the

Classroom Manual
page 161

Special Tools

Dust cover removal tool

Cotter pin remover

The wheel assembly fits onto the *hub* and the hub fits over the bearings on a spindle.

A *cotter key* is a piece of metal doubled back on itself. It is inserted into a hole and the ends bend out to hold it in place.

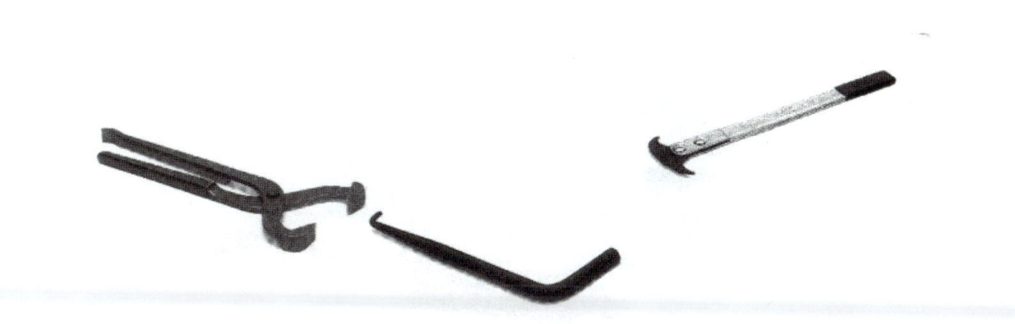

Figure 7-1 These tools can remove most automotive seals and dust covers. (Courtesy of Snap-on Tools Company)

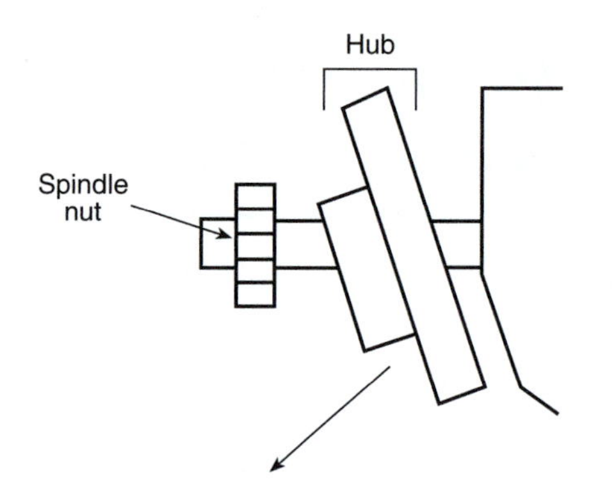

Figure 7-2 Screw the nut slightly onto the spindle and pull outward and downward on the hub assembly.

hub, or the seal and bearing do not come out, remove the hub and place it on the table with the seal facing upward.

Special Tools

Seal remover

Blow gun

Parts washer

A seal remover or large, flat-tipped screwdriver can be used to remove the seal. Place the tool or screwdriver across the hole in the seal (Figure 7-3). Hook the end of the tool or screwdriver under the inner edge of the seal and use the hub as a pivot to apply force to the seal. The seal should pop out fairly easy. The bearing can now be removed.

CAUTION: Always use protective gloves and face protection when cleaning bearings. Skin and eyes can be injured by the cleaning solution.

Before beginning work on the bearings, clean the inside of the hub completely. Dry it with compressed air and set it aside.

Classroom Manual
page 161

Inspecting and Repacking a Wheel Bearing

During a typical brake job, it is usual for the technician to service the **wheel bearings.** During the service, the bearings should be inspected. The inspection discussed below applies to every bearing but directly highlights ball and roller bearings. Engine bearing failures are made obvious by sound and engine operation.

Most shops have a standard fee for replacing the brakes on one axle. The fee includes repacking the *wheel bearing* and a new seal.

Inspection

Bearings fail because of a lack of proper lubrication, overloading, or improper installation. In most cases, the design of the total component prevents overloading. Few bearing failures result directly from improper installation, thereby leaving lubrication as the main cause of failure.

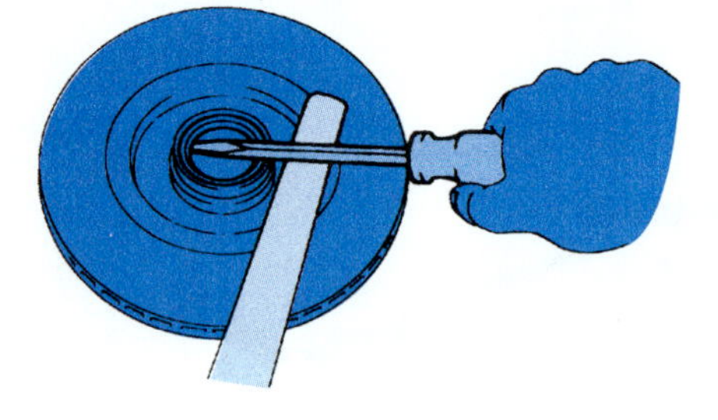

Figure 7-3 A screwdriver can be used to remove many automotive seals.

Bearing failure is indicated by a noise from the bearing area. At times, it is almost impossible to isolate the problem bearing. Axle bearings can be confused with wheel or differential bearings. A slow test drive around the lot may find the end of the vehicle from which the noise originates. Turning in tight circles may pinpoint the side of the car with the most noise. An experienced technician should perform the test drive if one is needed.

Bearing failure causes, two types of damage. When pieces break off the bearing, the damage is called **spalling** (Figure 7-4). Dents in the bearing or race are signs of **brinelling** (Figure 7-5). Either can cause more damage if not replaced. A dry bearing can weld itself to the races and the components by friction-produced heat.

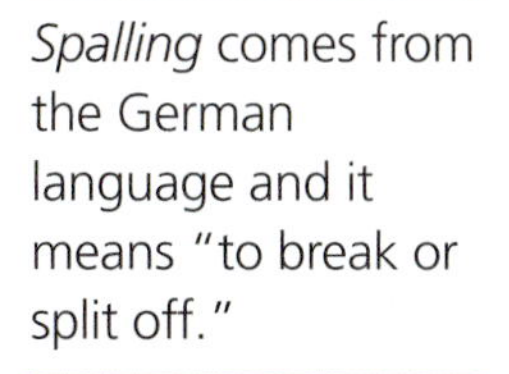

Spalling comes from the German language and it means "to break or split off."

Brinelling is a type of bearing failure where dents appear due to the roller hammering against the race.

CAUTION: Always use protective gloves and face protection when cleaning bearings. Skin and eyes can be injured by the cleaning solution.

CAUTION: Do not let the bearing spin while drying it with compressed air. The bearing may fly apart, thereby causing injury, or it could catch skin between the bearing and cage or race.

Remove the suspected bearing and clean it thoroughly with a parts cleaner. Ensure that all of the old lubricant is removed. The bearing can be air dried with compressed air. This will help remove the cleaning solution and any particles trapped in the bearing. However, when using compressed air to dry a bearing, there are some critical things that must be done. Hand protection must be used to protect the skin from the cleaner, compressed air, and any particles dislodged by the air (Figure 7-6). Eye and face protection should be always be worn during

Special Tools

Parts washer

Blow gun

Catch pan or vat

Figure 7-4 Note how the metal is flaking from this race (top) and bearing. (Courtesy of Chicago Rawhide)

Figure 7-5 Denting of the bearing metal is caused by the bearing moving up and down within its races. This may be a result of improper installation. (Courtesy of Chicago Rawhide)

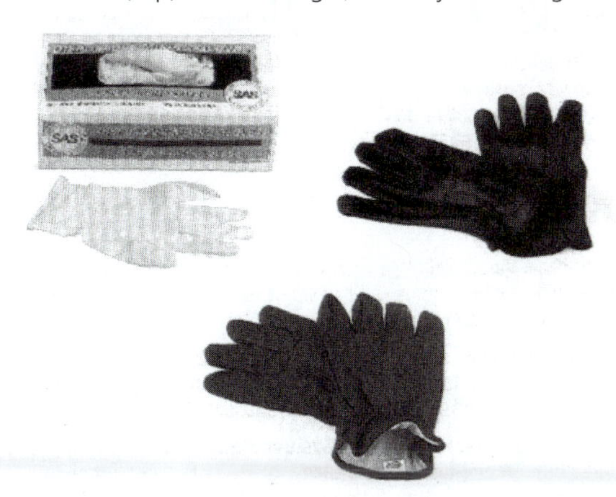

Figure 7-6 Protective gloves can be used by automotive technicians and they can still retain a good sense of touch. (Courtesy of Snap-on Tools Company)

the cleaning process. The bearing must be held in a manner to prevent any of the rollers or balls from spinning. A free-spinning bearing can disintegrate, thereby causing injuries. If at all possible, use a regulator to reduce air pressure. This will lessen the possibility of injury due to spinning bearings and loose particles within the bearing.

Most shops collect small amounts of *hazardous waste* in a large, sealed container for shipment later.

Dry the bearing over a vat or pan to catch the cleaner and particles blown off by the air (Figure 7-7). The waste will probably have to be treated as **hazardous waste** and stored accordingly. With the bearing cleaned, clean the races. Most can be cleaned while still installed in the component.

A *burned area* will be bluish in color.

With all of the bearing components clean and dry, inspect each ball or roller for dents, nicks, **burned areas,** or any other damage to the machined surfaces (Figure 7-8). Inspect the

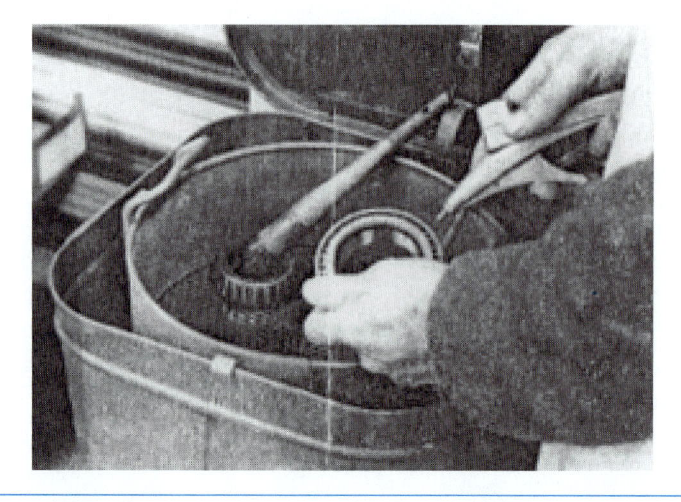

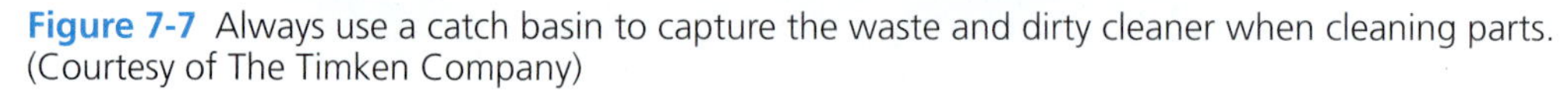

Figure 7-7 Always use a catch basin to capture the waste and dirty cleaner when cleaning parts. (Courtesy of The Timken Company)

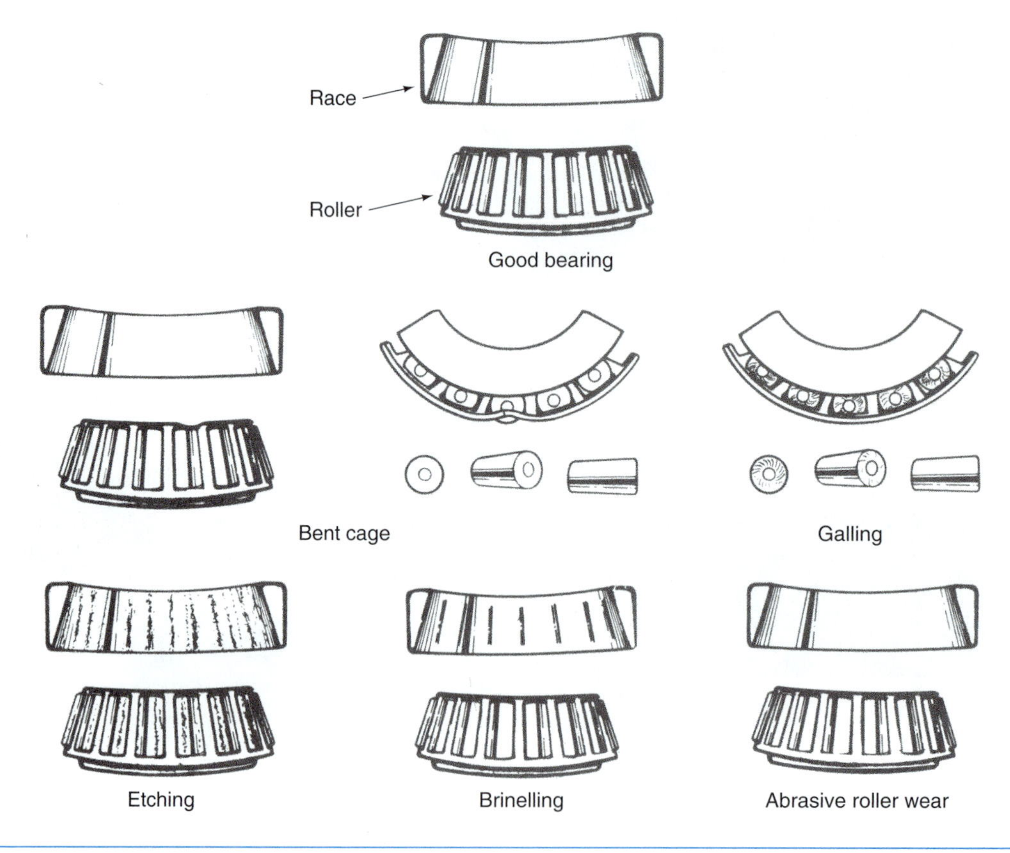

Figure 7-8 Note the different wear patterns on the three bottom bearings. (Reprinted with the permission of the Ford Motor Company)

cage for any bending or breakage. Inspect the races as much as possible. Most races can be inspected while installing a component. Also, inspect the component into which the bearing and races are fitted. If a race has been removed from the hub or **shaft,** carefully inspect the area where the race fits (Figure 7-9). This area must be as clean and undamaged as other parts of the bearing. Any type of damage to any part requires replacement of the complete bearing, including the races.

If the shaft or hub is damaged, consult the owner. Replacing or machining some shafts and hubs can be an expensive job. While there is not much choice concerning the repairs needed in this situation, the owner is the one paying and has the final word based on the technician's recommendation.

The *shaft* with the type bearings discussed in this section is called a "spindle." The shaft (or spindle) is part of a steering knuckle, which is replaced as a unit.

Repacking the Bearing

For a bearing to be completely lubricated, each roller or ball must be greased before installation on the vehicle. "Repacking" is usually the term used to indicate the procedure for prelubricating the bearing.

Special Tools

Bearing packer

There are several types of tools that can be used to repack bearings. Some operate by using compressed air while others are hand operated (Figure 7-10). The tools that have been around the longest and are sometimes the fastest and most accurate are, quite simply, the technician's hands. That is the procedure we will discuss here.

Before attempting to repack the bearing, ensure that it is clean and dry. Parts cleaner left on the bearing will break down the grease and lubrication will be reduced. Also ensure that the correct grease is being used. It should be multipurpose or a "G" classification. After repacking, make sure there is a clean strip or ribbon of grease extending upward and past the ends of each roller (Figure 7-11). Even with the best cleaning, a little dirty grease may sometimes be trapped within the bearing. The new grease will force it out.

Figure 7-9 The spindle or shaft must be as clean and as clear of damage as the bearings and races. (Courtesy of The Timken Company)

Figure 7-10 This is a hand-operated packer. The bearing fits under the black boot then the boot is forced downward. The grease is inserted into the packer under the red piston shown here at the bottom. (Courtesy of Snap-on Tools Company)

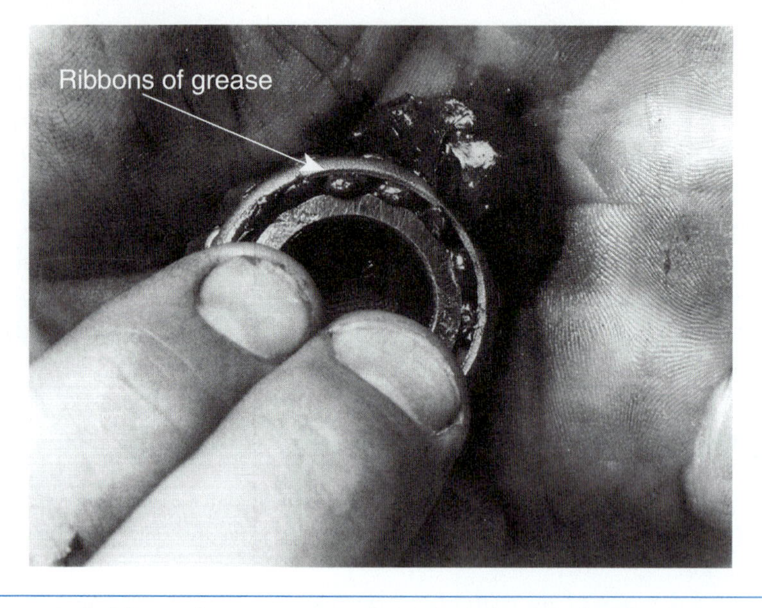

Figure 7-11 Until ribbons of grease appear above the rollers, the bearing is not properly packed. (Courtesy of The Timken Company)

A rough cloth will remove most of the *grease* from a technician's hands after the bearing has been repacked.

With the grease on hand and the bearing clean and serviceable, scoop some **grease** and place it in the palm of one hand. The amount of grease depends on the technician's hand of course, but it should be about the size of half a Mounds candy bar. Holding the bearing with the smaller cone (end) up, press a portion of the outer edge of the bearing down into the grease from the top of the mound (Figure 7-12). When the bearing edge touches the palm, press just slightly harder and drag that edge across the palm toward the wrist. It does not have to be dragged far. Repeat this step until clean grease ribbons are visible at the top end of the rollers. Rotate the bearing until the next rollers are aligned to be placed in the grease. Repeat the pressing, dragging, and rotating until all of the bearing rollers have clean grease between them. More grease may be needed to replenish what is in the palm. Before placing the bearing on a clean cloth or rag, spread a light coat of grease around the outside of the bearing. This will provide lubrication during the first several rotations of the bearing within its supported component. Repeat the procedures for any other roller bearing.

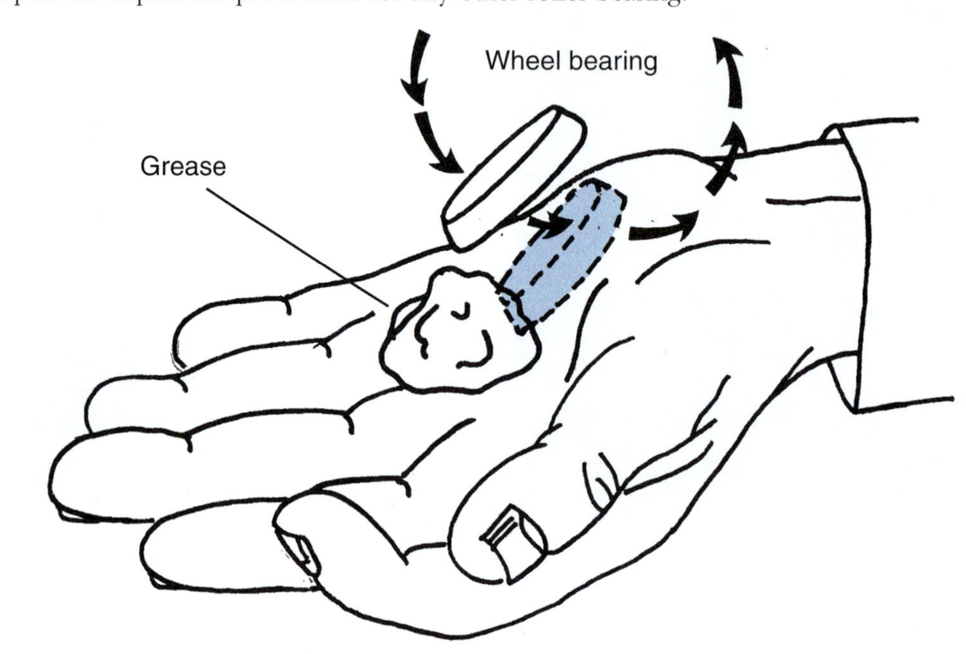

Figure 7-12 The bearing is moved down through the grease and rotated toward the heel of the palm.

Ball bearings can be repacked in a similar manner, but it may take a little longer. Most automotive ball bearing assemblies are not used in areas where they have to be repacked. Many ball bearing assemblies are pre-lubed and sealed at the factory. Some ball bearings are exposed to the lubricant within the component. However, new ball bearings used in these applications should be pre-lubed before installation. The best way is to drip the bearing in the same lubricant used in the component. A new bearing will run dry and be damaged if not pre-lubed in some manner. If the technician works on large vehicles, it is common practice to pre-lube the large bearings even if it is an old one being reinstalled.

Repacking a bearing with packing tools is fairly easy and much cleaner because each tool does what can be done by hand. The bearing is placed in a cone-shaped holder and another cone fitted over it (Figure 7-13). The grease is forced through the bearing rollers. Ensure clean ribbons of grease protrude past the rollers and some grease is applied to the outside by hand.

Removing and Installing a Race from a Hub

Classroom Manual
page 162

A race is removed from a hub or shaft with a hammer, punch, or a puller. On the hub we are working with in this chapter, the races can be removed with a hammer and pin punch. Set the hub on a solid surface like a metal workbench. If the race for the inner bearing is being removed, some type of **spacers** may have to be placed under the hub so the race has room to slide out. Two short pieces of 2 × 4 lumber work fine without damaging the hub. The punch will enter the hub from the opposite side of the race being removed.

It is best not to use a metal *spacer*. It would not absorb shock the way wood or other soft materials will.

CAUTION: Always use eye, face, and hand protection when using a hammer and punch. Injuries from flying particles or a missed aim with the hammer can result.

With the hub cleaned and in position, observe downward through the opening. A very small edge of the bottom race or two notches in the hub directly behind the race will be visible (Figure 7-14). In many cases, both are visible. The notches provide a place to rest the end of the punch against the race.

Tools are covered in Chapter 3, "Automotive Tools."

Special Tools

- 8-ounce ball peen hammer
- 6-inch punch
- Spacer material

Figure 7-13 This packer uses air pressure to force the grease into the bearing. (Courtesy of The Timken Company)

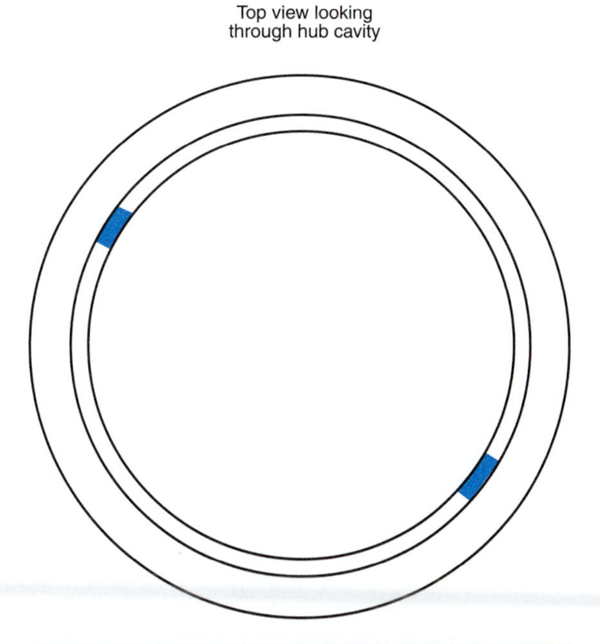

Figure 7-14 The backside (inner) of the race can be seen through the two notches at the lower edge of the hub's opening.

Set the punch in place and strike it sharply with the hammer (Figure 7-15). Move to the other notch and repeat. Do not strike the punch very hard unless absolutely necessary. Most races will pop out without too much trouble. It will be necessary to move the punch from notch to notch to keep the race fairly even as it is forced out. This is a simple task and there is no need to rush. Rushing usually causes a mistake with the hammer and results in a sore hand or finger. Repeat the procedure for the other bearing race.

Special Tools

8-ounce hammer

Bearing/seal driver

Installing the race is pretty much the opposite of the removal. A bearing/seal driver should always be used to install a race. The driver set comes with different sized drivers (Figure 7-16). Select a driver that fits snugly in the bearing side of the race and will slide inside the hub opening.

Set the race over the opening with the larger opening upward. Place the driver into the race and tap slightly with a hammer (Figure 7-17). The race and driver must be completely aligned with the opening. If the race is off just a little, it will not enter the opening. Once the race is started, harder strikes can be made with the hammer until the race bottoms out. This is noticeable because of the sound made when the race hits the bottom. The sound will be dull compared to the sounds made when the race is being driven in. Remove the tool and inspect the race for nicks or scratches. There should be no damage if the proper driver is used. Flip the hub over and install the other race.

Figure 7-15 A hammer and pin punch are used to drive the race from the hub. (Courtesy of Chicago Rawhide)

Figure 7-16 This driver set is used to seat seals and bearing races. (Courtesy of Snap-on Tools Company)

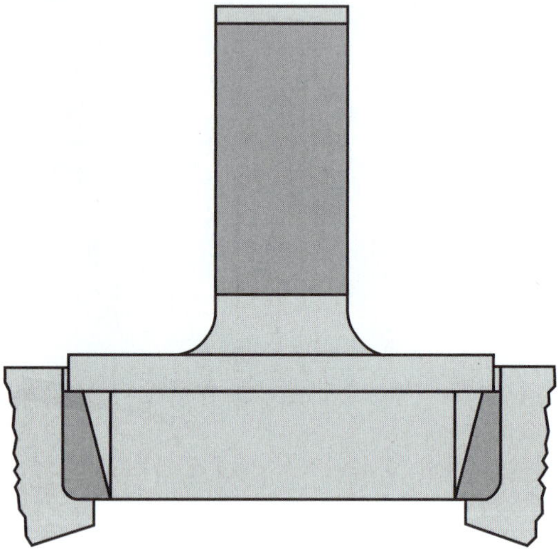

Figure 7-17 Note how this driver fits the bearing race. This prevents damage to the race. (Courtesy of Chicago Rawhide)

Installing a Wheel Bearing and Seal

Classroom Manual
page 174

With the race installation completed, the inner bearing and seal can be installed. Put some loose grease into the hub. It should fill the inner portion of the hub's opening but not overfill it (Figure 7-18). Excessive grease here will cause heat buildup which will melt the grease on the bearing, thereby causing damage to the bearings and seal. Place the hub with the inner bearing race facing up. This is normally the largest.

Set the greased inner bearing into its race. It should drop into the race far enough to be almost flush with the race. If it is not flush, it is probably in backward. The bearing should not block the seal.

With the bearing in place, align the wheel seal over the opening. Remember to place the seal's lip inward or toward the lubricant (Figure 7-19). The same driver used to install the race

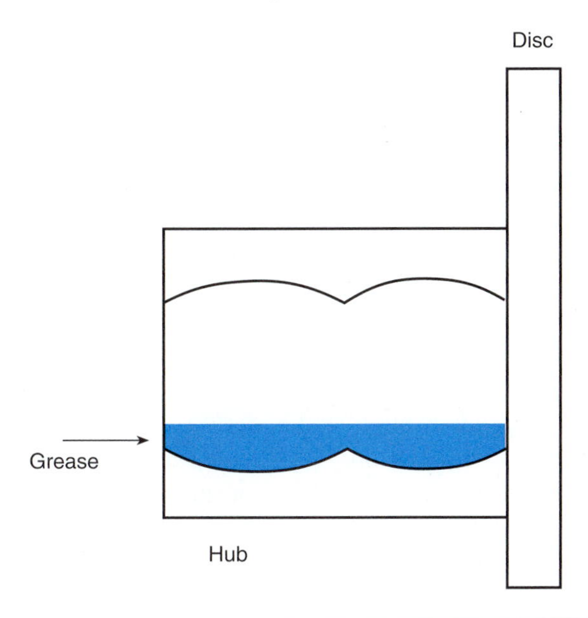

Figure 7-18 Only fill the center of the hub anywhere from a third to half full of grease. Too much could ruin the bearing.

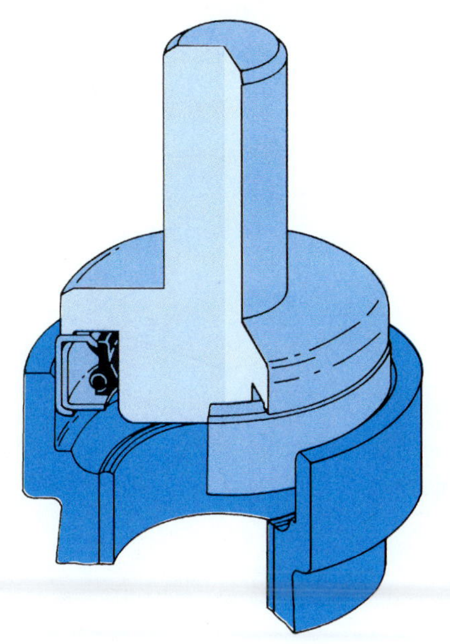

Figure 7-19 This tool is used to seat seals only. It will not work well for seating a bearing race. (Reproduced by permission of Deere & Company, © 1992, Deere & Company. All rights reserved.)

There are many ways used to adjust the wheel bearing tension. The only right way is to follow *service manual* instructions. The other ways rely on touch, feel, and guessing.

Special Tools

Torque wrench

can be used to install the seal, but normally a hammer is used. Hold the seal in place by hand and tap slightly around the edges of the seal until it is started evenly. Use the hammer to tap the seal the rest of the way into the cavity. If the seal has a flange that overlaps the hub, the sound will change as the flange mates with the hub. Many seals do not have a flange. They are driven in until the outer edge of the seal is even or flush with the outer edge of the hub.

Reinstall the hub and slide the outer bearing over the spindle and into the hub. Screw the nut on while turning the hub. This helps center the bearings on the spindle and in the hub. Consult the **service manual** for the torquing specification on the nut. Many times the nut is tightened to a torque and then backed off a certain amount. Over tightening may result in bearing failure because of excessive friction and heat. Under tightening will also damage the bearing and cause the wheel assembly to be loose on the spindle.

With the nut torqued, select a new cotter key and install it. Place some grease into the dust cup, about 1/3 full, and install it into the hub.

Classroom Manual
page 173

Installing a Chemical Sealant

This section is intended as a practice exercise. If cost is a factor, a tube of cheap toothpaste can be used instead of RTV. Find a scrap engine, transmission, differential cover, or a plain piece of flat metal. If a flat piece of metal is used, a marker can be used to track and simulate bolt holes. Any of the items can be cleaned after the practice is completed. The technician will have to use RTV in many applications.

Cutting the tube at an angle allows only a small portion of the tube to rest on the cover. This helps prevent spreading the RTV.

Remove the cover from the RTV tube and **cut it at an angle** until its opening is about 1/8 inch in diameter (Figure 7-20). Once the bead is started, do not break it. If it is necessary to move a hand, keep the tube end in place and begin to squeeze the tube again from that point. Also, when going by a bolt hole, lay the bead to the inside (lubricant side) of the hole (Figure 7-21). Do not run the bead over the hole.

Place the end of tube at a convenient starting point. The tube should be angled at 45 degrees to the cover (Figure 7-22). Squeeze from the bottom and continue squeezing as the tube is moved. The tube should be moved with an even spread. The RTV should come out in a smooth, even flow without humps or dips in the bead. Continue until the bead is laid completely around the cover. Finish the bead by rolling the tube tip into the starting point of the bead. This will mesh the two ends together and provides a leak resistant seal.

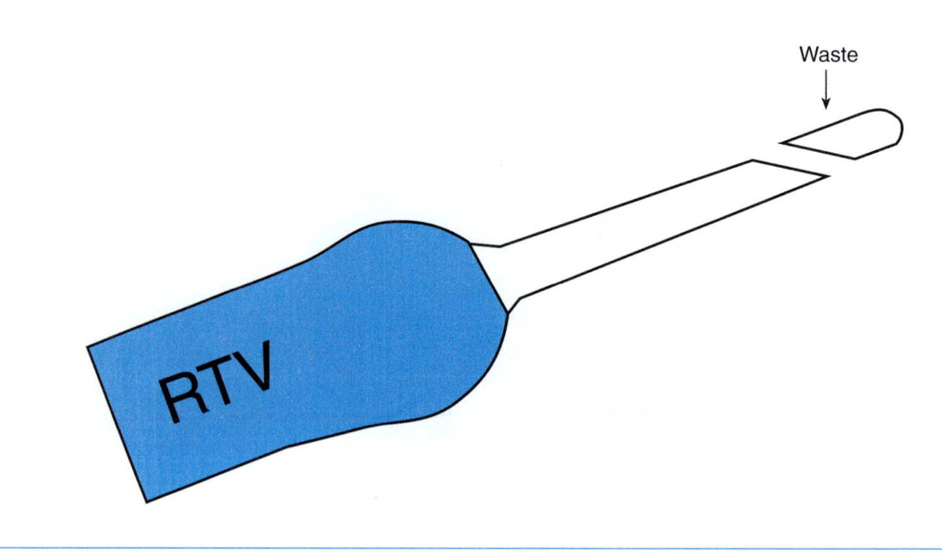

Figure 7-20 Cut the RTV tube at an angle.

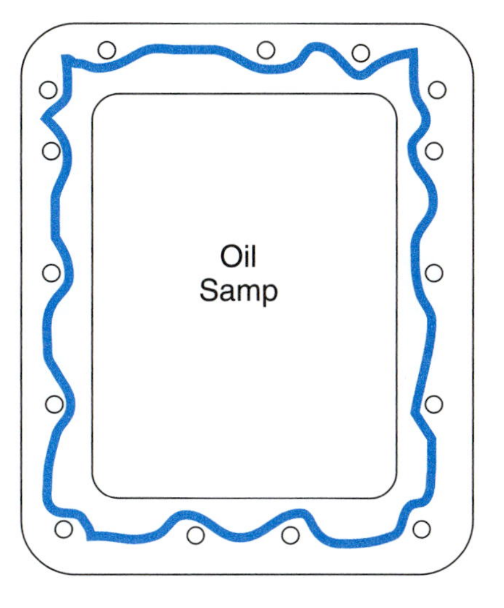

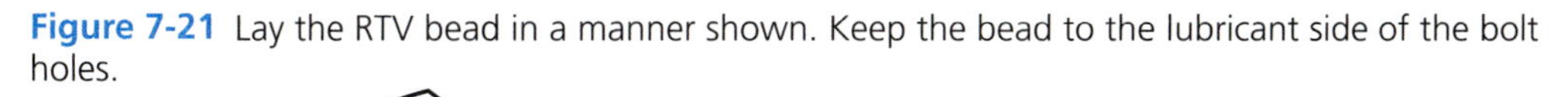

Figure 7-21 Lay the RTV bead in a manner shown. Keep the bead to the lubricant side of the bolt holes.

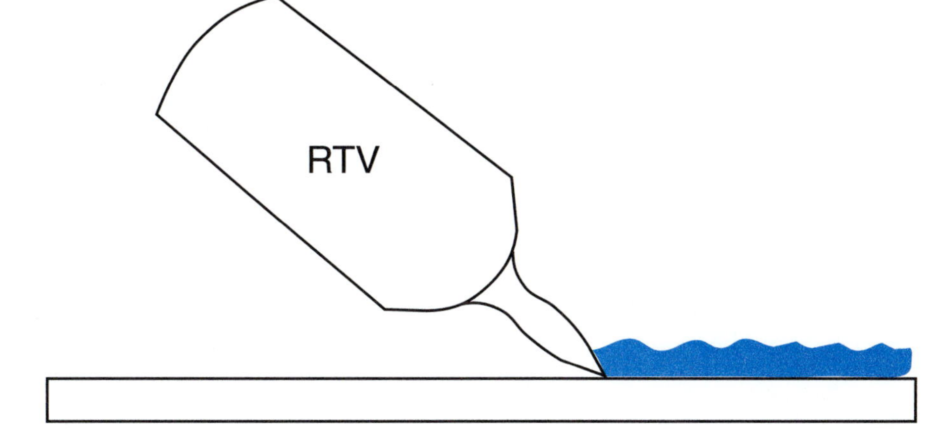

Figure 7-22 Keep the point of the tube to the metal at about a 45-degree angle. The angled cut will allow for a smooth bead.

CASE STUDY

A technician performed a routine job of replacing the differential fluid. He sealed the cover with RTV as recommended by the service manual. The customer returned some weeks later with a small but unsightly leak around the cover. The technician removed, cleaned, and installed the cover. RTV was used. Within a few days the customer returned with the same problem except the leak was worse. A careful inspection of the cover and housing revealed no visible problem. As a matter of curosity the technician screwed in some of the cover bolts. They bottomed out with about half of the bolt threads exposed. Digging into the holes with a pick revealed that every hole was filled about halfway full of RTV. The technician had laid the bead straight over the hole on the cover and the installed bolt pushed the RTV into the hole. The bolts were tightened with a quarter-inch-drive impact wrench so the technician did not have a feel for the bolt as it bottomed on the RTV. Cleaning the holes completely, installing RTV properly, and torquing the fasteners properly corrected the problem. Rushing the "simple" tasks on a routine job can result in a loss of profits and customers.

Terms to Know

Brinelling

Cotter key

Spalling

ASE Style Review Questions

1. Repacking a bearing is being discussed. *Technician A* says the grease should be an L classification. *Technician B* says the grease must extend through the rollers. Who is correct?
   A. A only
   B. B only
   C. Both A and B
   D. Neither A nor B

2. Laying a RTV bead is discussed. *Technician A* says RTV should cover the entire sealing surface. *Technician B* says to remove the tube from the work until the complete bead is laid. Who is correct?
   A. A only
   B. B only
   C. Both A and B
   D. Neither A nor B

3. Cleaning a bearing is being discussed. *Technician A* says compressed air at shop pressure should be used. *Technician B* says gloves should be used when drying the bearing. Who is correct?
   A. A only
   B. B only
   C. Both A and B
   D. Neither A nor B

4. Removing a wheel bearing is being discussed. *Technician A* says the seal must be removed before the outer bearing can be removed. *Technician B* says the nut or a screwdriver can be used to gain access to the outer bearing.
   A. A only
   B. B only
   C. Both A and B
   D. Neither A nor B

5. Installing a wheel bearing is being discussed. *Technician A* says the driving tool can be used to install the seal behind the inner bearing. *Technician B* says the outer bearing race may be installed with the same tool used to install the inner race. Who is correct?
   A. A only
   B. B only
   C. Both A and B
   D. Neither A nor B

6. Inspecting bearing components is being discussed. *Technician A* says the shaft is part of the bearing assembly. *Technician B* says the shaft should be inspected for the same problems as the bearing. Who is correct?
   A. A only
   B. B only
   C. Both A and B
   D. Neither A nor B

7. *Technician A* says spalling is when the bearing or race has dents in it. *Technician B* says brinelling causes pieces of the bearing to flake off. Who is correct?
   A. A only
   B. B only
   C. Both A and B
   D. Neither A nor B

8. *Technician A* says a special tool is used to remove the outer bearing. *Technician B* says a nut can be used to help remove the seal. Who is correct?
   A. A only
   B. B only
   C. Both A and B
   D. Neither A nor B

9. *Technician A* says the outer side of the bearing should receive a light coat of grease before it is installed in the hub.
Technician B says bearings that are lubricated by the component's lubricant should be pre-lubed.
Who is correct?
A. A only
B. B only
C. Both A and B
D. Neither A nor B

10. *Technician A* says the dust cap should be filled about three-quarters full of grease.
Technician B says the hub cavity should be full of grease before installing the bearings and seal. Who is correct?
A. A only
B. B only
C. Both A and B
D. Neither A nor B

Table 7-1 ASE Task

Inspect the camshaft bearing for unusual wear; remove and install camshaft bearings.

Problem Area	Symptoms	Possible Causes	Classroom Manual	Shop Manual
Engine performance	Oil consumption. Noisy engine.	**1.** Worn bearings.	162	170
		2. Damaged journals.	162	170
		3. Loose valve components.	165	170

SAFETY

Wear eye protection.

Table 7-2 ASE Task

Remove, clean, inspect, and repack wheel bearings, or repack wheel bearings and races.

Problem Area	Symptoms	Possible Causes	Classroom Manual	Shop Manual
Wheels	Vibration and loose steering.	**1.** Worn, damaged bearing or races.	162	161–167
		2. Bearing installed improperly.	162	169–170

SAFETY

Wear safety glasses.
Wear gloves.
Wear face protection.

Table 7-3 ASE Task

Inspect and replace external seals and gaskets.

Problem Area	Symptoms	Possible Causes	Classroom Manual	Shop Manual
Transmission, engine	Oil leaks.	**1.** Torn seal and gaskets.	170, 174	169–170
		2. Over-torque or under-torque gasket fasteners.	170	169–170

SAFETY

Wear safety glasses.

Table 7-4 ASE Task

Inspect and replace bushings.

Problem Area	Symptoms	Possible Causes	Classroom Manual	Shop Manual
Transmission	Oil leaks, noises.	**1.** Worn bushing.	169	—
		2. Worn shaft.	168	—
		3. Damaged or worn bearing.	168	—

SAFETY

Wear safety glasses.

Job Sheet 11

11

Name ______________________________ Date ______________

Removing, Repacking, and Installing Wheel Bearings

Upon completion and review of this job sheet, you should be able to remove, repack, and install a wheel bearing and seal.

Tools Needed

Seal remover
Bearing packer
8-ounce ball peen hammer
Seal driver
Hub with inner bearing and seal installed
Blocking materials
Catch basin

Procedures

■ **CAUTION:** Use eye and hand protection during parts cleaning. Injury or skin irritation may result if protective clothing is not worn.

1. Place the hub with the outside facing upward. If necessary, use blocks so clearance is available between the hub and bench.
2. Use the seal remover to force the seal from its cavity.
3. Remove the inner bearing.
4. Wash the bearing and hub cavity using a parts washers.

■ **CAUTION:** Do not allow the bearing to spin when using compressed air as a drying agent. Serious injury could result if the bearing comes apart.

5. Use reduced, pressure compressed air to dry the bearing. Use a catch basin to collect the cleaner and waste.
6. Place the bearing, small end down, into the bearing packer.
7. Force the grease into the bearing until clean ribbons of grease are visible at the top of the rollers.
8. Smear a light coat of grease on the outer sides of the rollers.
9. Install the bearing, small end first, into the cleaned and dried hub cavity.
10. Place the grease seal, lip first, over the cavity.
11. Hold the seal in place with the seal driver while using the hammer to drive the seal in place.
12. Drive the seal in until it is flushed with the edge of the hub.
13. Use some fingers to rotate the bearing within the hub. If it moves freely, the task is complete.

☑ **Instructor's Check** ______________________________

____

____

Engine Maintenance

Upon completion and review of this chapter, you should be able to:

- ❑ Perform an oil and filter change and fluid check.
- ❑ Change air and fuel filters.
- ❑ Test a battery.
- ❑ Inspect and replace secondary circuit components.
- ❑ Replace cooling system hoses.
- ❑ Inspect and replace drive belts.

Introduction

The technician's first jobs will probably be lubrication service and tires. While performing a routine oil change, the technician should observe the general condition of the vehicle and the engine. The customer may not realize that there is a potential problem. Driving the vehicle to the lift gives the technician a chance to get a feel for the brakes, steering, and engine operating condition. When a defect is suspected, notify the customer or the service writer.

Most entry-level technicians with sufficient supervision can perform the services covered in this chapter and the remainder of this book. Keep in mind that the shop manager or service writer will not assign more complicated jobs to a new technician until they feel comfortable with the technician's skill and knowledge.

SAFETY TIPS: The services performed in this chapter may require the technician to work on or with hot components or liquids. **Heat-resistant gloves** that cover the hands *and* forearms are available at local parts stores. Remember, heat absorbed by the outside of a material can burn the skin if touched. If the material is wet, it will have no resistance and the heat will travel straight through to the skin. Even when using protective clothing, the technician must be careful. Eye protection should always be worn. Always use the locks on a lift or use jack stands when working under the vehicle.

Heat-resistant gloves and forearm sleeves are usually made of woven kelvar. If the kelvar becomes wet, the heat will be absorbed by the liquid and transferred to the skin.

Changing the Engine Oil and Filter and Checking Fluid Levels

Classroom Manual
pages 207–209

Special Tools

Lift

Drain pan

Filter wrenches

Almost every technician starts his or her first job as an **oil changer** and will continue to perform this basic service throughout his or her career. If any task is considered to be a basic and very necessary vehicle service, it is the scheduled engine oil change.

Before starting the service, get the oil and filter. Usually the **type of oil** is determined by the engine specifications, but the customer may have a preference for a brand or type (Figure 8-1). Check the repair order for customer preference or any additional tasks the customer may have requested. The customer may have heard a noise or other symptom that she or he would like to have checked.

Move the vehicle to the bay. As the vehicle is operated, note the performance of the engine, brakes, and other systems. This short drive may point to something that requires a closer inspection. Raise the hood and make a quick inspection for the oil filter. Some engines have the filter mounted on the front of the engine and it is removed from under the hood instead of under the vehicle (Figure 8-2). Lift the vehicle to a comfortable working height.

Almost every shop offers *oil changes* and basic services.

The brand and *type of oil* are strictly a customer's choice. However, the technician may advise the customer on the type of oil required on a engine.

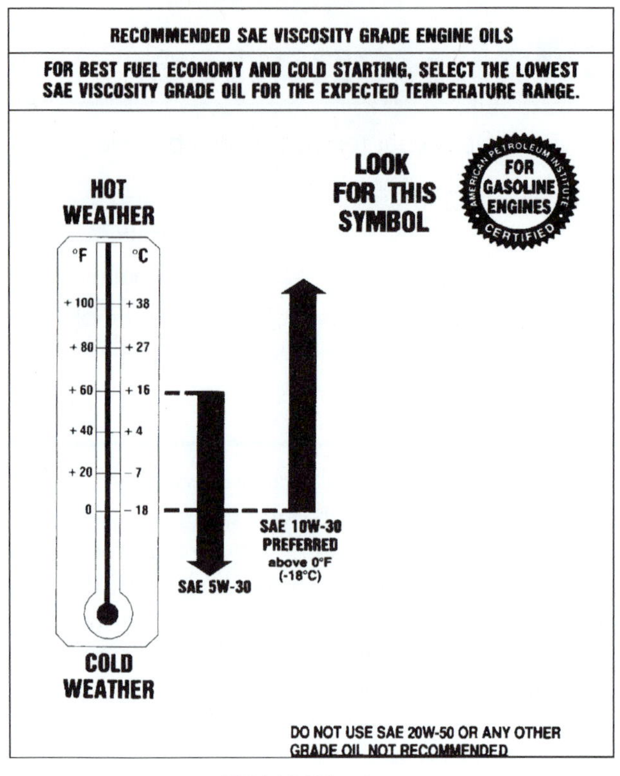

3800 V6 Engine

Figure 8-1 Oils are formulated to operate in different temperature zones. (Courtesy of Chevrolet Motor Division, General Motors Corporation)

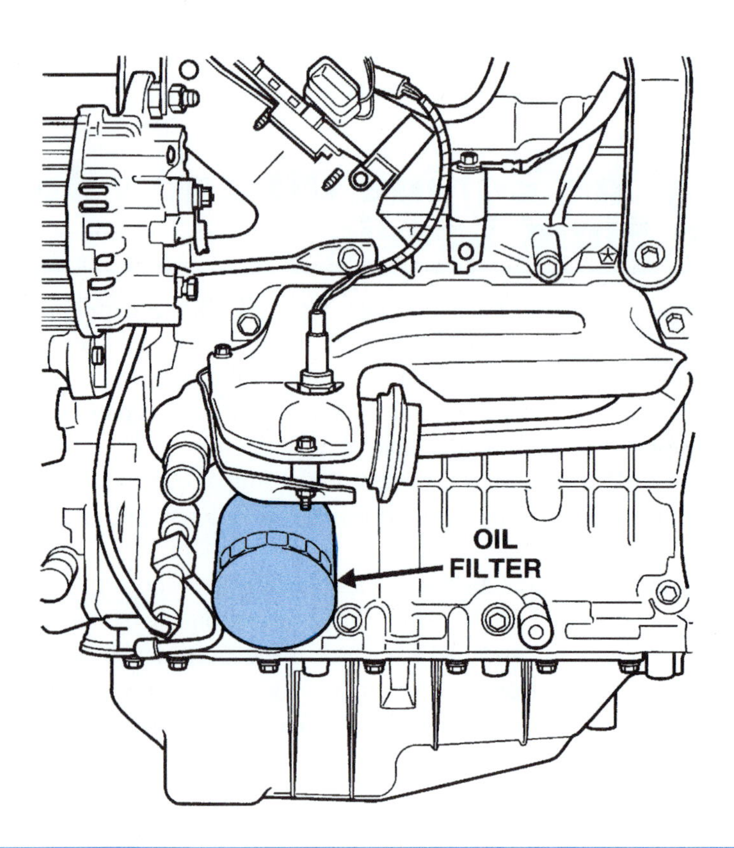

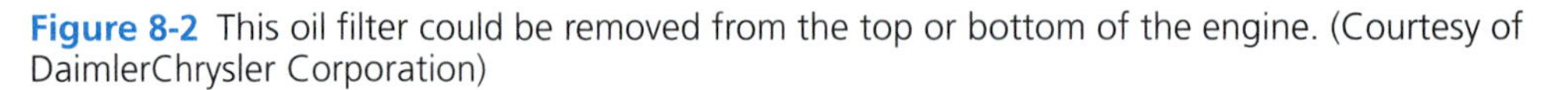

Figure 8-2 This oil filter could be removed from the top or bottom of the engine. (Courtesy of DaimlerChrysler Corporation)

Figure 8-3 The left drain pan has a hand pump to empty the container when it is full. (Courtesy of Snap-on Tools Company)

CAUTION: Use hand and forearm protection when changing oil on a hot engine. Burns could result from the oil, filter, and engine components.

WARNING: Do not remove the wrong drain plug. A few vehicles have a drain plug for the automatic transaxle. Accidentally draining the wrong fluid will cost the shop some money, but if the mistake is not caught before releasing the vehicle to the customer, damage to the engine and transaxle could occur.

CAUTION: Always capture the used oil from the engine. Clean any spills immediately. Failure to do so may result in an injury to the technician or charges by the EPA for illegal waste disposal.

INTERNET TIP: For proper disposal and control procedures for repair shop waste, information is available at the web site for the Coordinating Committee for Automotive Repair. The web site is http://www.greenlink.com.

In many cases, the vehicle is hot and the customer is waiting, so speed is essential. Do not, however, forget basic safety procedures. Use heat-resistance gloves and eye protection. With the vehicle raised, move the **drain pan** under the oil drain plug (Figure 8-3). Use the correct **filter wrench** to loosen the plug. A typical drain plug is not made of the strongest metal and can be rounded with a wrong wrench. A boxed-in wrench is suggested.

While the oil is draining, make a quick inspection of the **undercarriage** of the vehicle. Look for any leaks or bent/damaged components. On some vehicles, it is possible to see part of the brake disc and pads. Inspect the tire threads for wear. The thread should show even wear across the tire.

Check the **drain plug** for a gasket. Many plugs have a brass or plastic washer to help seal the drain hole (Figure 8-4). It is usually best to replace the gasket if one is used. Once the

A *drain pan* is usually mounted on a moveable mount.

Filter wrenches are specifically designed to remove oil filters.

The *undercarriage* of the vehicle includes everything that is mounted or visible from under the vehicle from bumper to bumper.

The *drain plug* is mounted in the lower portion of the oil pan and is used exclusively to drain the oil from the engine or another assembly.

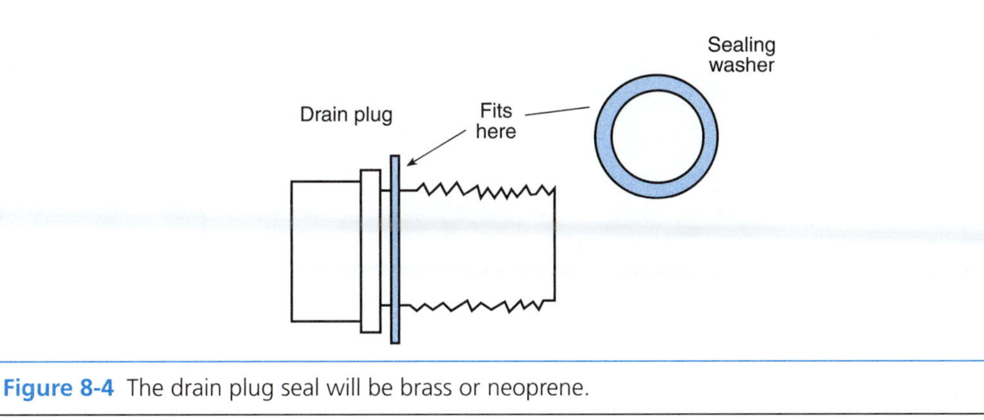

Figure 8-4 The drain plug seal will be brass or neoprene.

Photo Sequence 7
Typical Procedure for Changing the Oil and Oil Filter

P7-1 Always make sure the car is positioned safely on a lift before working under it.

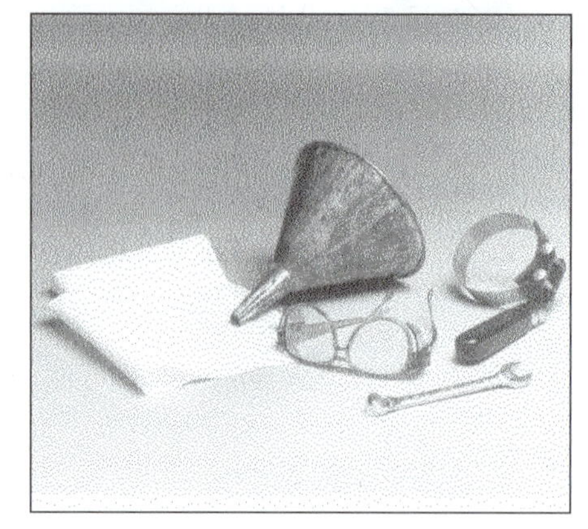

P7-2 The tools and other items needed to change the engine's oil and oil filter are rags, a funnel, an oil filter wrench, safety glasses, drain pan, and a wrench for the drain plug.

P7-3 Place the oil drain pan under the drain plug before beginning to drain the oil.

P7-4 Loosen the drain plug with the appropriate wrench. After the drain plug is loosened, quickly remove it so that the oil can freely drain from the pan.

P7-5 Make sure the drain pan is positioned so that it can catch all of the oil.

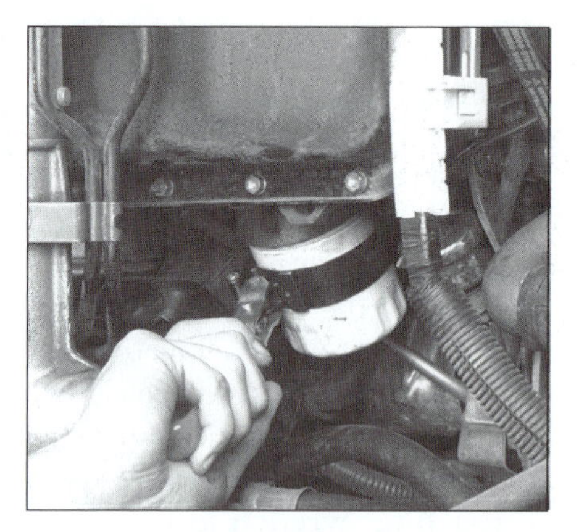

P7-6 While the oil is draining, use an oil filter wrench to loosen and remove the oil filter.

P7-7 Place the old filter in the drain pan so it can drain. Then discard the filter according to local regulations.

P7-8 After wiping the oil filter sealing area on the engine, apply a coat of clean engine oil on the new filter's seal.

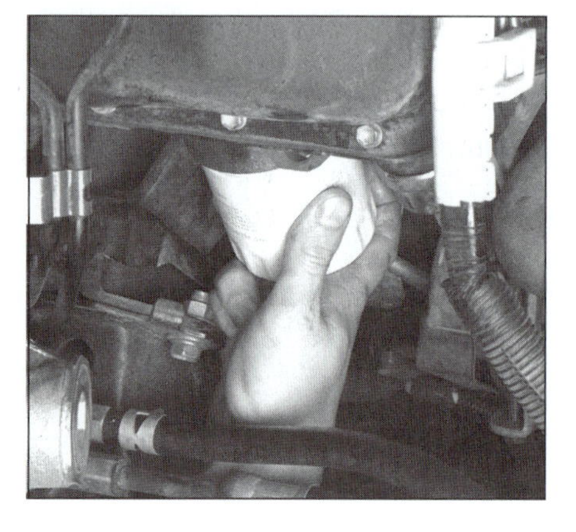

P7-9 Hand-tighten the filter to the block. Oil filters should be tightened according to the directions given on the filter.

Typical Procedure for Changing the Oil and Oil Filter (continued)

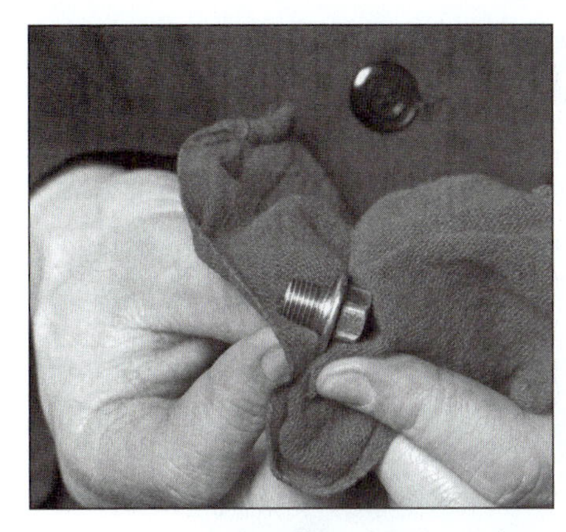

P7-10 Prior to reinstalling the drain plug, wipe off its threads and sealing surface with a clean rag.

P7-11 The drain plug should be tightened according to the manufacturer's recommendations. Overtightening can cause thread damage, and undertightening may result in oil leakage.

P7-12 With the oil filter and drain plug installed, lower the car to the ground and remove the oil filter cap.

P7-13 Carefully pour the oil into the engine. The use of a funnel usually keeps oil from spilling on the engine.

P7-14 After the recommended amount of oil has been put in the engine, check the oil level.

P7-15 Start the engine and allow it to reach operating temperature. While the engine is running, check the engine for leaks. Extra attention should be given to the seal of the oil filter. If there is evidence of a leak, turn off the engine and correct the problem.

P7-16 After the engine has been turned off, recheck the oil level and correct it if necessary.

Figure 8-5 A few different types of oil filter wrenches. (Courtesy of Snap-on Tools Company)

oil has completed draining, install and torque the drain plug. Do not over tighten the plug. The drain hole goes through thin metal and the threads can be stripped out.

Locate the filter and determine the best type of wrench to be used. There are several types of wrenches available and more than one may work (Figure 8-5). The location of the filter and the components around it will determine which wrench will be the easiest to use. Position the drain pan directly under the filter and loosen it. At this point, the technician may have to work around a very hot exhaust and other hot components. Many filters are placed so the draining oil will flow down over engine and vehicle components. Be alert with the drain pan and watch for splashing oil. Work carefully and safely.

While the filter is draining, secure the new filter and a container of oil. Open the oil and use a finger to smear a light coat of oil on the filter's gasket (Figure 8-6). This will help remove it during the next service.

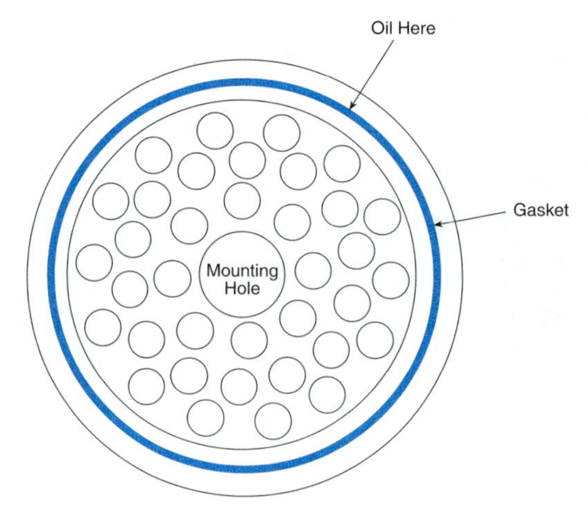

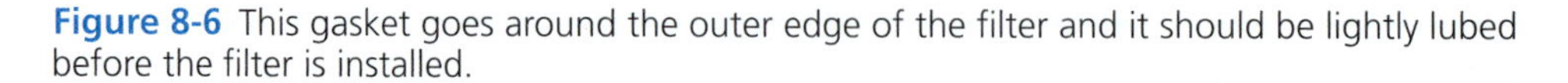

Figure 8-6 This gasket goes around the outer edge of the filter and it should be lightly lubed before the filter is installed.

Finish removing the old filter and inspect the end that fits against the mount. There should be a **gasket**. If not, the gasket is stuck to the mount and must be removed. Wipe the mount with a clean rag and install the new filter. Hand tighten the filter to the specifications shown on the filter. In almost all cases, the filter is hand tightened about one-half of a turn after the seal contacts the mount. Do not use a wrench to tighten the filter or it will be very difficult to remove during the next service.

The filter *gasket* is usually an O-ring.

With the plug and filter in place, check the area for tools and remove the drain pan from under the vehicle. Lower the vehicle to the floor.

There are different methods used to install oil into an engine. Some shops have overhead drops that are plumbed to large oil containers in a storeroom (Figure 8-7). Follow the equipment instructions and shop policy to select the amount and type of oil. Many shops use quart containers and the technician must pour each quart individually into a filler hole located in a valve cover.

Use a clean funnel to prevent oil from dripping on the engine. In the few situations when a funnel is not available, oil can be poured directly from the container. Most quart containers are shaped to reduce the amount of spillage. Notice where the pouring **spout** is attached to the container body (Figure 8-8). It is usually to one side at the top. Hold the container so the pouring spout is the highest point, placing the oil below the spout. The oil can be poured slowly and air can enter the container as the oil exits (Figure 8-9). Turning the container so the spout is down will cause a burping of the oil as air is drawn into the emptying container. Install the amount of oil specified and then replace the filler cap.

The offset position of the container's *spout* allows air to enter the container. It is suggested that the container not be squeezed to speed the process. This will block the airflow.

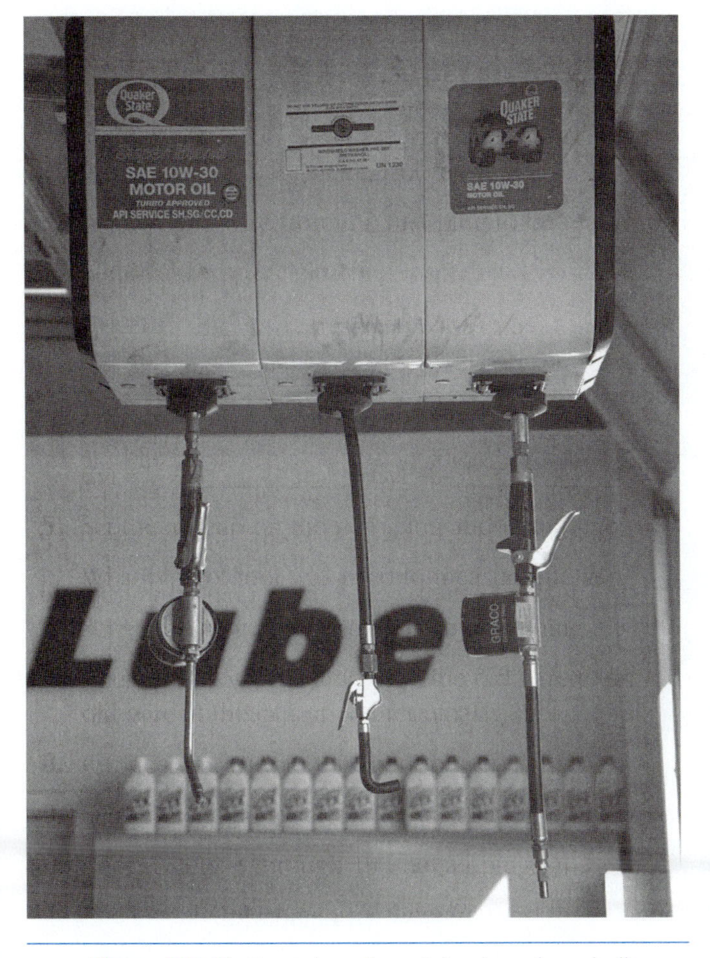

Figure 8-7 Fleet repair and specialty shops have bulk containers and the oil or grease is pumped to each bay as needed.

Figure 8-8 The spout of a quart container of engine oil is offset to facilitate pouring.

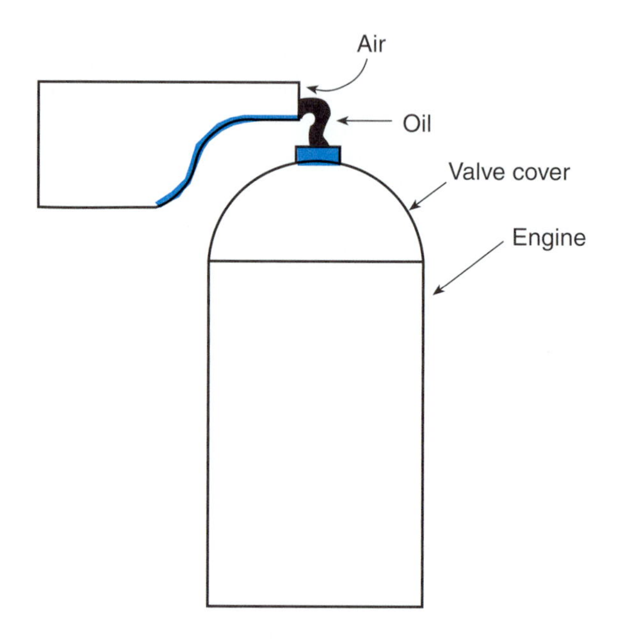

Figure 8-9 Keep the spout at the top as the oil is added to the engine.

The *windshield washer* reservoir is usually translucent and mounted along one of the fenders. However, on some vehicles, the only portion visible is the filler cap and the filler tube.

Conduct a check of the fluid levels available under the hood. The most common are the brake master cylinder, **windshield washer**, coolant, and transaxle or transmission (Figure 8-10). The transaxle and transmission must be checked with the engine running, but the fluid can be checked for burning or age. This fluid should be reddish. If it is burned or dark in color, a service recommendation may be made to the customer.

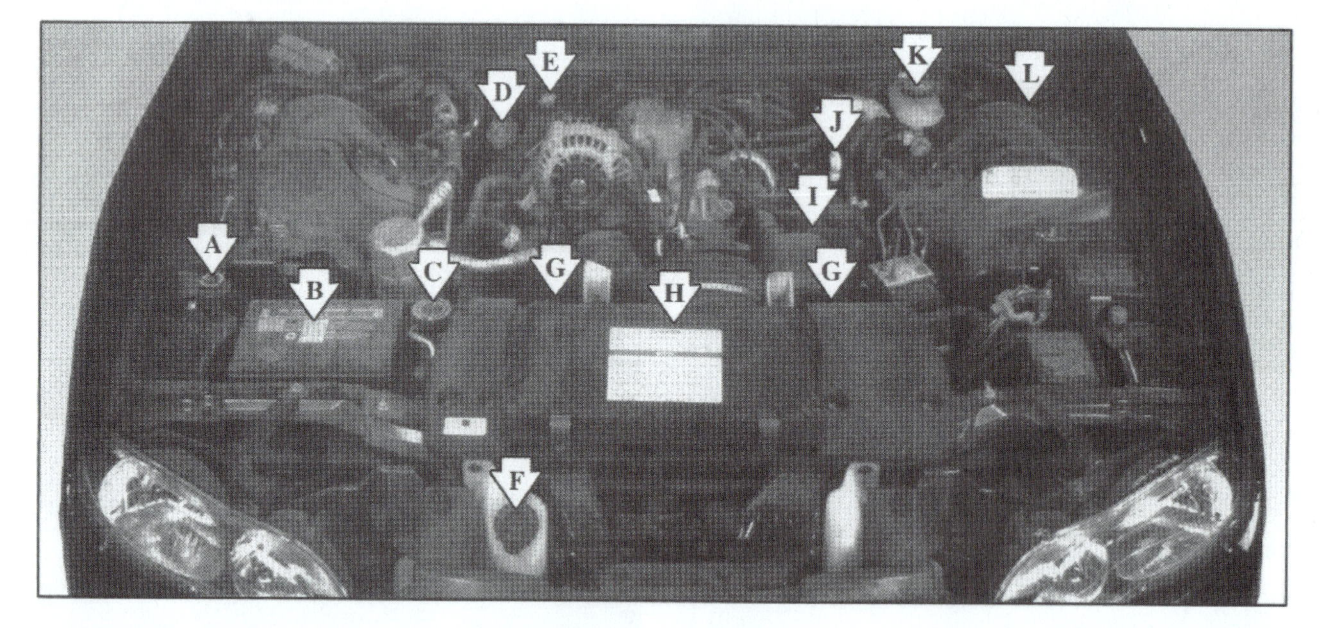

A. Engine Coolant Reservoir
B. Battery
C. Radiator Fill Cap
D. Engine Oil Fill Cap
E. Automatic Transmission Dipstick (If Equipped)
F. Windshield Washer Reservoir
G. Engine Cooling Fans
H. Air Cleaner
I. Power Steering Reservoir
J. Engine Oil Dipstick
K. Brake Fluid Reservoir
L. Clutch Fluid Reservoir (If Equipped)

Figure 8-10 In most shops, an oil change includes a check of the fluid level at each location shown here. (Courtesy of Chevrolet Motor Division, General Motors Corporation)

CAUTION: Do not open a hot coolant system. Serious burn injuries could result.

The **coolant** may be hot, so do not open the radiator. Check the coolant level in the overflow reservoir. It should be between the hot (high) and cold (low) marks (Figure 8-11). Removing a cap or observing the level through the translucent reservoirs provides a means to check the master cylinder and windshield washer fluids. A low brake fluid level may be correct based on the amount of brake wear. As the brakes wear, the fluid level will drop. The technician or service writer may consult with the customer if the brake fluid is low.

The *coolant* reservoir is similar to the one for the windshield washer. The filler cap will be labeled according to purpose of the container.

Some shops will **top off** some fluids without charge to the customer. Generally, antifreeze and brake fluids are charged. With all of the fluids checked, start the engine and observe the oil warning light. It should go out within a few seconds after the engine is running. If it does not, shut the engine down immediately and correct the problem. When the light is out, shut off the engine and recheck the engine oil level. Some of the oil is used to fill the filter and the level may drop slightly. The level, with the filter full, should be at the full mark or slightly below the mark. Top off the oil as needed. Check under the vehicle for leaks. When the oil level is correct and no leaks are apparent, clean the vehicle of any oil drips. Complete the repair order and return the vehicle to the customer. Properly dispose of waste oil and the old filter.

Topped off is a term used to indicate that fluid should be added to a system to bring it to the correct level.

Changing the Air and Fuel Filters

Classroom Manual
pages 210–214

Air Filter

The air filter is usually checked during an oil change service. However, it should be checked any time the customer complains of poor engine operation. The filter will be located in a housing on top of the engine or in an area near a fender behind a headlight. The housing cover may be bolted down or retained with spring clips (Figure 8-12).

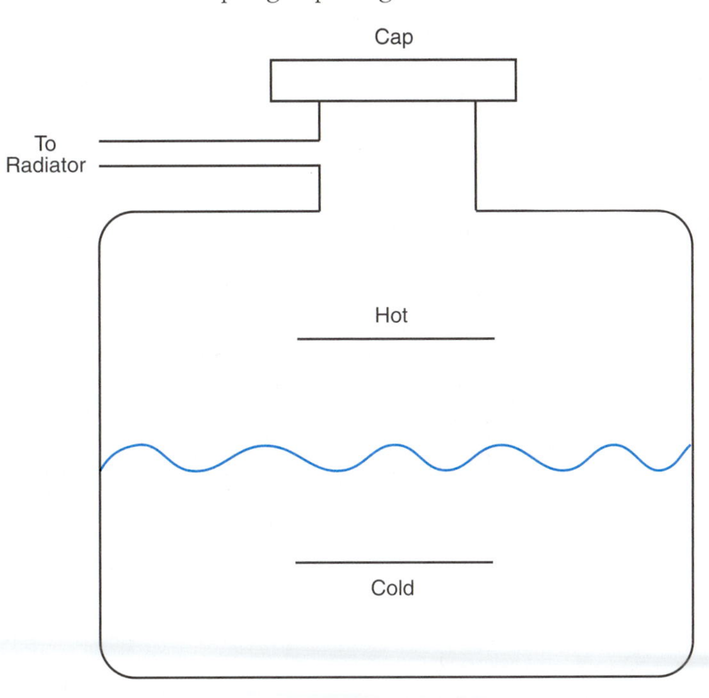

Figure 8-11 The coolant should be visible through the reservoir and between the hot and cold mark on a warm engine.

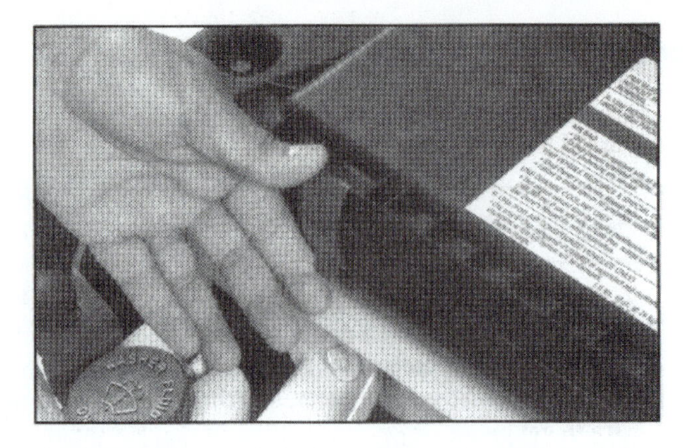

Figure 8-12 The air filter cover is retained by spring clips. Note the windshield washer reservoir cap to the lower left. (Courtesy of Chevrolet Motor Division, General Motors Corporation)

SERVICE TIP: Many of the vacuum hoses are made from hard plastic and break easily. Use both hands, one to pull the hose and one to hold the component, to pull the hose straight from the component. Do not bend or flex the hose. Use a small, pocketsize, flat-tipped screwdriver to unlock the electrical connection.

Select the correct replacement filter and open the housing. There may be electrical or vacuum connections that must be disconnected before the cover can be removed (Figure 8-13). Be sure that neither is damaged during removal. Remove the filter and inspect the housing for dirt, leaves, and oil. Dirt and leaves on the inlet side of the housing is normal. Clean the housing. If oil is present, this may indicate a problem with the PCV system. Further inspection of the PCV system should be accomplished.

In some vehicles, a separate air filter for the PCV is located in the engine air filter housing (Figure 8-14). This filter should also be replaced at this time. Install the new engine air filter and cover. If bolts are used, do not over tighten them. On the round units mounted directly to the top of the engine, the tightening of the bolts can be critical. In many cases, the cover on this type of housing applies force to the air filter. Over tightening can depress the filter and restrict airflow to the engine.

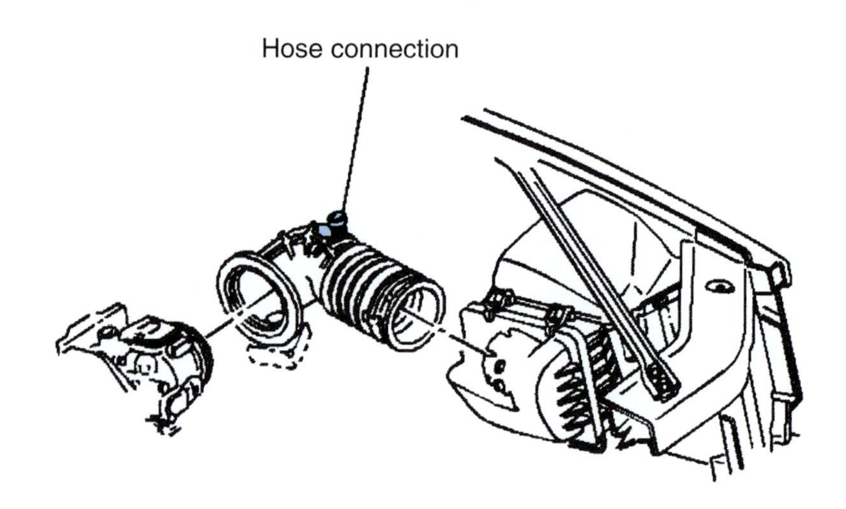

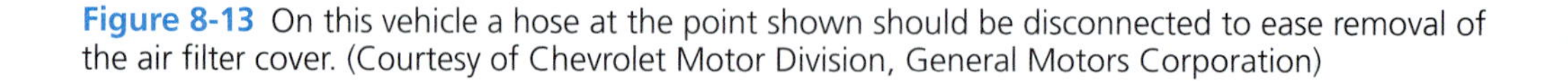

Figure 8-13 On this vehicle a hose at the point shown should be disconnected to ease removal of the air filter cover. (Courtesy of Chevrolet Motor Division, General Motors Corporation)

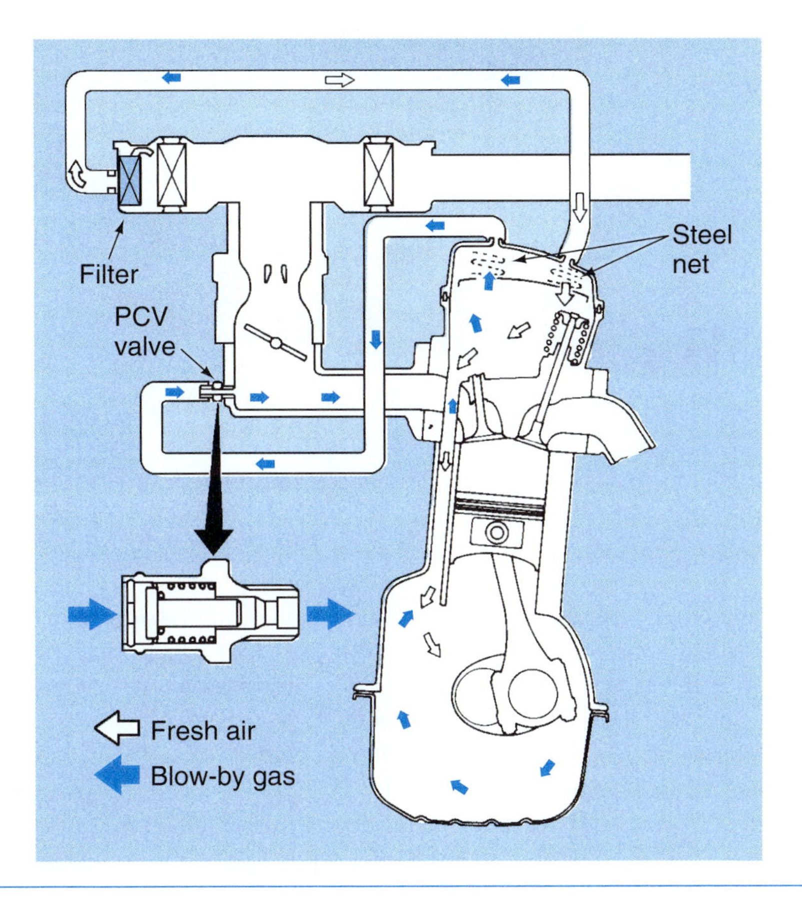

Figure 8-14 The small filter for the PCV system may be known a "crankcase breather filter." (Courtesy of Nissan North America)

Fuel Filters

CAUTION: Do not allow any hot components or flame-producing devices near the vehicle when the fuel system is being serviced. Dangerous fires could result.

Special Tools

- Line wrenches
- Quick disconnect tools
- Catch pan for fuel
- Hose clamping tools

Changing the fuel filter offers some safety and mechanical problems. Opening the line to remove the filter will allow some fuel to escape. A suitable catch container should be used. The oil drain pan should *not* be used, but the waste fuel must be stored in an appropriate container. Consult with a supervisor for instructions. The vehicle should be allowed to sit until the exhaust system is cool. It is preferable that the entire vehicle has cooled down.

The technical end involves relieving the fuel pressure on EFI systems and using tools to disconnect and close the lines. Consult the service manual to determine how the fuel pressure is relieved (Figure 8-15). It usually involves disabling the fuel pump and starting the engine. When the engine stalls, the fuel pressure has been removed but there will still be some fuel in the lines.

Figure 8-15 The filter should be twisted on the fuel line to loosen it before pulling them apart. (Courtesy of Chevrolet Motor Division, General Motors Corporation)

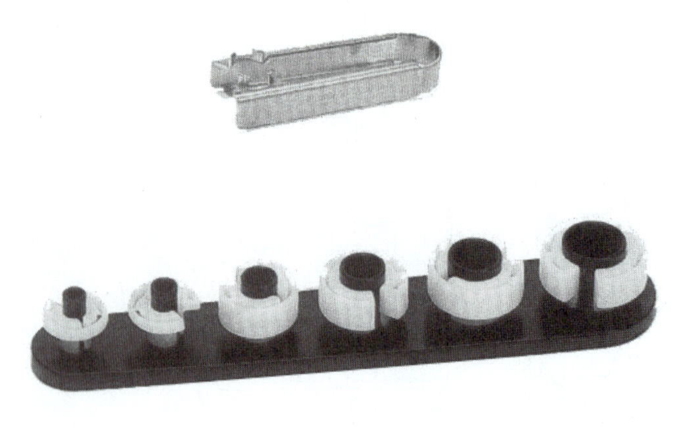

Figure 8-16 There are many types of quick disconnect tools available. (Courtesy of Snap-on Tools Company)

Some fuel filters are connected to the fuel lines with compression fittings requiring line wrenches for loosening. Others use quick-release fittings that can be unlocked with special tools or a small, flat-tipped screwdriver (Figure 8-16). Almost all filters have a small mounting bracket that is removed or loosened with a socket or screwdriver.

Before starting work, select the correct filter and check it for the bracket and replacement quick-release locks if used (Figure 8-17). Some types of quick-release locks are damaged during removal. Most EFI systems have fuel filters mounted on their frame rails and use quick-release connections used for this discussion. Other types follow the same procedures except for filter location and type of connection.

Follow the instructions to relieve the fuel pressure. Do not enable the fuel pump until the filter is replaced. Raise the vehicle to a good working height. Locate the filter and determine the tools needed. If the filter is connected to flexible hoses, the hoses can be clamped shut with special tools.

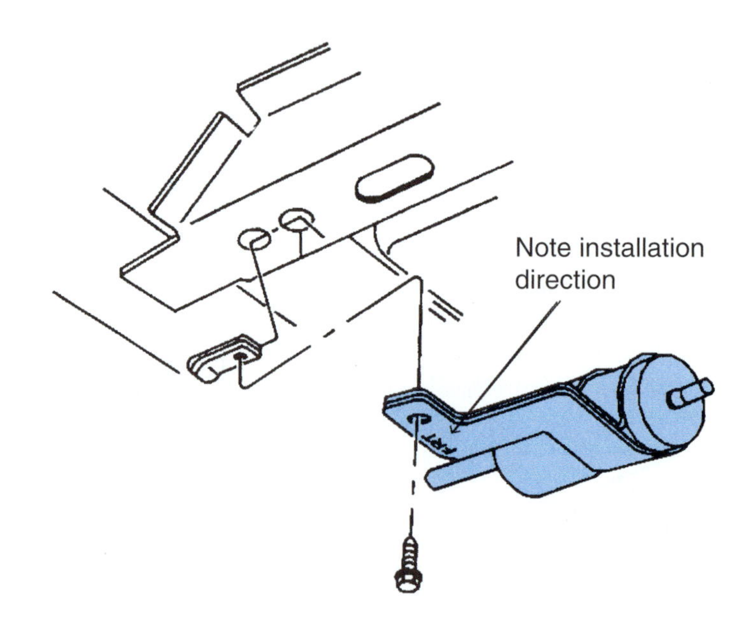

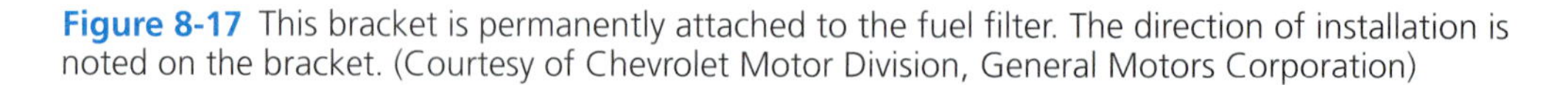

Figure 8-17 This bracket is permanently attached to the fuel filter. The direction of installation is noted on the bracket. (Courtesy of Chevrolet Motor Division, General Motors Corporation)

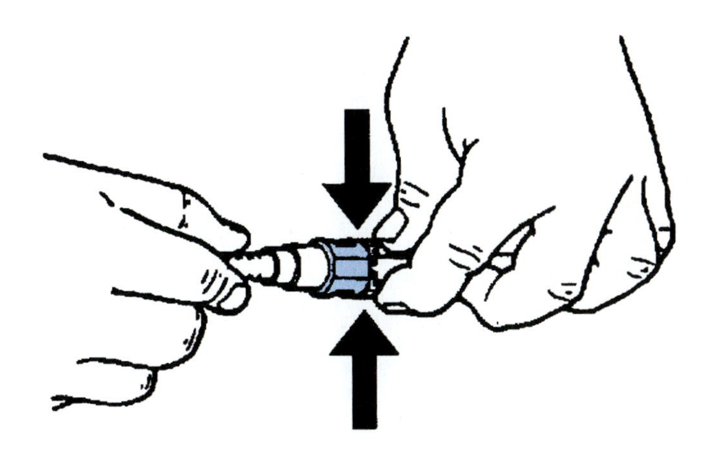

Figure 8-18 This type of quick release can be released by hand. It must be replaced when the filter is replaced. (Courtesy of Chevrolet Motor Division, General Motors Corporation)

Clamp the hoses if possible and position the pan securely as two hands will be required to remove the filter (Figure 8-18). Remove or loosen the bracket. Loosen the fittings by prying the locks open with the tool. Twist the filter once or twice to loosen it on the lines. **Fuel** should be allowed to drain at this point. Allow as much as possible to drain. Grasp a fuel line to hold it in place as the filter is pulled from the line. The locks will have to be held open while separating the filter and hose. Allow the fuel to drain before separating the filter from the other line. There will still be fuel in the filter so handle accordingly.

If rags are used to catch some of the *fuel*, ensure that they are stored separately from other rags.

Ensure that the filter is mounted in the correct direction. The filter should be marked "inlet" and "outlet" or may only have one end labeled (Figure 8-19). The label may be just an arrow showing flow. The filter is installed with the outlet end toward the engine. Install one end of the filter into the fuel line and be sure the lock slips in place. The short pipe at each end of the filter has a flange or collar. The lock must locked between the collar and filter body. The connection should be checked by attempting to pull the filter from the line. The filter may move a short distance. Repeat the procedures with the other end. Install and tighten the bracket. Remove any hose clamps installed.

Lower the vehicle just enough to gain access to the door. Do not start the engine at this time. Enable the fuel pump and turn the ignition key from OFF to RUN two or three times.

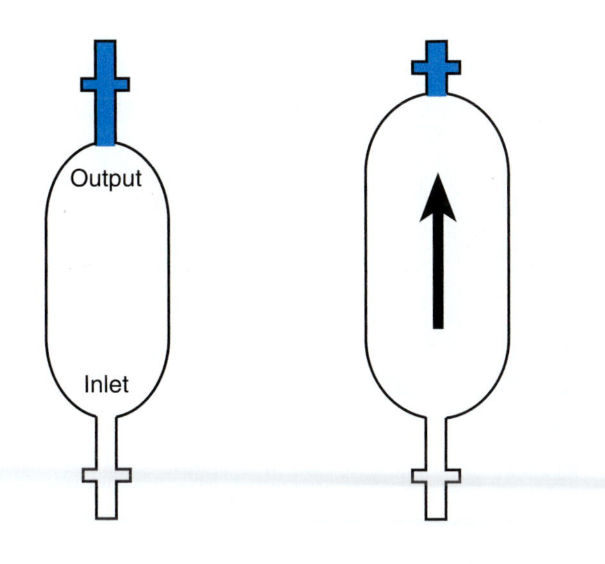

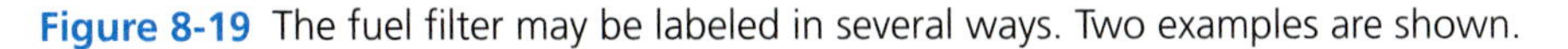

Figure 8-19 The fuel filter may be labeled in several ways. Two examples are shown.

This will pressurize the system. Check the fuel filter connections for leaks. If dry, start the engine and allow it to run long enough to settle down and operate smoothly. Shut off the engine and check the filter for leaks. This procedure is easier if a second person is available to check for leaks or operate the running engine while the filter area is checked. With the service complete, check for tools and lower the vehicle.

Drain all of the fuel from the old filter. Properly dispose of the filter and waste fuel. Complete the repair order and return the vehicle to the customer.

Inspecting, Servicing, and Testing the Battery

Classroom Manual
pages 179–182

Special Tools

Voltmeter or multimeter

VAT 40/60 or similar battery tester

WARNING: Do not load test the battery while it is connected to a computer-controlled alternator. Damage to the electronic circuits could result.

CAUTION: Prevent any hot component or flame from being near a discharging or discharged battery. A discharging battery produces hydrogen gas, which is highly flammable. An explosion could result.

A **voltage drop test** and cleaning can be done with the battery installed (Figure 8-20). The battery could be **load tested** while being installed and connected on older vehicles. Newer vehicles use computer-controlled alternators and the testing procedures are best done with the battery disconnected. A load test is used to determine the **capacity** of the battery. It is usually performed when the customer complains that the starter system is inoperative or the battery fails to charge. A spare or used battery could be used strictly for load-testing practice if desired.

A *voltage drop test* determines how much voltage is used between the points where the meter's leads are attached to the circuit.

A *load test* is the process of applying an electrical load to a battery to determine the electrical *capacity* of a battery.

Before beginning the test, determine if the installed battery is powerful enough for the vehicle and its electrical load (Figure 8-21). A fully loaded vehicle may function for a while on a battery that is too small, but the battery's life will be severely shortened. A service manual or battery guide will provide the battery capacity for each vehicle model. The capacity will be shown as ampere-hour or **cold cranking amps (CCA).** In most cases, the CCA will be the easiest thing to locate since it is identified by large letters on the label. Most vendors also advertise and sell their batteries by CCA.

Cold cranking ampere (CCA) is the amount of amperage available from a battery for 30 seconds at a temperature of 0 degrees F.

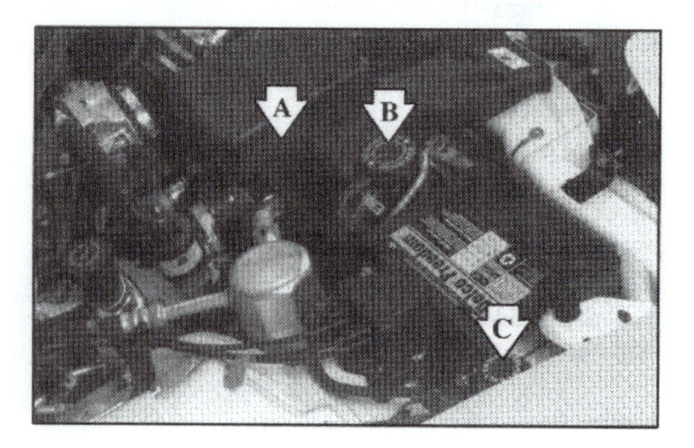

Figure 8-20 This battery could be load tested in the vehicle *if* it is disconnected before the test is started. (Courtesy of Chevrolet Motor Division, General Motors Corporation)

Service Data and Specifications (SDS) for a Battery

Applied area		USA	Canada
Type		GR21	GR24
Capacity	V-AH	12-49	12-63
Cold cranking current (For reference value)	A	490	550

Figure 8-21 Note the higher CCA for Canada. Canada has a lower average temperature than most of the United States and a stronger battery is required. (Courtesy of Nissan North America)

Begin the service by inspecting the battery. The top should be clean with no corrosion on the battery, terminals, posts, or cables. Before cleaning or load testing, the battery should have a voltage test and a voltage drop test. These tests may locate the problem without the load test. With a corroded battery or connections, it is expected that there will be a lost of voltage indicated by the tests. If voltage is lost or cannot leave the battery, voltage cannot enter the battery for charging.

Connect the red lead of a voltmeter to the positive post and the black lead to the negative post (Figure 8-22). *Do not connect to the cable terminals for this test.* The voltage should exceed 12 VDC. If it does not, tests on the charging system should be done by an experienced technician.

Assuming that the battery voltage exceeded 12 VDC, connect the red lead of a voltmeter to the positive post on the battery and touch the black lead to the top of the battery case (Figure 8-23). If there is a reading above 0.2 VDC, the battery is losing too much voltage across the case.

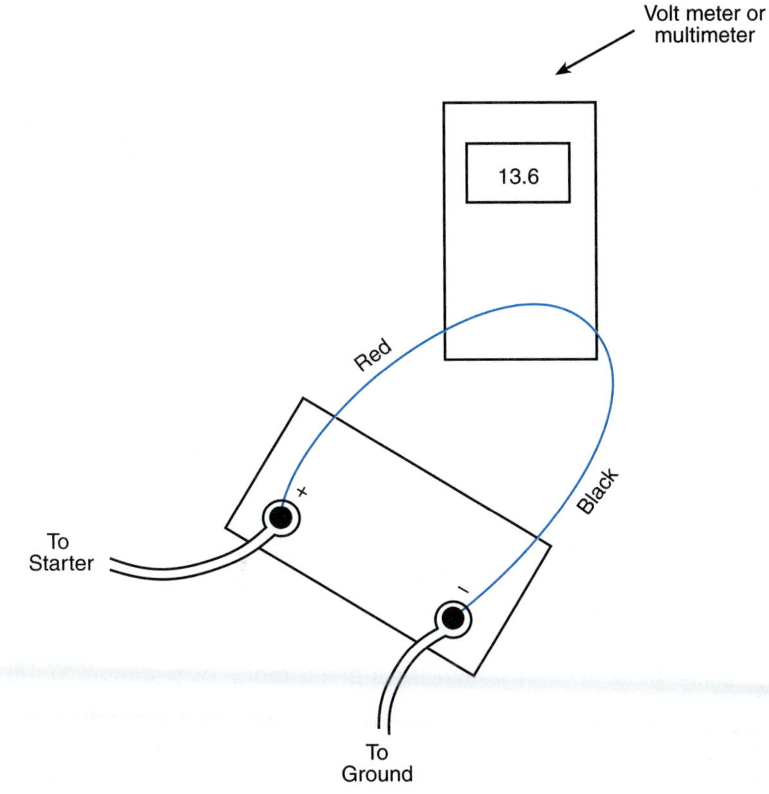

Figure 8-22 With the leads connected post-to-post, the battery voltage can be determined.

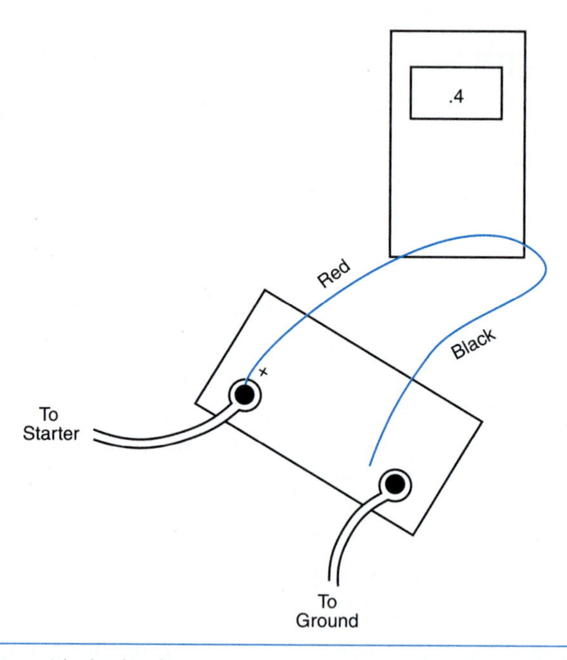

Figure 8-23 With the leads connected post to case any voltage loss because of corrosion or boilover can be measured. Note this measurement is above limits.

The next voltage drop test is done between the post and the terminals on each side of the battery. On the positive side, the red lead goes to the terminal and the black lead goes to the post. Do the opposite on the negative post. A reading over 0.2 VDC indicates a loss of too much voltage.

WARNING: Consult the service manual before disconnecting the battery. Some manufacturers require a small battery be attached to the cables as backup power. This maintains the PCM memory and operator's preferences. Do not use a vehicle battery for the backup. A small very low-ampere battery is required.

Even if the voltage drops are within limits, a dirty battery should be cleaned. Baking soda and a small amount of water will neutralize the acid and dissolve the corrosion. Use additional water to clean the waste from the battery and battery container. After the battery is clean, perform the voltage tests again. A marked improvement in the test results may indicate that the problem is probably repaired. Have an experienced technician check the charging system to ensure that the battery is being charged. Prepare the battery for load testing by disconnecting the cables, negative cable first.

Any technician performing a load test for the first time should be supervised for safety and proper procedures. Remove the battery from the vehicle and place it on a flat surface. Determine the battery's CCA and divide by two. For instance a battery with 550 CCA would be tested with a load of 275 amperes.

VAT 40/60 is a term used to designate a battery tester and is derived from a brand name. VAT stand for volt, ampere tester. Similar testers are sold under different brand names.

Study the service manual for the load tester. The **VAT 40** is a common tester and is fairly simple to operate. The **VAT 60** is digital version of the same machine and other testers are available. For this discussion we will use the VAT 40 since it is the simplest to explain, operate, and is still available in many shops. The procedures for all load testers are the same, but the controls and gauges are different.

Observe the gauges on the tester. The VAT 40 has two gauges, one for ampere and one for voltage (Figure 8-24). Each is color coded for a given range of readings. The top gauge is for amperage and the bottom for voltage. The ampere gauge has a red and blue scale. Note the range of the red scale on the ampere meter. This is the scale for load testing and it ranges from zero to 500 amperes, positive and negative. The voltage also has two scales: black and green. The green voltage scale will be used for this test.

Figure 8-24 The VAT 40 is an older but very reliable volt ampere tester.

To the right of the ampere gauge is the load knob. It should be positioned at zero or completely counterclockwise. Below the knob is the test selector switch (Figure 8-25). It should be positioned at the red/blue (#1) start position. Between the two gauges are two zero-adjust knobs and a red load indicator light. The zero-adjust knobs are used to calibrate the ampere needle to zero before starting the test. The red light will glow while the load is applied. Below the voltage gauge are the power selector, mechanical zero, and field selector switches. The power switch must be on INT.18V and the field in the OFF position. With all of the switches and selectors set properly, the leads can be connected.

WARNING: Do not exceed the test ampere limits or hold the load for more than 15 seconds. Damage to the battery or the tester could occur, or an explosion could occur with greater damage and injury.

Connect the red lead to the positive post and the black lead to the negative post (Figure 8-26). Adjust the ampere gauge to zero using the zero-adjust knob. Snap the large, green induction ampere lead around the black VAT 40 lead. The arrow should point away from the battery. Observe the voltage on the green scale. If the battery does not have at least 12 VDC, a load test would be invalid. The battery will have to be recharged before testing. During the test, observe the two gauges. Do not allow the amperage to exceed the ampere reading selected for this test. If the voltage drops below 9.6 VDC, stop the test. A charged battery that drops below 9.6 VDC during a load test is bad and should be replaced.

SERVICE TIP: If the gauge reads negative on the ampere scale, reverse the ampere lead and reconnect it with the arrow pointing in the opposite direction. If the voltage scale reads negative reverse the red and black leads.

Figure 8-25 The test switch must be set to the #1 (start) position before proceeding with the test.

We will use a 550 CCA battery for this test discussion. The test ampere would be 275 amperes for this battery. With the battery connected to the tester, rotate the load knob clockwise until the red ampere gauge is registering 275 amperes (Figure 8-27). Hold the knob in position for no more than 15 seconds while observing the voltage scale. A battery that maintains at least 9.6 VDC is good. If necessary, recharge the battery before installing it in the vehicle.

A vehicle may operate with a bad or weak battery if it can be started. The battery is the key to cranking and starting the engine. It may also be so bad or weak that the alternator cannot supply power to the rest of the electrical circuits. Poor maintenance is a major cause of battery failure.

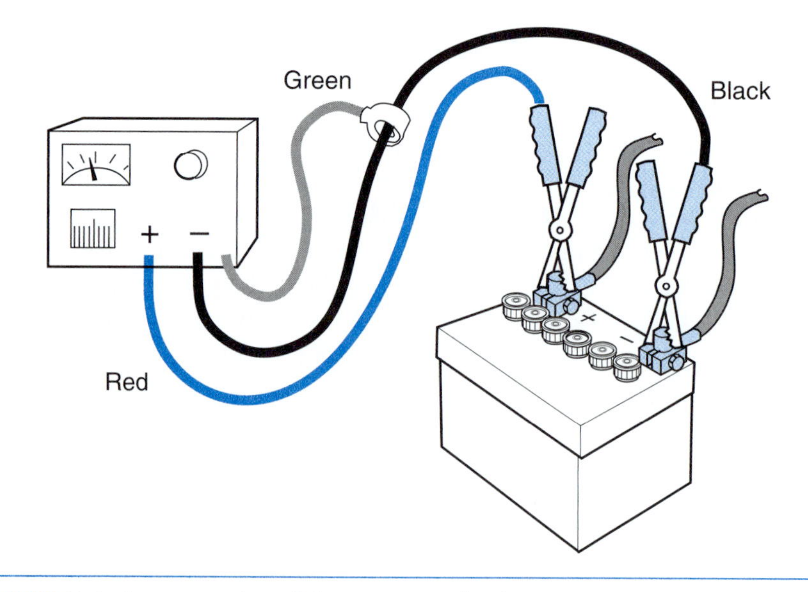

Figure 8-26 Note the connection of the green amp lead.

Figure 8-27 Do not exceed the calculated ampere load and do not load the battery for more than 15 seconds.

Inspecting and Replacing Secondary Circuit Components

Classroom Manual
pages 214–218

Special Tools

Spark plug socket

Spark plug boot puller

Spark plug feeler gauge

There are numerous components associated with the vehicle's secondary components. The number and type depend on the type of ignition system. We will discuss the inspection and replacement of the spark plugs and plug wires since almost every vehicle uses them. Before starting the service, use the service manual to determine the type of spark plug, type of wires, and the spark gap specification (Figure 8-28). Determine if any covers or special procedures must be followed to remove the spark plug or wire.

Replacement Parts

Air Cleaner Filter	A1163C
Battery	75-60
Engine Oil Filter	
VIN Engine Code K	PF47
VIN Engine Code G	PF25
Fuel Filter	GF578
PCV Valve	
VIN Engine Code K	CV892C
VIN Engine Code G	CV895C
Radiator Cap	RC24
Spark Plug	
VIN Engine Code K	41-921 (0.060 inch Gap)
VIN Engine Code G	41-931 (0.060 inch Gap)

Figure 8-28 Two different General Motors engines, K and G, require two different spark plugs but have the same gap. (Courtesy of Chevrolet Motor Division, General Motors Corporation)

Locate a spark plug in the engine. Some plugs are difficult to access and may require the removal or movement of a cover, hose, or other component. Applying force with an adequate hand and tool position may require some ingenuity on the part of the technician. Observe the area around the plug and determine the best method for access. Before removing the cover or other components, observe how and where they are installed. Any hoses or electrical connections should be marked before disconnection. Basic tools are usually the only ones needed for this task: ratchet, spark plug socket, extension, and universal joint. The only type of socket that should be used is a spark plug socket (Figure 8-29). Check the inside of the socket to ensure that the rubber boot is in place. Spark plug sockets are made to hold the spark plug during removal and installation procedures, thereby preventing breakage of the plug.

WARNING: Do not attempt to start an engine with coil packs when a spark plug wire is disconnected or not grounded. Damage to the coil pack may occur.

SERVICE TIP: Remove only one spark plug and one wire at a time. Start the engine after connecting each new wire. It only needs to run for a few seconds. Removing all of the wires at once can lead to additional time trying to decide where which each wire is supposed to be connected. Starting the engine gives an instant report on the installation of the last plug and wire installed.

Plug wires tend to adhere to the spark plugs as they age, making removal difficult without a wire remover. Most plug wire sets come with a small container of dielectric grease. A small dab of the grease is placed in the boot before installation on the plug.

Grasp the **spark plug wire** as close as possible to the plug. Twist the wire boot back and forth on the spark plug. This will loosen the boot and made it easier to pull from the plug. Do not pull on the wire. It will pull apart. If there is difficulty in removing the boot, use the boot puller (Figure 8-30). It hooks the bottom of the boot and will easily remove the most difficult ones.

Attach the socket to the ratchet or extension and place it over the spark plug. Rotate the socket by hand. It should not turn. If it does, wiggle the socket and push down to force it onto the plug. Use the ratchet to remove the plug from its hole. Lay out the spark plugs by cylinder number.

Use the round wire feeler gauge to set the gap on the new plug to specifications. Screw in the new plug and torque to specifications (Figure 8-31). Do not over tighten.

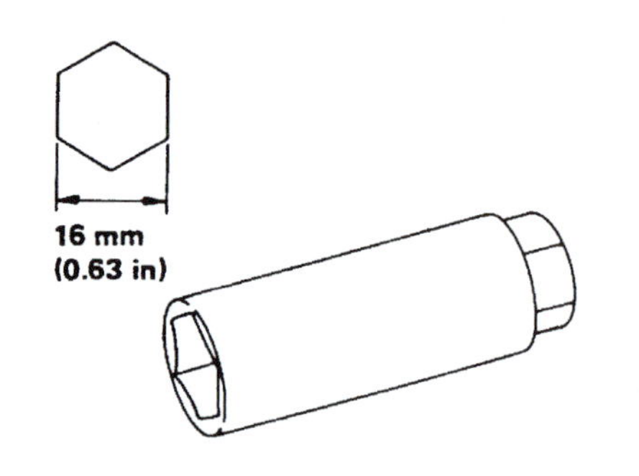

Figure 8-29 The right end of this plug socket has a square hole for the ratchet. The hex on the right end can be used with an open-end, box-end, or another socket wrench to assist in turning the spark plug socket. (Courtesy of Nissan North America)

Figure 8-30 Spark plug boot pliers should be used to remove difficult boots. (Courtesy of Snap-on Tools Company)

TORQUE SPECIFICATION

Part tightened	N·m	kgf·cm	ft·lbf
Spark plug x Cylinder head	18	180	13
Distributor x Cylinder head	19	195	14
Crankshaft position sensor x Oil pump	9.5	97	84 in.·lbf

Figure 8-31 Note how low the torque is for this spark plug. This torque is typical of most vehicles with aluminum cylinder heads. (Reprinted with permission)

Remove the disconnected spark plug wire from each holddown clip or bracket all the way back to the distributor or coil pack. Route the new wire into the areas where the old one was laid. Ensure that the new wire is placed into the holders or clamps and is not touching any engine component that may get hot. A spark plug wire touching the exhaust manifold or pipe will burn through quickly. Remove the old wire from the distributor or coil pack and replace it with the new one (Figure 8-32). Connect the other end to the spark plug. Ensure that the wire is securely connected and the boot is pushed on as far as it will go at each end. Start the engine. Repeat the procedures for the other spark plugs. Recheck the wires to ensure that they are properly routed and not touching hot components.

Inspect the **spark plugs** as they are removed (Figure 8-33). A black, sooty plug indicates a rich air/fuel mixture. A burned plug or one with brown deposits may mean the mixture is too lean. Oil or oily deposits show excessive engine oil consumption. Worn electrodes usually mean the plugs have been in the engine too long. A typical "good" used spark plug will have tan streaks, no deposits, and even though the gap may be wider than specified, both electrodes will show minimum wear. Consult with the customer or service writer if abnormal spark plug conditions are detected.

Spark plugs can help determine the engine's operations. Engine builders for race cars use the spark plug to determine how well, or how bad, the engine is performing.

If the repair is satisfactory, clean the area and vehicle, store all tools, dispose of the waste wires and plugs, complete the repair order, and return the vehicle to the customer.

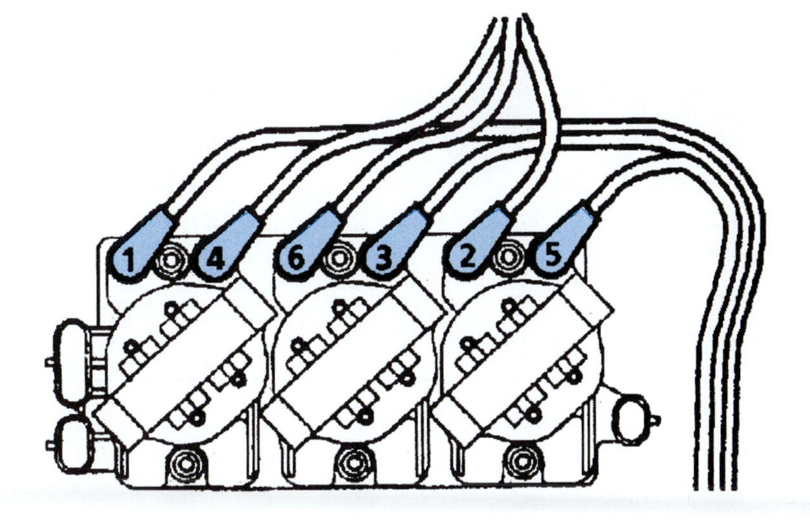

Figure 8-32 Note the cylinder numbers on this General Motors coil pack. (Courtesy of Chevrolet Motor Division, General Motors Corporation)

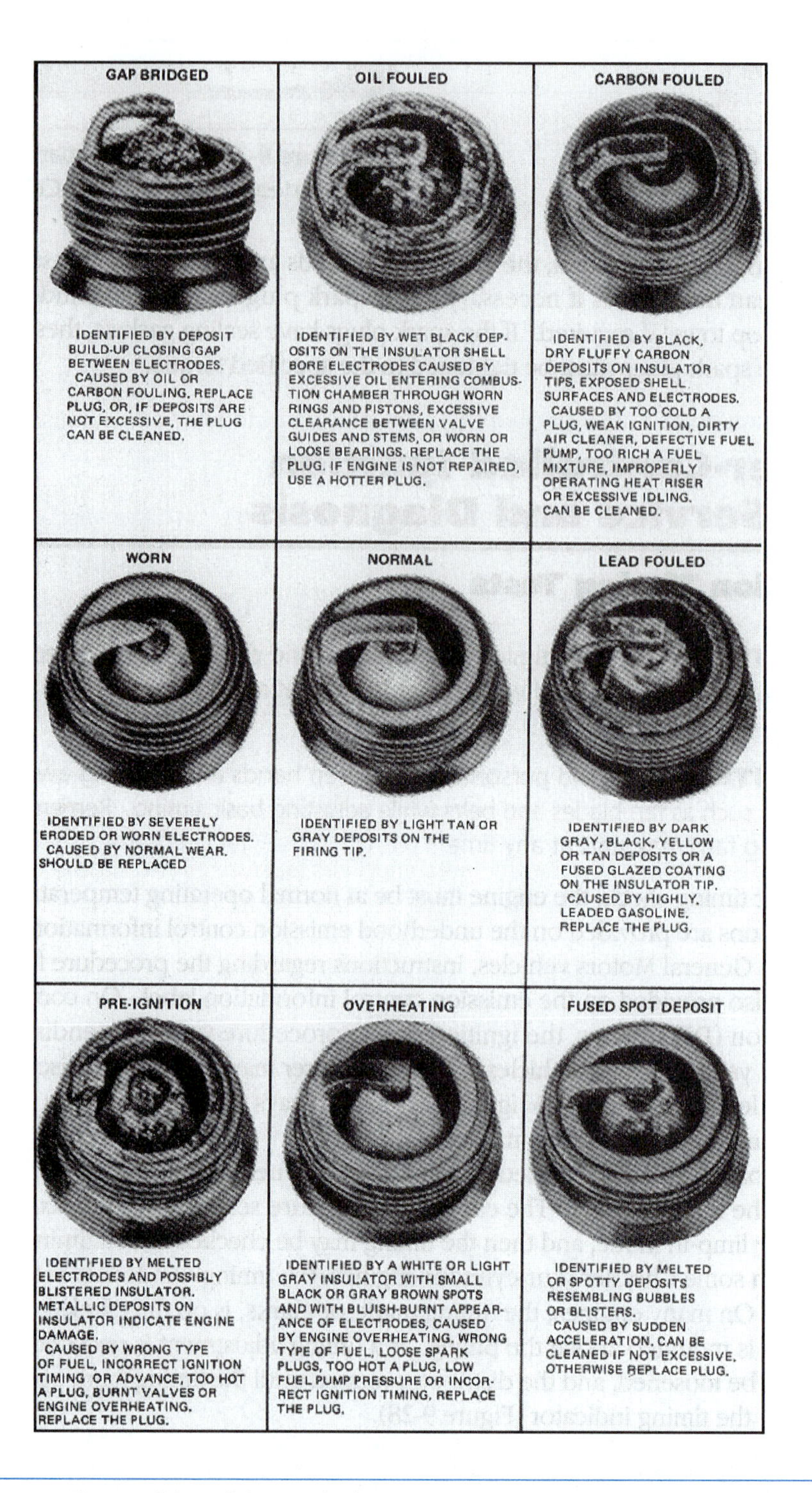

Figure 8-33 The condition of the spark plugs can tell the technician a great deal about how the engine is performing. (Courtesy of Ford Motor Company)

Cooling System Inspection and Service

Classroom Manual
pages 218–221

Inspections

CAUTION: Do not open, service, or repair a hot coolant system. Serious injury could result.

Special Tools

Drain pan

Hose clamp pliers

Inspection of the cooling system consists of checking for external coolant leaks, hoses, belts if applicable, and the visual condition of the coolant. The coolant should be a 50/50 mixture of water and **antifreeze** and should be a bright green color. Check the inside of the cap for rust, corrosion, or **slime**. If any of these three conditions is present, the cooling should be flushed and new coolant installed.

Antifreeze is a chemical mixture use to cool, reduce corrosion, and clean the cooling system.

External cooling leaks are easy to spot and can be easily identified by the green color of the antifreeze. An experienced technician should locate and repair internal cooling leaks. Check under the vehicle and inside the engine compartment for coolant. Do not forget to check the heater hoses and heater. A leaking heater will deposit coolant in the passenger compartment on the floorboard. The owner will usually complain of a coolant smell.

Antifreeze is very slick and oily, but not *slimy*. Metal dissolving in the coolant causes it to feel slimy.

Hoses and Coolant Changes

WARNING: Some new engines require special procedures for refilling the coolant systems. Follow the manufacturer's instructions. Damage to an engine may occur if the system has air or is not completely full.

CAUTION: Collect and dispose of the waste coolant according to shop policy. Wrongful disposal of waste coolant could damage the environment, injure or kill wild or domestic animals, and bring legal charges filed by the EPA against the shop owner.

Most radiator hoses are easily replaced. The coolant must be drained, captured, and recycled or disposed of properly. Before starting the replacement process, read the service manual to determine the correct method for draining the system, refilling the system, and choosing the amount and type of antifreeze required (Figure 8-34). Some engines, including diesels, require special antifreeze.

NOTICE:

When adding coolant, it is important that you use only DEX-COOL® (silicate-free) coolant. If coolant other than DEX-COOL is added to the system, premature engine, heater core or radiator corrosion may result. In addition, the engine coolant will require change sooner -- at 30,000 miles (50 000 km) or 24 months, whichever occurs first. Damage caused by the use of coolant other than DEX-COOL® is not covered by your new vehicle warranty.

Figure 8-34 Some vehicles require a special antifreeze. Note the last sentence in the notice. (Courtesy of Chevrolet Motor Division, General Motors Corporation)

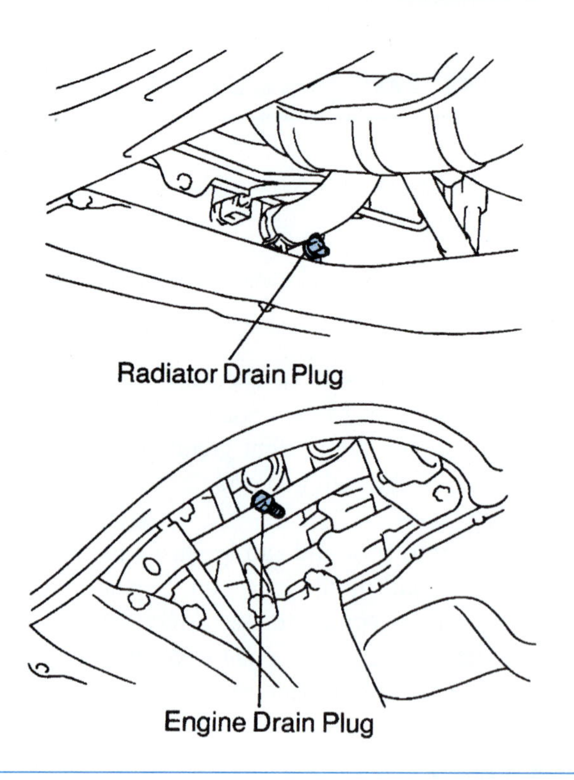

Figure 8-35 Many engines do not have a drain plug or peacock on the engine block. (Reprinted with permission)

A *peacock* is a drain plug for the coolant system.

Drain the coolant into a suitable container using the **peacock** located on the lower side of the radiator (Figure 8-35). Do not loosen the cap at this time. Some vehicles have peacocks on the engine block. If this is to be a complete coolant change, those peacocks will have to be opened to drain all of the coolant. When the amount of draining coolant slows, remove the radiator cap. Initially leaving the cap on allows the coolant recovery tank to drain through the radiator.

Release the hose clamp using pliers, wrenches, or a screwdriver. The hose will probably be stuck to the mounts and will have to be twisted and pulled off. A possible solution is to split the hose where it fits over the mounts and slide a flat-tipped screwdriver between the hose and mount (Figure 8-36). Use a small piece of sandpaper or scraper to remove any corrosion from

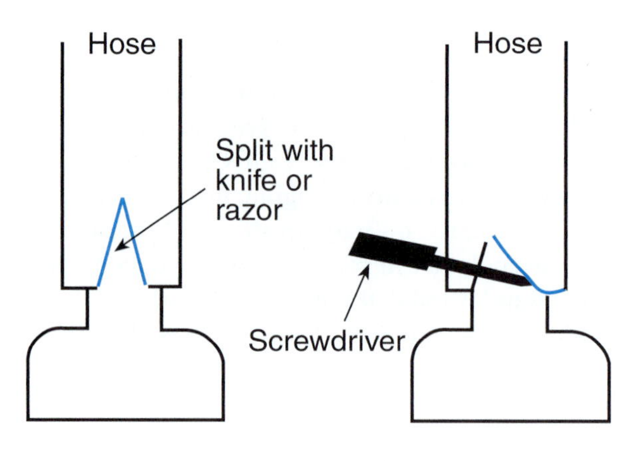

Figure 8-36 Split the hose and pry it open for removal. Do not do this if the hose is to be reused.

the mounts. Slip a new clamp over the new hose and slide the hose on the mount. Place the clamp about halfway between the end of the mount (feel it through the hose) and the end of the hose. Tighten the clamp until bits of the hose extend into the clamp's screw slots. Repeat the procedures for the other end and any other hoses. The same basic procedures can be used to replace the heater hoses also. Do not use excessive force to remove a hose from the heater. Heater hose connections are thin and can be damaged while trying to remove the hoses. Use a knife or another blade to slice the hose as mentioned above.

Close the peacocks and pour in the correct amount of antifreeze. Completely fill the cooling system with water. Follow the manufacturer's instructions to bleed air from the system. Some vehicles have a small bleeder screw at the upper radiator hose/engine connection for this purpose. Others require increasing the engine speed to 1,200–1,500 rpms, adding additional coolant, and capping the radiator before reducing the engine speed to idle. The heater temperature control should be placed in the HOT position during refilling. This will flush air from the heater and completely fill the system with coolant.

Once the system is full, allow the engine to idle while checking for leaks. If the repair is satisfactory, complete the repair order, dispose of the waste, clean the area, and return the vehicle to the customer.

Replacing the Serpentine Drive Belt

Replacing a serpentine belt can be complicated and simple. The method of removing the belt tension or even locating the **tensioner** is usually the most difficult. Before beginning the repair, locate a drawing of the belt's routing (Figure 8-37). If a drawing is not available, it may be advisable to draw a diagram of the old belt's routing. It would appear to be a simple operation, but a single-belt system has many bends as the belt goes around each pulley. Without a drawing, errors in routing the new belt may cost the technician valuable time.

Classroom Manual
page 181

Special Tools

Serpentine belt tensioner tool

A *tensioner* is a spring-loaded device used to maintain preset tension on a belt during engine operation.

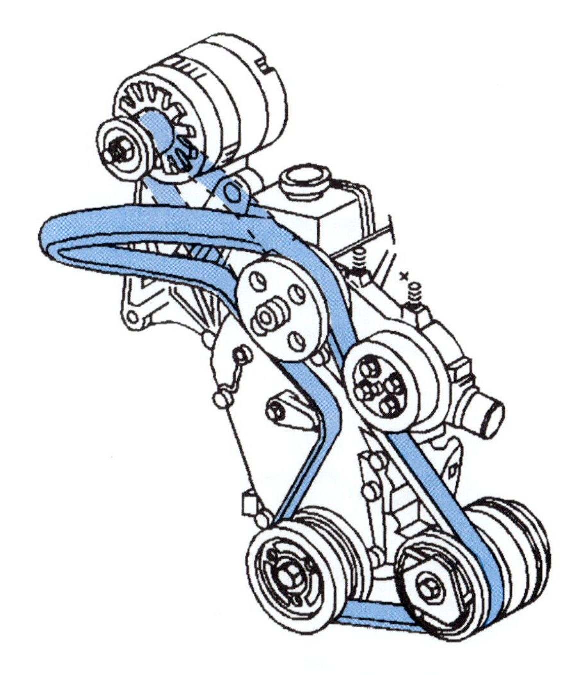

Figure 8-37 This serpentine belt would be simple to install compared to many. The tensioner on this belt would be forced down and away to relieve the tension. (Courtesy of Chevrolet Motor Division, General Motors Corporation)

Locate the tensioner and determine the type of tool or wrench needed to remove tension from the belt. In some cases, a box-end wrench is all that is needed. A special tool is available for this task and will work in most cases (Figure 8-38).

CAUTION: Ensure that the tensioner and tool are securely mated to each other before attempting to relieve tension from the belt. The tensioner is under a great deal of force. A tool that is not placed properly could slip and cause injury or damage.

SERVICE TIP: Before attempting to move the tensioner, note how the belt is threaded over it. If the belt contacts the top of the tensioner pulley, the tensioner will probably have to be swung down and away from the belt. Just the opposite is true if the belt touches the pulley on the lower side.

Use a tool to force the tensioner back from the belt. A great deal of force is usually required. With the tension released, remove the belt from the pulley that is the easiest to access. The alternator, is possible, is the best option. This pulley is fairly small and is best suited for removing and installing the belt. Once the belt is loose, slowly allow the tensioner to move back into place. Sometimes the tool will hang on an engine component and need to be worked loose. Do not allow the tool to hold the tensioner as the belt is removed. It could come loose and cause an injury.

Remove the belt from the rest of the pulleys and from the vehicle. Install the new belt starting at the crankshaft pulley. Maintain hand tension on the belt as it is routed and installed over all but the last pulley. The last pulley should be the easiest to access. Secure the belt while the tool is replaced on the tensioner.

CAUTION: Do not stand in the rotating plane of a belt. It can come off and cause injury.

This step is easier and safer to do with two people, but one person can do it. Move the tensioner until it is at the furthest end of travel. Pull the belt over the last pulley. Before allowing the tensioner to move, make a quick check to ensure that the belt is properly positioned

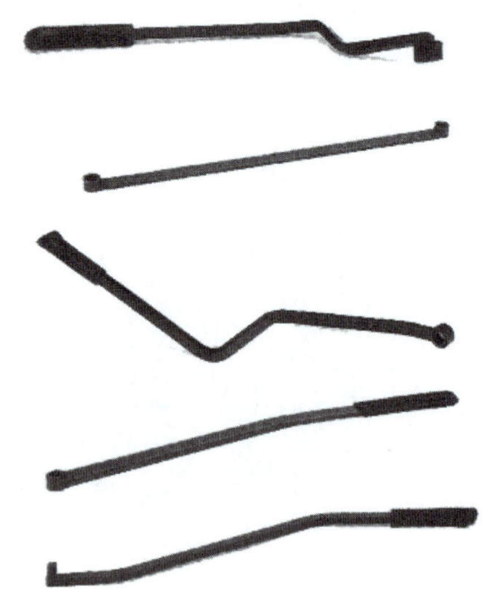

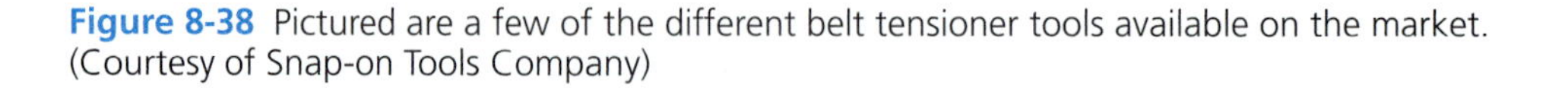

Figure 8-38 Pictured are a few of the different belt tensioner tools available on the market. (Courtesy of Snap-on Tools Company)

on each pulley. Release the tensioner and recheck the belt. If the belt is not positioned properly on any pulley, remove the tension and place the belt properly. Have a second technician crank the engine once or twice while observing the belt from the side. If the belt routing is correct, start the engine and observe the belt operation. If the repair is correct, remove and store the tools, complete the repair order, and return the vehicle to the customer.

Inspecting the Presence and Condition of Emission Control Devices

Classroom Manual
pages 221–225

Each of the emissions systems outlined in the Classroom Manual needs periodic servicing and replacement. In state and local areas requiring emissions testing, a vehicle will not pass inspection if emissions components are missing or damaged, even if the exhaust emissions are within standards. In this section, basic inspection procedures on some emissions systems are discussed. Systems not listed usually require an experienced technician and electronic testers. All systems will require an experienced technician for detailed testing.

Secondary Air Injection

CAUTION: Do not attempt to touch or repair hot engine components. Serious burn injury could result. It is best to allow the vehicle to cool before beginning the tasks in this section.

The drive belt for the secondary air injection is inspected as other belts. As the engine is operated, there should be no exhaust noises in or near the secondary air diverter valve. If there is exhaust noise, check the flexible hose running from the valve to the metal line going to the exhaust manifold or catalytic converter (Figure 8-39). There is a check valve inside the metal line, which prevents exhaust from entering the flexible hose and valve. It may become stuck in the open position over a period of time and gases can enter the hose. If it allows exhaust to back up, pulsing or heat may be felt in the flexible hose. The check valve requires replacing if

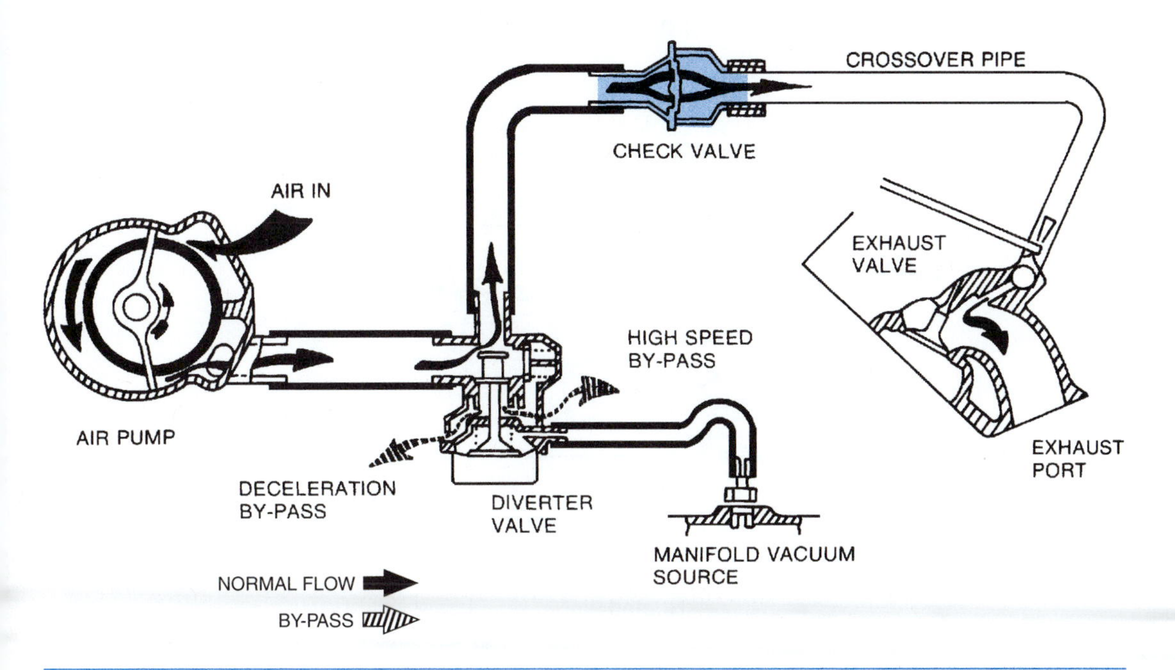

Figure 8-39 A bad check valve would allow hot exhaust gases into the flexible hose and the diverter valve. (Courtesy of General Motors Corporation, Service Operations)

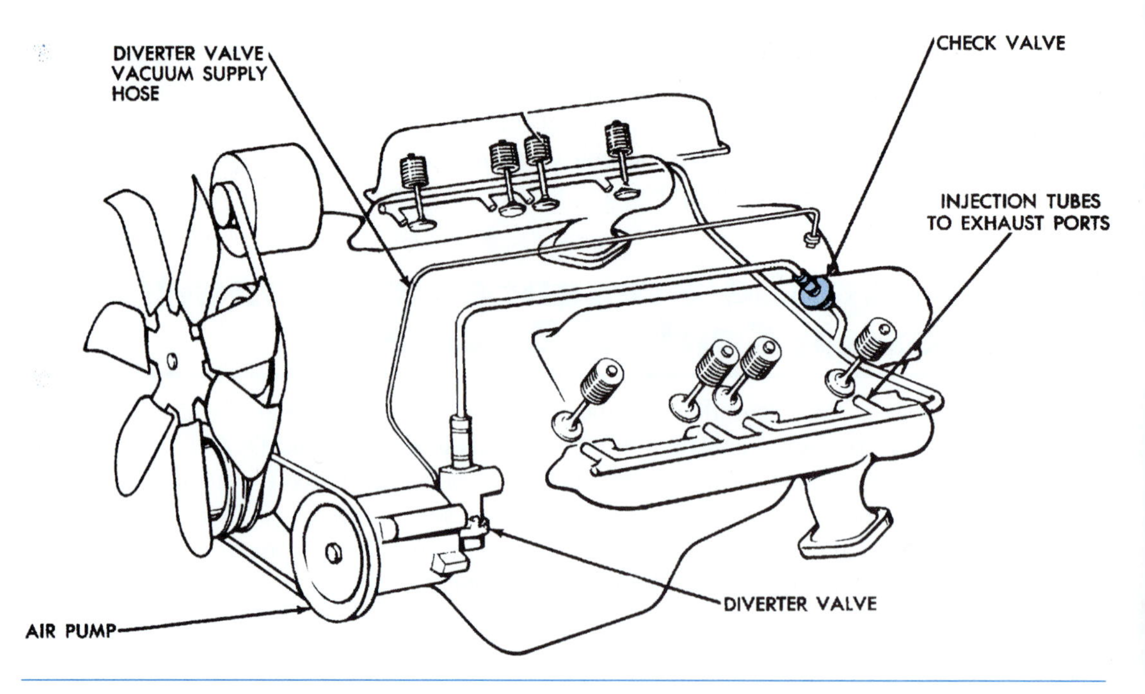

Figure 8-40 Many air heater hoses disappear after several services. The air cannot be heated in cold weather without the hose. (Courtesy of DaimlerChrysler Corporation)

this condition is noted. Accelerate the engine several times and quickly release the throttle. If there is a backfire through the exhaust, the diverter valve may not be shutting off the airflow during deceleration (Figure 8-40). Vacuum hoses leading to the diverter valve should be flexible, crack-free, and respond to squeezing much like any other rubber-type hoses. Replace any that show signs of wear or dry rotting.

Thermostatic Air Cleaners

The TAC system should have a large flexible hose leading from the air intake snorkel to a stove around the exhaust manifold (Figure 8-41). There will be one or two vacuum lines from the air filter housing to the intake manifold or a control device. Inspect them the same as other vacuum hoses and replace as necessary.

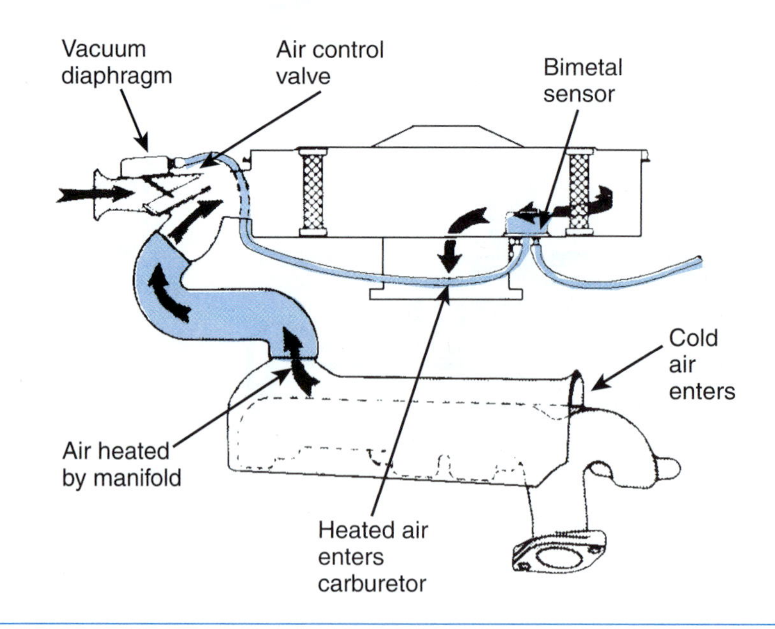

Figure 8-41 There are only a few hoses on the EVAP system that can be inspected visibly. (Courtesy of Nissan North America)

Evaporative Systems

Removing the fuel tank filler cap is the first check of the EVAP system. If the sound of air moving occurs as it is loosened, the cap is probably working. The other end of the system is located in the engine compartment and has vacuum hoses that need to be checked (Figure 8-42). The remainder of the system can only be inspected for leaks. Further tests to diagnose the EVAP system require electronic test equipment. Always replace the fuel filler cap with one of identical design and operation.

Catalytic Converters

CAUTION: Use care when working around an operating converter. The catalytic converter may reach temperatures of 1,400 degrees Fahrenheit. Second- and third-degree burns on hands and arms could result.

The catalytic converter cannot be checked completely, but some simple checks can be made. Start the engine and allow it to warm. While the engine is operating, check the end of the tailpipe. If there is soot in the pipe or the exhaust gas is black, an overly rich air/fuel mixture is indicated. Too much fuel in the exhaust will clog a converter and restrict the exhaust flow. After the engine has warmed to operating temperature, attempt to observe the exhaust pipe between the converter and exhaust manifold. A pipe that is beginning to turn or has turned red also indicates a clogged exhaust system. Do not touch the pipe or any other component of the exhaust. A noise from near or inside the converter may indicate a loose heat shield or the material in the converter may be coming apart or loose.

The remainder of the emissions systems are checked at this point for their presence. Some of the components may not be visible and may require testing equipment. As mentioned in the Classroom Manual, some of the systems have been replaced with better technology.

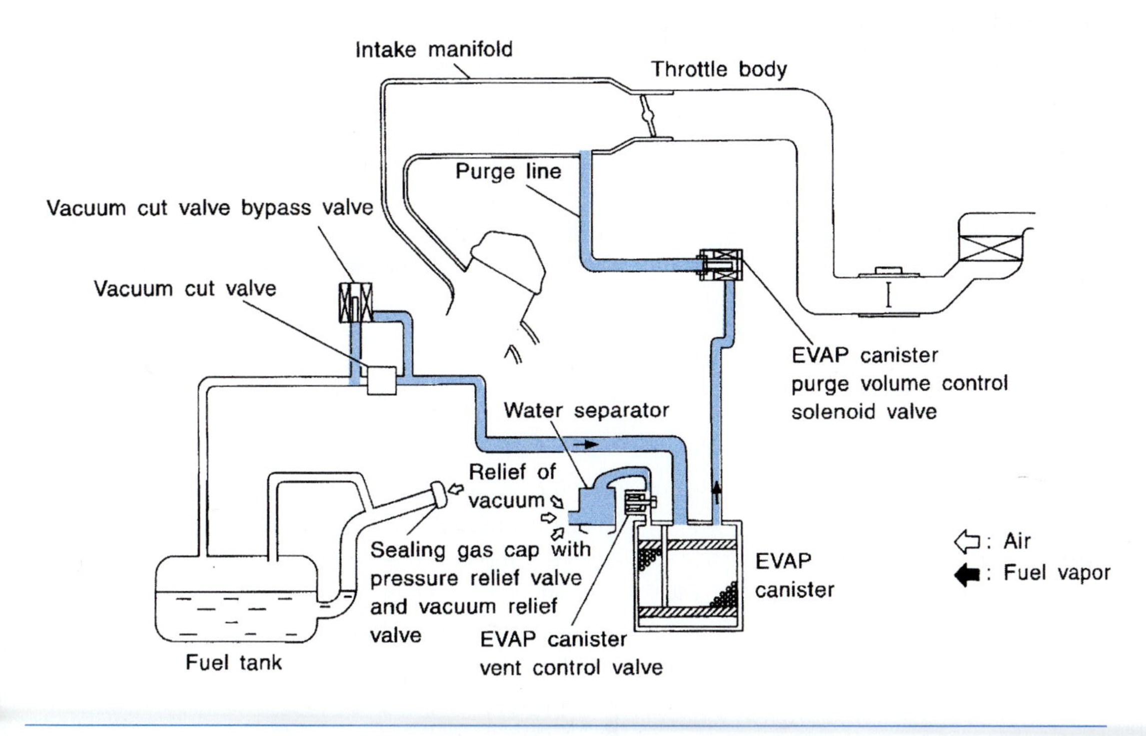

Figure 8-42 The EVAP system. (Courtesy of Nissan North America)

CASE STUDY

A vehicle came in with a cooling problem. The vehicle cooled correctly until stopped in traffic. The temperature gauge would read high and the air conditioning would shut down. After performing routine checks on the cooling fan and fan controls, the only problem found was that the coolant fan switch would not switch on at the specified temperature. A continuity (resistance) check of the fan switch showed an open circuit even at high temperature. A switch was ordered and the engine allowed to cool completely. While draining the coolant, the technician noticed that the coolant was fairly clear, indicating low antifreeze content. The coolant contained a great deal of corrosion and it was slimy to the touch. The technician recommended and gained approval for a complete cooling system flush. When the switch was removed, the portion that was immersed in the coolant was completely corroded over. Out of curiosity, the technician tested the old switch using boiling water. The corroded switch remained open even at 220 degrees Fahrenheit (100 centigrade). The switch was cleaned with baking soda and fine sandpaper and retested. The switch now closed at 218 degrees Fahrenheit. A system flush, new coolant, and a new switch completed the repairs.

Terms to Know

Antifreeze
Cold cranking ampere (CCA)
Diverter valve
Drain pan
Drain plug
Load test
Peacock
Top off
Undercarriage
Voltage drop
Windshield washer

ASE Style Review Questions

1. *Technician A* says a plugged PCV may be indicated by oil in the air filter housing.
   Technician B says a broken vacuum hose on the TAC system may cause poor cold driveablilty. Who is correct?
   A. A only
   B. B only
   C. Both A and B
   D. Neither A nor B

2. The ignition system is being discussed. *Technician A* says an oil-fouled spark plug may also indicate a possible problem in the exhaust.
   Technician B says a spark plug with the electrodes worn down and a very wide gap indicates the plug is being changed on schedule. Who is correct?
   A. A only
   B. B only
   C. Both A and B
   D. Neither A nor B

3. The charging system is being discussed. *Technician A* says the battery must be disconnected before removing and checking it for voltage drop.
   Technician B says the only way to properly load test the battery is with the battery installed and connected. Who is correct?
   A. A only
   B. B only
   C. Both A and B
   D. Neither A nor B

4. The fuel system is being discussed. *Technician A* says the pressure must be released on all vehicles before removing the fuel filter.
   Technician B says poor engine operation may result if the inlet end of the fuel filter is toward the fuel tank. Who is correct?
   A. A only
   B. B only
   C. Both A and B
   D. Neither A nor B

5. A spark plug tip that has a tan color indicates that:
   A. the plug wear is normal.
   B. the plug has been overheated.
   C. there is rust in the cooling system.
   D. none of the above.

6. The lubrication system is being discussed.
   Technician A says an oil filter gasket should have a light coat of RTV applied.
   Technician B says the old gasket may stick to the filter or the filter mounting. Who is correct?
   A. A only
   B. B only
   C. Both A and B
   D. Neither A nor B

7. The cooling system is being discussed. *Technician A* says proper installation of a serpentine belt requires two people.
   Technician B says a slimy feel to the coolant indicates the coolant should be changed and the system flushed. Who is correct?
   A. A only
   B. B only
   C. Both A and B
   D. Neither A nor B

8. *Technician A* says the oil filter must be torqued to specifications.
   Technician B says the oil pan plug may required a new gasket. Who is correct?
   A. A only
   B. B only
   C. Both A and B
   D. Neither A nor B

9. Load testing a battery is being discussed.
   Technician A says the battery should be tested at one-third of its CCA.
   Technician B says the green amp lead of the VAT 40 should be placed around the meter's red lead. Who is correct?
   A. A only
   B. B only
   C. Both A and B
   D. Neither A nor B

10. The VAT 60 is being discussed. *Technician A* says the VAT 60 is an analog version of the 40 model.
    Technician B says the basic operation of the two models is very similar. Who is correct?
    A. A only
    B. B only
    C. Both A and B
    D. Neither A nor B

Table 8-1 ASE Task

This is not a listed ASE task: Change the engine oil and oil filter.

Problem Area	Symptoms	Possible Causes	Classroom Manual	Shop Manual
Engine	Mileage/ time.	**1.** Routine service.	207–209	177–183

SAFETY

Wear safety glasses.
Allow engine to cool.
Use hand protection.
Dispose of waste.

Table 8-2 ASE Task

Inspect, service, or replace fuel and air induction system components, intake manifold, and gasket.

Problem Area	Symptoms	Possible Causes	Classroom Manual	Shop Manual
Engine	Mileage/time.	**1.** Routine service.	210	185
	Poor performance.	**2.** Restricted fuel filter.	—	187–190
	Oil in air filter	**3.** Restricted air filter.	—	185–187
	housing.	**4.** restricted PCV filter.	—	186

SAFETY

Wear safety glasses.
Allow engine and transaxle to cool.

Table 8-3 ASE Task

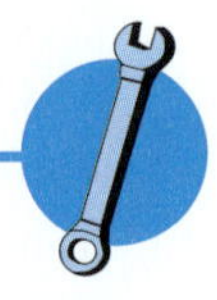

Inspect, service, or replace air filters, filter housings, and intake ductwork.

Problem Area	Symptoms	Possible Causes	Classroom Manual	Shop Manual
Engine	Mileage/time.	**1.** Routine service.	210–214	185–190
	Poor performance.	**2.** Restricted air filter.	210–214	185
	Poor fuel mileage.			

SAFETY

Wear safety glasses.

Table 8-4 ASE Task

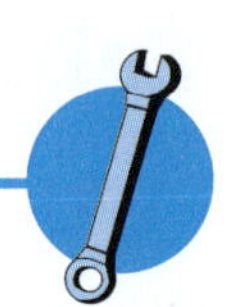

Inspect and replace positive crankcase ventilation (PCV) system components.

Problem Area	Symptoms	Possible Causes	Classroom Manual	Shop Manual
Engine	Mileage/time.	**1.** Routine service.	222	186
	Poor performance.	**2.** Restricted PCV filter.	—	186
	Oil in air filter	**3.** Restricted PCV system.	—	186
	housing.			

SAFETY

Wear eye protection.

Table 8-5 ASE Task

Inspect, replace, and adjust drive belts, tensioners, and pulleys.

Problem Area	Symptoms	Possible Causes	Classroom Manual	Shop Manual
Performance	Belt squealing.	**1.** Belt loose.	181	201–203
	Chirping.	**2.** Belt worn.	—	201–203
	Poor accessory performance.	**3.** Tensioner or pulley worn.	—	201–203
		4. Accessory damaged.	—	201–203

SAFETY

Wear safety glasses.

Table 8-6 ASE Task

Inspect, test, service, repair or replace ignition system secondary circuit wiring and components.

Problem Area	Symptoms	Possible Causes	Classroom Manual	Shop Manual
Performance	Engine hard to start.	**1.** Worn spark plugs.	214	196–197
		2. Corroded, damaged spark plug wires.	214	196–197
	Poor idle. Time/mileage.	**3.** Routine service.	214	195-198

SAFETY

Wear safety glasses.
Allow engine to cool.

Table 8-7 ASE Task

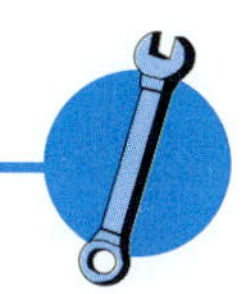

Inspect, service, and replace battery, battery cables, clamps, and hold-down devices.
Perform battery capacity load, high-rate discharge test; determine needed repairs.

Problem Area	Symptoms	Possible Causes	Classroom Manual	Shop Manual
Engine	Will not start.	**1.** Corroded cables, terminals.	181	191
	Battery will not hold charge.	**2.** Battery shorted, open.	181	191
		3. Low electrolyte.	181	192
		4. Inoperative charging system.	181	—

SAFETY

Wear safety glasses.
Allow engine to cool.
Disconnect negative cable first and reconnect last.

Table 8-8 ASE Task

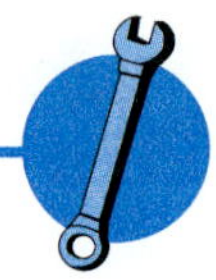

Inspect, clean, fill, or replace battery.

Problem Area	Symptoms	Possible Causes	Classroom Manual	Shop Manual
Electrical	Poor starting.	**1.** Corrosion on cables, terminals.	181	191
	Poor charging.	**2.** Broken/loose wires, connection.	181	191
	Dim or no lights.	**3.** Weak, damaged, old battery.	181	191
		4. Loose alternator drive belt.	181	201–203
		5. Bad alternator.	181	—

SAFETY

Wear safety glasses.
Allow engine to cool.
Disconnect battery's negative cable first before servicing battery or alternator.

Table 8-9 ASE Task

Inspect and replace engine cooling and heater hoses.

Problem Area	Symptoms	Possible Causes	Classroom Manual	Shop Manual
Engine	Overheating.	**1.** Dried, cracked hoses.	218	199
	Leaks.	**2.** Hose collapsed.	218	199
	Age.	**3.** Broken, loose clamps.	218	199
		4. Routine service	—	199–201

SAFETY

Wear safety glasses.
Allow engine to cool completely.

Table 8-10 ASE Task

Inspect coolant; drain, flush, and refill cooling system with recommended coolant; bleed air as required.

Problem Area	Symptoms	Possible Causes	Classroom Manual	Shop Manual
Engine	Overheating.	**1.** Low coolant level.	—	199
	Leaks.	**2.** Deteriorated/aged coolant.	—	199
	Age.	**3.** Contaminated coolant.	—	199
		4. Low antifreeze content.	—	199

SAFETY

Wear eye protection.
Contain waste coolant.

Table 8-11 ASE Task

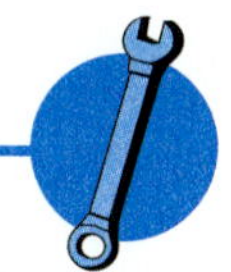

Inspect, service, and replace positive crankcase ventilation (PCV) filter/breather cap, valve, tubes, orifices, and hoses.

Problem Area	Symptoms	Possible Causes	Classroom Manual	Shop Manual
Emissions	Oil in air filter housing.	**1.** Plugged PCV system.	222–223	—
		2. Worn engine.	222–223	—

SAFETY

Wear eye protection.

Table 8-12 ASE Task

Inspect fuel tank, tank filter, and gas cap; inspect and replace fuel lines, fittings, and hoses; check fuel for contaminants.

Problem Area	Symptoms	Possible Causes	Classroom Manual	Shop Manual
Emissions	Engine will not start.	**1.** Fuel tank over pressurized.	223	205

SAFETY

Wear safety glasses.
Provide for fire prevention.

Job Sheet 12

12

Name ______________________________ Date ______________

Performing an Engine Lubrication Service

Upon completion of this job sheet, you should be able to perform a routine engine lubrication service.

Tools Needed

Drain pan
Filter wrenches
Box-end wrenches
Service manual

Procedures

1. Move the vehicle to the lift. Record any noise or obvious defects found during the movement. ______________________________

   ____

   ____

   ____

2. Engine Identification:

   Displacement ______________ Block ______________ Valve train ______________

3. Determine the type and amount of oil and the filter:

   Oil ______________ Amount __________

   Filter/customer preferred brand ______________________________

4. Raise the vehicle and place the drain pan under the oil pan plug.

5. Remove the plug and drain the oil. Note the condition of the oil. ______________

   ____

6. Check the plug gasket and replace if necessary.

7. Install the plug after the oil has drained. Torque to specifications.

8. Move the drain pan if necessary. Remove and drain the oil filter. Did the gasket come off with the old filter? ______________________________

9. Clean the oil filter mount, lube the new gasket, and install the new filter.

10. Lower the vehicle.

11. Install the proper amount of oil into the engine.

12. Start the engine. Watch the light or gauge. If no oil pressure is indicated within a few seconds, shut down the engine and locate the problem.

13. If the oil pressure is correct, shut down the engine.

14. Check for leaks at the oil filter and oil pan plug. ______________________________

____

15. Check the oil level and top off as needed.

Was oil required? ____________ If yes, how much? ______________________________

Instructor's Check ____

____

Job Sheet 13

13

Name ____________________ Date ____________

Removing, Inspecting, and Replacing Spark Plugs and Wires

Upon completion of this job sheet, you should be able to replace the spark plugs and wires.

Tools Needed

Hand tool set
Service manual

Procedures

1. Locate and record the specifications for the engine.

   Engine ID (type, size, valve)____________

   Spark plug type ____________

   Spark plug torque____________

   Spark plug gap____________

   Spark plug wire type ____________
2. Determine the best method of reaching the spark plugs.
3. Remove the wire from the first plug to be removed.
4. Remove the spark plug and inspect it. Record all findings.

   Cyl #1 ____________ Cyl #2 ____________

   Cyl #3 ____________ Cyl #4 ____________

   Cyl #5 ____________ Cyl #6 ____________

   Cyl #7 ____________ Cyl #8 ____________
5. Set the air gap on the new plug and install into the cylinder head.

   What type of tool was used to measure the gap?____________
6. Select the new spark plug wire by matching it to the old one.
7. Route the new wire alongside the old wire to the distributor cap or coil pack. Replace the old wire at the cap or coil terminal
8. Connect the wire to the spark plug.
9. Start the engine to check the installation of this spark plug.
10. Repeat steps 3 through 10 until all plugs and wires are replaced.
11. Recheck the placement of the wires and their connections at the plug and distributor/coil pack.

12. Start the engine. Is the engine operating smoothly? ______________________

If no, why? ______________________

____

13. Record your diagnosis of the engine based on your inspection of the old plugs and wires.

____

____

____

____

____

____

____

Driveline Service

Upon completion and review of this chapter, you should be able to:

- ❑ Diagnose driveline noise.
- ❑ Adjust a manual linkage clutch.
- ❑ Service a manual transmission or transaxle.
- ❑ Service an automatic transmission or transaxle.
- ❑ Service a drive shaft or axle.
- ❑ Service a differential.

Introduction

Like the engine, most of the powertrain components need routine servicing for the best performance. Many times a driver will routinely change the engine oil and filter, but fail to do the same for the transmission/transaxle and differential. Noise is sometimes the first indication that the driveline is failing.

Driveline Diagnosis by Noise

Classroom Manual
page 229

Noises are sometimes one of the first things that a customer recognizes as a possible problem.

Most driveline **noises** are easy to separate from the noise of an engine failure. However, the noise can echo or vibrate through the vehicle leading the technician to look in the wrong place. If the noise can be associated with a type of component, it can help eliminate some of the possible sources. Some of the driving conditions needed during a test drive can be produced on a lift in the shop. Only an experienced technician operating within the shop's policy should perform a test drive on the road or on the lift.

WARNING: Care must be taken when operating the vehicle in an attempt to isolate a driveline noise. The vehicle has to be operated to load, unload, and stress the driveline. It is best to use an empty parking lot or a low traffic road. Failure to pay attention to other motorists or traffic conditions can result in serious consequences.

WARNING: Do not conduct a test drive without permission of the owner, managers, and only in accordance with shop policy. An experienced technician should perform any test drive. Failure to follow proper procedures could result in damage to the vehicle, injury, and legal actions.

WARNING: Use extra care in placing lift pads, lifting, and operating a vehicle when the vehicle is operated while raised on the lift. Do not raise the vehicle any more than necessary to clear the wheels of the floor. Clear the areas of equipment and people in front of and behind the vehicle. Damage to the vehicle or injury could occur if the operation is not done in a safe and cautious manner.

Manual Transmissions/Transaxles

Noises from the transmission and transaxle usually include growling from the gears or whining from the bearings. Gear noise will change or disappear when another gear is selected. Sometimes a damaged gear, synchronizer, or maladjusted clutch will result in grinding or grating noises as the damaged teeth engage teeth on the mating gear (Figure 9-1). Bearing whine will change in pitch as the vehicle speed changes.

Figure 9-1 Each gear should be inspected for damage or excessive wearing. (Courtesy of Nissan North America)

Noises can be further diagnosed by operating the vehicle under different conditions. Usually, transmission noise will not change because of road condition or the turning of the vehicle. It will appear as the load is applied and released. If the clutch is disengaged as the vehicle is driven and the noise does not change, it is probably caused by driveline components behind the transmission.

Disengaging the clutch unloads the driveline and the noise will change in pitch, volume, or both. Placing the transmission in neutral as the vehicle is moving will cause most transmission noise to disappear or at least change drastically. A vibration may be felt in the shift lever as a certain gear is engaged.

Automatic Transmissions/Transaxles

All of the internal components of an *automatic transmission/transaxle* are heavy, continuously lubed, and tend not to fail unless abused.

Normally, **automatic transmissions/transaxles** do not produce major noises. The most common noise is a whine from the hydraulic pump or possibly the torque converter. If there is a clunking or other noise coming from the transmission area, the flexplate may be cracked (Figure 9-2). A cracked flexplate will also result in noise during cranking as the starter tries to turn

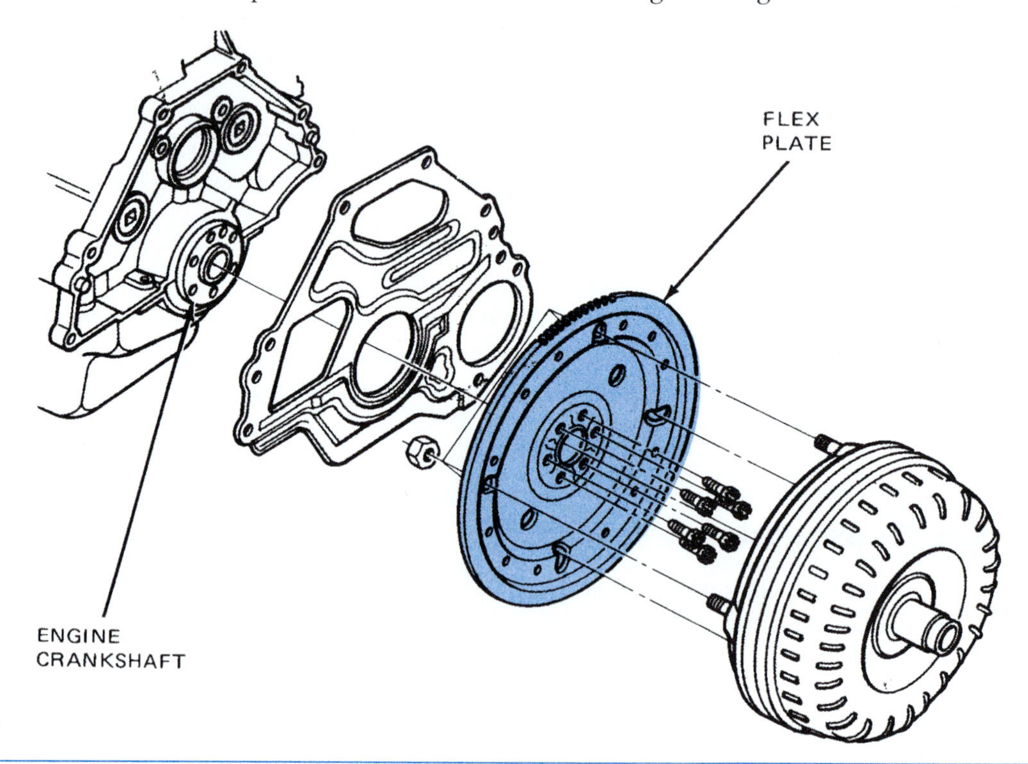

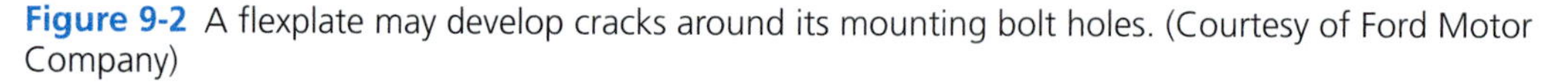

Figure 9-2 A flexplate may develop cracks around its mounting bolt holes. (Courtesy of Ford Motor Company)

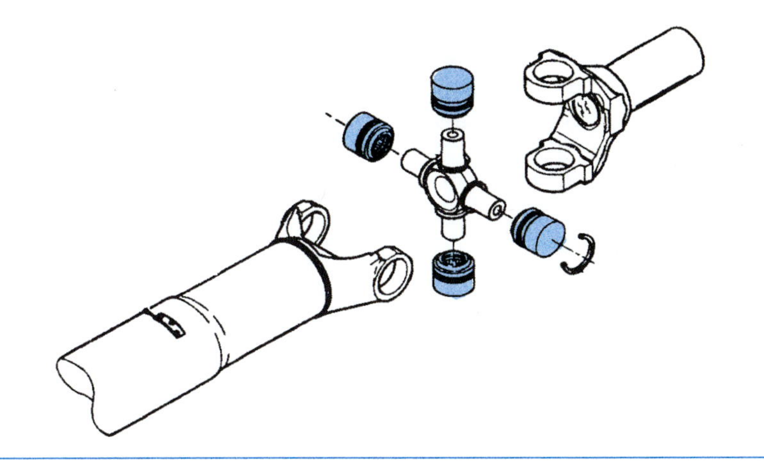

Figure 9-3 Lack of lubrication on the needle bearing may result in a squeaking noise. (Courtesy of Chevrolet Motor Division, General Motors Corporation)

the flexplate and torque converter. Any other noise from the automatic transmission/transaxle usually means it has to be replaced or rebuilt.

Drive Shafts

A bent or unbalanced drive shaft will cause a continuous vibration, which gets worse as the shaft's speed is increased. The vibration may lessen, but will not go away completely as the load is changed. This vibration should not be confused with bad or unbalanced tires. Tires will usually produce a vibration on an axle. Front tires will cause a vibration or jerking of the steering while rear tire vibration is felt in the seat. Drive shaft vibration may resound throughout the vehicle.

U-joints can create a clunking noise as the drive shaft is loaded and unloaded (Figure 9-3). This very closely resembles noise from worn differential gears. Closer inspection is required. A squeak at speeds of 3 to 6 mph usually means the U-joints are dry. If a **grease fitting** is present on the joint, lubricating it may help. It is best to replace a squeaking U-joint as soon as possible. If the joint breaks apart during vehicle operation, the drive shaft assembly can come through the floorboard causing injury to the passenger. A drive shaft that drops to the road and digs into the road surface can flip a car over.

Most original-equipment U-joints do not have *grease fittings.* Replacement U-joints will include a fitting with the joint.

Differentials

As stated above, clunking noise from under the car may be worn U-joints or worn differential gears. Other differential noises can be just as difficult to isolate. A high-pitched whine from the differential can mean the pinion gear and ring gear are set too tightly or one of the bearings may be going bad (Figure 9-4). Operating the vehicle in tight, slow turns can isolate axle bearing

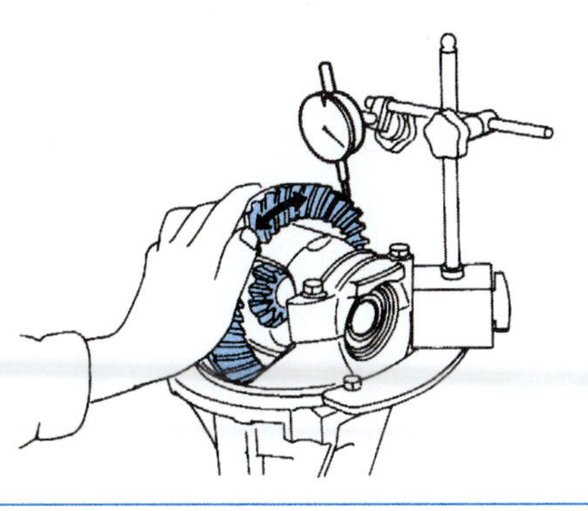

Figure 9-4 Improper backlash between the differential gears may produce a whining or roaring sound. (Courtesy of Nissan North America)

noises. As the vehicle leans, its weight loads and unloads the bearings sited near the drive wheels. The bad bearing will make more noise when it is on the outside of the turn and loaded. Rear-wheel-drive axles usually do not make any noise until they break. Then there is no drive or noise.

FWD Drive Axles

Most *front-wheel-drive (FWD) axle bearings* are one-piece, double ball bearings that are permanently lubed.

The noise from **front-wheel-drive axles** will originate in the CV joints. They are fairly easy to diagnose and isolate. From a dead stop, accelerate hard straight ahead. A fast, clicking noise is usually related to a bad inner CV joint. Operating the vehicle in a tight, slow circle will produce a slow-to-fast clicking noise from bad outer CV joints. The noise will be worse when the bad joint is loaded or on the outside of the turn. Bad outside CV joints may also be felt in the steering wheel as a turn is made.

A bearing mounted in the wheel hub supports the outer end of the drive axle (Figure 9-5). A common noise from a damaged bearing is a loud crushing or cracking sound. This may also produce a noticeable feel in the steering wheel and increases when the bearing is loaded.

Clutches

Clutch problems can be easy to diagnose because the system is so simple. Clutch noises have to be separated from transmission noises.

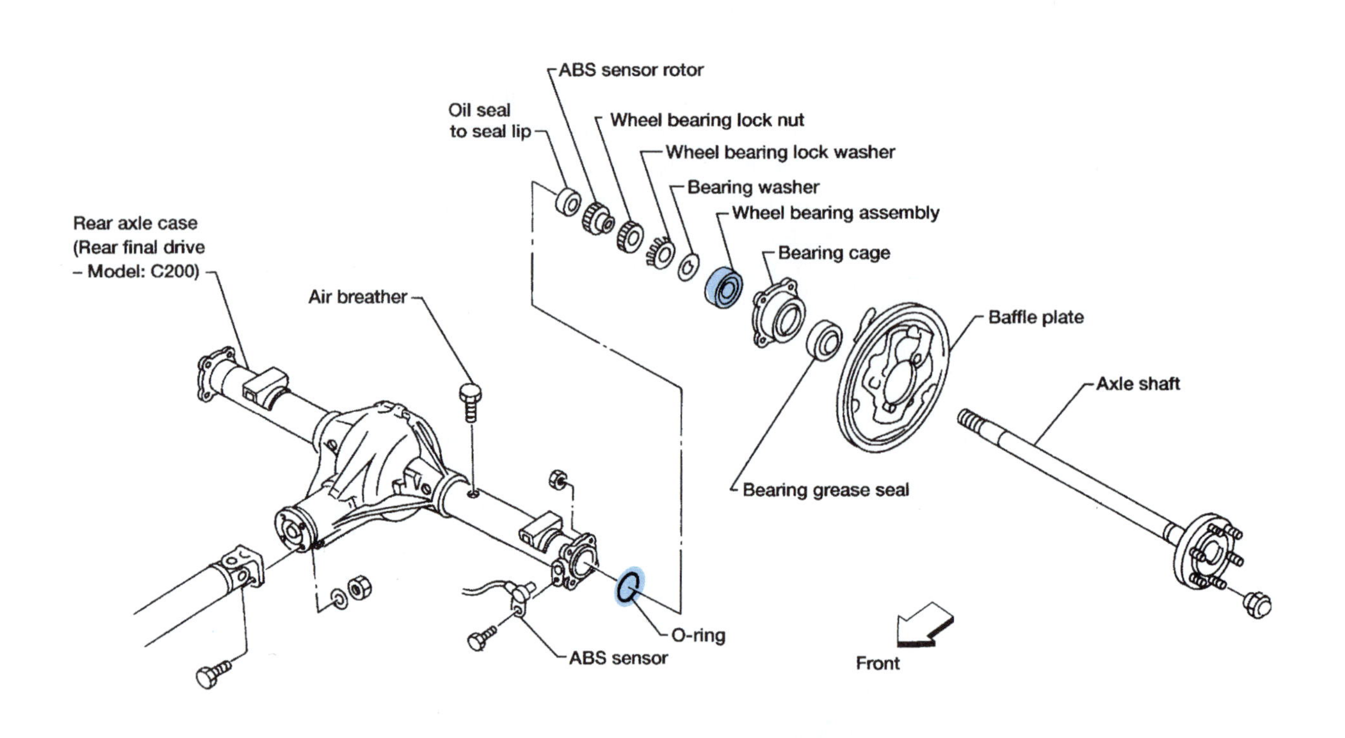

Figure 9-5 The rear axle bearings can produce a roaring sound which may change when the vehicle changes direction. (Courtesy of Nissan North America)

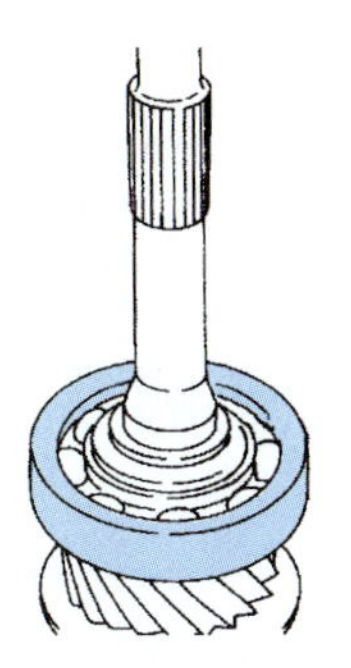

Figure 9-6 Noise from the transmission input bearing are most noticeable with the vehicle under load. The noise will disappear with the clutch disengaged and the transmission in neutral. (Courtesy of Nissan North America)

Diagnosis

A rattling or whining sound with the **clutch disengaged** can be traced to the release bearing. If the noise is audible at all times but changes in pitch or volume as the clutch is operated with the transmission in neutral, the noise points to a continuously-running release bearing that is dry of lubrication. A transmission input shaft's bearing noise will be most prominent with the clutch engaged and the transmission in neutral (Figure 9-6). It will disappear within a short time when the clutch is disengaged and held that way for a few seconds. A chattering noise and jerking as the clutch is being engaged usually indicates a bad clutch disc, pressure plate (clutch cover), or flywheel. A leak in the engine's rear main or transmission input seal will also cause chattering and jerking during clutch operation. The clutch assembly must be removed for an exact diagnosis.

Disengaging the clutch allows the transmission gears to stop turning, thereby eliminating any noise.

A worn clutch disc will slip during load conditions, most notably in the higher gears. A racing engine as the engine load goes up in high gear is the first sign of a worn clutch disc. Mechanical clutch linkage that has not been properly adjusted will not allow the pressure plate to clamp correctly and will cause the same problem. This type of linkage requires periodic adjustment as the clutch disc wears (Figure 9-7). An early sign of pressure plate wear or maladjusted linkage is a difficulty in shifting into first or reverse gears while the vehicle is halted. The pressure plate in this instance is not completely releasing the clutch disc. There may not be a slipping condition, but other gears may also be hard to select when the vehicle is under load.

Since most **hydraulic clutch linkages** are self-adjusting, slipping normally indicates bad components in the clutch assembly. If the hydraulic clutch system should leak, the pressure

The fluid in a *hydraulic clutch* is usually brake fluid and will deteriorate over time, thereby causing leaks in the master and slave cylinder.

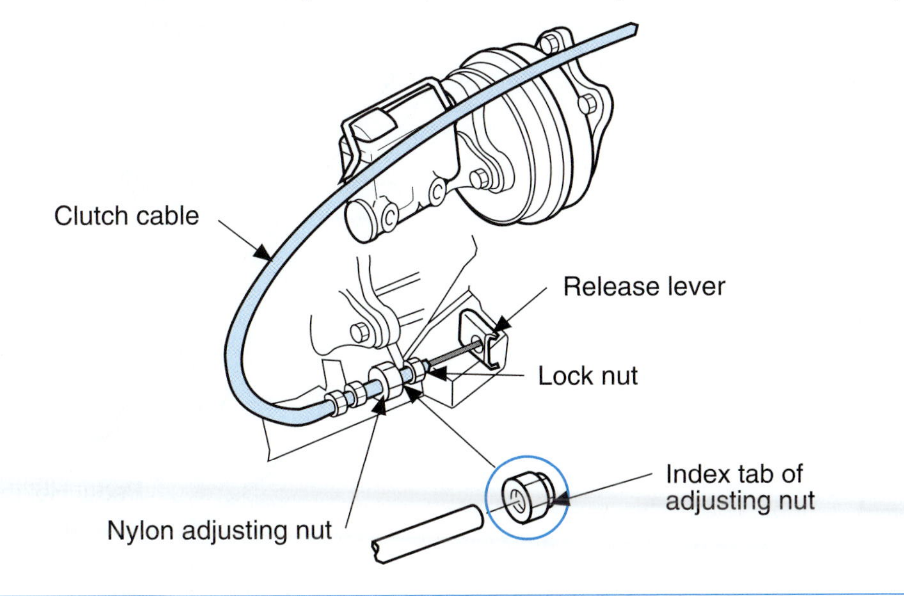

Figure 9-7 This clutch is adjusted at the nuts located near the clutch fork. (Courtesy of Ford Motor Company)

plate will not release the clutch disc. Topping off and bleeding the system may temporarily correct the problem, but the only proper repair is to replace the leaking component.

Classroom Manual
pages 231–236

Special Tools

Lift OR

Floor jack and jack stands

A *bleeder screw* is a hollow fitting that, when loosened, allows air or fluid to escape from the component. It is used to bleed air out of hydraulic systems.

Very few hydraulic control clutch systems have pedal *free travel*. Almost all manual clutch linkage systems have a specific free travel.

Adjusting the Manual Linkage for a Clutch

Mechanical clutch linkages must be adjusted periodically. Before discussing the mechanical type adjustment we will discuss changing the fluid in a hydraulic clutch. The fluid can be replaced easily by attaching a rubber hose to the slave cylinder **bleeder screw** and draining the old fluid into a container, preferably a clear one (Figure 9-8). Add fluid to the master cylinder and operate the pedal until clean fluid comes out of the slave. Keep the end of the hose below the level of fluid in the container and watch for air bubbles as the system is bled. Bleed the air out of the system by closing the bleeder screw and pumping the pedal several times. Hold the pedal to the floor and slightly open the bleeder. Continue this process until there are no air bubbles in the fluid. This process should be done at about the same interval as that for the coolant and brake fluid, about every two years. Consult the service or operator manual for the maintenance schedule.

Many newer models of cars and light trucks have hydraulically-controlled clutches, but there are many older vehicles still on the road. A heavy-truck or equipment fleet technician will perform this repair as a routine maintenance service.

Adjusting the pedal's **free travel** sets the mechanical linkage engagement point for the clutch. Free travel is the distance the pedal moves before the release bearing contacts the release levers of the pressure plate (Figure 9-9). The vehicle will probably have to be lifted some amount to gain access to the adjustment. If the vehicle is lifted to a normal (stand up) working height, a second person will have to ride the vehicle up the lift to operate the clutch and measure the pedal free travel. If only one technician is available, it is probably quickest and easiest to jack and block the front of the vehicle. This will allow the technician to use a creeper to make the adjustment, slide out, measure the free travel, and if necessary slide back under the vehicle to make additional adjustment.

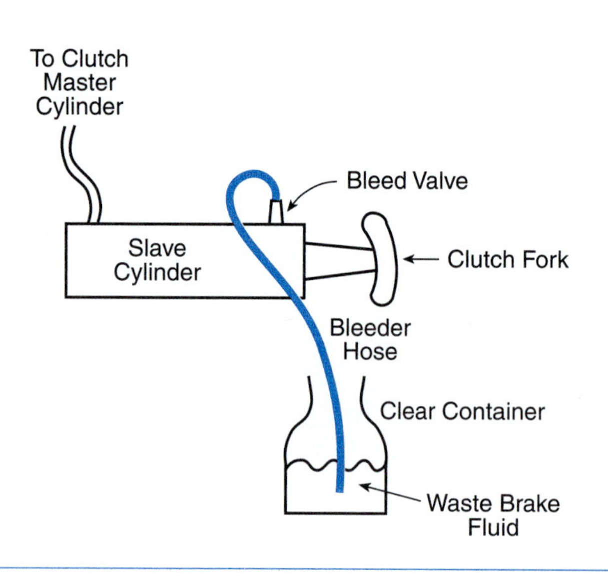

Figure 9-8 The hydraulic clutch system can be bled using a rubber hose and a catch container.

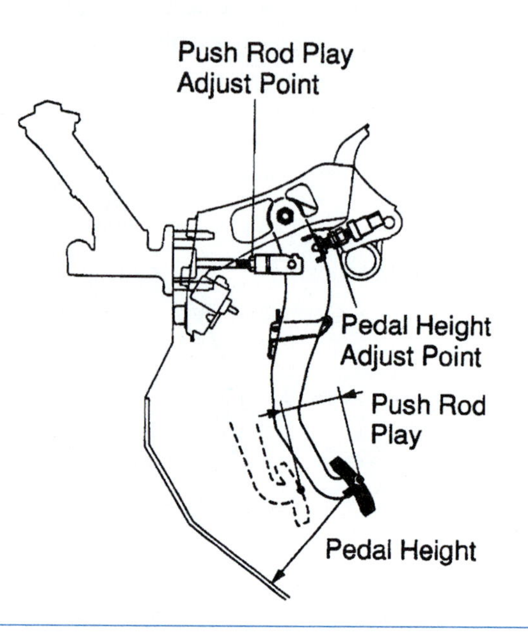

Figure 9-9 Proper clutch pedal free travel is essential for clutch life and operation. (Reprinted with permission)

The adjustment device will be located on the clutch linkage at the left side of the bell housing near the release bearing's fork. It usually consists of one or two nuts that are screwed on a threaded rod. Adjust the linkage using the adjusting nuts until the pedal free travel is the correct distance. Most clutches have a free travel of 3/4 inch to 1 inch. Consult the service manual for the correct specifications.

WARNING: Ensuring that the parking brake is set and the areas to the front and rear of the vehicle are cleared before running the engine to check clutch operation. Damage to the vehicle, shop equipment, or injury to persons may result if the clutch fully engages unintentionally.

With the adjustments completed, check the clutch and transmission operation on the lift. All gears should be checked at the halt with the engine running. First and reverse should be selected easily at the halt. All other gears should be smooth with no grinding or hanging. The vehicle will have to be tested on the road to completely check the adjustment. Consult the instructor or supervisor.

Transmission, Transaxle, and Differential Service

Classroom Manual
pages 236–266

Both manual and automatic transmissions and transaxles need routine servicing. Most of the time it is only a fluid and filter change. The service manual specifies the interval.

Manual Transmissions/Transaxles

Special Tools

Drain Pan

Air- or hand-operated pump OR

Suction gun

This type of transmission does not need very much service other than a fluid change. Very few manual systems use a filter. The fluid can be drained through a drain plug located at the bottom of the transmission housing. Manual transmission fluid comes in different blends. Older vehicles use a 85W-90 thick oil with few additives, but newer systems require lighter, thinner fluid. The current classification is listed as GL-4 or GL-5 (Figure 9-10). GL fluids may have

Figure 9-10 Manual transmission and differential fluids range from a thick 85W90 to thin automatic transmission fluid.

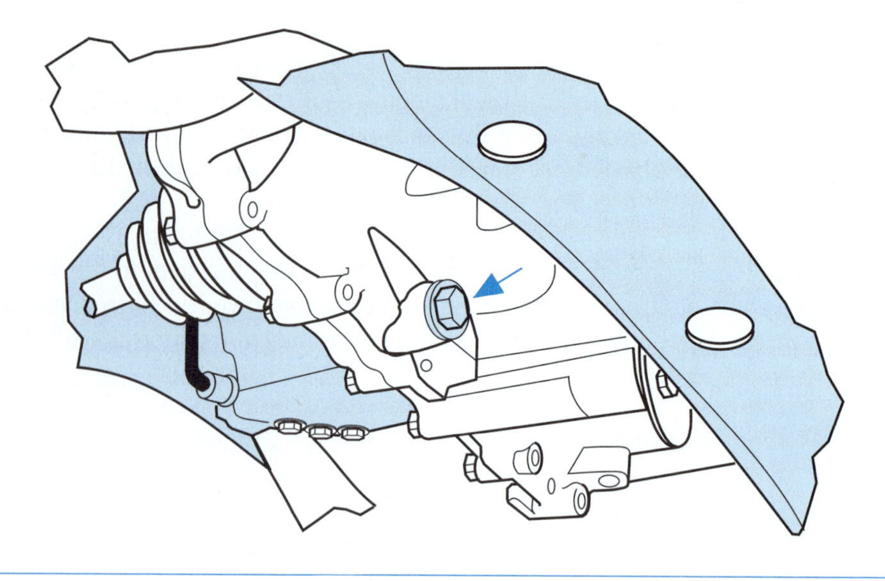

Figure 9-11 A drain plug for a manual transaxle. (Courtesy of Subaru of America, Inc.)

additives that lubricate, clean better, and are designed for specific temperature ranges. Many manufacturers also used special lubricants for their manual transmissions. Newer Chevrolet and GMC trucks, for instance, require a special fluid for their manual transmissions. Other manufacturers use automatic transmission fluid or a light-weight engine oil for their manual transmission or transaxle systems. Most manual transmissions require three to five quarts of fluid.

Attempting to service a *transmission* that is not at the same angle as an operating vehicle may not allow all of the fluid to drain.

To change the fluid in a manual **transmission**, lift the vehicle to gain access to the drain plug (Figure 9-11). The transmission should be hot enough for the thick fluid to flow freely. Cold transmission fluid is very slow to flow and will not remove all of the debris that may be inside the bottom of the transmission. The transmission should be at the same angle as it is when the vehicle is setting on its tires. Remove the drain plug, capture the old fluid, and dispose of it according to the shop's policies. While the fluid is draining, check for leaks at the transmission's extension housing seal where the drive shaft mounts or the drive axle seals on a transaxle. Repair as necessary.

The *suction pump* is used to draw fluid from a container and force the fluid into the transmission/transaxle by hand.

Reinstall and torque the drain plug once all of the fluid has drained. Fluid can be installed with an air-operated pump or using a hand **suction pump**. The suction pump is slow and can be messy.

WARNING: Overfilling a transmission could cause high pressure within the housing and rupture a seal or gasket. Damage could result from an overfilled transmission.

Locate the fill plug. It is normally about halfway up on the side of the transmission. Remove the plug and insert the pump's hose tip into the hole. If an air- or hand-operated pump is used, the amount being installed is measured with a dial on the pump (Figure 9-12). Suction

Figure 9-12 This type of lubrication gun is attached to a large bulk container of lubricant. (Courtesy of Snap-on Tools Company)

Figure 9-13 A suction gun can be slow and messy when transferring fluids. (Courtesy of Snap-on Tools Company)

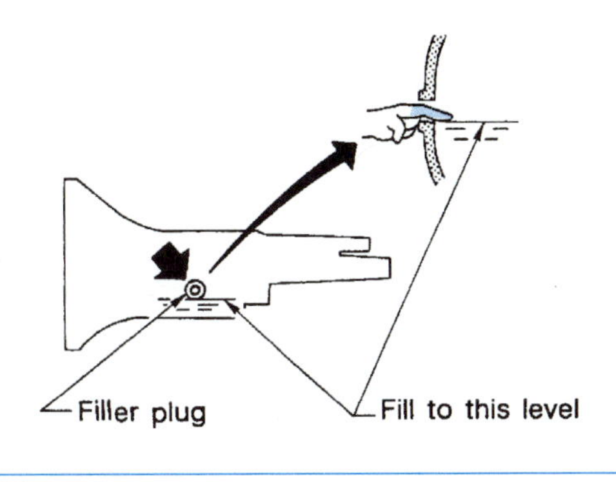

Figure 9-14 Manual transmission and differential fluid levels can be checked using a finger. (Courtesy of Nissan North America)

pumps will hold about a pint of fluid and will have to be refilled until the transmission is full (Figure 9-13). Most manual transmissions do not have a dipstick to check fluid level. The most common method of checking the fluid involves the use of the technician's finger (Figure 9-14). Insert a finger into the fill hole to the first knuckle. Bent the fingertip downward inside the transmission. If the fluid touches the finger, the level is correct. There are special tools that can be brought or made to check the fluid level. A quick tool can be made from a welding rod. Bend the rod 90 degrees, about the length of the last little finger joint from one end (Figure 9-15). About an inch from this bend, curve the rod so the bent end can be fitted into the fill hole. Place the bent end into the fill hole and the rod against the bottom of the hole (Figure 9-16). If the fluid level is correct, there will be fluid on the end of the rod inserted into the fluid. A tool is not an absolute necessity unless the technician performs this type of service several times a day or week.

During filling, the fluid is added until it barely begins to run from the fill hole. If it is overfull, drain and capture the excess. Reinstall and torque the fill plug.

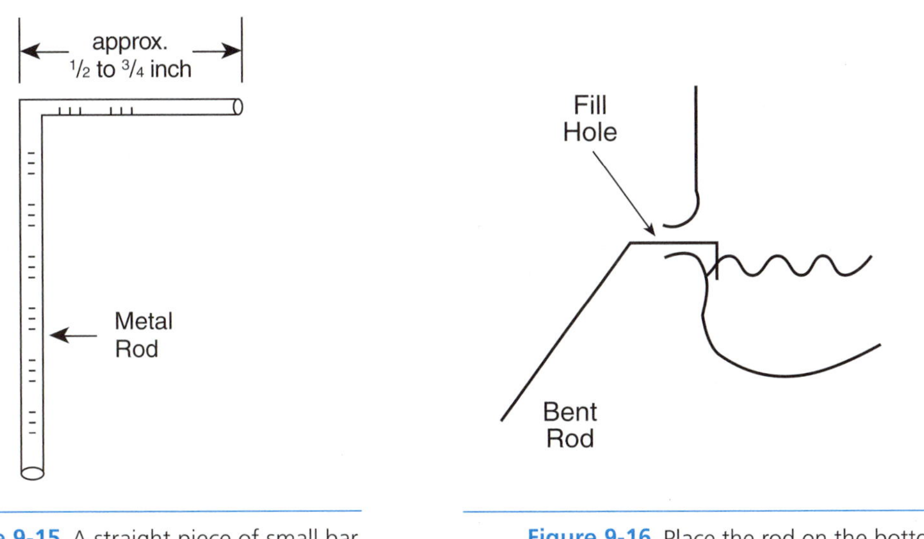

Figure 9-15 A straight piece of small bar stock can be made to check the fluid level in a manual transmission or differential.

Figure 9-16 Place the rod on the bottom of the fill hole and allow the short portion to extend down into the fluid.

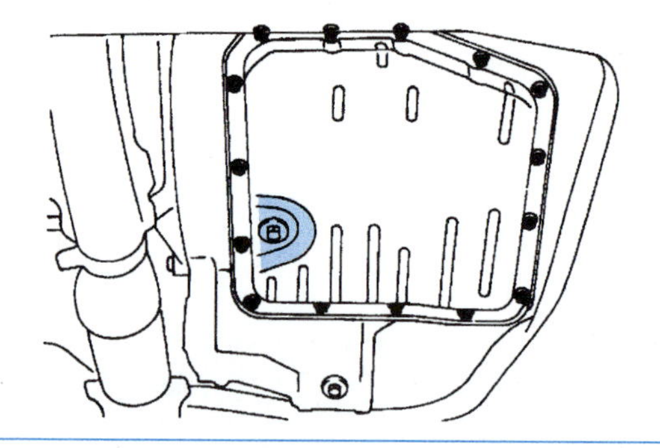

Figure 9-17 This automatic transaxle has a drain plug to drain the fluid. (Reprinted with permission)

Special Tools

Large drain pan

Funnel

Automatic Transmissions/Transaxles

Automatic transmission/transaxle service includes a fluid and filter change. The fluid removed will not be all of the fluid in the system. A typical fluid change will only remove the fluid in the transmission. The torque converter and pump will retain some fluid. Before starting the service, check the manual for the correct amount of fluid used for a not-for-the-full-amount servoce. It is best if the transmission is warm. Transmission and transaxle services are performed using similar procedures.

Before proceeding compare the pan gasket in the filter kit with the pan. There are many different pan shapes, none of which are particular to a certain brand or model. If the pan is removed and the gasket is incorrect, the lift and bay will be tied up until another gasket is available. In some extreme cases that may be one or two days. In most cases, if the gasket is correct, the filter will be correct.

Lift the vehicle to gain access to the transmission's oil pan. Most automatics do not have drain plugs. The pan has to be removed (Figure 9-17). Care must be taken to prevent spills. Remove the pan bolts from the lowest side of the pan. Place a wide drain pan under the transmission and remove bolts along each side, starting at the point where the first bolts were removed. At some point, the pan will tilt and fluid will begin to drain. Loosen the other pan bolts, but do not remove all of them until most of the fluid is removed. Keep the last two or three bolts loose until the pan can be supported by hand. Hold the pan against the transmission and remove the last several bolts. Control the oil pan and pour the remaining fluid into the drain. There will be fluid dripping from the filter and transmission. Keep the drain pan in place until the oil pan is reinstalled. Note how the filter is installed. Automatic transmission filters come in many different shapes and their installation location may not be readily apparent. Remove the filter bolts and filter.

Usually, there is a small magnet in the bottom of the oil pan (Figure 9-18) with a few metal shavings on the magnet. This is normal. Excessive shavings indicate gear or bearing damage. Look for black friction material in the bottom of the pan as well. There will probably be

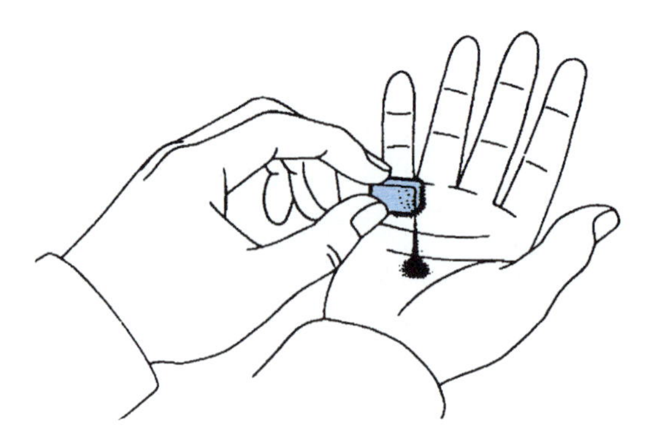

Figure 9-18 The magnet lies in the transmission oil pan to collect and hold metal shavings. (Reprinted with permission)

Photo Sequence 8
Typical Procedure for Performing an Automatic Transmission Fluid and Filter Change

P8-1 To begin the procedure for changing automatic transmission fluid and filter, raise the car on a lift. Before working under the car, make sure it is safely positioned on the lift.

P8-2 Locate the transmission pan. Remove any part that may interfere with the removal of the pan.

P8-3 Position the oil drain under the transmission pan. A large diameter drain pan helps prevent spills.

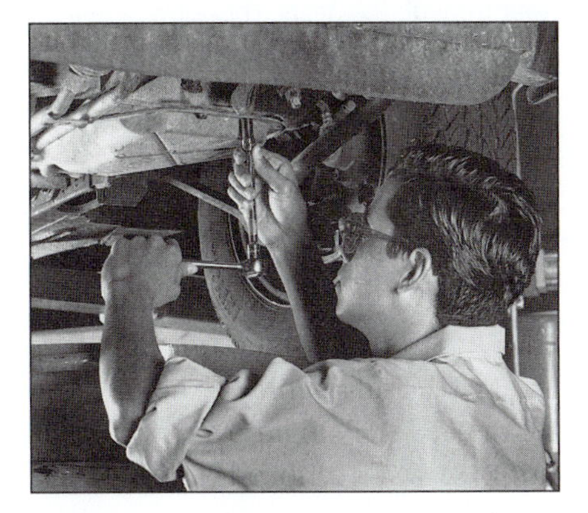

P8-4 Loosen all of the pan bolts and remove all but three at one end. This will cause fluid to flow out around the pan into the drain pan.

P8-5 Supporting the pan with one hand, remove the remaining bolts and pour the fluid from the transmission pan into the drain pan.

P8-6 Carefully inspect the transmission pan and the residue in it. The condition of the fluid is often an indication of the condition of the transmission and serves as a valuable diagnostic tool.

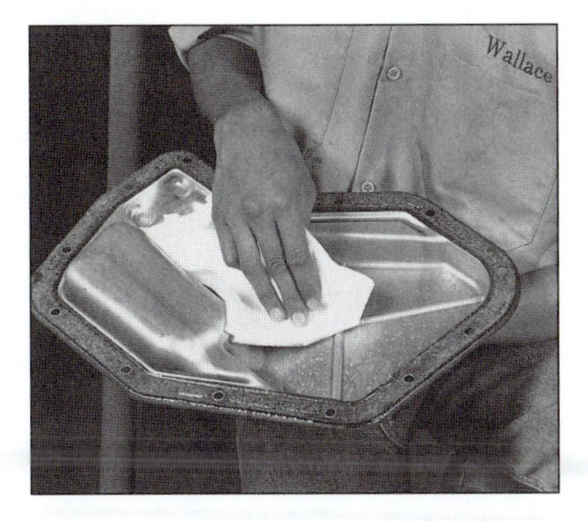

P8-7 Remove the old pan gasket and wipe the transmission pan clean with a lint-free rag.

P8-8 Unbolt the fluid filter from the transmission's valve body. Keep the drain pan under the transmission while doing this. Additional fluid may leak out of the filter or the valve body.

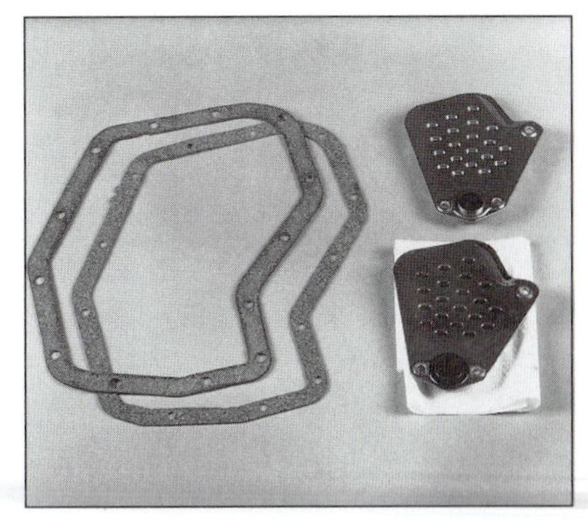

P8-9 Before continuing, compare the old filter and pan gasket with the new ones.

Typical Procedure for Performing an Automatic Transmission Fluid and Filter Change (continued)

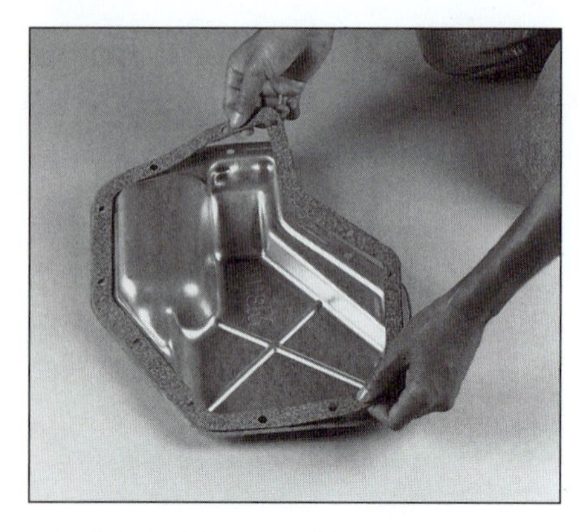

P8-10 Install the new filter onto the valve body and tighten the attaching bolts to proper specifications. Then lay the new pan gasket over the sealing surface of the pan and move the pan into place.

P8-11 Install the attaching bolts and tighten them to the recommended specifications. Carefully read the specifications for installing the filter and pan. Some transmissions require torque specifications of inch-pounds rather than the typical foot-pounds.

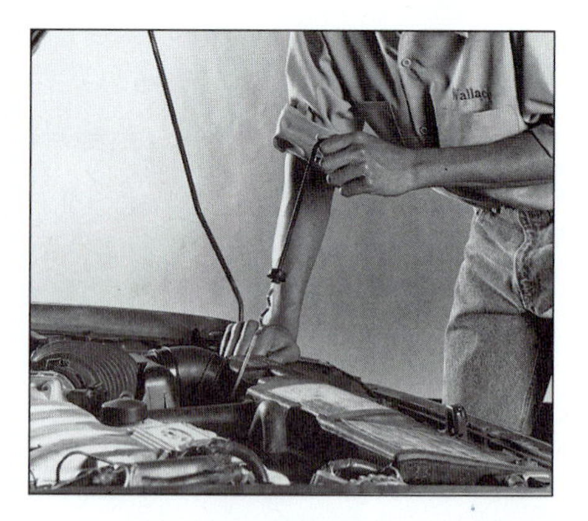

P8-12 With the pan tightened, lower the car. Pour ATF into the transmission. Start the engine. With the parking brake applied and a foot on the brake pedal, move the shift lever through the gears. This allows the fluid to circulate throughout the entire transmission. After the engine reaches normal operating temperature, check the transmission fluid level and correct it if necessary.

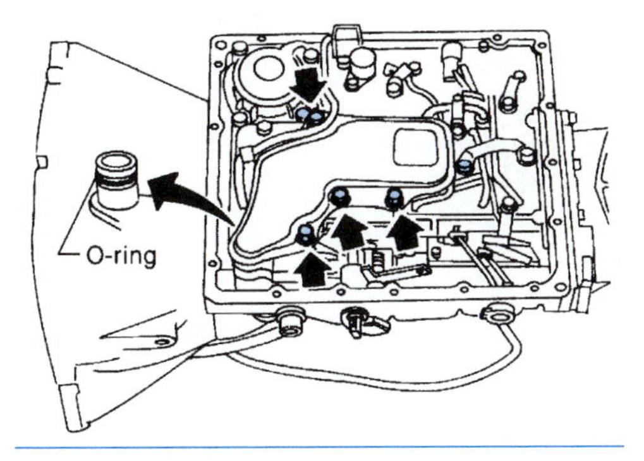

Figure 9-19 The transmission/transaxle filter bolts should be torqued. (Courtesy of Nissan North America)

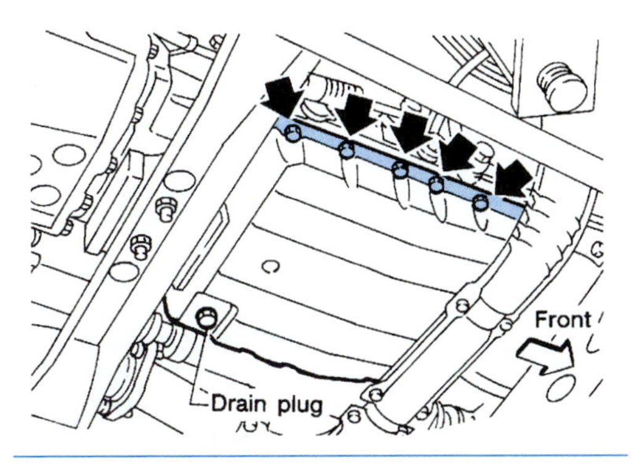

Figure 9-20 The oil pan bolts must also be torqued for proper sealing of the pan. (Courtesy of Nissan North America)

some, but like the metal shavings, this is normal. Pieces or chucks of friction material indicate that the hydraulic clutches are bad. The lower side of the filter needs to be checked for the same items. Excessive shavings or friction material normally require a transmission overhaul or rebuild. Consult the instructor, supervisor, or service manual for advice as needed.

Clean the oil pan of all materials, including any gasket material. The oil pan mounting area on the transmission also has to be cleaned of all gasket materials. Be sure to check the bolt holes in the pan and transmission for damage or gasket material. Some manufacturers use RTV instead of a gasket to seal the pan, but RTV can enter and block the hole. Like the manual system, take the time now to check the seals at the drive shaft or drive axles.

WARNING: Overfilling a transmission could cause high pressure within the housing and could rupture a seal or gasket. Damage could result from an overfilled transmission.

Install the filter and torque the filter bolts (Figure 9-19). If RTV was the original sealing material but an aftermarket gasket is available, use the gasket. Place the gasket on the oil pan. Do not use RTV on the gasket unless the gasket manufacturer recommends it. If RTV has to be used in place of a gasket, it should be laid in a 1/8-inch bead and applied evenly around the pan. Remember to lay the bead on the fluid side of the bolt hole and *do not* allow RTV to cover the hole. Ensure that the RTV meets the vehicle manufacturer's specifications and it is allowed to cure 10 to 15 minutes before mounting the pan to the transmission. Before installing the oil pan, wipe the fluid from the mounting area on the transmission. Place the pan in place and start the bolts. Overtightening the bolts can strip the hole threads very easily. It is important to use a good torque wrench and tighten the bolts, in sequence, to torque specification (Figure 9-20). With the pan installed, lower the vehicle and add the proper amount of fluid. The fluid should register at the cold level mark or slightly below the full mark on the **dipstick.**

A *dipstick* is a device for measuring fluid in a component. Most common are dipsticks for the engine, automatic transmission, and power steering fluids. Some dipsticks have cold and hot measurement markings.

With the fluid installed, start the engine and select each gear in turn to fill the pump and clutches. The fluid in most automatic transmissions/transaxles is checked with the engine running and the gear selector in park or neutral and hot. The first check after a service will be with cold fluid, but the level should be nearly correct but not be overfull (Figure 9-21). With

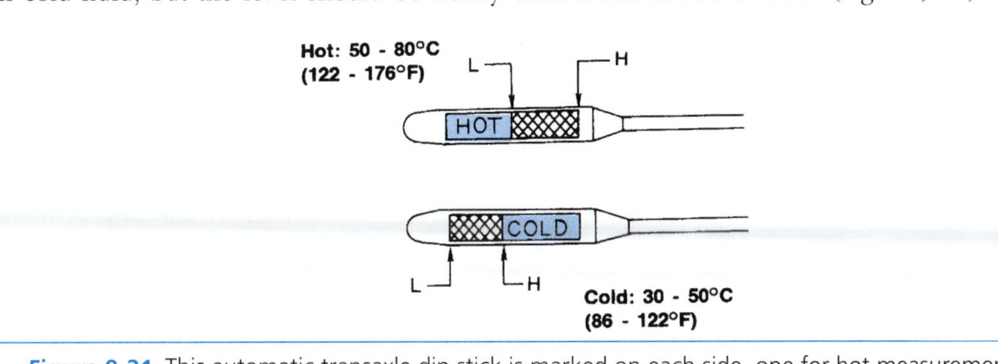

Figure 9-21 This automatic transaxle dip stick is marked on each side, one for hot measurements and the other for cold. (Courtesy of Nissan North America)

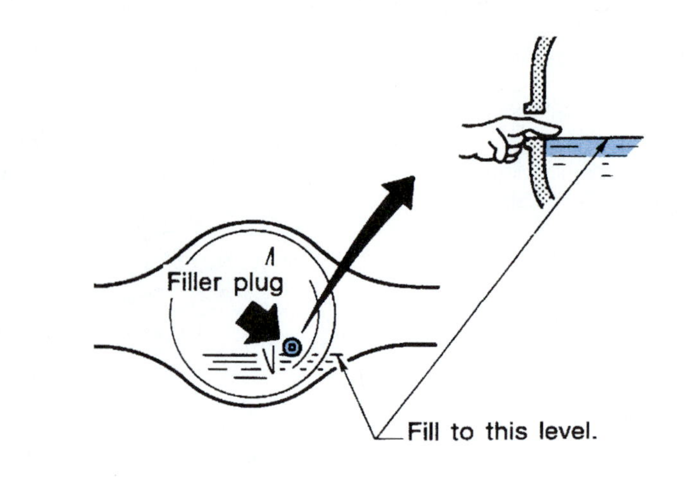

Figure 9-22 The differential fluid level is checked using a finger. (Courtesy of Nissan North America)

the proper level for cold fluid, the vehicle may be driven to heat the fluid so it can be checked at the hot fluid level. Next, check the oil pan for leaks. Remove any fluid over the full mark and repair any leaks if necessary.

Special Tools

Drain pan

Air- or hand-operated pump OR

Suction gun

Lock-up or *limited slip* differentials are used to transfer power from a spinning wheel to the wheel that has traction.

Differentials

The differential is serviced by changing the oil. Some differentials have a drain plug while others have a rear cover that must be removed. The fill plug is usually about half way up the side similar to the manual transmission. The fluid is drained, refilled, and checked the same as a manual transmission and uses the same type of fluid on older differentials (Figure 9-22). Lock-up or **limited slip** differentials use special additives for the clutches. Using regular fluid can seriously damage a lock-up differential. Newer differentials use special blended fluids. Final drives on transaxles are serviced as part of the transaxle.

Classroom Manual
pages 260–261, 263–266

Drive Shaft and Drive Axle Repair

Drive shaft repair usually consists of replacing a U-joint or replacing the entire drive shaft assembly. On the front-wheel-drive axle, CV joints or boots can be replaced, but the most common task is to replace the entire axle. CV joint boots can be replaced, but because of the CV joint's highly machined and matching mechanism, damage to the joint can occur very quickly when a boot is torn and the grease is flung outward. This is usually the cheapest and quickest method for the customer and shop.

Replacing just a boot or one joint may save some time or the cost of the part, but the labor will offset most of that savings. In addition, the shop will have to warrant the repair component and the labor. Replacing the entire drive axle allows the shop to give the customer a new axle or one that was rebuilt in a specialty driveline shop at roughly the same cost and a better warranty.

Drive Shafts

We will cover a basic U-joint replacement. Some vehicles may have a center bearing on the drive shaft and more than two U-joints. The procedures will be the same except for removing the center bearing. Before removing the drive shaft, mark the position of the shaft to its companion flange on the differential (Figure 9-23). The companion flange is the part where the driveshaft's rear U-joint mounts to the differential. There may be another companion flange at the center bearing or the transmission depending on vehicle design. Use a **sealing tool** to slide over the transmission's output shaft once the drive shaft is removed. This will stop any fluid from leaking.

WARNING: Do not over tighten the vise on the drive shaft and do not place the vise over the balancing weight. It may bend or dent the tube, knock a weight off, and unbalance the assembly.

WARNING: Do not use tools for purposes for which they were not designed. Damage to components, tools, or injury to a person may occur.

Place the drive shaft in a vise. Remove the rear U-joint by first removing the retaining clips and pressing the bearing cups out. The clips may be internal or external. Some original U-joints do not use clips, but have plastic retainers that can be sheared when the cups are pressed out. The replacement U-joints for this type will have internal clips. When pressing the cups, use an adapter to support the drive shaft's yoke. This prevents the yoke from spreading. A common practice is to use sockets that are slightly larger and slightly smaller than the cup. The smaller socket is used to push the cups away from the yoke while the larger socket acts as a standoff between the yoke and vise. However, in Chapter 3, "Automotive Tools," there is no mention of using sockets as pressing or standoff tools. Another common practice is to use a hammer and punch. While the hammer and punch can be used as drivers in some instances, this is not one of them. There are tools designed for removing and installing U-joints (Figure 9-24).

Special Tools

Vise

Marker

U-joint bearing press

Grease gun

Drip pan

A *sealing tool* is a pliable cover shaped like the slip joint. The plug slides over the output shaft to prevent oil from leaking out during servicing.

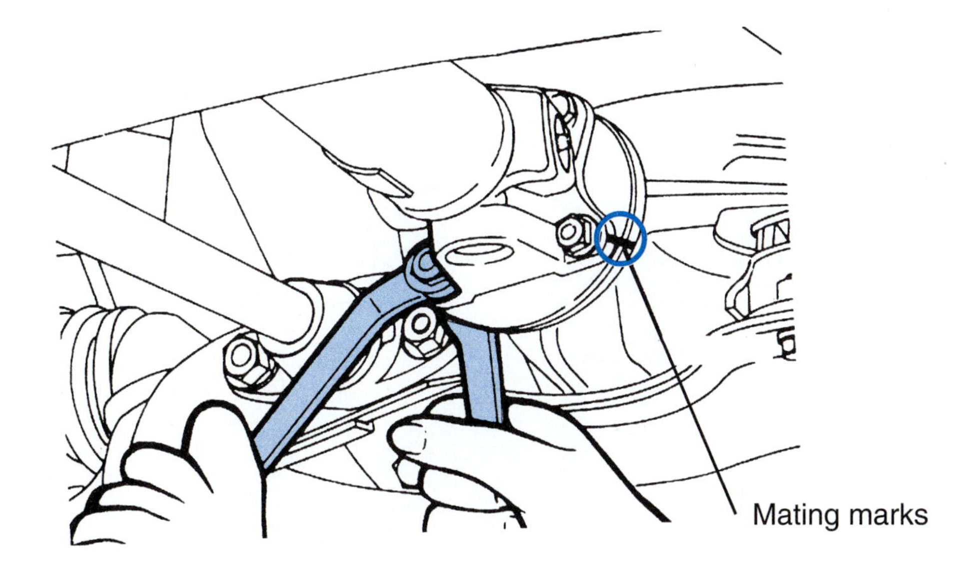

Figure 9-23 Always mark the drive shaft flange with the differential flange before separating the two. (Courtesy of DaimlerChrysler Corporation)

Figure 9-24 Special tools are available for pressing the U-joints in and out of the drive shaft yoke. (Courtesy of Snap-on Tools Company)

The *yoke* of a light drive shaft (small car) can be easily distorted. The correct use of the proper tool will help prevent this distortion.

The *grease fitting* will usually be installed on the drive shaft side of the U-joint.

Pressing on one cup will force the opposite cup out of the **yoke**. With the first cup out, try to remove the U-joint cross by moving it into the empty bore and twisting it to the side. If the space is not sufficient, the yoke will have to be rotated and force applied to the end of the cross. This will move the second cup outward enough to remove the cross. The second cup can be knocked out with a rod installed through the empty bore once the cross is removed.

During the installation of the U-joints, the bearing cups must be kept squarely in the bore as force is being applied. The needle bearings can be displaced if the cup is dropped or is not aligned correctly on the cross (Figure 9-25). The cross has a grease fitting hole that must be placed correctly as well. The hole is positioned to allow the insertion of the **grease fitting** and allow the connection of a grease gun with the drive shaft on the vehicle.

Place the cross in a bore and align a cup on the cross to keep the bearings in place as the cup is pressed. Press the first cup just enough to position the other end of the cross to align the second cup. With both cups being guided by the cross, apply force to each side so the cups are pressed into the yoke. One cup will probably be in position before the other.

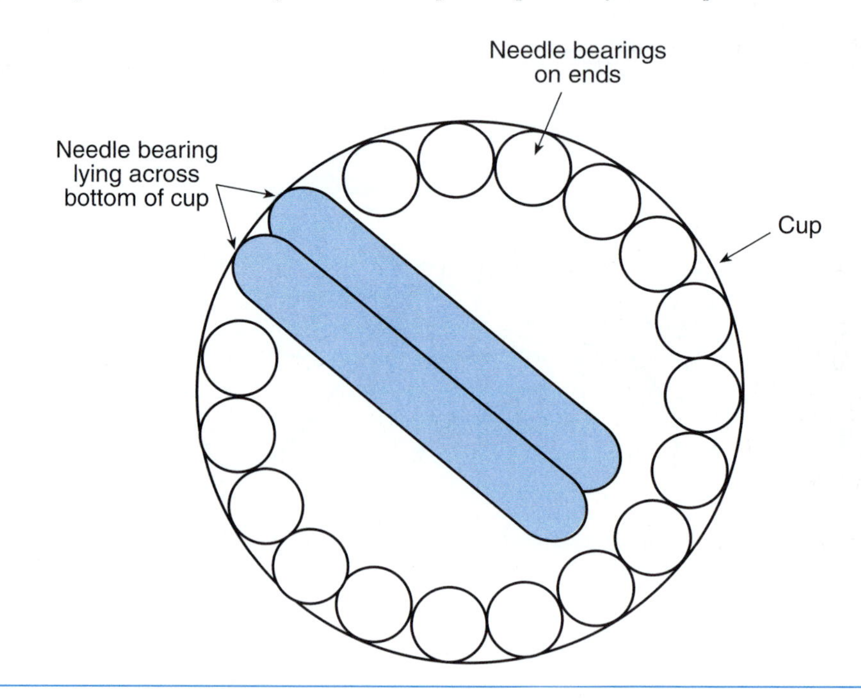

Figure 9-25 Needle bearings stand on end and can be easily knocked over during installation.

Photo Sequence 9
Typical Procedure for Disassembling a Single Universal Joint

P9-1 Clamp the slip yoke in a vise and support the other end of the drive shaft.

P9-2 Then remove the lock rings on the tops of the bearing caps. Make index marks in the yokes so that the joint can be assembled with the correct phasing.

P9-3 Select a socket that has an inside diameter large enough for the bearing cap to fit into, usually a 1 1/4-inch socket will suffice.

P9-4 Select a second socket that can slide into the shaft's bearing cap bore, usually a 9/16-inch socket.

P9-5 Place the large socket against one vise jaw. Position the drive shaft yoke so that the socket is around a bearing cap.

P9-6 Position the other socket to the center of the bearing cap opposite the one in line with the large socket.

P9-7 Carefully tighten the vise to press the bearing cap out of the yoke and into the large socket.

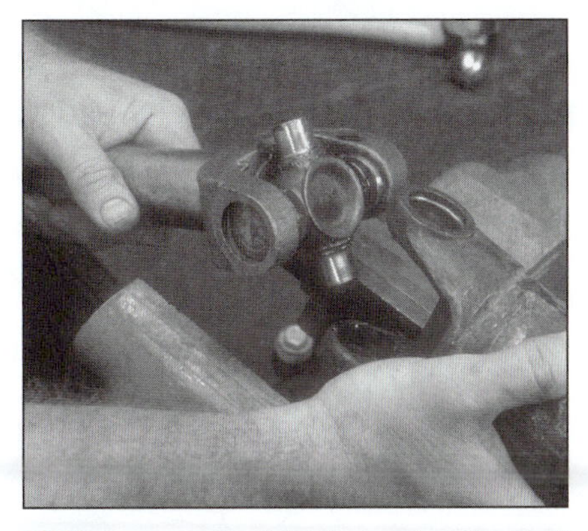

P9-8 Separate the joint by turning the shaft over in the vise and driving the spider and remaining bearing cap down through the yoke with a brass drift punch and hammer.

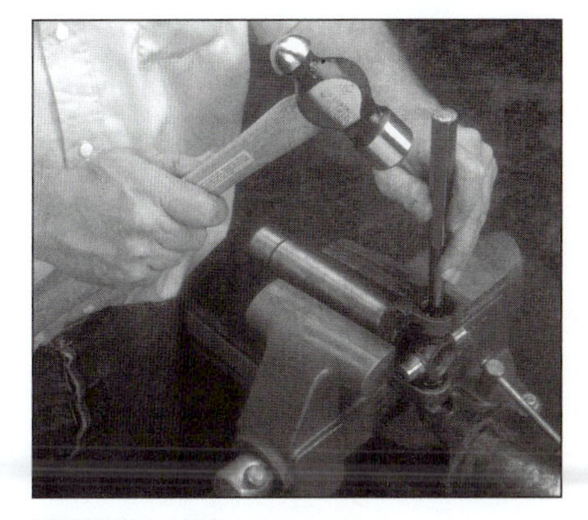

P9-9 Use a drift and hammer to drive the joint out of the other yokes.

Install the retaining clip on the cup then press and clip the second cup into position. If the second cup does not go all the way in, one or more of the needle bearings may have dropped to the bottom of one of the cups. The joint will have to be removed and checked. Install the grease fitting.

Some rear U-joints have the second set of cups in a flange, which bolts to the companion flange (Figure 9-26). Replace the cups on this type of arrangement at this time. Many systems have outer cups that connect to the companion flange by U-bolts. If this is the case, move the two new cups out of the way. They will not be needed until the drive shaft is reinstalled. Rotate the U-joint to ensure that it is free.

Mark the position of the slip joint to the drive shaft. This U-joint is removed and installed the same as the rear one, however, it may require several attempts to clamp the shaft in the proper position in the vise. Ensure that the slip joint is positioned correctly as the U-joint is being installed. Install the grease fitting and grease the joint. If all four cups are installed on the rear joint, it can also be greased. If only two cups are installed, the joint cannot be greased correctly until the drive shaft is installed into the vehicle.

Remove the sealing tool from the transmission and slide the slip joint into position. Place the two rear bearing cups on the cross and in position on the companion flange. Ensure that the marks are aligned. Install and torque the bolts. If necessary, grease the rear U-joint. The vehicle can now be lowered, the area cleaned, the repair order completed, and the vehicle returned to the customer.

Drive Axles

Replacing a front-wheel-drive axle is not normally assigned to an entry-level technician, but he or she may be asked to assist in the replacement. The replacement process is fairly simple and can be learned quickly.

Front-wheel-drive axles consist of two CV joints and a shaft. If the shaft is bent, the entire axle is *always* replaced. If a CV joint is damaged or the boot is torn, either can be replaced, but as mentioned above, it is usually cheaper in the long run for the customer to replace the axle as an assembly. We will discuss the replacement of a typical drive axle. The following is typical of manual or automatic transaxles and should be supervised by the instructor or an experienced technician until proficiency is achieved by the new technician.

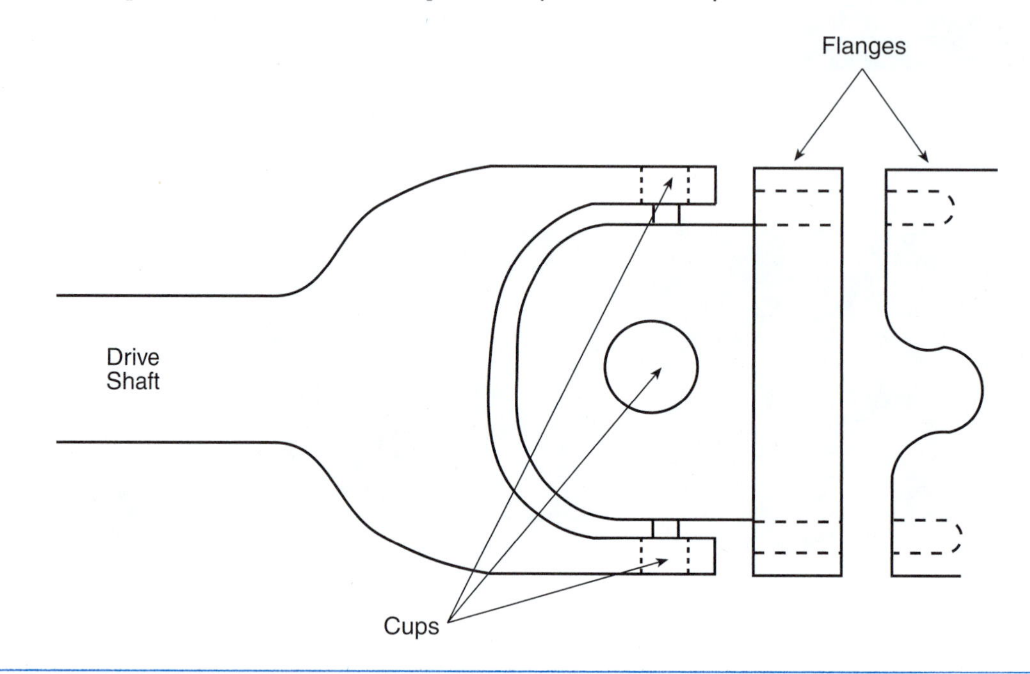

Figure 9-26 Flanges like the two shown must be bolted together.

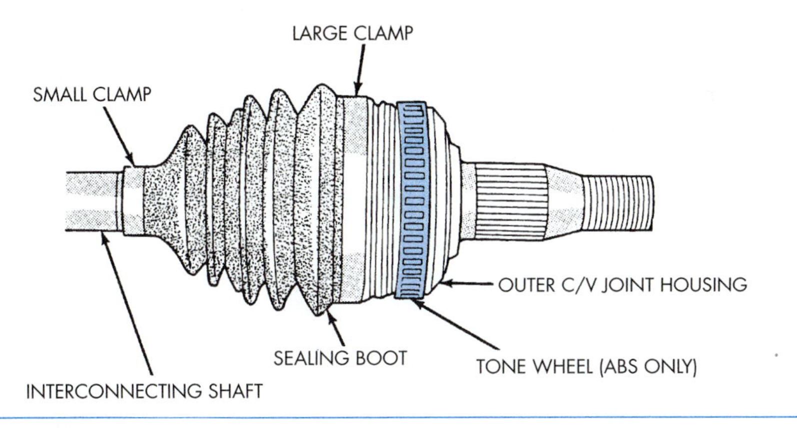

Figure 9-27 The tone ring on this drive axle signals the speed of the axle and wheel to the ABS control module. (Courtesy of DaimlerChrysler Corporation)

WARNING: Use extreme care when working on vehicles with antilock brake systems. Damage to the system or components may occur if improper tools or procedures are used.

Lift the vehicle to gain access to the inner end of the shaft. A drain pan should be available to catch any fluid that may drain from the transaxle's final drive housing. Remove the wheel assembly. Locate, remove, and hang the brake caliper and rotor out of the way. Before removing any other components, inspect the area around the **steering knuckle**. Closely inspect for antilock system components. If a sensor is present, consult the service manual for procedures to protect it during axle replacement. Also check the outer CV joint for a **tone ring** (Figure 9-27).

There are several methods used to attach the hub to the suspension. If a wheel alignment is performed at the strut, the vehicle will have to be checked for alignment before returning it to the customer.

The first step after inspection is to remove the axle nut. The **cotter key** and locking plate or tab are removed followed by the nut. There are two ways to set up the knuckle for removal of the axle. The first is to remove the steering linkage from the knuckle and then remove the strut/knuckle mounting bolts. The free knuckle, swinging on the lower **ball joint**, can be swung out of the way as the axle stub is pushed from the **hub** (Figure 9-28). The wheel alignment will probably have to be checked in this case.

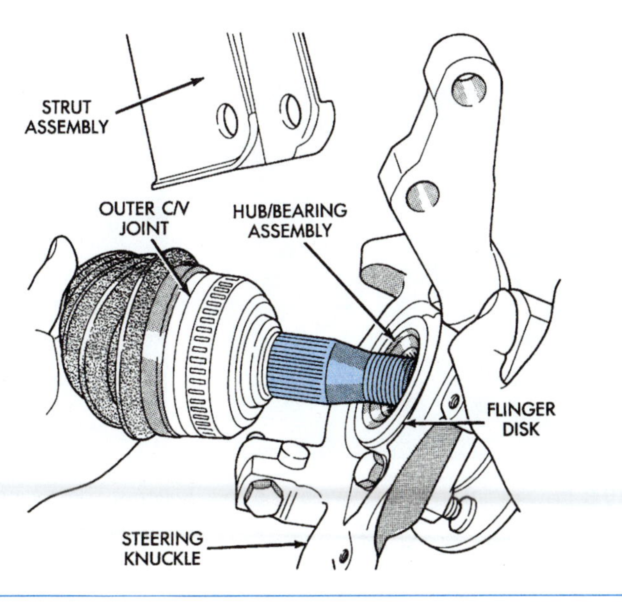

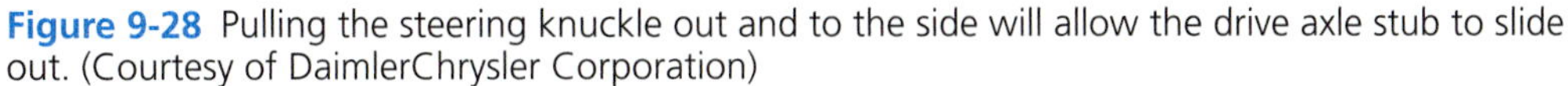
Figure 9-28 Pulling the steering knuckle out and to the side will allow the drive axle stub to slide out. (Courtesy of DaimlerChrysler Corporation)

Special Tools

Torque wrench, $^3/_8$ and $^1/_2$ inch

Small pry bar

The *steering knuckle* supports the wheel assembly and hub. The hub fits into the steering and supports the axle with a bearing.

A *tone ring* is a metal ring around the CV joint. It has humps and valleys that create a signal for the antilock wheel sensor. Damage to one hump will cause the antilock system to malfunction.

A *cotter key* is a piece of metal bent double and fitted into a slot or hole. It is used in most cases to prevent a nut or bolt from working loose.

A *ball joint* is used between the steering knuckle and the control arms. The joint allows the knuckle to pivot as the wheel assembly moves during steering and road conditions.

A *hub* provides the connection component between the drive axle and the wheel assembly. It houses the necessary bearings and has mounting surfaces and hardware for the axle, brake rotor, and wheel.

Photo Sequence 10
Typical Procedure for Removing and Installing Drive Axles

P10-1 With the tire on the ground, remove the axle hub nut and loosen the wheel nuts.

P10-2 Raise the vehicle and remove the tire and wheel assembly.

P10-3 Remove the brake caliper and rotor. Be sure to support the caliper so that it does not hang by its weight.

P10-4 Remove the bolts that attach the lower ball joint to the steering knuckle. Then pull the steering-knuckle assembly from the ball joint.

P10-5 Remove the drive axle from the transaxle by using the special tool.

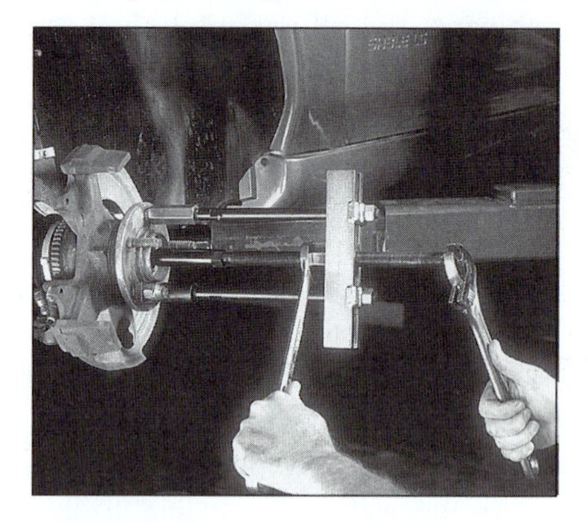

P10-6 Remove the drive axle from the hub and bearing assembly using a spindle remover tool.

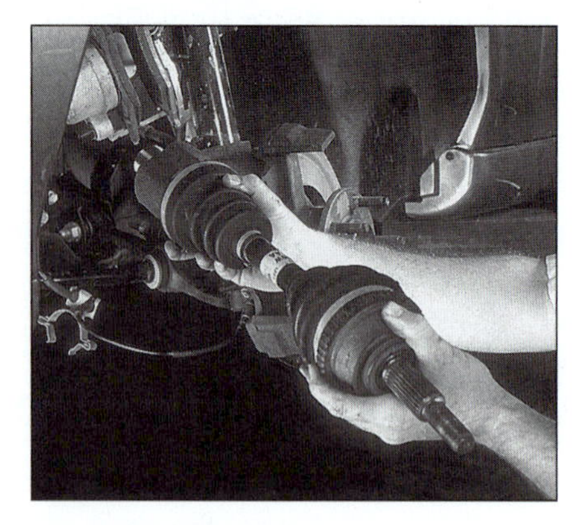

P10-7 Remove the drive axle from the car.

P10-8 Install joint in a soft-jawed vise to make necessary repairs to the shaft and joints and prepare to reinstall the shaft.

P10-9 Loosely install the halfshaft into the steering knuckle and transaxle.

Typical Procedure for Removing and Installing Drive Axles (continued)

P10-10 Tap the splined end into the transaxle or install the flange bolts and tighten them to specifications.

P10-11 Pull the steering knuckle back and slide the splined outer joint into the knuckle.

P10-12 Fit the ball joint to the steering-knuckle assembly. Install the pinch bolt and tighten it.

P10-13 Install the axle hut nut and tighten it by hand.

P10-14 Install the rotor and brake caliper.

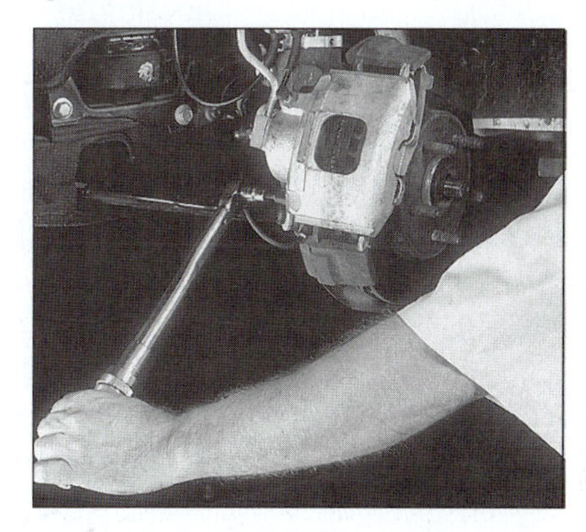

P10-15 Torque the caliper bolts to specifications.

P10-16 Install the wheel and tire assembly.

P10-17 Lower the car.

P10-18 Torque the hub nut to the specified amount.

A second method is to separate the steering linkage from the knuckle and unbolt the lower control arm. When this is done, the knuckle will hang from the strut (Figure 9-29). This is the best method because it will not change the wheel alignment.

WARNING: Do not drive directly on the end of the axle. Damage to the axle threads will occur. Use the axle nut or a piece of wood between the drive and the axle.

Axle stubs are the outer ends of the axle. The stubs fit into the transaxle or the hub. Each stub will have spines to fit either the hub or final drive gears. The hub end will be threaded for the axle nut.

Separate the axle from the hub by pulling out the steering knuckle to the side. In most cases, the axle will slide from the hub. If the axle will not slide out, press the **axle stub** from the hub using a special puller tool. Once loosened, the axle should slide out. Allow the hub and knuckle to hang to one side as the axle is separated from the transaxle.

WARNING: Always use a drive axle that has the correct locking device on the transaxle end of the axle. An undersized clip may allow the axle to come out of the transaxle during operation. An oversized clip may damage transaxle gears during axle installation and hinder removal if it does not lock in place.

WARNING: Do not pry or hit the CV boot. Damage to the boot may occur and allow the CV grease to be flung out during operation.

A small pry bar is fitted between the inner end of the inner CV joint and the transaxle housing to unlock the inner end of the axle (Figure 9-30). A special tool is also available to unlock the axle. Slide the axle out. Check the end of the axle's inner stub for a retaining clip or lock. Ensure that the replacement axle has the same type and size of lock or clip.

Place the new axle into the transaxle until it locks or is fully positioned against the transaxle housing. The axle must not slide out when using only hand force. A brass hammer can be used to tap the axle in place. If it does not lock in, inspect and correct the problem. Line the outer axle stub with the hub and slide the hub onto the axle. Install the washer and loosely tighten the axle nut. Install the lower ball joint and the steering linkage. As the ball joint and steering linkage nuts are tightened to torque, line up the nuts to the cotter key's hole, if used. If the nut passes the hole, do not loosen the nut. Tighten it until the hole aligns again. If the nut has to be loosened, it must be removed completely and the two components broken apart before reinstalling.

Figure 9-29 Separating the steering knuckle from the lower control arm prevents alignment of the steering following a drive axle or CV boot installation.

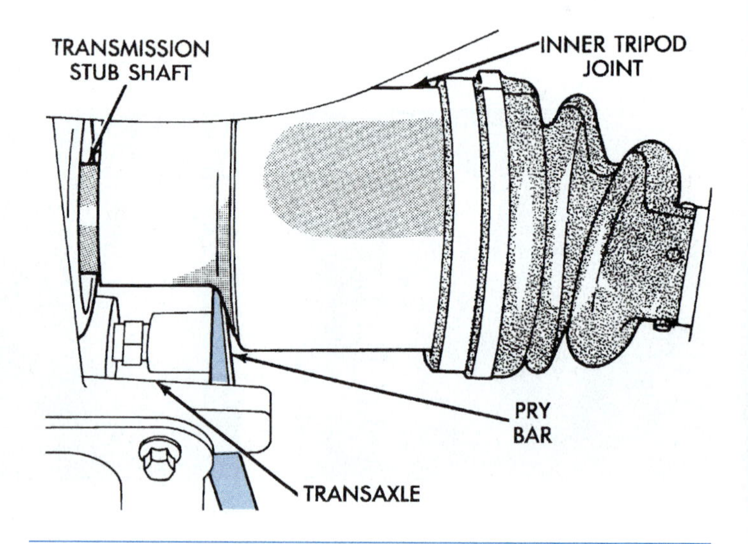

Figure 9-30 A small pry bar can be used to unlock the drive axle from the final drive gears. (Courtesy of DaimlerChrysler Corporation)

Install the brake components and wheel assembly. Lower the vehicle until the tires are resting on the floor. Torque the axle nut and install the locking collar and cotter key. If locking tabs are used, tap or crimp the tabs in place. Operate the brakes to reseat the pads. Finish the repair by completing the repair order, cleaning the area, storing the tools, and returning the vehicle to the owner.

CASE STUDY

An automatic transaxle was rebuilt and installed in a vehicle. A rebuilt pump and torque converter were also installed. During the test drive, a serious fluid leak was detected at the bottom of the bell housing. Knowing the only possible transaxle components that could leak in this area were the pump gasket, pump seal, or torque converter, the technician removed the transaxle. Inspection of the seal and torque converter showed no damage to either part. Fluid was present around the inside of the bell housing and over the torque converter. The technician removed the pump and found the gasket to be dry. A closer inspection of the suspected components was conducted. When the pump seal was removed, it was found that the pump bushing behind the seal cavity was missing. This allowed the converter pump lug to move within the pump and break contact with the seal. A replacement pump was supplied and the transaxle reinstalled. A test drive revealed no leaks and the vehicle was returned to the customer.

Terms to Know

Ball joint
Bleeder screw
Cotter key
Dipstick
Free travel
Grease fitting
Hub
Limited slip
Power pack
Steering knuckle
Suction pump

ASE Style Review Questions

1. Drive axles are being discussed. *Technician A* says some drive axles have center bearings. *Technician B* says the CV joint or boot may be replaced individually. Who is correct?
   A. A only
   B. B only
   C. Both A and B
   D. Neither A nor B

2. Drive shafts are being discussed. *Technician A* says the drive shaft must be marked before removal. *Technician B* says the U-joint does not need to be marked before removing. Who is correct?
   A. A only
   B. B only
   C. Both A and B
   D. Neither A nor B

3. Transmission fluid is being discussed. *Technician A* says most manual transmissions use automatic transmission fluid.
   Technician B says the level of fluid in an automatic transaxle should be checked while hot for an accurate measurement. Who is correct?
   A. A only
   B. B only
   C. Both A and B
   D. Neither A nor B

4. Transmission servicing is being discussed. *Technician A* says the oil pan bolts must be torqued.
   Technician B says the fill plug is on the side of a manual transmission. Who is correct?
   A. A only
   B. B only
   C. Both A and B
   D. Neither A nor B

5. A vehicle has a rattling noise from the area of the clutch. The noise goes away when the clutch is disengaged. *Technician A* says the problem is in the release bearing.
   Technician B says the transmission input shaft bearing is probably the cause. Who is correct?
   A. A only
   B. B only
   C. Both A and B
   D. Neither A nor B

6. A whining noise is heard as the engine is running and the transmission is in neutral. *Technician A* says this may be caused by a torque converter.
   Technician B says the output shaft bearing may be the cause. Who is correct?
   A. A only
   B. B only
   C. Both A and B
   D. Neither A nor B

7. During servicing of an automatic transmission, metal shavings are found on the magnet.
   Technician A says a small amount of shavings is normal.
   Technician B says excessive shavings may indicate a problem with the overrunning clutch. Who is correct?
   A. A only
   B. B only
   C. Both A and B
   D. Neither A nor B

8. Differential servicing is being discussed. *Technician A* says the fluid may be the same as the manual transmission.
   Technician B says lockup differentials use regular fluid. Who is correct?
   A. A only
   B. B only
   C. Both A and B
   D. Neither A nor B

9. U-joint replacement is being discussed. *Technician A* says the bearing cups are pressed out.
   Technician B says a cup that fails to seat completely may be caused by a misplaced needle bearing. Who is correct?
   A. A only
   B. B only
   C. Both A and B
   D. Neither A nor B

10. A clunking noise is heard during gear changing. *Technician A* says the noise is probably in the differential or U-joints on an automatic-transmission-equipped vehicle.
    Technician B says the noise may be caused by a bad inner CV joint. Who is correct?
    A. A only
    B. B only
    C. Both A and B
    D. Neither A nor B

Table 9-1 ASE Task

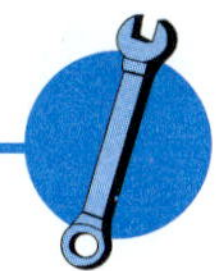

Remove and install antilock brake system (ABS) components following manufacturers' procedures and specifications.

Problem Area	Symptoms	Possible Causes	Classroom Manual	Shop Manual
Brakes	Axle/brake service.	**1.** Replace drive axle.	261	220, 234–239

SAFETY

Wear safety glasses.
Use proper procedures and tools.

Table 9-2 ASE Task

Diagnose noise and vibration problems; determine necessary action.

Problem Area	Symptoms	Possible Causes	Classroom Manual	Shop Manual
Automatic transaxle	Vibration during operation.	**1.** Bearings and gears worn or damaged.	252–256	218–219
		2. Pump worn.	247, 252	218

SAFETY

Wear safety glasses.
Allow engine and transaxle to cool.

Table 9-3 ASE Task

Replace fluid and filter.

Problem Area	Symptoms	Possible Causes	Classroom Manual	Shop Manual
Automatic transaxle, transmission	Scheduled service.	**1.** Mileage.	—	226–230
	Slipping, leakage.	**2.** Seals worn, damaged.	—	226–230
		3. Worn clutches, bands.	253–254	226–230

SAFETY

Wear safety glasses.
Allow transmission/transaxle to cool.

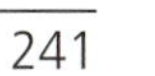

Table 9-4 ASE Task

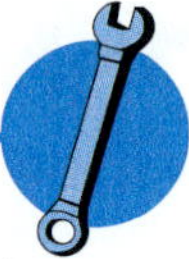

Diagnose clutch noise, binding, slippage, pulsation, chatter, pedal feel/effort, and release problems; determine needed repairs.

Inspect, adjust, and replace clutch pedal linkage, cables and automatic adjuster mechanisms, brackets, bushings, pivots, and springs.

Inspect, adjust, replace, and bleed hydraulic clutch slave and master cylinders, lines, and hoses.

Problem Area	Symptoms	Possible Causes	Classroom Manual	Shop Manual
Manual drive train	Hard shifting. Slipping clutch. Leakage.	**1.** Worn clutch disc.	233	220–222
		2. Improper linkage adjustment.	234	222–223
		3. Low clutch fluid.	235	221
		4. Leaking master or slave cylinder.	235	221
		5. Air in hydraulic system.	235	221

SAFETY

Wear safety glasses.
Allow engine to cool.

Table 9-5 ASE Task

Diagnose transaxle noise, hard shifting, jumping out of gear, and fluid leakage problems; determine needed repairs.

Problem Area	Symptoms	Possible Causes	Classroom Manual	Shop Manual
Manual drive lines	Hard shifting. Noise during shifting/operation.	**1.** Worn gears/synchronizers.	236–246	217–218
		2. Low transmission fluid.	—	—
		3. Clutch linkage misadjusted.	234–236	218, 221–223
		4. Clutch disc worn.	233	221
		5. Clutch pressure plate worn.	232	221

SAFETY

Wear safety glasses.

Table 9-6 ASE Task

Diagnose shaft and universal/CV joint noise and vibration problems; determine needed repairs.

Inspect, service, and replace shafts, yokes, boots, and universal/CV joints.

Inspect, service, and replace shaft center support bearing.

Problem Area	Symptoms	Possible Causes	Classroom Manual	Shop Manual
Drivelines	Noise.	**1.** Worn/damaged CV joints.	263	220, 234–239
	Vibration.	**2.** Worn/damaged U-joints.	266	219, 231–234
		3. Bent drive shaft.	260	219, 231–234
		4. Bent drive axle.	261	220, 234–239
		5. Worn/damaged center bearing.	260	219, 231–234

SAFETY

Wear safety glasses.

Table 9-7 ASE Task

Inspect lubrication devices; check fluid level and refill with proper fluid.

Inspect, flush, and refill with correct lubricant.

Problem Area	Symptoms	Possible Causes	Classroom Manual	Shop Manual
Transmission, transaxle differential. Limited slip differential.	Schedule service.	**1.** Mileage/time.	—	—
	Leaks, noise.	**2.** Leaking seals.	—	223–225, 236
	Limited slip inoperative.	**3.** Low fluid level.	—	223–225, 236
		4. Worn/damaged limited slip clutches.	261–262	230

SAFETY

Wear safety glasses.

Job Sheet 14

14

Name ______________________ Date ______________

Changing Fluid in an Automatic Transmission

Upon completion of this job sheet, you should be able to change the fluid in an automatic transmission or transaxle.

Tools Needed

Drain pan
Hand tools
Service manual
Torque wrench $^1/_4$-inch drive, inch-pounds

Procedures

1. Identify the transmission and specifications

   Vehicle make __________ Model __________ Engine __________

   Transmission ______________________________

   Type of fluid ______________________________

   Amount of fluid __________ L/Qt

   Fluid to be checked ________ hot/cold __________ park/neutral

   Filter bolt torque __________ inch/foot-pounds

   Pan bolt torque __________ inch/foot-pounds

2. Lift the vehicle to gain access to the oil or fluid pan.
3. Position the drain pan and remove the lowest pan bolts.
4. Remove other bolts slowly until the fluid begins to drain.
5. Support the pan and remove all bolts. Drain any remaining fluid.
6. Inspect the magnet. Describe any materials that adhered to the magnet.______________

   ____

   ____

   ____

7. Remove the filter. Remove any filter gasket or O-ring if present.
8. Remove all gaskets or RTVs from the pan and pan mounting area on the transmission.
9. Install the filter and gaskets or O-ring. Torque the filter bolts.
10. Place the gasket or RTV on the pan. Wipe the pan mounting area on the transmission clean.

11. Position the pan and gasket and install the bolts.
12. Torque the pan bolts in a sequence.
13. Lower the vehicle. Keep the wheels off the floor.
14. Add the specified type and amount of fluid. Conduct an initial fluid level check. Top off as needed.
15. Set the parking brake if it has not already been set. Start the engine and engage each gear for five to ten seconds. Ensure that the transmission engages reverse and drive.
16. Select the park (neutral) gear.
17. Check the transmission fluid for cold level. Top off as necessary.
18. Test drive the vehicle until the engine is at operating temperature.
19. Check the fluid level. Is the fluid hot? ________ Cold? ________
20. Top off as necessary.

____

____

Job Sheet 15

15

Name ______________________________ Date ______________

Servicing a Differential

Upon completion of this job sheet, you should be able to service a differential on a rear-wheel-drive vehicle.

Tools Needed

Hand tools
Service manual
Torque wrench, 3/8-inch or 1/4-inch drive
Drain pan with screen or magnet or wide flat pan
Air- or hand-operated pump OR
Suction gun

Procedures

1. Identify vehicle and specifications:

   Vehicle make __________ Model __________ Engine __________

   Transmission ______________________________ manual/automatic

   Type/amount of fluid ____________________

   Cover bolt torque ____________________

2. What is your reason for servicing this differential? (diagnosis, symptoms) ______________

   ____

3. Lift the vehicle to a comfortable working height.
4. Position the drain pan under the differential. If possible, use a pan that will capture and hold metal shavings with a screen or magnet, or use a wide, flat pan in which fluid can be examined after draining.
5. If the differential has a drain plug, remove it and drain the fluid.
6. Once the fluid is completely drained, install and torque the plug to specifications. Go to step 21.
7. Remove the lower bolts on the cover. Loosen the side bolts starting at the bottom on each side. Do not remove bolts. Fluid may start to drain as the bolts are loosened.
8. If necessary, use a screwdriver to pry the bottom of the cover from the housing. Use caution since fluid will begin to flow as soon as the seal is broken.
9. Allow the fluid to drain.
10. With most of the fluid drained, remove the remainder of the bolts and the cover.

11. Examine the waste fluid for any metal shavings. There should be none. Inspection results:

____

____

12. If shavings are found, further diagnosis of the gears and bearings is required.
13. Clean the cover and housing of all sealing material. Clean the inside of the housing as much as possible.
14. If a gasket is available, position it on the cover and insert a bolt through the cover.
15. Place the gasket and cover on the housing. Start two bolts to hold them in place. Start the remaining bolts and hand tighten all of them.
16. Torque the cover bolts to specifications. Go to step 21.
17. If a gasket is not used, lay a bead of RTV on the cover. Ensure that the bead is about 1/8-inch thick, evenly applied around the cover edges, and attached to the fluid side of the bolt holes. Let stand for 10 to 15 minutes.
18. Align the cover's bolt holes with the holes in the housing.
19. Align the cover with the housing. Ensure that the cover does not slide as the bolts are started.
20. Start and hand tighten all bolts. Torque all bolts to specifications.
21. Remove the fill plug. It is normally about halfway up on the side of the housing just behind the drive shaft connection.
22. Secure the correct fluid and install it through the fill plug. Use the available fluid pump to install the fluid.
23. As the fluid nears specified capacity, reduce the fluid flow.
24. Cease filling just as soon as the fluid begins to flow from the fill hole.
25. Check the fluid level using the first joint of the little finger or a measuring tool. Add or drain fluid as necessary to the correct level.
26. Check the cover for leaks.
27. If the repair is complete, clean the area of tools and waste fluid and lower the vehicle.
28. Complete the repair order and return the vehicle to the owner.

____

____

Job Sheet 16

16

Name ______________________________ Date ______________

Replacing a U-Joint

Upon completion this job sheet, you should be able to replace a U-joint in a drive shaft assembly.

Tools Needed

Lift OR
Floor jack and stands
U-joint bearing press
Grease gun
Marker
Transmission sealing plug
Drain pan
Vise

Procedures

1. Identify the vehicle.

   Vehicle make __________ Model __________ Engine __________

   Transmission type____________U-joint bolt torque ____________

   Are all of the U-joints being replaced? ____________

   Is there a center bearing on this shaft? ____________

   Is it being replaced? ____________

   If yes, what is the bearing bolt torque? ____________
2. Lift the vehicle to a good working height.
3. Compare the replacement U-joint with the installed joint.

   Are they the same in size of bearing cup? ____________

   Do they have the same locking devices? ____________
4. Mark the rear end of the drive shaft to the differential companion flange.
5. Position a drip pan under the rear of the transmission.
6. If present, remove the center bearing fasteners. Support the shaft.
7. Remove U-joint/flange fasteners at the differential. Support the shaft.
8. Slide the slip joint from the transmission. Set the shaft aside.
9. Install the sealing plug.
10. Position the shaft in a vise with the vise jaws near the U-joint being removed.
11. Remove all locking clips, if present, from the joint.
12. Position the U-joint bearing press.
13. Tighten the press until one bearing cup is forced out.
14. Remove the press, twist the joint cross sideways, and remove the cross. If it will not clear, position the press in an opposite direction to the first setup. Force the second bearing cup out. Remove the cross.
15. Rotate the shaft in the vise to more fully expose the second set of bearing cups.

16. Repeat steps 12 through 14.
17. Remove the bearing cups from the replacement joint and place them aside.

■ **NOTE:** Read steps 18 through 25 before starting the installation.

18. Start one cup into one hole of the yoke.
19. Insert the cross into the yoke by placing one end of the cross into the seated cup.
20. Twist the cross until the opposite end is fitted into the second and opposite hole of the yoke. Hold in place.
21. Start a second bearing cup over the exposed cross end and into the yoke.
22. Position the cross so the two ends are started slightly into their respective cups and the grease fitting is facing in the correct direction. Hold in place.

WARNING: Do not apply excessive force while pressing the cups. They will require some force, but if it becomes difficult, the bearing cups will have to be removed to check for dropped needle bearing. If a bearing is dropped, excessive force could damage it, the cup, and the yoke.

23. Position the bearing press and slowly press the cups into the yoke. Rotate the cross as the cups are pressed.
24. At about the halfway point, stop pressing the cup. Check the position of the cross within the cups. If the cross is positioned properly, continue pressing the cups.
25. When the cups are in position, move the cross as far as it will go in each direction. Install the locking clips if the cross can be moved by hand. If it will not move, the cups and cross will have to be removed to check for dropped bearings.
26. Install the grease fitting.
27. Repeat steps 18 through 25 to remove and install the other cups on this joint.
28. Repeat steps 8 through 25 to remove and install the other U-joints.
29. Grease all U-joints that have all bearing cups installed and locked. The other U-joints will be greased after the drive shaft has been installed on the vehicle.
30. Move the shaft to the vehicle.
31. If required, replace the center bearing onto the shaft.
32. Remove the sealing plug and slide the slip joint into the transmission.
33. If this is a two-piece drive shaft (with a center bearing), support the shaft as the rear bolts are being installed.
34. Install the bearing cups, if necessary, onto the cross.
35. Position the rear U-joint into the companion flange and start the bolts. Align the marks by turning one rear wheel to rotate the flange.
36. Torque the fasteners.
37. Install the center bearing, if present, and torque the fasteners.
38. If necessary, grease the U-joints that have not been lubricated.
39. If necessary, top off the transmission fluid at this time (manual) or after the vehicle has been lowered (automatic).
40. Secure all tools and clean the area of any oil spots.
41. Lower the vehicle, complete the repair order, and return the vehicle to the customer.

Suspension and Steering Service

Upon completion and review of this chapter, you should be able to:

- ❑ Diagnose suspension and steering malfunctions.
- ❑ Replace a shock absorber.
- ❑ Replace a power-steering pump drive belt.
- ❑ Replace inner tie rod on a rack and pinion system.
- ❑ Diagnose tire wear.
- ❑ Remove and install wheel assembly.
- ❑ Remove, balance, and mount tires on wheels.
- ❑ Set up alignment lifts and machines.

Introduction

The driver may not realize a steering or suspension problem exists until a part fails. The technician should make at least a cursory inspection of any vehicle he or she may have serviced. The inspection may only be a quick ride into the shop or a check of the thread on the tires. Spotting a potential hazard may bring business to the shop, but more important is the recognition and correction of a fault that could prevent accidents and injury.

Checking Frame and Suspension Mounting Points

Classroom Manual
page 271

As stated in the Classroom Manual, the condition of the frame affects the suspension and steering operations. A technician performing suspension and steering work should make an inspection of the frame. The frame should not show any usual bends or cracks. Mounting fasteners that show any sign of movement indicate a portion of the frame has shifted.

Attention should also be directed to the point where brackets are welded to the frame. Any break in a weld bead could indicate a twisted or bent frame. Abuse of the vehicle through overloading or driving can result in distorted frame members. Check the door, hood, and trunk alignment and hinge fittings. A misaligned door, hood, and trunk does not necessarily mean the frame is bent, but a bent or twisted frame could deform the body.

Diagnosing Suspension Problems

Classroom Manual
pages 272–273

At some time in their travels, everyone has seen obvious indications of poor suspension and steering components in vehicles on the road. The compact car with six passengers, the overloaded pickup, worn tires, and other indicators point to faults or potential faults. Each could eventually result in a visit to a shop. Most suspension problems can be prevented with some maintenance and timely service. The components discussed in this section are control arms bushings, ball joints, springs, and shock absorbers. Torsion bars are checked in the same manner as springs and struts are checked in the same manner as shock absorbers.

Driver Complaints

The customer's *complaint* should be shown on the repair order.

The typical driver's **complaint** on suspension systems involves noise, swaying, and sagging of a corner or end of the vehicle. Most noise complaints are stated as a heavy thump, grating, or grinding noise as the vehicle wheels moves over the road surface. This may indicate a weak spring that would allow the control arm to bottom out against the rebound bumper on the frame, a bad ball joint, a missing or worn control arm bushing, or other component failure

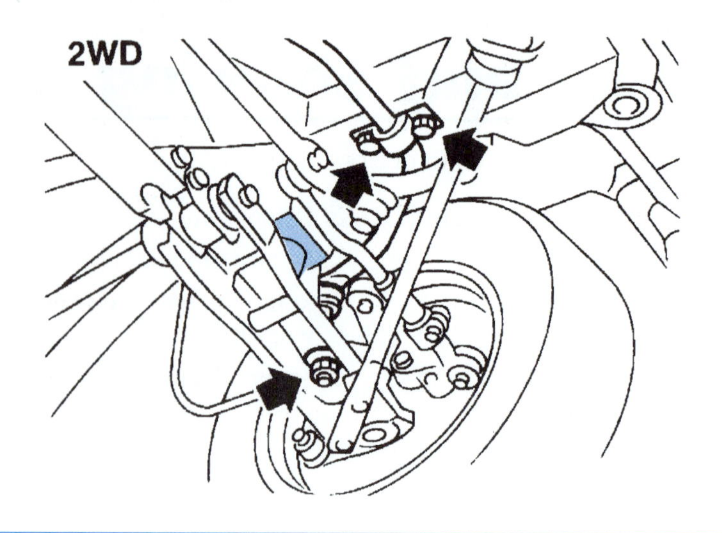

Figure 10-1 The bumper prevents the control from contacting the frame when the spring "bottoms" out. (Courtesy of Nissan North America)

(Figure 10-1). The driver may also complain that the vehicle seems to roll or lean excessively during a turning maneuver. This could be caused by a weak spring, a worn or missing sway bar bushing, or mountings that are worn or missing. A technician should test drive the vehicle with the driver if possible. Test drive results along with the following inspections will isolate the problem component within a minimum time.

Control Arms

Most control arms do not suffer damage unless abused, but the bushings supporting the control arms do erode. The outer ends of the bushings are visible under the vehicle (Figure 10-2). The bushings should not be cracked, dry, or oil soaked. Pushing with a flat-ended rod or screwdriver can check the bushings' flexibility. The bushing material should flex and return to its orginal condition. If the material is very soft or does not return, it should be replaced. Replace all of the bushings on a control arm as a set. Worn bushings will allow the control arm to shift and may create a thud as the vehicle accelerateds, brakes, or turns.

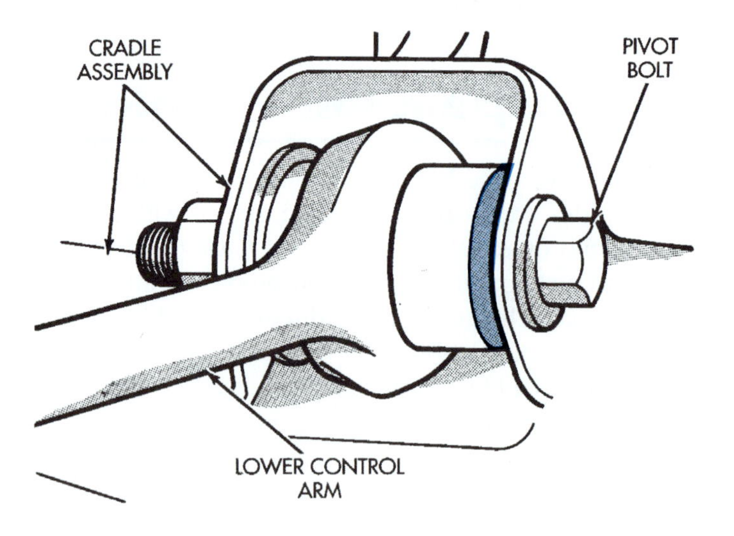

Figure 10-2 The edges of the bushings can be seen between the control arm and the frame. (Courtesy of DaimlerChrysler Corporation)

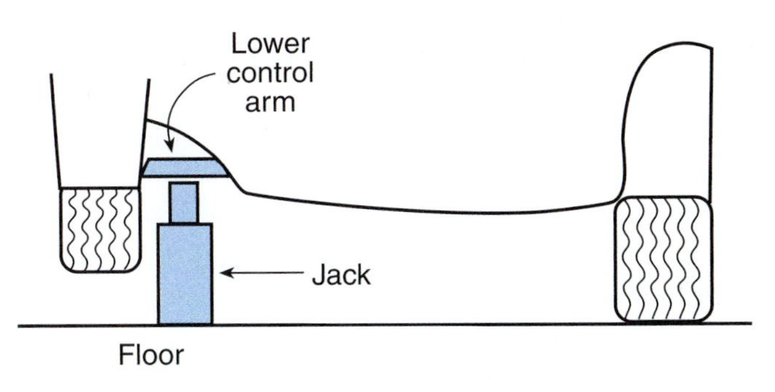

Figure 10-3 The jack should lift directly under the ball joint or as near as possible.

Ball Joints

Ball joints attach the control arm to the steering knuckle and must not have excessive movement between the joint **stud** and the steering knuckle. Excessive movement is any movement more than specified. There are several means to check ball joint wear. The service manual will provide exact procedures for performing this inspection. Some general guides are listed here.

Studs on ball joints and steering linkages are tapered and are interference fitted into a tapered hole.

Some ball joints have an indicator that is visible when the joint is worn. The grease fitting is designed and fitted into a floating or moveable retainer. As the joint wears, the grease fitting can be moved by hand. In some vehicles, any movement is cause for replacement of the ball joint. Some ball joints have the fitting's retainer extending above the ball joint. If the grease fitting shoulder should sink and become flush with the ball joint, the joint must be replaced.

WARNING: Do not replace a ball joint relying on hand and visual inspection alone. The quick test is used to make an initial diagnosis. Always refer to the service manual for specific inspection and tolerances.

Special Tools

Pry bar

Jack

Still other vehicles have ball joints that cannot be measured until the weight is removed from the joint. The weight of the vehicle must be on the control arm for a proper inspection. On a vehicle with the spring anchored on the lower control arm, place a jack under the control arm as close as possible to the ball joint (Figure 10-3). Operate the jack until the tire is an inch or two off the ground. Use a pry bar under the center of the tire and attempt to lift the tire (Figure 10-4). A good method is to place one hand at the top of the tire. This will give the technician a feel for any vertical movement of the wheel. If there is any movement, the ball joint needs to be checked further with a dial indicator or other procedures shown in the service manual.

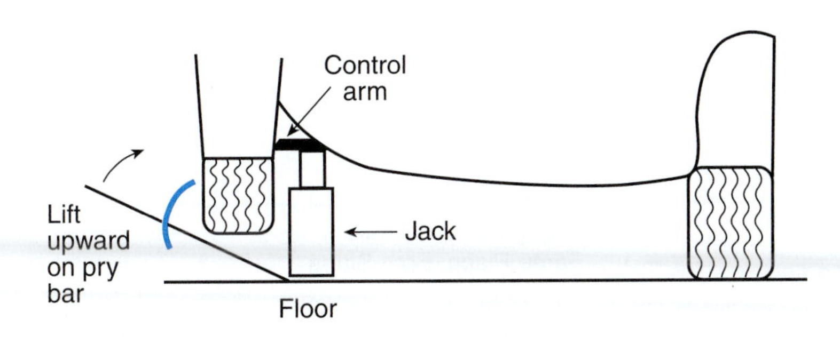

Figure 10-4 The wheel can be shifted by the pry bar if there is wear in the ball joint or the wheel bearing is loose.

Some vehicles use *rubberized joints* that have a great deal of movement even when new. The service manual has specific instructions for the placement of the dial indicator and the allowable vertical movement of the joint. Also, do not confuse a loose wheel bearing with ball joint failure.

An SLA system with the spring on the upper control arm must have the upper ball joint checked in a similar manner. Systems that use torsion bars and struts also have specific inspection procedures to check the ball joints. Always check both wheels during the inspection.

Springs and Torsion Bars

A worn or broken spring or torsion bar will not support its share of the vehicle's weight. A corner of the vehicle that is lower than the others indicates a weak or sagging spring. Some fairly quick measurements can confirm the presence of a weak spring or torsion bar. Locate a point that can be found on each side of the vehicle. An example could be permanent point on each front fender. The distance is measured from each point vertically to the floor and compared to each other (Figure 10-5). Figure 10-6 shows the measurement points on one vehicle. Some vehicles require measurements to be made for checking the ride height before a proper wheel alignment can be accomplished. This diagnosis and test applies to all types of springs.

Broken springs will cause the vehicle to drop in ride height, will not absorb shock, and will likely cause some noise to be produced. Any time the wheel with a broken spring encounters a dip or hump, the shock will be transmitted to the steering linkage and the vehicle frame. It will be readily noticeable by the passenger and particularly by the driver. A broken spring is replaced not repaired. Some shops recommend that both springs on an axle be replaced together. That may be a valid recommendation in some cases, but it not usually required by most manufacturers of vehicles or suspension components.

Shock Absorbers and Struts

Contrary to belief, shock absorbers and struts do not support the vehicle's weight. A weak spring does not necessarily mean the shock absorber is bad. However, a worn or broken spring may damage the shocks. Always check the shocks if a spring problem is found. Worn shocks

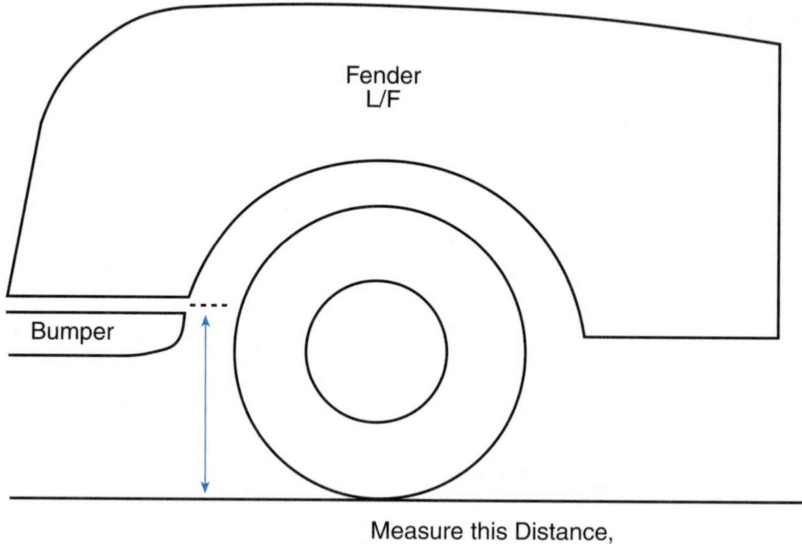

Figure 10-5 Pick identical points on each side of the vehicle and measure to the ground or floor.

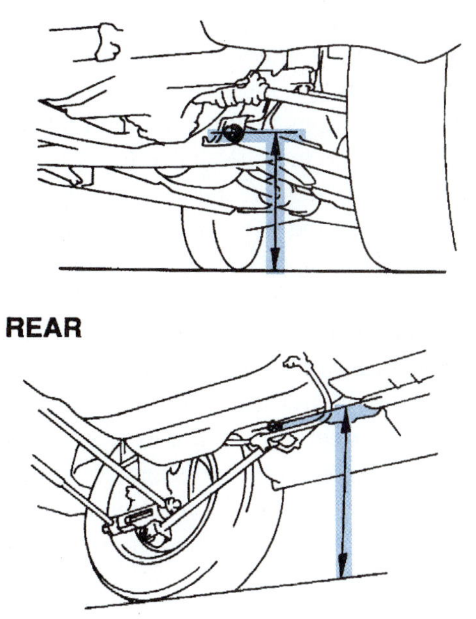

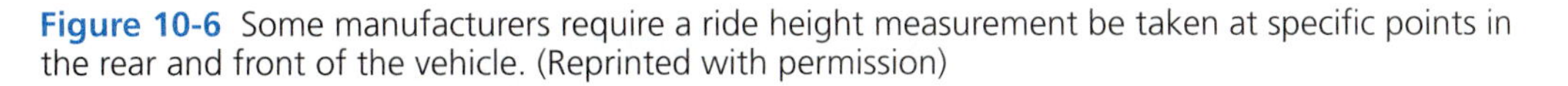
Figure 10-6 Some manufacturers require a ride height measurement be taken at specific points in the rear and front of the vehicle. (Reprinted with permission)

and struts tend to make the vehicle **wave** or bounce as the wheels encounter road hazards. As the spring jounces and rebounds, the weak shock cannot control it. The end or corner of the vehicle will continue to move up and down until the spring's energy is exhausted.

To test a shock absorber or strut, forcibly press downward on the corner with the suspected shock or strut several times. Watch for the vertical movement of the vehicle. A good, functioning shock or strut should allow the typical spring to jounce and rebound no more than once before stopping the vehicle at its ride height. Struts and shocks like springs can only be replaced.

Shocks or struts can also be broken. The broken part is usually the rod end (Figure 10-7). It will also produce a banging or knocking noise as the wheel encounters road irregularities.

As the vehicle "*waves*" from poor shocks or struts, the alignment angle changes on the front wheels and can cause steering control problems.

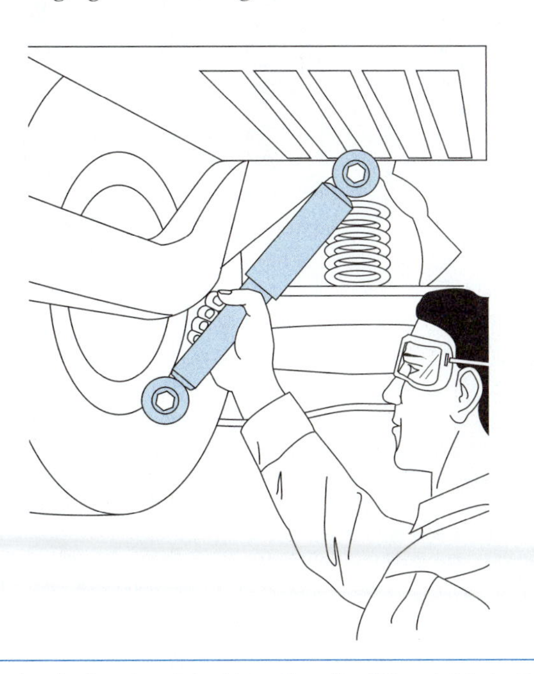

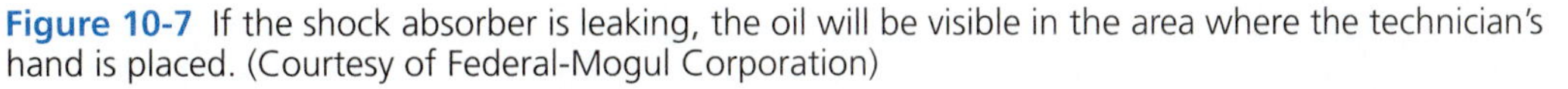
Figure 10-7 If the shock absorber is leaking, the oil will be visible in the area where the technician's hand is placed. (Courtesy of Federal-Mogul Corporation)

Seepage is the very small amount of fluid that seeps past the seal. Most seals are designed to allow some lubrication seepage around the seal's contact point with the shaft.

Naturally, there will be no shock absorber action. A more common condition with older shock or struts is oil leakage. There may be a very light indication of oil on the outside of the housing. This is considered to be **seepage** and is a normal condition. However, any visible liquid oil or indications of heavy leaking is a sign that the seals are bad and the shock or strut need to be replaced (Figure 10-7). Again, some shops may recommend replacement of the shocks or struts on the same axle. With struts, it may be more desirable to replace both since front struts are part of the steering system also. It is the customer's choice, the overall condition of the vehicle, and the advice of the technician that determine the course of action.

Additional malfunctions of the suspension may initially be thought to be steering problems. Weak springs, shocks, or struts could result in poor steering and ride complaints. Poor springs on the rear axle can result in a vehicle that wants to wander from side to side.

Bumping or grinding noise can indicate poor spring *isolators*. The same is true of shock and strut mountings. Damaged upper strut mountings and bearings could result in crushing or grinding noises as the wheels are steered or a hump is encountered (Figure 10-8). Worn or missing body bushings can result in a noise that is similar to that of worn shock bushings. A thorough investigation must be conducted to locate noises that seem to be suspension problems. Another fact to consider is that worn or broken suspension components can damage other suspension and steering components.

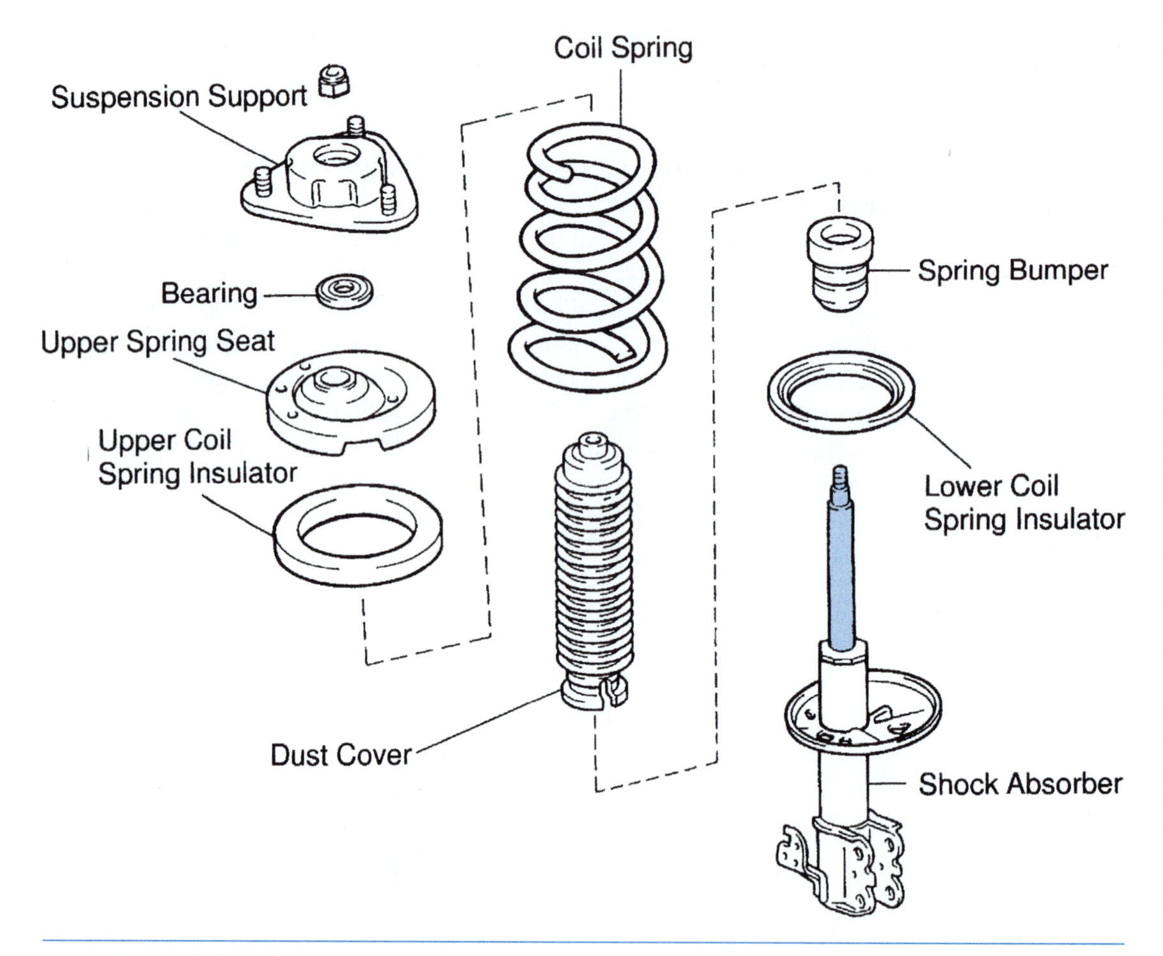

Figure 10-8 A shock absorber or strut rod usually breaks at the stepped-down area at the top of the rod. (Reprinted with permission)

Replacing Suspension Components

Front Shocks

Classroom Manual pages 278–288

Overall, shocks are very simple to replace. The wheel assembly may be removed for easier access. Remove the bolts or nuts that hold the shock in place. On some vehicles, the upper fastener is accessed through the engine compartment. Remove the top fastener on this type before raising the vehicle. The weight of the vehicle will help hold the rod while the nut is being removed. If the top nut and shock rod turn together, a special tool can be used to break off the nut (Figure 10-9), but in most cases, an impact wrench or air ratchet will remove the nut easily. Raise the vehicle and remove the lower fasteners. Remove the shock from the vehicle. Check to make sure all of the old bushings are also removed. The bushings on the rod type are two-piece with metal retainers.

Gas-charged shocks hold the rod downward in the cylinder by a wire or lock. The rod must be released during installation. Inspect the mounting area and decide the best method of installing and releasing the rod. Many times, the best method is to fasten the lower end of the shock and then release the rod. The rod is not under a great deal of pressure, and extension is fairly slow and can be controlled. The rod's tip is guided into place. Other than the automatic extension of the rod, the installation is the same as non-gas shocks.

The wire on a *gas-charged shock absorber* is soft and easily cut.

WARNING: Do not use pliers or other unpadded tools to hold the strut or shock absorber rod still while the nut is installed. A damaged rod can damage the strut or shock absorber's upper seal.

Install the rod's lower bushing retainer and bushing. Insert the rod into its mounting hole. Install the upper bushing, install the retainer, and loosely tighten the nut. Mount the lower end of the shock and install its bushings and fasteners. The shocks may have *rod-type* or *plate-type* mountings at each end or one of each (Figure 10-10). The rod-type, where the nut screws

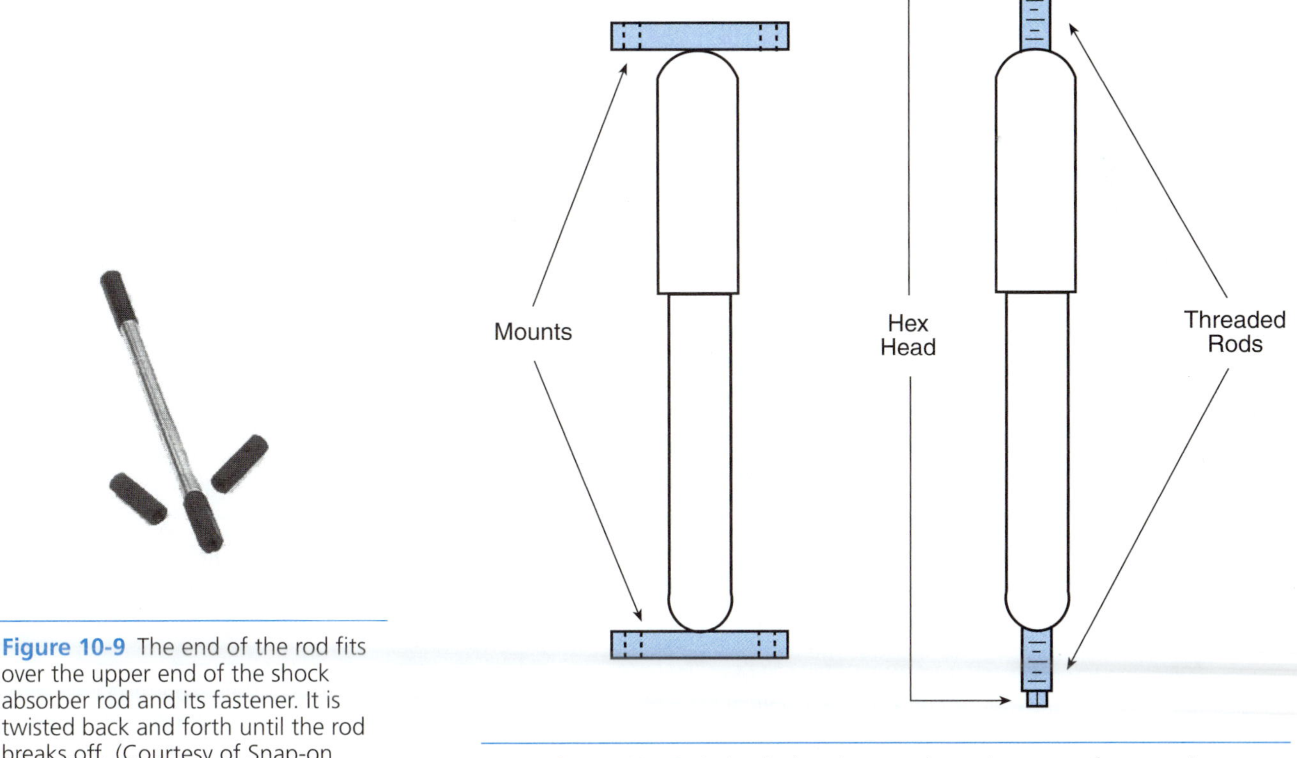

Figure 10-9 The end of the rod fits over the upper end of the shock absorber rod and its fastener. It is twisted back and forth until the rod breaks off. (Courtesy of Snap-on Tools Company)

Figure 10-10 A shock absorber may have the types of mount shown or a combination of the two.

directly onto the rod, always has bushings. The bar-type usually does not have bushings. A special wrench is available that will hold the rod while the nut is tightened (Figure 10-11). Torque all fasteners. Lower the vehicle, clean the work area, and complete the repair order.

Rear Shocks

Special Tools

Jack stand, short or tall as needed

CAUTION: Removing shocks on some solid rear axles will release their coil springs. Failure to support this type of axle could cause injury or damage.

Replacing the rear shocks is similar to replacing the front ones, and some rear struts can be replaced following the procedures for rear shocks. Do not remove rear struts if they are also used to mount coil springs. A service manual is required to remove this type of strut correctly and safety.

Before removing any fasteners, inspect the suspension to ensure that the spring will not come loose when the shock is disconnected. This is especially true of rigid axles with coil springs. The shock absorber helps hold the coil in place by preventing the axle from dropping too far. An axle with a leaf spring should also be supported. When the shock absorber is removed, the axle will drop and the new shock absorber can be connected without raising the axle. The upper shock absorber fastener on some cars is accessible through the trunk.

If the upper mounting is in the trunk, raise the vehicle enough to gain access to both ends of the shock. The vehicle can be raised to a good working height if both mounts are under the vehicle. Remove the upper fastener and then their lower ones. Remove the shock and bushing. Most rear shocks have bushings pressed into the shock absorber mounting. Others use bushings similar to the ones on the front (Figure 10-12).

If a gas-filled shock is used, decide on the best method to remove the holddown wire and install the shocks. Usually, the easiest method is usually to place the upper end in position, loosely tighten the fasteners, and cut the wire. The shock absorber rod can be guided to the lower mount as it extends. Mounting the lower end first and then guiding the rod upward into place is another method. It is usually easiest to guide the shock into place instead of extending it and then forcing it into place.

Figure 10-11 This special socket is made of two connected pieces. The bottom fits over the fastener and the top fits over the hex end of the rod. The top is then held while the bottom is turned by an open-end wrench. (Courtesy of Snap-on Tools Company)

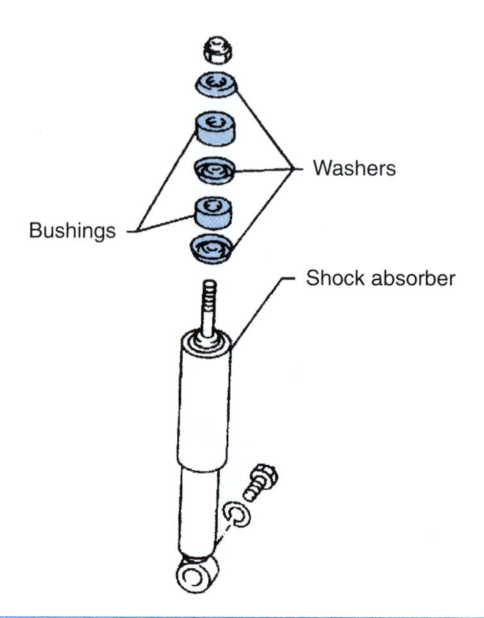

Figure 10-12 The bushings on this shock are held in place by three washers/retainers and a nut. (Courtesy of Nissan North America)

Mount any necessary bushings and retainers. Install the shocks and torque the fasteners. Lower the vehicle, clean the work area, complete the repair order, and, if all repairs are complete, return the vehicle to the customer.

Steering Component Diagnosis

Classroom Manual
page 288

CAUTION: Do not work on the steering column or steering wheel without disabling the air bag system. Serious injury or death could result if the air bag functions and the technician is working in the vicinity of the steering wheel or column. Consult the service manual for disarming procedures.

Many times, the first indication of a steering component failure is the customer buying new tires or complaining of poor steering. The technician can identify some problems by doing a quick inspection while performing an oil change or some other repair work. Each item of the steering system should be checked each time the vehicle has any work performed on the steering or suspension systems including tire replacement or repair. Most of the parts can be easily checked without any special tools except for lifting equipment. Tires will be addressed in a later section.

Driver Complaints

Like the suspension system, the customer may complain of noise or improper movement of the vehicle that may indicate a steering component failure. A worn linkage socket or inner tie rod may produce a popping or snapping noise as the steering wheel is turned. A loose or worn idler arm may be stated by the customer as a "wander"on a straight road or excessive play in the steering linkage. Excessive steering wheel play may be the result of worn gears in the steering gear box, but this is more common with recirculating ball and nut than with rack and pinion systems.

A symptom that came to be known as "morning sickness" may be present on older rack and pinion steering. This is a result of poor power-steering boost after the vehicle has been shut down overnight. The system leaks pressure until the fluid warms up and the rack's piston seals swell. The driver has to apply more force than usual to steer the vehicle during cold drive away. The steering unit has to be replaced or rebuilt to correct this problem. Since rack and steering units are rebuilt by aftermarket vendors, it is usually best to replace the steering unit with a rebuilt model instead of trying to rebuild it at the shop. The overall cost will be about the same, but the time will be much less. Like suspension problems, the technician may have to road test the vehicle to further isolate the failed component.

Linkage

CAUTION: Do not work on the steering column or steering wheel without disabling the air bag system. Serious injury or death could result if the air bag functions and the technician is working in the vicinity of the steering wheel or column. Consult the service manual for disarming procedures.

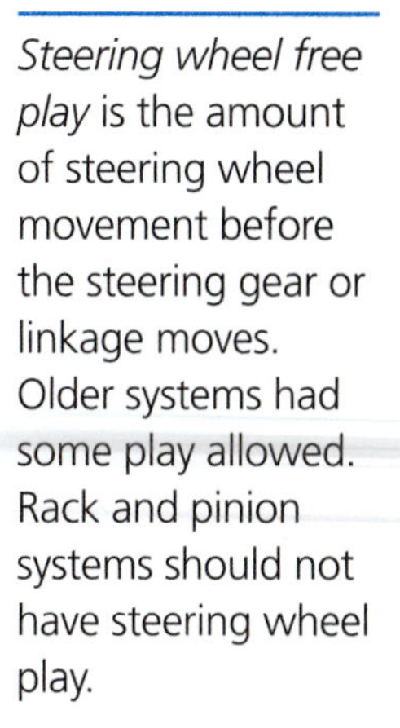

Steering wheel free play is the amount of steering wheel movement before the steering gear or linkage moves. Older systems had some play allowed. Rack and pinion systems should not have steering wheel play.

Before lifting the vehicle, check the linkage for **steering wheel free play**. With the engine off, move the steering wheel in each direction while watching the front wheels. Any movement of the steering wheel should result in an immediate corresponding movement of both wheels. If one wheel does not follow the steering wheel, there is probably a problem with the linkage to that wheel. A loose or worn steering gear will result in erratic action of both wheels. If there is excessive free play between the steering wheel and the steering linkage, the steering

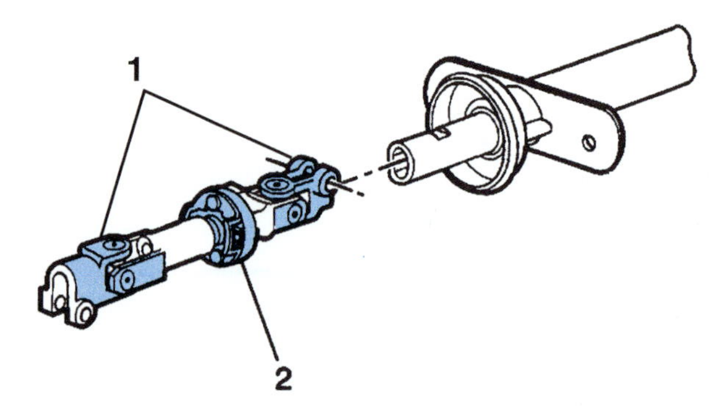

Figure 10-13 Item #1 is a set of U-joint-type couplings. Item #2 has a neoprene or rubber type coupling. (Courtesy of Chevrolet Motor Division, General Motors Corporation)

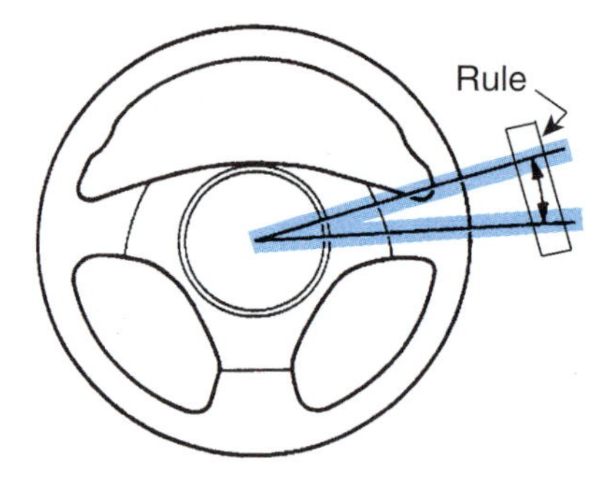

Figure 10-14 The distance is measured by using a rule or similar device to determine the amount of free play of the steering wheel. (Reprinted with permission)

The *coupler* is a flexible connection between the steering column and the gear or the upper and lower shafts. Some systems use a connection similar to a U-joint.

gear or **coupler** must be replacd (Figure 10-13). Neither can be repaired even though the steering gear can be resealed.

A more accurate method of checking free play is to measure the amount of travel of one spot on the steering wheel. Slowly, move the steering wheel in one direction until all of the free play is removed. Align a graduation on a rule or other linear measuring device with a spot on the wheel. Hold the measuring device in place and parallel to the wheel's direction of rotation (Figure 10-14). Turn the wheel in the opposite direction to the first movement until the free play is removed again. Note where spot on the wheel aligns on the measuring device. The distance between the two graduations is the amount of free play.

Make a check of the power-steering system by starting the engine and turning the wheels completely from side to side. The complete movement should be smooth and continuous. The engine should not stall at idle nor should there be any belt squealing, pump noise, or jerks of the steering wheel. Replace or adjust belt if it is worn or loose. Add power-steering fluid if there is pump noise. If the power-steering pump emits a moaning sound, check its fluid level. If the level is correct, the pump may need to be replaced.

CAUTION: Do not work on the steering column or steering wheel without disabling the air bag system. Serious injury or death could result if the air bag functions and the technician is working in the vicinity of the steering wheel or column. Consult the service manual for disarming procedures.

There are specific tests that can used to diagnose the power-steering systems. One uses a beam-type torque wrench. Only an experience technician should perform this test because the center pad of the steering wheel must be removed. The driver's air bag, if equipped, is part of the pad. Once the air bag is disarmed and the pad is removed, the torque wrench and socket are placed over the steering-wheel retaining nut (Figure 10-15). The steering wheel is turned using the torque wrench, and the amount of force is registered on the wrench. Another test is done with a pressure gauge set up for use with power-steering units. While any pressure gauge reading over 300 psi can be used, finding the correct adapters may be a problem. Most power-steering test gauges come in sets with the correct adapters and can be used fairly easily.

Most power-steering systems have a pressure switch. The switch sends a signal to the PCM to adjust idle speed during parking. A faulty switch may cause the engine to stall or stumble when parking. The engine may also have some problem in the fuel or air idle systems that may caused it to stall during parking. The service manual lists the steps to test the switch and its circuits. However, the noted repairs do not have to be accomplished before the remainder of the inspection is performed.

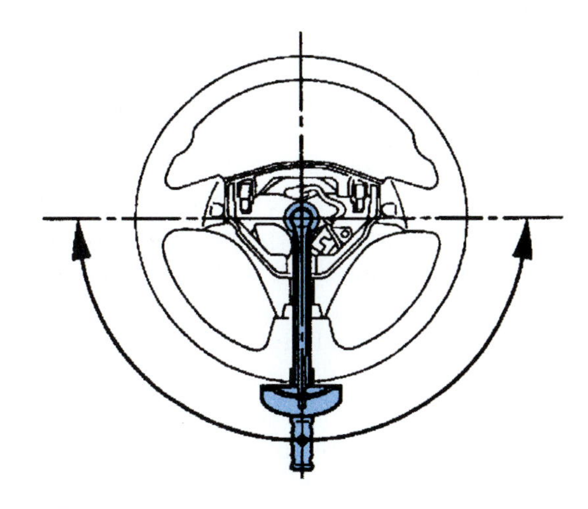

Figure 10-15 Using a beam-type torque wrench to measure power steering assist. (Reprinted with permission)

WARNING: Do not condemn "rubberized" type steering connections by feel alone. This type of socket and ball component will have some movement even when new. Consult the service manual for specifications and test procedures.

The linkage has to be checked for worn or broken connections or parts. A drive-on lift is best suited for the inspection. It will keep the weight of the vehicle on the wheels and apply normal operating tension on the steering and suspension systems. Raise the vehicle to a good working height. With the vehicle lifted, apply linear force against the connections of each of the sockets in the linkage (Figure 10-16). Apply the force back and forth in line with the socket stud. There should be absolutely no movement between the stud and the hole it fits through. Any movement of the stud within its mating hole requires the component be replaced. Do not

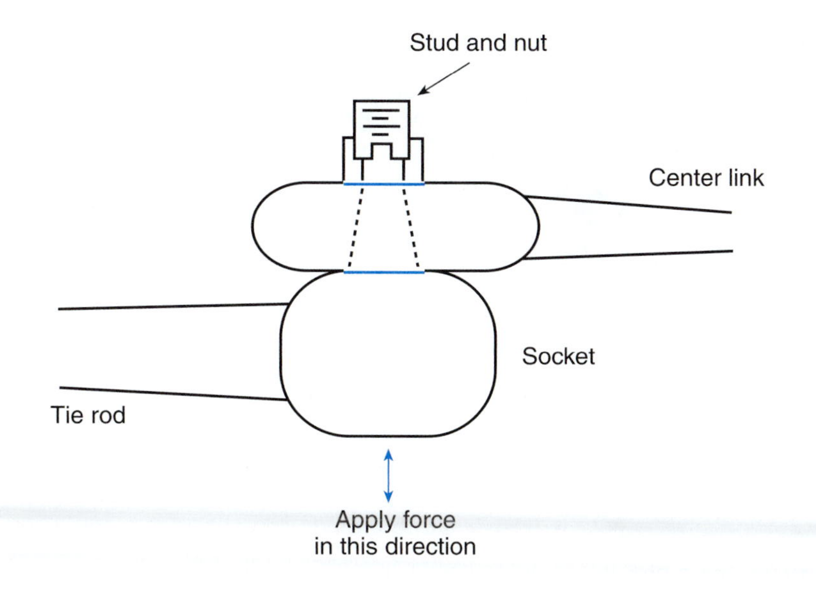

Figure 10-16 There should be no excessive movement between the stud and the tapered hole.

confuse the side-to-side movement of the ball within the socket. The ball should move from side to side and some manufacturers allow for some vertical movement of the ball within the socket. Consult the service manual for specific tests.

The pitman arm located at the bottom of a recirculating ball and nut system is splined onto the sector shaft and held in place by a large nut and lock washer (Figure 10-17). There should be absolutely no movement between the pitman arm and shaft and the nut should be completely tight. Here, any movement requires the replacement of the pitman arm, sector shaft, or both. If the sector shaft is the problem, it is recommended that the entire steering gear be replaced along with the pitman arm. It would be extremely unusual to find a problem at this connection. A nut that was not properly torqued is the only reasonable cause of slack between the arm and shaft. If there is a loose connection here, it advisable to closely inspect the entire system for similar loose fasteners or improperly installed components.

The idle arm, if equipped, must be checked for mounting and arm movement. The mounting bolts must be tight and the bracket flush against its mount, which is usually the frame rail (Figure 10-18). As the connection between the center link and idler arm is checked, notice the movement of the idler arm on its pivot. Some manufacturers allow for a limited amount of movement at this point; others allow no movement. A few idler arms can be adjusted to remove some movement. Always check the vehicle's service manual for specific procedures and specifications.

Have someone move the steering wheel side to side while watching the movement of the linkage. The entire system should move together. On a rack and pinion system, one common problem is a worn inner tie-rod socket. A short movement of the rack with no movement

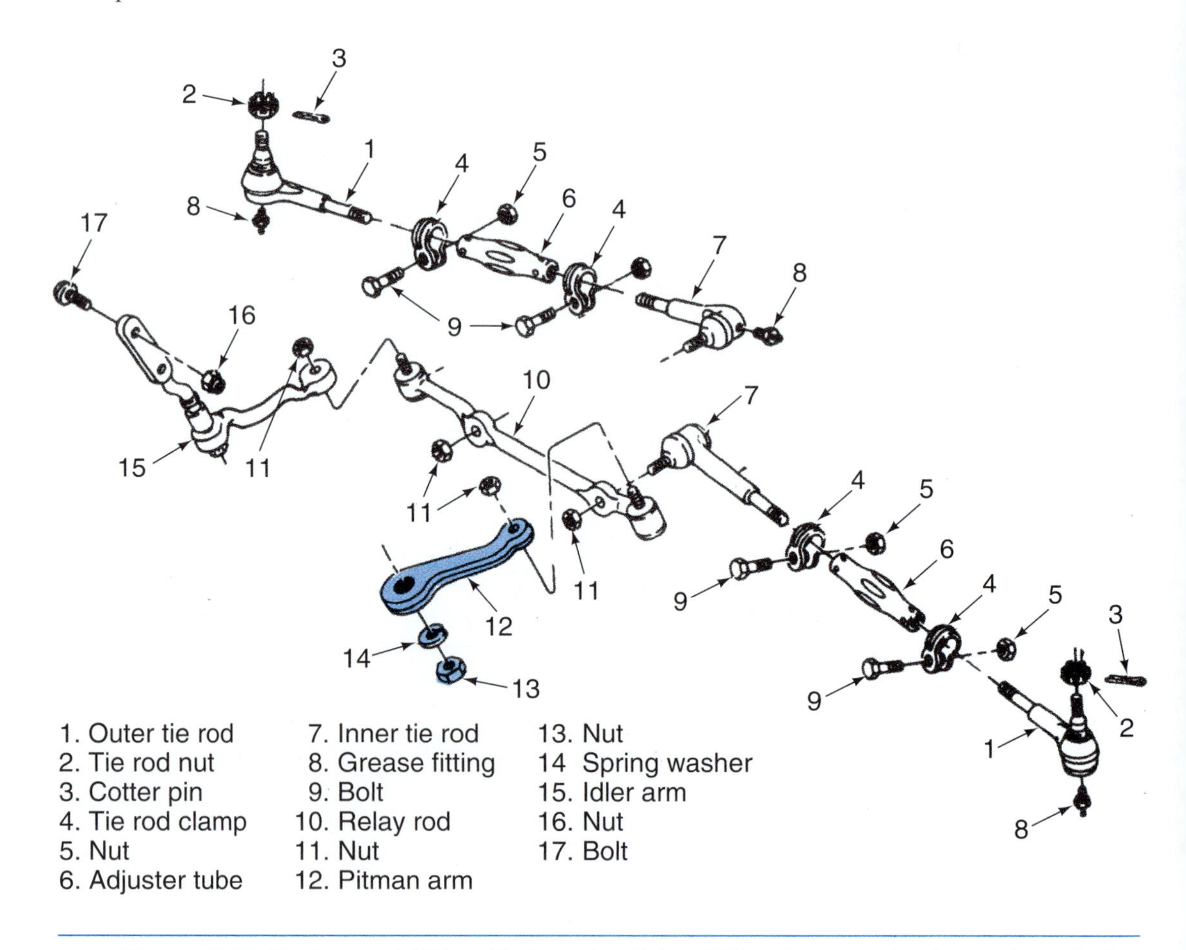

Figure 10-17 The large nut (13) and lock washer (14) hold the pitman arm in place and usually do not loosen under normal conditions. (Courtesy of Chevrolet Motor Division, General Motors Corporation)

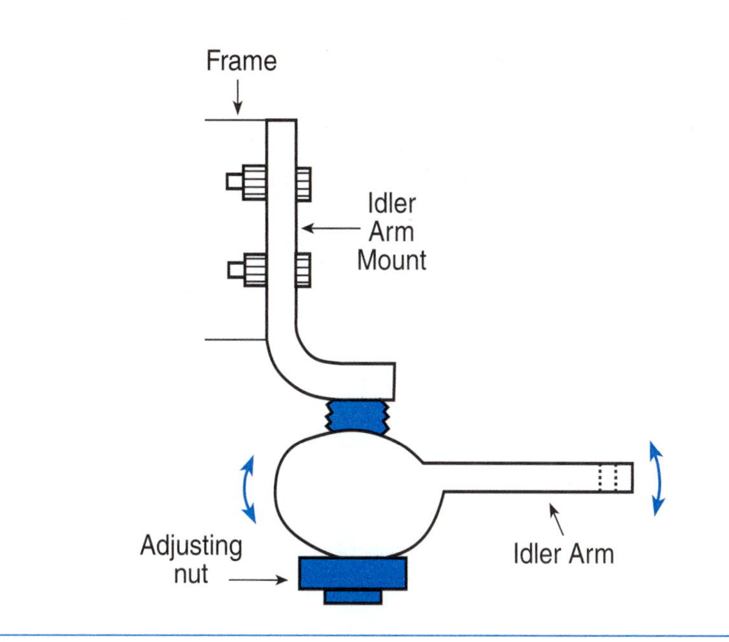

Figure 10-18 Some idler arms have an adjustment to remove excessive movement.

of the tie rod indicates a worn socket. It can be verified by locking the steering wheel and trying to move the affected wheel by hand. If there is a socket problem, the wheel and tie rod will move slightly. The same procedure can be applied to test the linkage on all types of steering systems. Remember, there is no movement allowed at the steering linkage connections and the steering should be smooth from one side to the other.

Replacing Steering Linkage Components

Classroom Manual
pages 288–294

If an alignment adjustment is made by *heating* or bending a steering or suspension component and an accident or damage results, the vehicle manufacturer will not accept responsibly under product warranty.

WARNING: Do not apply **heat** to remove any steering linkage or fastener. Heat will change the temper of the component and could result in damage or injury.

CAUTION: Do not work on the steering column or steering wheel without disabling the air bag system. Serious injury or death could result if the air bag functions and the technician is working in the vicinity of the steering wheel or column. Consult the service manual for disarming procedures.

The power-steering belt is replaced by loosening the tensioner. The pump may be driven by a single serpentine belt or it may have a separate belt. The replacement of a serpentine belt was covered in Chapter 8, "Engine Maintenance."

A single belt, or belts used to drive the power-steering pump, usually employs the pump as the belt tensioner. To replace the power-steering belts, it may be necessary to remove other belts. It is recommended that all of the belts be replaced at one time.

In most vehicles, there will be at least two fasteners that must be loosened so the pump can move. One will be located near the front of the pump through a slot in the bracket (Figure 10-19). Another will be located at the bottom of the pump bracket and will extend through the pump to the rear part of the bracket. Other types may have two slots, one at the front and one at the rear. Each will have a fastener. There may be two fasteners at the bottom instead of a single, long one. Inspect the pump mounting brackets to determine which fasteners to loosen. Each of the fasteners will have to be loosened before the pump can be moved.

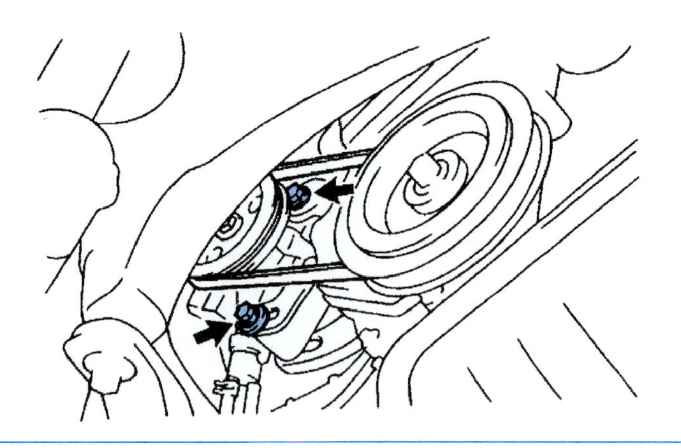

Figure 10-19 Note the end of the slot at the lower fastener. (Reprinted with permission)

Special Tools

Pry bar

WARNING: Do not pry on the power-steering reservoir or pulley to move it. Damage to the reservoir or pulley may result.

Some brackets have threaded adjusting bolts and brackets to adjust belt tension (Figure 10-20). The pump will have to be pushed or pulled on other types. Some pumps can be moved by hand while others have to be forced with a pry bar. If necessary, place the pry bar between the pump and the engine or other stationary component. Move the pump toward the engine until the belt is loose enough to be removed. Remove the belt and install the new one.

Position the pry bar to apply outward force to the pump. Pull the belt to the correct tension and hold in place. Tighten the easiest fastener to access. Release the pry bar and check the belt tension. If it is not adjusted properly, loosen the first fastener and tighten the belt again. If the tension is correct, torque all the fasteners.

Special Tools

Ball joint or tie-rod end press

Impact wrench

Torque wrench

Most steering linkage parts are replaced in similar methods by forcing the stud from its mounting hole. After lifting the vehicle, remove the cotter key, if used, and the nut from the stud. A special tool is then used to remove the stud from its mounting hole (Figure 10-21). Press the stud out and remove the steering linkage from the vehicle. In most cases, more than one connection will have to be removed to free the damaged components.

Tie-rod ends can be removed from the steering knuckle using the same tool. The end is then unscrewed from the tie rod or adjusting sleeve. Count the number of turns as the end is being screwed off. Install the replacement end the same number of turns. This will approximately set the toe of the front wheels until an alignment is completed.

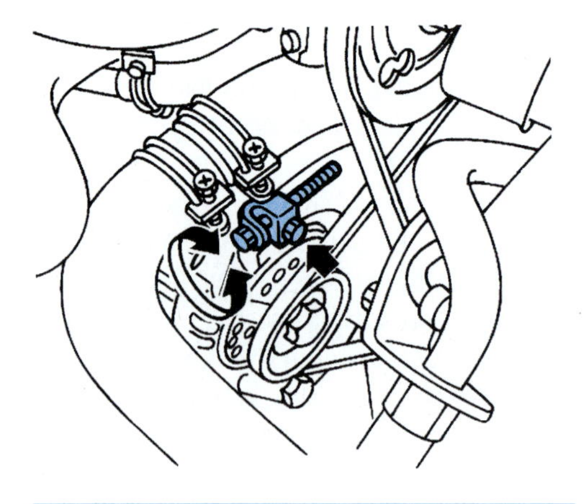

Figure 10-20 This type of belt adjustment fixture is used on many vehicles, light and heavy. (Courtesy of Nissan North America)

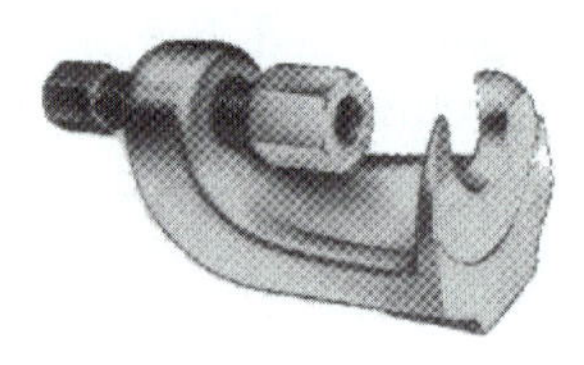

Figure 10-21 This tool works similar to a press. (Courtesy of MATCO Tool Company)

Installation is basically the reverse of removal. Before inserting the stud into its mating hole, ensure that the rubber boot is placed over the socket and a grease fitting is installed. Insert the stud and hand tighten the nut. Do the same with any other connection that needs to be done on this component.

The new part should come with a new nut and cotter key if used. The nut will be a self-locking type or a castellated nut (Figure 10-22). The **castellated nut** is used with a cotter key. A standard torque wrench and socket cannot be used because of space and the arrangement of the components. If special torque wrenches are not available, tighten the nut as much as possible using a standard box-end wrench. This is not really the proper procedure, but it is an accepted practice.

A *castellated nut* is so named because of the turret-type projections on one side.

With a castellated nut, tighten it the same way but line up a slot in the nut with the cotter keyhole in the stud. Do not loosen the nut to align the hole or slot. Continue to tighten it a little at time until they are aligned. If the nut has to be loosened, the complete connection has to be broken and the installation must be started from the beginning. The nut will have to be removed and the stud will have to be forced from the hole and reinstalled. This is due to the tapered shape of the stud and hole. Once they are forced or pressed into place, they must be locked at that point. Loosing the nut relieves force on the taper and the two mating components will not have the correct fit.

Special Tools

Inner tie-rod socket wrench

Ball joint or tie-rod end press

On rack and pinion systems, the inner tie rod and socket are removed in a little different manner. Disconnect the tie-rod end from the steering knuckle as stated above. On the inner end, remove both boot clamps and slide the boot outward toward the end of the rod. It may be necessary to turn the steering wheel until the boot and socket are better visible. Inspect the socket connection. There should be a small hole or the end of a pin visible toward the inner end of the socket (rack end). The pin locks the socket to the rack and needs to be removed. A pin punch and small hammer will accomplish the task. Other systems may use a setscrew or other device to lock the socket. They must be removed before proceeding.

There is an equalizing tube that runs between the boots on each end of the rack (Figure 10-23). As a boot is removed, ensure that the tube remains in place in the other tube. When the boot is installed, ensure that the tube is properly positioned in each boot.

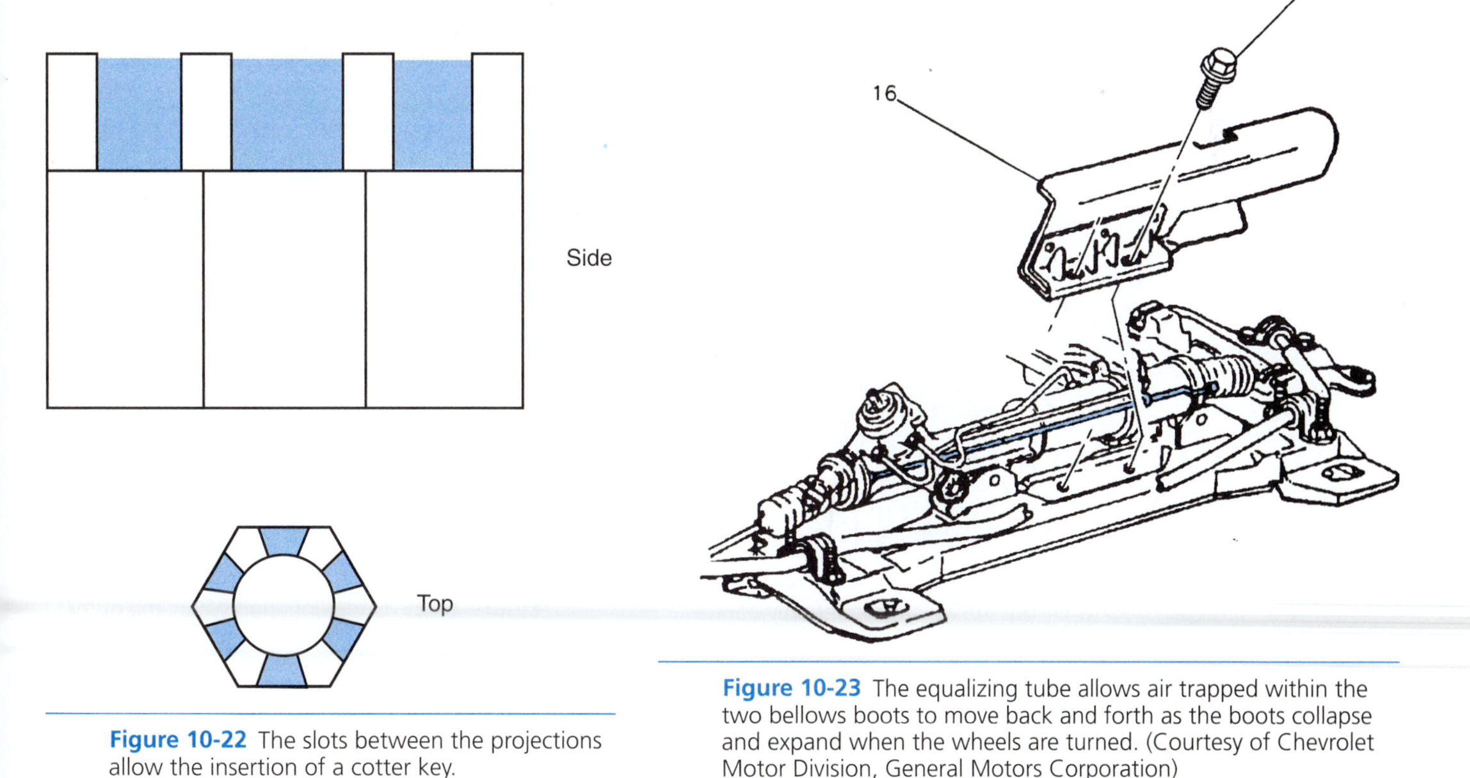

Figure 10-22 The slots between the projections allow the insertion of a cotter key.

Figure 10-23 The equalizing tube allows air trapped within the two bellows boots to move back and forth as the boots collapse and expand when the wheels are turned. (Courtesy of Chevrolet Motor Division, General Motors Corporation)

With the lock removed, locate two or more flat sides on the socket. There are special extended sockets available or an open-end wrench to remove or install the inner tie rod. Remove the tie-rod end, lock nut, clamps, and boot from the inner tie rod. Remove the inner tie rod from the rack. Slide the boot, clamps, and screw nut onto the new tie rod. Screw and torque the inner tie rod onto the rack and install the locking device. Move the boot over the rack housing and install the inner clamp. Leave the outer clamp loose for now since the toe must be checked and adjusted. Install the tie-rod end onto the tie rod the same number of turns it took to remove it. Connect the end to the steering knuckle as stated above. Hand tighten the lock nut until the toe is checked. The alignment can now be checked and the repair completed.

Classroom Manual
pages 294–300

A good *tire wear* diagnosis can save the technician time and labor.

Tire Wear Diagnosis

Correctly interpreted, **tire wear** can point quickly to a steering or suspension malfunction. Abnormal tire wear is any one part of the tire that is worn more than a matching part of the same tire. It is primary thread wear, but may include damage to the sidewall or bead. Also, individual tires may show different types of wear.

A driver may complain of the vehicle pulling or drifting to one side on a straight road. This is usually a result of poor wheel alignment, and corrective actions can be taken at a shop. Usually, a vibration felt at certain speeds can be traced to unbalanced tires, but it could possibly be a driveline problem as well. If tire rotation and balance have been recently performed and a pulling to one side starts after the work, radial tire belt movement is usually the problem. Switching the two front tires to opposite sides may correct the pull, but at times, the tires must be placed back in their orginal condition to eliminate the pull.

An inspection of the tire's sidewall may show sloping ridges and valleys, and most radial tires will be affected to some extent. Normally, this is not a concern, but the driver may question the presence of the ridges. If the ridges seem to affect steering, ride, or durability, the tire needs to be replaced.

Tire Pressure Wear

The tread should wear evenly around and across the tread area. The most common fault is incorrect tire pressure. An over-inflated tire will ride on its center tread and not wear the other parts of the tire (Figure 10-24). An under-inflated tire will ride on its edges with little wear to the inside area (Figure 10-25). Scruff marks on the sidewall may also be apparent on an under-inflated tire. This results when the air pressure is not high enough to hold up the sidewall when cornering. The customer should be advised to check tire air pressure at least once a month or as the operator manual states. Under-inflated tires also result in a loss of fuel mileage and poor steering control.

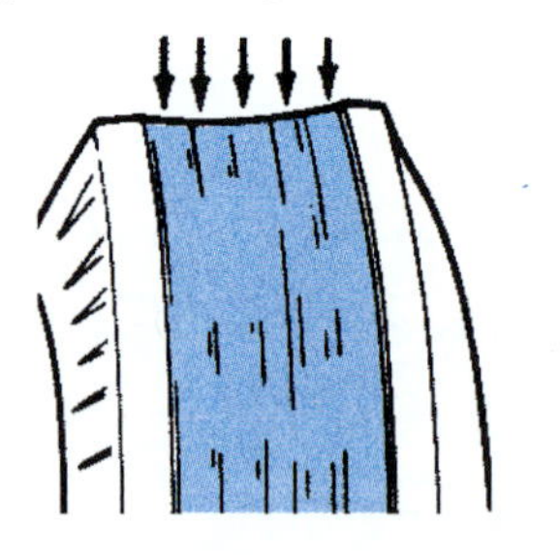

Figure 10-24 Over-inflation wears the center part of the tread. (Courtesy of DaimlerChrysler Corporation)

Figure 10-25 Under-inflation wears the outer edges of the tread. (Courtesy of DaimlerChrysler Corporation)

Camber Wear

Camber wear is indicated by excessive wear on the edge of the tire with little wear on the center or opposite edge (Figure 10-26). Too much **positive camber** results in wear on the outer three or four treads, and excessive **negative camber** shows up as inner tread wear. The remainder of the tread will show even wear. Camber is corrected using an *alignment machine*. Some vehicles do not have camber adjustment, while others have adjustments on each wheel. If there is no adjustment mechanism, the suspension, frame, and body have to be checked and damaged or worn components must be replaced.

Camber wear and *toe wear* can be confusing. Remember that toe wear will create a saw-tooth pattern across all or most of the tread.

A *positive camber* is the inward tilt of the wheel at the top.

Negative camber is the outward tilt of the wheel.

Toe Wear

Toe wear will also show the most wear at the tire's edge, but it will be spread over most of the tread width. Toe wear is indicated by a saw-tooth pattern across the tread (Figure 10-27). The pattern is obvious when the technician's hand, palm down, is drawn across the tread width. The edge of each tread will be sharpened as though it had been cut with a sharp object. The sharpened edges point in the direction of **toe**. If the edges point to the outside of the tire, the tire is *toed-out* too much. In this case, the amount of greatest wear will be on the inside treads, which is similar to camber wear. However, camber wear will not create the saw-tooth pattern.

Toe-in means that the front edges of the wheels are closer than the trailing edges of the same wheels.

Wear Caused by Balance and Rotation

Balancing and rotating tires are important for long tire life. They should be part of routine tire maintenance and be accomplished about every 5,000 miles. Failure to do so could result in poor riding and steering comfort.

Tires that are not kept in balance may develop **cupping** of the treads at irregular intervals around the tire (Figure 10-28). The cupping occurs as the heavy spot of the tire hits the road and causing a scruffing of the tire against the pavement. Tires that have not been rotated may also cup the tread. When determining if balance or rotation causes the wear, remove the tire assembly and check its balance on a tire balance machine. Even at this point the technician may not be able to completely eliminate rotation as the problem, so the best solution is to balance and rotate all tires on the vehicle.

Cupping wear caused by balance or rotation is not as clearly distinguishable as other types of tire wear. Normally, balance wear will be in one spot while rotation will be all around the tire.

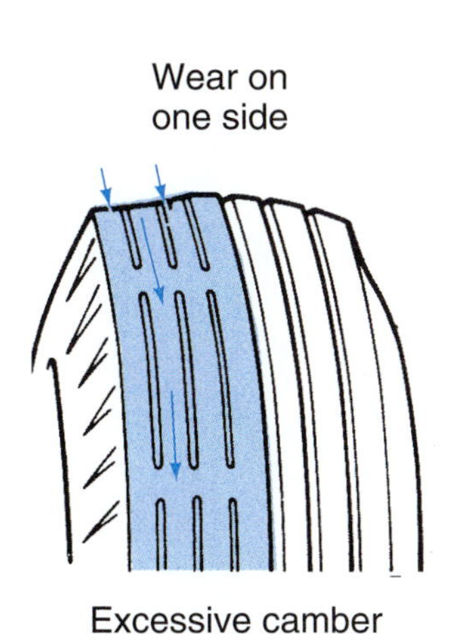

Figure 10-26 Camber will wear one edge of the tread. (Courtesy of DaimlerChrysler Corporation)

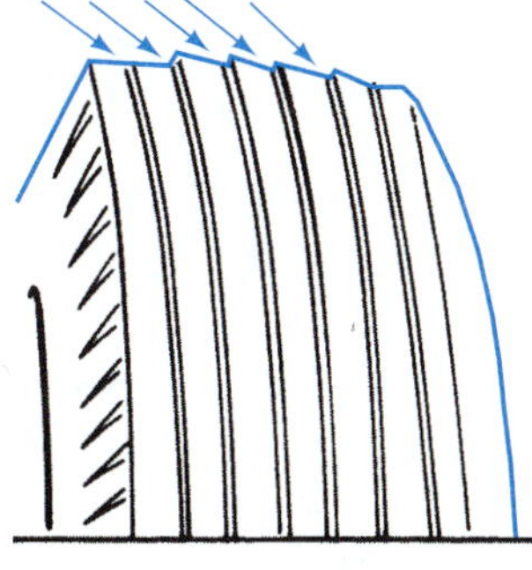

Figure 10-27 Toe produces a saw-tooth or feathered edge to the tread. (Courtesy of DaimlerChrysler Corporation)

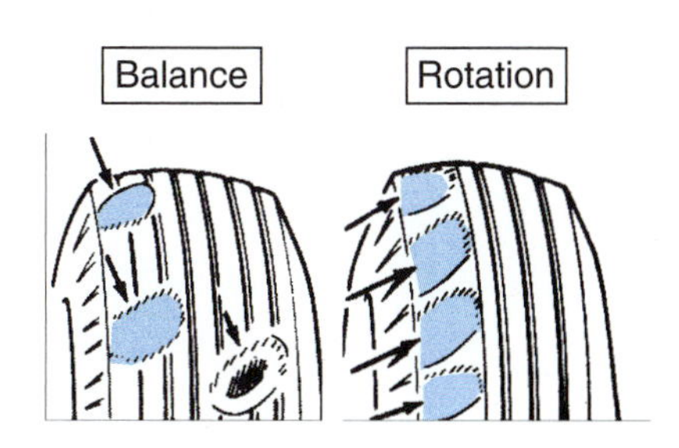

Figure 10-28 Lack of rotation may produce a regular cupping of the tread. (Courtesy of DaimlerChrysler Corporation)

Wheel and Tire Service

WARNING: The lug nuts on a wheel should never be tightened with a straight impact wrench. They must be torqued according to the manufacturer's specifications using a torque wrench, torque stick, or a reliable adjustable impact wrench. Damage to the wheel, studs, hub, or brake disc could result.

Some of the more common problems with wheels and tires can be traced to poor or no service. The driver or the technician may also cause a problem. Wheel and tire service is limited to replacement, proper torquing, or repairing a hole in the tire.

Wheel Assembly Removal and Installation

WARNING: Apply extra care when working with chrome or magnesium wheels. Improper torquing of the lug nuts could result in damage to the wheel.

Special Tools

Jack and jack stands or lift

Impact wrench

Torque wrench or wheel torque wrench

Service manual

Almost everyone at some point in their driving career must remove and install a wheel. Most flat tires do not occur at a shop but on a dark, lonely road or in the middle of rush-hour traffic. Drivers may carry some safety and repair equipment in their vehicles, but a torque wrench (available at the automotive shop) is the one thing they do not have To change a tire, place a jack under the vehicle's lift point and raise it until the tire is clear of the ground or floor. If in a shop, a 1/2-inch impact wrench can be used to remove the lug nuts. Otherwise, the wrench that comes with the vehicle is used. Perform the necessary work on the wheel and tire or replace it with another assembly. Start the lug nuts on the studs and hand-tighten them. Use a hand wrench to tighten the nuts as much as possible. If in a shop, use a torque wrench of some type to tighten the nuts to specification. A cross-pattern should be followed to torque or tighten the nuts so the tightening process does not cock the rim on the hub (Figure 10-29). A cocked rim will prevent proper torquing of some of the nuts and will be obvious once the vehicle is on the road.

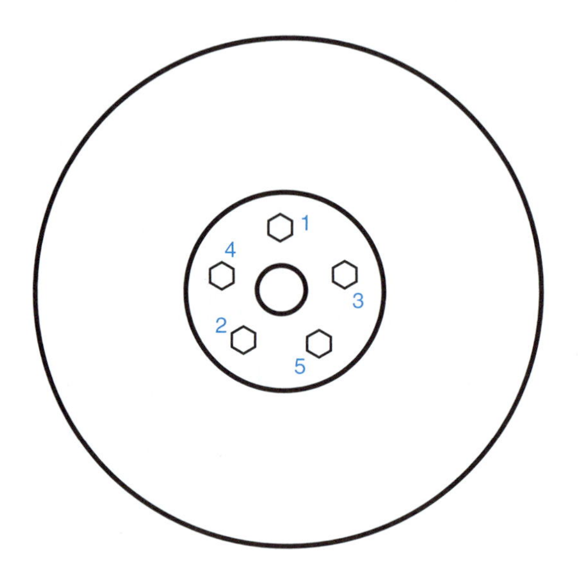

Figure 10-29 Lug nuts should be tightened in a crosshatch pattern similar to the pictured sequence 1–5.

Tire Replacement

The following instructions are general guidelines for a tire changer with a side breaker. The instructions for the tire changer *must* be followed. Remove the wheel assembly and release all of the air from the tire. Select the proper mounting tools to install the wheel assembly on the tire changer. Chrome and magnesium wheels require special mounts to prevent the wheel from damage or marring. Place the deflated tire and wheel assembly onto the bead breaking portion of the machine. Ensure that the clamp does not lock over the rim but presses against the tire at a point near the rim. Operate the clamp until the tire bead breaks away from the rim. Turn the tire around and break the bead on the other side.

Special Tools

Tire changer

Service manual

CAUTION: Keep hands and fingers clear of the bead area as the tire is being forced onto the wheel. Serious injury could result to hands or fingers if they get caught between the wheel and tire bead.

Mount the wheel assembly onto the tire changer and lock it in place. Use the changer's bead guide to remove one side and then the other side from the rim. Before installing the tire, lube the beads with tire lubricant. Position one side of the tire on the rim and use the bead guide to install the tire. With the tire mounted, inflate it to the proper pressure. Some tires may require a band or air tool to initially seat the seat. Do not inflate the tire more than the pressure noted on the sidewall. Once the tire is properly inflated, remove it from the changer.

Wheel Balancing

WARNING: Before using any machinery, read and study the operating manual. Failure to do so could result in damage to the machine or injury to the operator.

Special Tools

Tire balancer

Wheel weight hammer/tool

To properly balance a wheel assembly it must be correctly mounted to the tire balancer. Various centering cones and fasteners are available on most tire balancers. Select a cone that fits approximately halfway through the wheel's large center hole. In most cases, all of the old weights are removed before mounting the assembly onto the balancer. The following instructions may differ between balancer brands and models.

Place the cone onto the balancer's spindle. There should be a spring already in place on the spindle. If not, install the spring and then the cone. The spring keeps the cone pressed into the wheel's center. Install the wheel assembly over the spindle and onto the cone. Install the protection devices for magnesium or aluminum wheels, if required, followed by the retainer nut. There are usually three measurements that must be entered into the balancer's computer. They are *rim diameter, rim width,* and *the distance from the edge of the rim to the balancer.* There is a measuring bar that slides out of the balancer and the end is placed against the rim. The bar will be numbered, and the number showing at the edge of the balancer is entered into the computer (Figure 10-30). This is the distance between the wheel assembly and the balancer's point of reference.

The rim's width is measured with a tong-like device. The ends are placed at the portion of the rim where the weights are installed (Figure 10-31). The diameter of the tire or rim is stamped into the sidewall of the tire. The method of entering the measurements into the computer varies from push buttons to rotating knobs. Follow the instructions in the machine's operating manual.

Photo Sequence 11
Typical Procedure for Dismounting and Mounting a Tire on a Wheel Assembly

P11-1 Dismounting the tire from the wheel begins with releasing the air, removing the valve stem core, and unseating the tire from its rim. The machine does the unseating. The technician merely guides the operating lever.

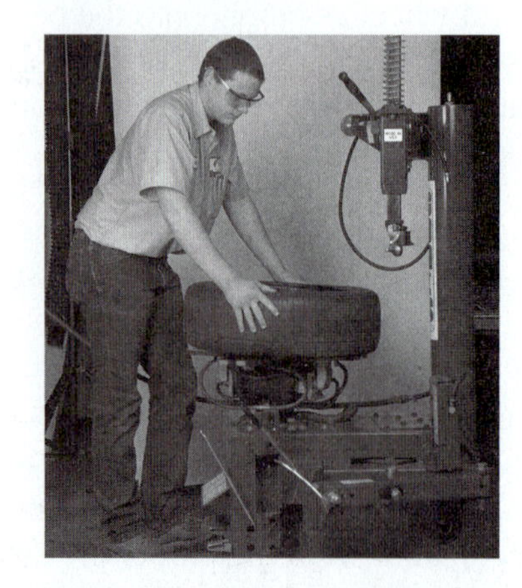

P11-2 Once both sides of the tire are unseated, place the tire and wheel onto the machine. Then depress the pedal that clamps the wheel to the tire machine.

P11-3 Lower the machine's arm into position on the tire and wheel assembly.

P11-4 Insert the tire iron between the upper bead of the tire and the wheel. Depress the pedal that causes the wheel to rotate. Do the same with the lower bead.

P11-5 After the tire is totally free from the rim, remove the tire.

P11-6 Prepare the wheel for the mounting of the tire by using a wire brush to remove all dirt and rust from the sealing surface. Apply rubber compound to the bead area of the tire.

P11-7 Place the tire onto the wheel and lower the arm into place. As the machine rotates the wheel, the arm will force the tire over the rim.

P11-8 Reinstall the valve stem core and infalte the tire to the recommended inflation.

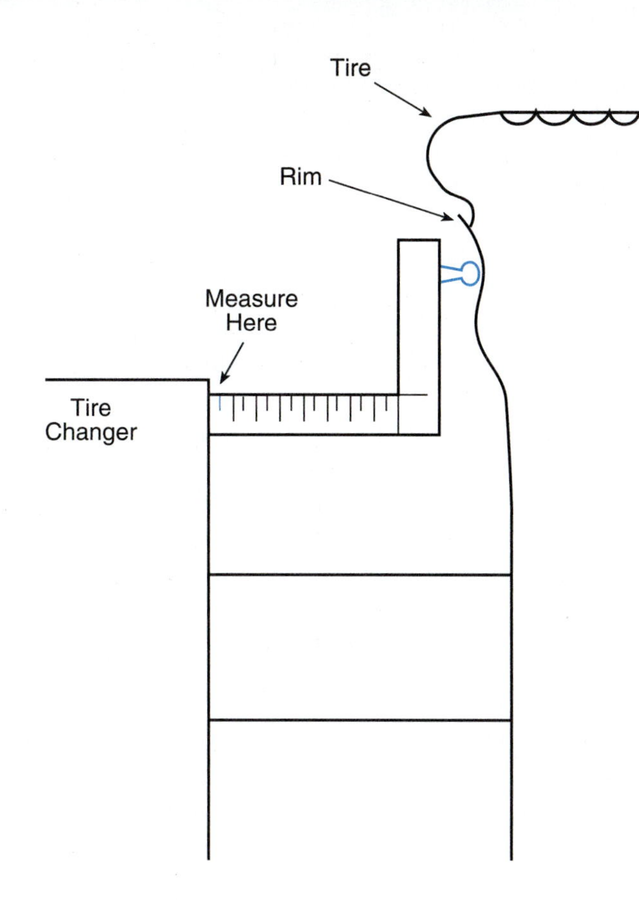

Figure 10-30 The bar may be marked in inches or millimeters.

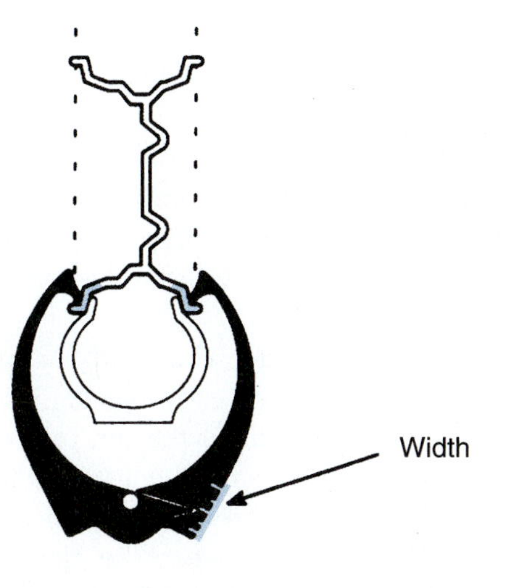

Figure 10-31 The rim's width is measured with a tong-like device similar to the one shown. (Courtesy of Hunter Engineering Company)

SERVICE TIP: At times, the wheel assembly may not balance without adding several ounces of weight at various points. This may be corrected by removing all of the weight, deflating the tire, breaking the beads loose, and rotating the tire 90 degrees or 180 degrees on the wheel. The tire is inflated and balanced. This procedure may cause heavy spots on the rim and tire to counteract each other, which will require less weight to balance the wheel assembly.

With the measurements entered, close the hood and press the SPIN button. The balancer will spin the wheel assembly and bring it to a halt. The display or scale will indicate where weights are to be placed and the size of each weight. Usually, the wheel assembly is rotated until the point of the new weight is straight up. A light or the alignment of two or more lines shown on the balancer's display will indicate that point. There will be two points selected for weights: one for the inside edge of the wheel, another for the outside edge. They are usually not directly across from each other. Install the proper weights, close the hood, and spin the tire. If everything is correct, the balancer will indicate zero (0) or by some other means indicate the tire and wheel are balanced. If the assembly is still out of balance, additional weights will be indicated.

Some customers with magnesium or chrome wheels desire that no weights be showing on the outside of the rim. Most new balancers have procedures for measuring and marking points for adhesive weight to the inside of the rims. The balancer has to be set up differently. Follow the instruction manual for the special procedures. There are also special rim weights for chrome wheels.

Setting Up an Alignment Machine

Classroom Manual
pages 300–303

Special Tools

Alignment machine

Alignment lift

There are several brands and models of alignment machines in use today. They range from units that measure a white light (very few left) to fully computerized units that do practically everything except turn the wrench. The following section covers a typical setup for a computerized system based on the Hunter WinAlign. Other brands are very similar in their setup procedures.

Switch on the alignment machine so it warms up while the vehicle is positioned. Raise the vehicle on the lift and lower the levering legs. Unlock the lift and lower it slowly until it is resting completely on four legs. The vehicle will have to be lifted from the rack at this point. Before lifting the vehicle, however, block the rear wheels on both sides, place the vehicle in neutral, be sure the parking brake is off, and unlock the steering wheel.

Position the lift's auxiliary jack under the front axle, and move the pads directly under or as close as possible to the lower ball joint. The jack should lift the end of the vehicle evenly without one wheel being substantially lower then the other. Raise the front of the vehicle and lock the jack in place. Repeat the lifting procedures with the rear of the vehicle. The pads at the rear must be placed correctly to prevent damage to the vehicle and allow room for the technician to access the adjustment devices. Consult the service manual as needed. Ensure that the jack is locked.

Select the rear wheel alignment head. The head is composed of the *mounting bracket* and the *sensor*. The knob at the top of the head's bracket is used to open or close the clamping device. The clamps can be attached under the rim (wheel weight area) or on the outside (tire bead area). It is suggested that the clamps be attached to the outside whenever possible. The clamping knob should be positioned at the top with the head installed onto the wheel. Rotate the tire so that the valve stem is between the eleven- and one-o'clock positions. Remove the valve stem cap. Look under the clamping knob and inside the alignment head bracket. There should be a short cable with a connector similar to the valve stem cap (Figure 10-32). Screw this onto the valve stem to prevent the alignment head from falling completely to the floor if it dropped or becomes unclamped from the rim.

Place the lower clamps between the tire and rim. Turn the adjusting knob until the upper clamps can be positioned onto the rim (Figure 10-33). Ensure that all four (two top and two lower) are correctly positioned; then tighten the clamping knob. It is not necessary to apply a great deal of torque. The clamps should be tightened just enough to hold the head in place. They can be tested by lightly pulling out the head. Again, it is not necessary to apply a great deal of force. The head can be jerked off, and over tightening the clamping knob may distort

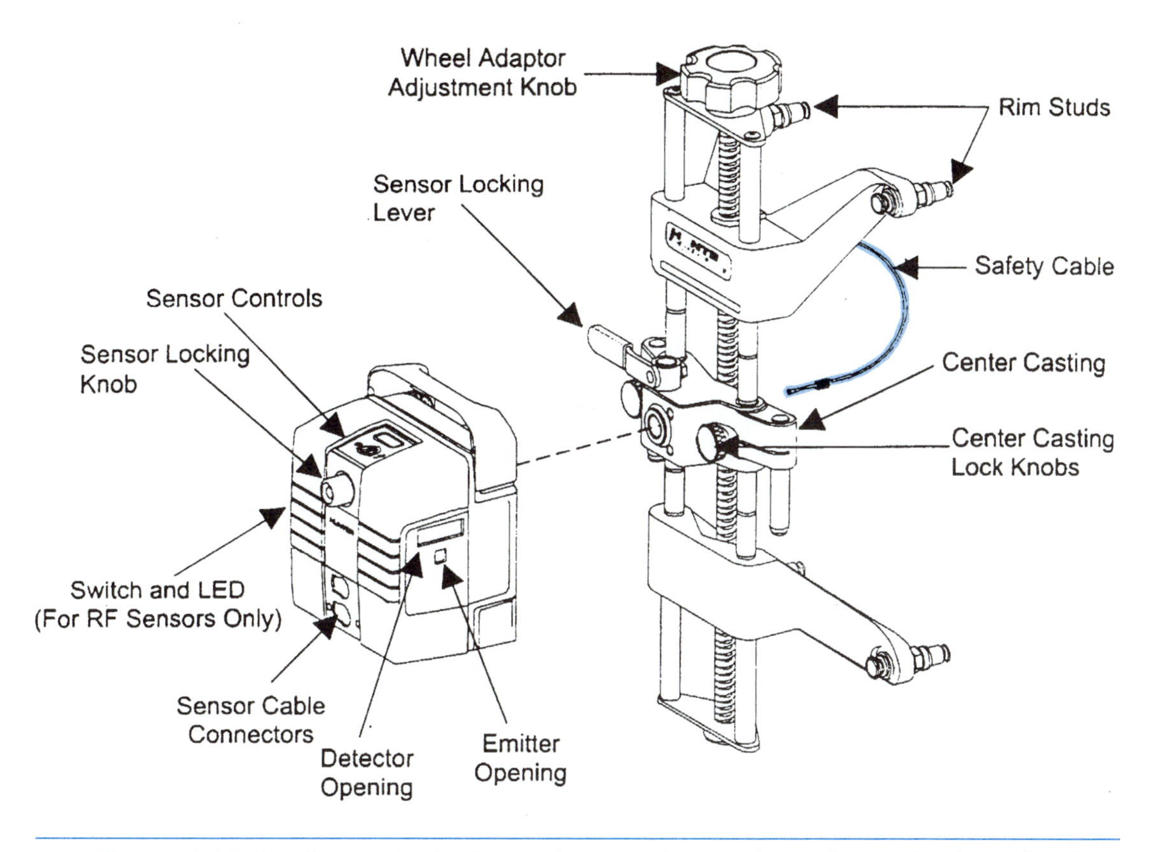

Figure 10-32 The alignment's wheel sensor is mounted to an adapter that can be clamped to the wheel. (Courtesy of Hunter Engineering Company)

the head's mounting bracket. If the wheel attempts to turn with the head installed, place a wheel block to prevent the turning. Repeat the head installation procedures for the other three wheels. If the heads and machine require cable connections, install the cables and turn on all alignment heads. Most will have a light or some type of indicator to show that the power is connected and on.

With all heads installed, the computer can be set up for the vehicle. The machine's manufacturer assumes that the technician has some knowledge of data entry and retrieval. Most alignment machines are user-friendly, and technicians need to know only the basic steps. Usually, the most popular machines have two methods of entering or retrieving data: *keyboard* or *light*

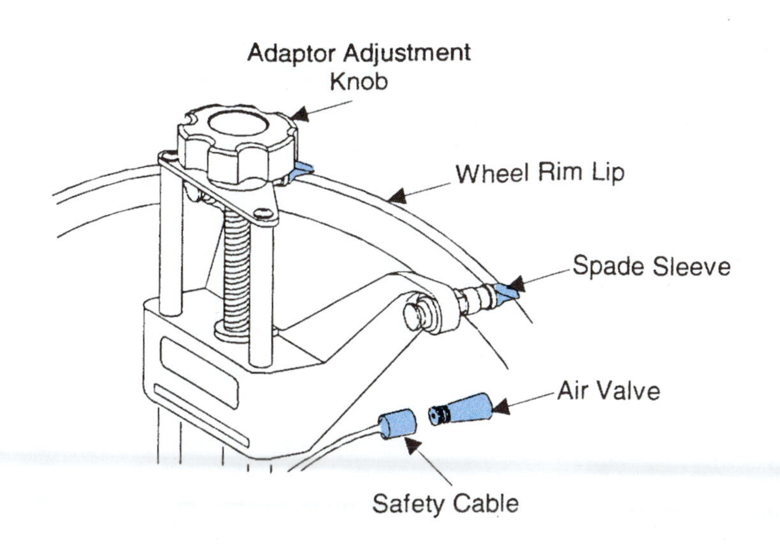

Figure 10-33 Attach the adapter clamps to the rim by the spade sleeve. Note the safety cable. (Courtesy of Hunter Engineering Company)

pin. The keyboard will have several designated *function keys*. Function keys allow the technician to move between areas of information or screens. Each screen will have more detail based on that area. The Hunter machine has four function keys that are shown at the bottom of the entry screen: K1, K2, K3, and K4. Depending on the information desired, each key will open a different screen and the function of each K key will change to match the new screen. It is best to practice, using the machine's instruction manual, until the technician is comfortable with the keyboard arrangement.

A *light pen* is basically a small flashlight wired to the computer that works like a computer's mouse. Point the pen at the desired title or listing on the screen and click the button. The computer will then display the information requested. The following training section will use a 1994 Honda Civic for the test vehicle and the light pen.

On the opening screen there should be a key or section that is titled "Begin Alignment" or similar terminology (Figure 10-34). Point the pen at that listing and click. The screen should display a list of vehicle brands. A point to remember here is to select the *brand* not the manufacturer (Figure 10-35). For instance, Ford and Chrysler manufacture several brands, including Ford and Chrysler. Selecting Ford will not show data for a Lincoln Mark V. Lincoln would have to be selected for that vehicle. In this case, use the arrow keys on the keyboard to scroll down until "Honda" appears. Point the pen at Honda and click. The next screen displays the Honda models. Scroll up or down until "Civic" appears. Point and click the pen. On the next screen, select the year, "1994," as the date of manufacture. The next screen recommends a four-wheel alignment and compensating procedures for the four sensors. Two methods are offered on the Hunter machine: *three-point* and *rolling*. Since three-point is recommended, point the pen at "use 3-point procedure" (lower left corner of screen) and click (Figure 10-36). The alignment machine is now set for a four-wheel alignment on a 1994 Honda Civic. Almost every computerized alignment machine is set up in a very similar manner. The primary differences are the use of the keyboard, light pen, and the exact terminology displayed on the screen. From this point on, all machines will display alignment specifications and measurements data in almost identical form.

Most alignment machines have software to enter customer, vehicle, and other repair order information. They can also print out the data and include some diagnostic and repair information. The amount, style, and display methods differ on each brand and model. The machine's operating instructions will detail the exact procedures.

Figure 10-34 Use the light pin to click the "Begin Alignment" button or use the K4 key on the keyboard (not shown). (Courtesy of Hunter Engineering Company)

Figure 10-35 Use the down arrow button by clicking with the light pen or the K3 key (not shown). (Courtesy of Hunter Engineering Company)

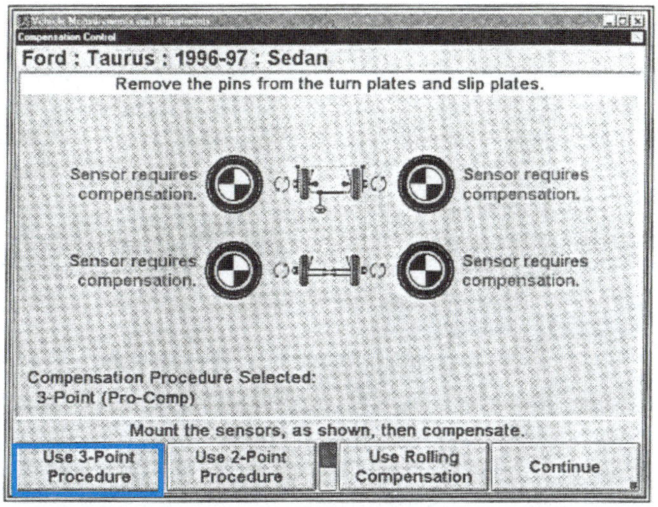

Figure 10-36 The "3-point" button or K1 (not shown) selects the type of sensor compensation procedures. (Courtesy of Hunter Engineering Company)

Alignment procedures taken after the setup will not be covered here, but for the reader's general information, the following steps are taken to complete an alignment:

- Sensors are compensated individually.
- Locking pins are removed from turntables and skid plates.
- The vehicle is lowered to lift.
- Caster measurement is taken.
- Rear caster, camber, and toe adjustments are made.
- Front caster, camber, and toe adjustments are made.
- Heads are removed and stored.
- The vehicle is removed from the lift and the pins are reinstalled.

Detailed information on suspension, steering, and alignment components and procedures can be found in *Today's Technician Automotive Suspension and Steering Systems* available from Delmar Publishers. Other sources include school, local, and Internet libraries. Suspension and steering parts manufacturers like MOOG and TRW have Internet sites with information on design and function of their products.

CASE STUDY

A technician was performing a routine tire rotation and balance along with other repairs. One tire proved difficult to balance and required excessive weights in the technician's opinion. The tire was deflated, broken down, and slid 180 degrees on the wheel. The amount of weight was reduced and the tire finally balanced out. Due to the other repairs, the shop foreman tested the vehicle and it had a pronounced vibration similar to an unbalanced tire. The technician rechecked the difficult tire and found it to be over two ounces off. The technician removed the tire from the rim. Inside were several large lumps of chassis grease and a new patch. Based on a conversation with the owner, it was determined that the tire had been patched several days earlier. The owner also stated that the vibration started after the tire was patched. The tire was low-profile mounted onto a wide rim. It was deduced that the shop performing the patch could not seat the tire and had used chassis grease to fill the large gap between the tire and rim so air could be held long enough for the tire to seat. Apparently, some of the grease got into the tire. The technician cleaned the interior of the tire and balanced it with no further problems.

Terms to Know

Camber wear
Castellated nut
Isolators
Joint stud
Pinch bolt
Toe wear
U-bolts

ASE Style Review Questions

1. *Technician A* says the upper bearing on a shock absorber can cause a grinding noise during turns.
   Technican B says a broken coil spring may allow the frame to connect the bumper on the lower arm. Who is correct?
   A. A only
   B. B only
   C. Both A and B
   D. Neither A nor B

2. Coil springs are being discussed. *Technician A* says a broken spring may affect steering angles.
   Technican B says a weak spring on a SLA system may affect steering angles. Who is correct?
   A. A only
   B. B only
   C. Both A and B
   D. Neither A nor B

3. A rack and pinion system is being discussed. *Technician A* says a worn pinion gear may cause erratic movement between the steering wheel and wheels.
   Technican B says a worn inner tie-rod socket will cause both wheels to respond slowly to steering wheel movement. Who is correct?
   A. A only
   B. B only
   C. Both A and B
   D. Neither A nor B

4. The steering system is being discussed. *Technician A* says any movement in the socket/ball requires a replacement of that steering component.
   Technican B says any movement between the stud and its mating hole requires a replacement for that component. Who is correct?
   A. A only
   B. B only
   C. Both A and B
   D. Neither A nor B

5. A vehicle with a lowered corner is being discussed. *Technician A* says this could be caused by a broken shock.
   Technican B says a broken torsion bar mount could be the cause. Who is correct?
   A. A only
   B. B only
   C. Both A and B
   D. Neither A nor B

6. Steering systems are being discussed. *Technician A* says the excessive steering wheel play on a recirculating ball and nut could be caused by a worn worm gear.
   Technican B says a bad coupler could cause steering wheel play. Who is correct?
   A. A only
   B. B only
   C. Both A and B
   D. Neither A nor B

7. Inspection of the steering linkage is being discussed. *Technician A* says sideways movement of the socket is normal.
   Technican B says force should be applied against the centerline of the socket stud to check for movement between the stud and hole. Who is correct?
   A. A only
   B. B only
   C. Both A and B
   D. Neither A nor B

8. A tire with abnormal wear is being discussed. *Technician A* says cupping of the tread could be a sign of poor maintenance.
   Technican B says balancing and rotation may help correct the problem. Who is correct?
   A. A only
   B. B only
   C. Both A and B
   D. Neither A nor B

9. *Technician A* says a saw-tooth wear pattern across the tread and excessive wear on a tire's edge may indicate that both the toe and camber are incorrect. *Technican B* says excessive wear on the outer edge of a tire indicates the caster is too negative. Who is correct?
   A. A only
   B. B only
   C. Both A and B
   D. Neither A nor B

10. Wheel alignment is being discussed. *Technican A* says A cable is used to align the sensor with the wheel assembly.
   Technican B says that vehicle and customer data can be printed out by most computerized alignment machines. Who is correct?
   A. A only
   B. B only
   C. Both A and B
   D. Neither A nor B

Table 10-1 ASE Task

Diagnose manual steering gear (non-rack and pinion type) noises, binding, uneven turning effort, looseness, hard steering, and lubricant leakage problems; determine needed repairs.

Problem Area	Symptoms	Possible Causes	Classroom Manual	Shop Manual
Steering	Noise during operation.	**1.** Broken/worn components.	251, 288–294	251, 259–263
	Loose steering action.	**2.** Worn worm or sector gears.	289–290	259–263

SAFETY

Wear safety glasses.
Disable air bag system.
Use wheel blocks.
Use lift locks or jack stands.

Table 10-2 ASE Task

Diagnose rack and pinion steering gear noises, vibration, looseness, and hard steering problems; determine needed repairs.

Problem Area	Symptoms	Possible Causes	Classroom Manual	Shop Manual
Steering	Noise during steering operation.	**1.** Broken/worn components.	292–293	259
		2. Worn rack or pinion gear.	292–293	259

SAFETY

Wear safety glasses.
Disable air bag system.
Use wheel blocks.
Lock lift or use jack stands.

Table 10-3 ASE Task

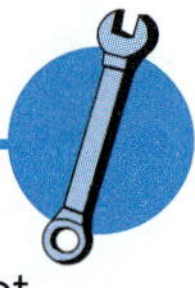

Inspect and replace rack and pinion steering gear inner tie-rod ends (sockets) and bellows boot.

Problem Area	Symptoms	Possible Causes	Classroom Manual	Shop Manual
Steering	Noise during steering operations. Popping or bumping noise at steering wheel during steering.	**1.** Worn inner tie rod.	292–293	262, 265
		2. Worn tie-rod ends.	292–293	262, 265

SAFETY

Wear safety glasses.
Use wheel blocks.
Lock lift or use jack stands.

Table 10-4 ASE Task

Diagnose power steering gear (non-rack and pinion) noises, vibration, looseness, hard steering, and fluid leakage problems; determine needed repairs.

Problem Area	Symptoms	Possible Causes	Classroom Manual	Shop Manual
Leaks and increased steering efforts.	Red, pink oil spots on floor.	**1.** Power steering pump leaks.	293–294	263–265
		2. Power steering connections loose or worn.	293–294	—
		3. Power steering lines cracked or split.	293–294	—
		4. Worn pump.	293–294	—
		5. Loose/worn drive belt.	293	263–265

SAFETY

Wear safety glasses.
Use wheel blocks.
Lock lift or use jack stands.

Table 10-5 ASE Task

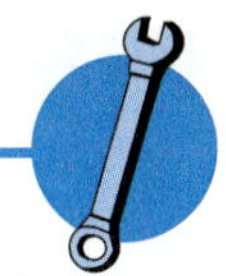

Diagnose power rack and pinion steering gear noises, vibration, looseness, hard steering, and fluid leakage problems; determine needed repairs.

Problem Area	Symptoms	Possible Causes	Classroom Manual	Shop Manual
Leaks	Red, pink oil spots on floor.	**1.** Power-steering pump leaking.	292–294	—
		2. Power-steering connections loose or worn.	292–294	260
		3. Power-steering lines cracked or split.	292–294	—
		4. Fluid over full.	292–294	—
		5. Pump worn.	293	259, 260
		6. Belt loose or worn.	293	260, 263–265

SAFETY

Wear safety glasses.
Use wheel blocks.
Lock lift or use jack stands.

Table 10-6 ASE Task

Inspect power-steering fluid level and condition; adjust level in accordance with vehicle manufacturers' recommendations.

Problem Area	Symptoms	Possible Causes	Classroom Manual	Shop Manual
Leaks	Red, pink oil spots on floor.	**1.** Power-steering pump/lines leaking.	293	260
		2. Fluid deteriorated from age/over heating.	293	—

SAFETY

Wear safety glasses.
Use wheel blocks.
Lock lift or use jack stands.

Table 10-7 ASE Task

Inspect, adjust tension and alignment, and replace power-steering pump belt(s).

Problem Area	Symptoms	Possible Causes	Classroom Manual	Shop Manual
Power steering system	Poor power assist. Belt noise.	**1.** Loose belt. **2.** Worn belt. **3.** Worn pump.	293 293 293	263–265 263–265 260

SAFETY

Wear safety glasses.

Table 10-8 ASE Task

Inspect pitman arm.

Problem Area	Symptoms	Possible Causes	Classroom Manual	Shop Manual
Steering	Rough steering movement.	**1.** Worn/broken components. **2.** Loose fasteners.	289–290 289–290	262 262

SAFETY

Wear safety glasses.
Use wheel blocks.
Lock lift or use jack stands.

Table 10-9 ASE Task

Inspect relay rod (center link/drag link/intermediate rod).

Problem Area	Symptoms	Possible Causes	Classroom Manual	Shop Manual
Steering	Rough steering movement. Noise during steering.	**1.** Worn/damaged components. **2.** Loose fasteners.	290–291 290–291	262 262–263

SAFETY

Wear safety glasses.
Use wheel blocks.
Lock lift or use jack stands.

Table 10-10 ASE Task

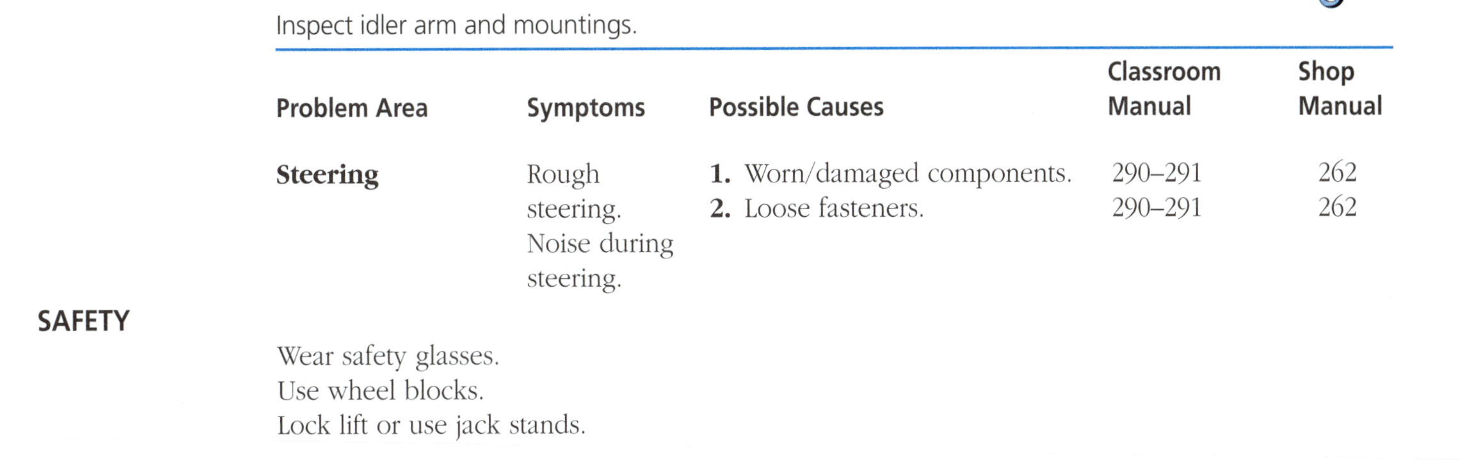

Inspect idler arm and mountings.

Problem Area	Symptoms	Possible Causes	Classroom Manual	Shop Manual
Steering	Rough steering. Noise during steering.	**1.** Worn/damaged components.	290–291	262
		2. Loose fasteners.	290–291	262

SAFETY

Wear safety glasses.
Use wheel blocks.
Lock lift or use jack stands.

Table 10-11 ASE Task

Inspect tie rods, tie-rod sleeves, clamps, and tie-rod ends.

Problem Area	Symptoms	Possible Causes	Classroom Manual	Shop Manual
Steering	Rough steering. Noise during steering.	**1.** Worn/damaged components.	290–291	262–263
		2. Loose fasteners.	290–291	262–263

SAFETY

Wear safety glasses.
Use wheel blocks.
Lock lift or use jack stands.

Table 10-12 ASE Task

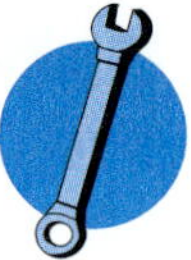

Diagnose front suspension systems noises, body sway/roll, and ride height problems; determine needed repairs.

Problem Area	Symptoms	Possible Causes	Classroom Manual	Shop Manual
Steering	Noise during steering. Excessive lean during steering.	**1.** Worn/damaged components.	271	251–256
		2. Missing/damaged bushings.	272–273	251
		3. Broken components.	272–273	251–256

SAFETY

Wear safety glasses.
Use wheel blocks.
Lock lift or use jack stands.

Table 10-13 ASE Task

Diagnose rear suspension system noises, rods, tie-rod sleeves, clamps, and tie-rod ends.

Problem Area	Symptoms	Possible Causes	Classroom Manual	Shop Manual
Steering	Rough steering.	**1.** Worn/damaged components.	278–281	251, 254–256
	Noise during steering.	**2.** Loose fasteners.	278–281	251, 254–256

SAFETY

Wear safety glasses.
Use wheel blocks.
Lock lift or use jack stands.

Table 10-14 ASE Task

Diagnose tire wear patterns; determine needed repairs.

Problem Area	Symptoms	Possible Causes	Classroom Manual	Shop Manual
Steering	Vibration. Wandering during operation.	**1.** Worn/damaged components.	294–299	259–263
		2. Tire inflation incorrect.	294	266
		3. Wheel alignment incorrect.	300–303	267

SAFETY

Wear safety glasses.
Use wheel blocks.
Lock lift or use jack stands.

Job Sheet 17

Name ______________________________ Date ______________

Lubricating Suspension and Steering Systems

Upon completion of this job sheet, you should be able to complete a lubrication on the suspension and steering systems and other systems' components.

Tools Needed

Service manual
Lift or floor jack with stands
Grease gun, powered or hand operated

Procedures

1. Raise the vehicle to a good working height.
2. Locate each grease fitting on the suspension:

Component	Fitting (y/n)
Upper ball joint (2 if used)	________
Lower ball joint (2 if used)	________
Steering knuckle (used on solid front axle without control arms; one top, one bottom; 4 total)	________ ________
Tie-rod ends (2 if used)	________
Tie-rod/center link (2 if used)	________
Idler arm mount (1 if used)	________
Center link/idler (1 if used)	________
Center link/pitman arm (1 if used)	________
Front mount of rear leaf spring (rarely used, but may be 2)	
If the vehicle is RWD, check drive shaft U-joints, usually included as part of routine service. (1 per joint)	________

Inspect other components for fittings.
Are there additional fittings? ______________

If so, list. (Possible locations: trailer hitch, U-joint on rear-drive axle like some Corvettes; power takeoff and shaft; mounts on front leaf spring; U-joint-type coupling on steering shaft; other visible points where two or more components move/work together.)

____

____

____

____

____

____

3. Lube each fitting in turn, starting at one front wheel and moving across the vehicle to the other wheel. It may be necessary to turn the wheel left or right to access the fittings.
4. Do not over lube. Stop lubing when a very small amount of grease is forced from the boot.
5. Lube any fittings at the rear of the vehicle.
6. Lube any other fittings found under the vehicle.
7. Lower the vehicle.
8. Lube any suspension/steering components not accessible from under the vehicle.
9. Check the power-steering fluid level.
10. Clean the work area, complete the repair order, and return the vehicle to the customer.

____

____

Job Sheet 18

18

Name ______________________ Date ______________

Replacing Inner Tie-Rod Sockets on a Rack and Pinion Steering System

Upon completion of this job sheet, you should be able to replace the inner tie-rod socket on a rack and pinion steering system.

Tools Required

Inner tie-rod removal socket/wrench
Tie-rod end removal tool
Standard hand tools
Lift or jack and jack stands
Service manual

Procedures

1. Use the service manual to obtain torque specification for:

   Wheel nuts

   Inner tie-rod socket

   Tie-rod end/steering knuckle fastener

   Tie-rod end locking nut

2. Lift the vehicle to a good working height or until the tire is clear of the floor.
3. Remove the wheel assembly.
4. Loosen the tie-rod end lock nut.
5. Remove the tie-rod end nut and use a tie-end removal tool to separate the end from the steering knuckle.

   ■ **NOTE:** Steps 6 and 7 and complete removal of the boot can be done on the bench *IF* an open-end wrench is used to unscrew the socket. If the special socket wrench is used, steps 6, 7, and 8 must be performed as listed.

6. Unscrew the end from the tie rod while counting the turns.

   Number of turns ______________ left or right threads

7. Remove the locking nut from the inner tie rod.
8. Remove the boot clamps and remove the boot.
9. Remove the locking device for the inner tie-rod socket and move it to the rack.
10. Slide the socket or wrench over the inner tie rod and unscrew the rod. Do not let the equalizing tube drop from the opposite boot.

11. Screw and torque the new inner tie rod onto the rack.
12. Install the inner tie rod to the rack locking pin.
13. Install the boot while guiding the equalizing tube into its cavity in the boot. Install the boot's inner clamp.
14. Install the locking nut onto the inner tie rod.
15. Screw the tie-rod end onto the inner tie rod with the same number of turns used during removal. Hand tighten the lock nut.
16.. Install the tie-rod end stud into the steering knuckle. Torque the fastener.
17. Install the wheel assembly and torque the lug nuts.
18. Use an alignment machine to adjust toe.

Job Sheet 19

19

Name ______________________________ Date ____________________

Dismounting and Mounting a Tire

Upon completion of this job sheet, you should be able to remove and mount a tire on a wheel.

Tools Required

Tire changer
$^{1}/_{2}$-inch drive impact wrench
Tire valve removal tool
$^{1}/_{2}$-inch drive torque wrench

Procedures

1. Locate the tire data and lug nut torque.

Tire size ____________________

Tire type ____________________

Recommended tire inflation ____________________

Maximum pressure on sidewall ________________

Lug nut torque ________________

2. Study all operating instructions for the tire changer.

3. Inspect the tire changer and operate each of the control valves. The valves should be spring loaded to the OFF position. If any valve does not return automatically to the OFF position immediately after being released, do not use the machine until repairs have been completed.

4. Lift the vehicle and remove the wheel assembly.

5. Remove the valve from the tire's valve stem.

6. Once the tire is completely deflated, remove all wheel weights.

■ **WARNING:** Do not place a magnesium or chrome rim onto the tire changer without using special adapters. Damage to the wheel may result if protective adapters are not used. Consult the machine's operating instructions.

7. Place the tire in the bead breaker.

A. If the breaker is on the side of the changer, move the tire until the clamp reaches the *edge* of the rim, but not directly on the rim.

1. Guide the clamp to the tire and operate the power valve.

2. Maintain the valve in operating mode until the tire breaks from the bead. Release the valve.

3. Turn the tire until the other side is under the clamp.

4. Repeat steps 7.A.1 and 7.A.2 until the second side breaks loose.

B. If the breaker is mounted on a turntable, place the tire on top of the turntable.

1. Operate the power valve to lock the tire assembly to the platform.

2. Position top and bottom clamps, as required, near the rim's edge.

3. Operate the clamp-operating valve until the tire breaks loose from the rim on both sides.

8. If the tire is not on the turntable, place it there and clamp it in place using the changer's clamping device.

9. Position the tire removal/installing guide and lock it in place.

10. Use the changer's pry bar to lift a section of the tire onto the guide. Do only one side of the tire at a time.

■ **CAUTION:** Use caution when the pry bar is inserted between the tire and guide or wheel while the turntable is rotating. Guide the bar until it can be lifted from the tire. Injury can occur if the pry bar jerks loose or becomes dislodged and hangs in the wheel assembly.

11. Operate the turntable rotation valve until the pry bar can be removed and this side of the tire is separated from the rim. Turn off the rotation valve.

12. Pull up on the tire near the guide.

13. Insert the pry bar between the lower side of the tire and the wheel. Lift the tire over the guide.

14. Operate the turntable rotation valve until the pry bar can be removed and this side of the tire is separated from the rim. Turn off the rotation valve.

15. Move the guide to the side and remove the tire.

16. Lightly lubricate the bead area of the new tire with tire lubricant.

17. Place the tire over the wheel.

18. Position the tire guide within the circle of the tire's beads and lock in place onto the rim. The tire should circle the guide.

19. Start the front of the tire's lower bead down over the edge of the rim and over the guide.

■ **CAUTION:** Keep hands and fingers clear of the bead area as the tire is being forced onto the wheel. Serious injury could result if they get caught between the wheel and tire bead.

20. Operate the rotation valve and apply downward pressure on the tire as the guide forces one side of the tire onto the wheel.

21. Turn off the rotation valve.

22. Position the top of the tire over the edge of the rim and the guide.

23. Operate the rotation valve and apply downward pressure on the tire as the guide forces one side of the tire onto the wheel.

24. Turn off the rotation valve.

25. Connect the changer's tire inflation line to the tire's valve stem.

■ **CAUTION:** Do not allow tire pressure to exceed the maximum pressure listed on the sidewalls. Serious injury or death could result from an exploding tire.

26. Inflate the tire until the bead "pops" into place on the wheel.

27. With the tire inflated, disconnect the air line and release the changer's locking clamps.

28. Install the valve into the valve stem.

29. Reconnect the air line to the valve stem and inflate it to specified pressure.

30. Disconnect the air line and install the valve stem cap.

■ **NOTE:** Wheel assembly should be balanced before completing step 31.

31. Install the wheel assembly onto the hub; install and torque lug nuts.

☑ **Instructor's Check** ______________________________

____

____

Job Sheet 20

20

Name ______________________________ Date ______________

Balancing a Wheel Assembly

Upon completion of this job sheet, you should be able to properly balance a wheel assembly.

Tools Required

Tire balancer
Wheel weight hammer/tool

Procedures

1. Study the balancer's operating instructions.
2. Select special adapters if the wheel is made of magnesium or chrome.
3. As required, select and install the centering cone onto the balancer's spindle. The cone may be installed on the outside of the wheel on some balancer's or certain types of wheels.
4. Position the wheel assembly onto the cone or spindle as required.
5. Install the lock nut onto the spindle and screw it into place.
6. If the balancer is not powered, turn on the power switch.
7. Select the type of balancing procedures desired: static or dynamic. Some balancers have special procedures that can be used.
8. Measure the distance from the balancer to the edge of the wheel. Measure at the part of the rim where the weights are installed. Use the measuring device attached to the balancer.
9. Enter the measurement using the buttons or knobs on the balancer.
10. Enter the wheel's width into the balancer.
11. Enter the wheel's diameter into the balancer.
12. Ensure that all wheel weights are removed from the wheel.
13. Close the hood and operate the balancer to spin the wheel assembly.
14. After the wheel assembly has stopped, observe the reading. If it is not balanced, rotate the wheel by hand until the balancer's alignment marks are matched for the inside of the wheel.
15. Attach the proper size weight, as determined by the balancer, to the inside of the wheel.
16. With the balancer's marks aligned, install the weight onto the rim. Use the weight hammer.

17. Rotate the wheel by hand until the marks are aligned for the outside of the wheel.

18. Select and install the weight on the outside of the wheel.

19. Close the hood and operate the balancer.

20. If the wheel assembly is balanced, remove the wheel assembly from the balancer.

21. If the wheel assembly is not balanced, repeat steps 13 through 19 as necessary to complete balancing.

■ **NOTE:** If the wheel assembly cannot be balanced on the second attempt, recheck the measurements and the mounting of the wheel assembly on the balancer. The tire may have to be removed from the wheel and repositioned.

22. Once the tire is balanced, remove the wheel assembly from the balancer and install it on the vehicle.

____

Servicing Brake Systems

Upon completion and review of this chapter, you should be able to:

- ❑ Discuss automotive brake fluids.
- ❑ Diagnose brake systems.
- ❑ Discuss safety precautions associated with ABS and air bag systems.
- ❑ Replace brake power boosters.
- ❑ Replace master cylinders.
- ❑ Replace disc brake pads.
- ❑ Discuss general replacement procedures for brake components.

Introduction

The brake system provides safe, controllable slowing and stopping of the vehicle. Considering that most vehicles weigh between one and two tons and may be traveling at interstate highway speeds, major failure of a component can lead to a life-threatening or at least an unnerving situation. While fairly simple to repair, brake system service requires the technician to accurately install and test components to prevent a possible accident.

This chapter will cover basic brake diagnosis and repairs. Since ABS and traction control systems generally require the use of a scan tool and special procedures for bleeding and installing some components, repairs to those systems will not be discussed. For specific information on ABS and traction control systems, the technician may refer to *Today's Technician Automotive Brake Systems* and *Automotive Computer Systems* for detailed instruction on the two systems' operation, testing, and repair.

Brake Fluid Types and Cautions

Classroom Manual
page 320

Glycol is a member of a group of alcohols. It should not be confused with "glyco" which is a combined form of glycerol, a sugar.

Before beginning the diagnosis of a brake system, it would be wise to consult the service manual for the type of brake fluid that is to be used. There have been instances where adding a few ounces of fluid to top off the system have led to expensive repairs done at the shop's expense.

There are three basic compounds used for brake fluids: ***glycol****-based, silicone-based,* and *petroleum-based.* Installing the wrong fluid into a system results in quick and complete deterioration of brake components. European vehicle manufacturers used a petroleum-based fluid in some of their older vehicles. Petroleum-based brake fluid has been discontinued for many years, but a classic vehicle refurbisher may encounter it.

Silicone-based brake fluid is used on a few recent vehicle makes and models. This type of fluid works very well as a brake fluid, but causes the most problem because there are so few vehicles using it (Figure 11-1). An inattentive technician may assume the brake fluid is glycol-based, top it off without checking, and cause serious problems in the brake system. Using the wrong type of fluid will cause deterioration of the sealing components and degrade the braking operation greatly. Even if the mistake is caught before the vehicle leaves the shop, the entire system must be taken down and all seals and flexible hoses must be replaced. This should be done at the shop's expense and the technician's expense. Always consult the service manual.

Glycol-based brake fluids are used in almost every brake system sold. Their popularity and frequent use are the reasons why mistakes, like the one listed above, occur. This brake fluid performs well and is fairly durable.

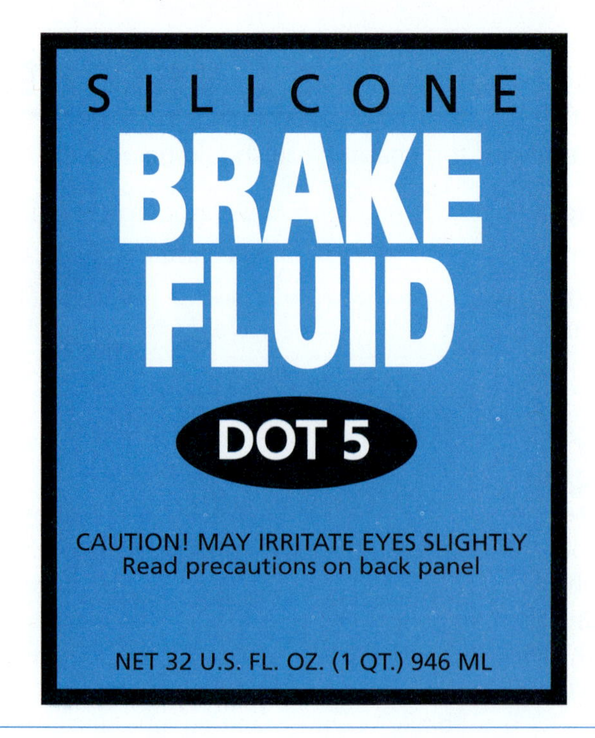

Figure 11-1 Silicone-based brake fluid cannot be mixed successfully with glycol-based fluids.

DOT stands for the U.S. Department of Transportation.

Brake fluids are classified by their resistance to boiling. The classes are **DOT 3**, **DOT 4,** and **DOT 5** with the highest, 5, having the highest boiling point (Figure 11-2). DOT 3 is the most common brake fluid. If a system is being flushed, a lower-rated brake fluid *cannot* replace a higher-rated one (DOT 3 *cannot* replace DOT 4 and DOT 5), but a higher-rated brake fluid *can* replace lower-rated ones (DOT 5 *can* replace DOT 4 and DOT 3). Consult the service manual for the proper type and classification.

The following precautionary statements should be as much a part of the technician's professionalism as the tools she or he uses every day.

WARNING: Do not mix glycol-based and silicone-based brake fluids. The mixture could cause a loss of braking efficiency.

WARNING: Protect the vehicle's paint surface from glycol-based brake fluids. This fluid can damage and remove the finish very quickly. If brake fluid gets on the vehicle's finish, wash the area immediately with cold, running water.

	STANDARDS		
	DOT 3	DOT 4	DOT 5
Boiling Point	min. 205°C	min. 230°C	min. 260°C
Wet Boiling Point*	min. 140°C	min. 155°C	min. 180°C

*Indicates decline in boiling point caused by increasing water content.

Figure 11-2 The higher the DOT number, the higher the boiling point of the brake fluid. (Courtesy of Nissan North America)

WARNING: Store brake fluid in a sealed, clean container. Brake fluid will absorb water and its boiling point will be lowered. Exposure to petroleum products such as grease and oil may cause deterioration of brake components when the contaminated fluid is installed.

WARNING: Do not use petroleum-based (gasoline, kerosene, motor oil, transmission fluid) or mineral-oil-based products to clean brake components. These types of liquids will cause damage to the seals and rubber cups and decrease braking efficiency.

CAUTION: Wear eye protection when dealing with brake fluid because it can cause permanent eye damage. If brake fluid gets in the eyes, flush with cold, running water and see a doctor immediately.

Diagnosing Brake Systems

Classroom Manual
pages 315–323

Brake system failures can be classed as either *hydraulic* or *mechanical* failures. Both will create symptoms that can be readily connected to a component or assembly. Mechanical failures will be discussed by common symptoms and their probable causes.

There are two conditions that constitute hydraulic failure: *external leaks* and *internal leaks*. Although mechanical conditions will create the leaks, they will be addressed as hydraulic problems to clarify the operational symptoms. An external leak is defined as "brake fluid exiting the system completely." An internal leak is defined as "a fluid leak within the system that does not exit the system."

External Leaks

A driver may complain that when the brakes were applied, the pedal dropped drastically and it took much more distance to stop the vehicle. The most common cause of this problem is an external leak somewhere in the system. A complete leak where the fluid exits the system results in a complete loss of pressure. The pedal will drop much lower than normal or go to the floorboard immediately upon brake application. If a single-piston master cylinder is being used, the pedal will hit the floorboard and cause a complete loss of all brakes. Pumping the pedal in an effort to regain control only compounds the problem. On a dual-piston master cylinder, the pedal will drop much lower than normal, but the driver will have brakes on one of the split systems and can stop the car. Again, pumping the pedal will not improve braking. An external leak of this type is noticeable when performing a visual inspection. The brake fluid level will also drop in the reservoir.

If the master cylinder is leaking externally, brake fluid will be leaking between the rear of the master cylinder and power booster or firewall (Figure 11-3). Usually, an external master

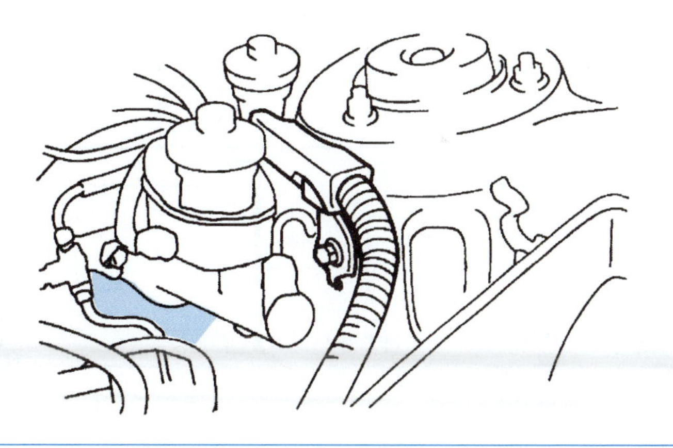

Figure 11-3 Brake fluid between the power booster and master cylinder indicates an external leak of the master cylinder. (Reprinted with permission)

cylinder leak will not result in complete brake failure until the fluid in the reservoir is exhausted. An external leak at a line, wheel cylinder, or caliper will cause a total loss of brakes on that system. The leaking brake fluid will be visible on the inside of the wheel assembly and may be dripping from suspension or steering components. Small, external leaks may be present but not readily visible. The leak is indicated when the reservoir must be topped off regularly. This type of leak is usually found at wheel cylinders during routine brake service. During inspection, the dust boots of the wheel cylinder are opened just enough to check for fluid inside the boots. The inside of the boots should be completely dry.

Internal Leaks

Drivers of older vehicles often complain of slow brake pedal drop while sitting at a stoplight. This situation is not as common on newer vehicles because of advances in machinery and materials, but it still occurs. The symptoms will usually occur as the car ages. Notice that the driver can stop the vehicle normally, but the braking action seems to lessen or disappear when the pedal is depressed for a time. Since the vehicle will stop with normal brake pedal movement, an external leak is eliminated. However, a pedal will drop when there is a loss of pressure, which points to an internal leak. An internal leak is almost always found in the master cylinder. Any other component that leaks will result in an external leak with the possible exception of ABS modulators. The modulators can leak internally, causing the same symptoms noted with a leaking master cylinder. However, because of regulations and new technology, this problem is found on the isolation/dump solenoids on older RWAL systems.

An internal leak in the master cylinder allows fluid to flow from the compression chamber, past the seal or cup, and back into the reservoir (Figure 11-4). High pressure created during stopping may prevent the seals from leaking. However, as the driver reduces force to hold the vehicle in place, the pressure drops slightly. This pressure reduction is enough to relax the seals and allow fluid to seep. The fluid returns to the reservoir and the master cylinder pistons move forward, dropping the pedal. Eventually, the condition could cause the pedal to reach the floor if the brakes are applied long enough. The fluid level in the reservoir will remain the same. Usually, the cheapest, shortest, and best repair is the replacement of the master cylinder. There are rebuild kits for some master cylinders, but the additional labor cost may exceed the cost of a replacement component.

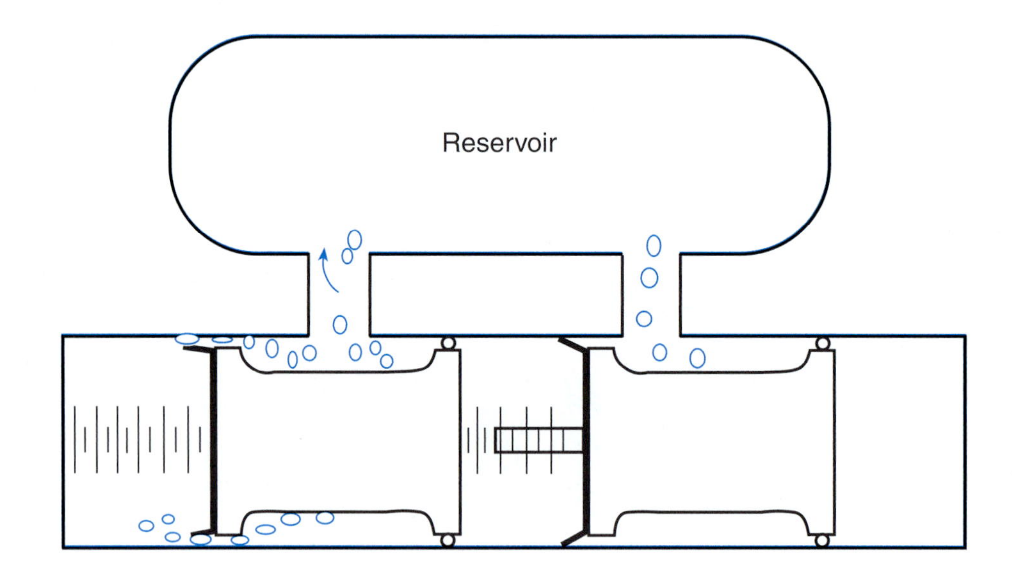

Figure 11-4 If the piston cup is corroded or damaged, pressurized fluid can escape back to the reservoir causing the brake pedal to drop.

Grabbing or Locking

A **grabbing brake** occurs when one or more wheel brakes grab the rotor or drum and then quickly release it. The vehicle jerks as its slows. This situation could be caused by oil or brake fluid on the pads or shoes. Hot spots on the drum or rotor can also cause the problem as will worn or weak retaining hardware. Grabbing brakes are more common with drum brakes.

A *grabbing brake* is one that tries to lock (grab) the wheel on each rotation and is usually caused by broken parts, wrong shoes, or hot spots.

A **locking brake** is one that will cause the wheel to stop turning and slide when the brakes are applied normally. This is caused by the friction material on that pad or shoe engaging with more friction than the other wheels. This can be caused by oil or brake fluid on the friction material or a brake that is adjusted too tightly. Brake fluid in particular will soften the friction material, which tends not to slide on the drum or rotor. Shoes and pads that have been contaminated by brake fluid or oil must be replaced and the rotor or drum cleaned. A second cause could be the opposing brake's failure to work correctly. For instance, if the left, front wheel consistently locks, the first thing to check is the right wheel brake to make sure it is working. Locking brakes are more common with drum brakes because leaking brake fluid flows directly on the shoes. A locking disc brake can be traced to a failure of the brake on the opposite wheel.

A *locking brake* can throw the vehicle out of control because of the loss of traction between the sliding tire and the road.

Pulling to One Side

A locking or grabbing brake may cause the the vehicle to pull to one side during braking. In this case, a steady, consistant, smooth pull caused by brake failure on the opposite side of the pull is being discussed. The driver must hold the steering wheel to one side to counteract the pull. Correcting the problem can be accomplished by adjusting the brake shoes or cleaning the slide pins, assuming there is no mechanical damage or bad parts. A pull to the side is often caused by a problem on the front axle, but an overly tight rear brake may cause a steady drift (light pull) to that side during slow to moderate braking. Excessive position camber may also cause pulling to one side during braking, but the driver will probably report this as a steering symptom since it will be present without braking.

Another pulling condition can be caused by internal damage to a brake hose. The internal lining of the flexible hose can come apart, and every time the brake is applied, the torn lining moves and blocks the fluid flow (Figure 11-5). The result is good braking on one wheel and little or none on the damaged side. Usually, there are no external signs of the blockage. Over a period of time, the hose could become weak enough for a bulge to form during braking, but the driver will probably have the brakes checked before this happens. If a thorough inspection is completed without finding the cause of the locking brake, the technician should suspect the brake hose on the non-braking side. This hose can be removed and the interior can be inspected by straightening the hose and looking at a light through the hole.

Fading

Fading occurs when the brakes are applied for long distances such as driving down a long grade. The brakes will appear normal at first, but braking action will be reduced and pedal feel will become firmer the longer the brakes are applied.

When the brakes are applied for long distances, they heat and lose friction. As the friction between rotor/pad or drum/shoe decreases, the braking action decreases. The driver applies more force and the heating becomes more intense. The eventual result is a complete loss of braking and a very hard pedal. Releasing the brakes will allow them to cool, but damage to the brake friction components may have occurred. When a light-vehicle driver complains of a fading condition, the technician should check for a trailer hitch. Pulling a trailer with a light vehicle can cause a fading condition.

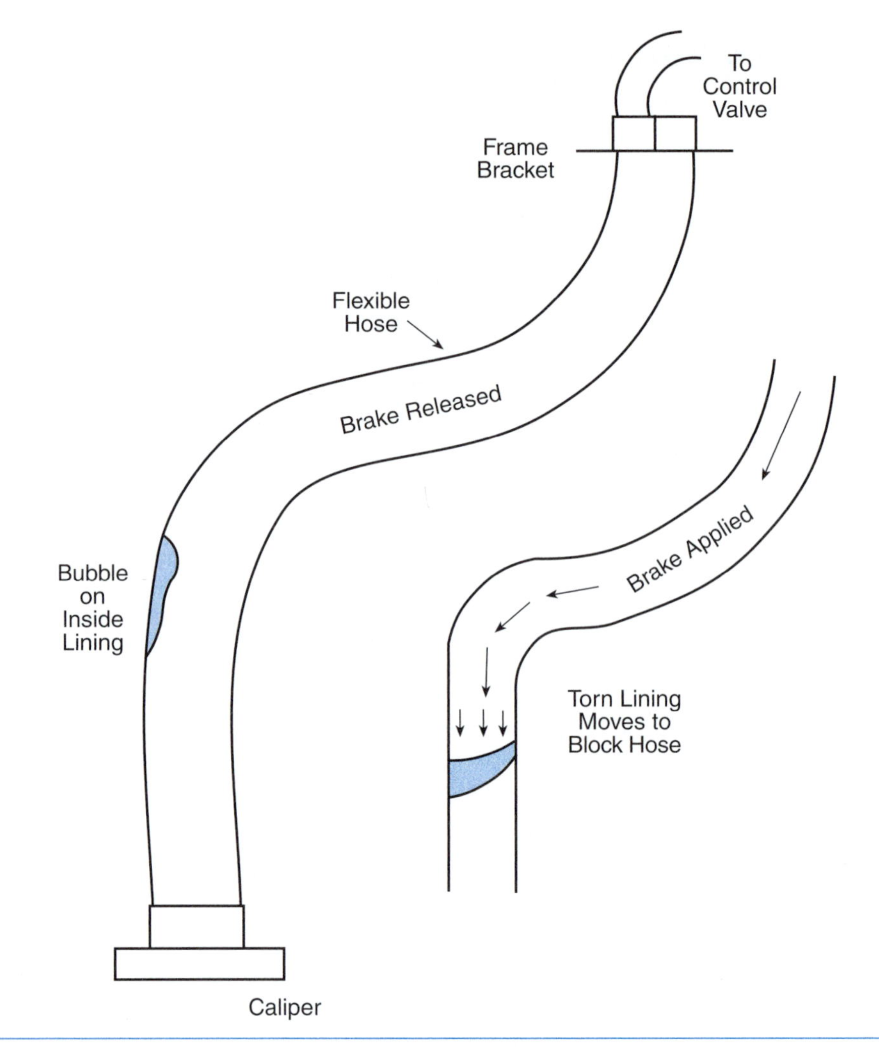

Figure 11-5 A torn or damaged interior lining can move and block the hose during brake application.

Many times, signs are posted at the top of steep slopes warning drivers, particularly truck drivers, to select a lower gear before going down the hill (Figure 11-6). The vehicle's driveline and engine are used to slow the vehicle and reduce brake usage.

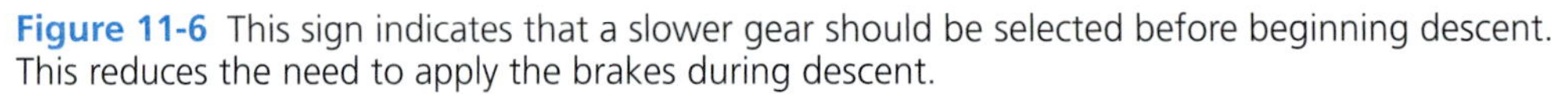

Figure 11-6 This sign indicates that a slower gear should be selected before beginning descent. This reduces the need to apply the brakes during descent.

Hard Pedal but Little Braking

When a driver applies the brake and has a **hard pedal,** the vehicle may not slow as it should. Water between the rotor and pad or drum and shoes will cause this condition. The water will burn off quickly and the brakes will return to normal, but repairs are required if the same condition occurs every time the brakes are applied.

In most cases, the condition stems from brakes that have been overheated. Failure to completely release the parking brake, repeated hard or long braking, pulling a trailer, misadjusted drum brakes, or a stuck caliper will cause the friction material to glaze from heat. The result is loss of friction between the pads/shoes and the disc/rotor. The pad or shoe slides over the machined surfaces without gripping. In most cases, the cause is a driver who consistently waits until the last moment to apply the brakes. Hand sanding the friction material might provide a quick fix to remove the glaze, but the correct repair procedure is to replace the shoes/pad and machine or replace the drums or rotors.

An overheated or **glazing** on one wheel may cause the vehicle to pull to one side during braking. Like other braking faults, pad or shoe glazing may not be obvious on first inspection. A glazed pad or shoe normally produces a very shiny, hard look to the rotor or drum. Bluish-colored spots on the machined surface are usually a sign that the brakes have been overheated and **hot spots** have formed on the braking surface.

The wrong shoes or pads can cause a *hard pedal*. They can also create almost all of the brake symptoms discussed in this section. A hard pedal is defined as a pedal that will not depress past a certain point, but the vehicle does not slow as it should. The driver will feel like he or she is pressing on an immovable metal block.

Glazing is a film over the friction surface caused by overheating of the friction material. The glaze will be very slick and greatly reduces the braking effect.

Pedal Pulsation and Vibration

A correctly machined drum and rotor, along with good, clean shoes and pads, will bring the vehicle to a smooth stop without vibration. During the life of the brake pad or shoes, heat attacks the working components during every braking action. This continuous heating and cooling, sometimes under poor weather and road conditions, causes the rotor or drum to warp or wear unevenly. Repeated hard braking increases the likelihood of damage. Disc brake systems are most susceptible to **pedal pulsation** and vibration during braking. Vibration on drum brake systems can be traced to **hot spots** on the drum, damaged shoes, or worn/damaged mounting components. For the most part, drum brakes do not produce pedal pulsation but they do tend to grab.

On a disc brake rotor, braking surfaces that are not parallel will cause pedal pulsation. The rotor has two machined braking or friction surfaces that should be parallel to each other. If one surface has a high or low area compared to its companion, or if the rotor is warped, the brake pad and piston will be pushed in and out of the caliper during braking (Figure 11-7). This action is transmitted back through the brake system, tries to kick up the brake pedal, and is repeated on every revolution of the rotor. In addition, there will be several points on the rotor that are not parallel or warped. The result is a light-to-heavy pulsation of the pedal and vibration of the vehicle. The damage could be severe enough to jerk the steering wheel back and forth as the two front brakes pulsate at different times.

Customers who are not familiar with the action of ABS may complain that at times the brakes seem to vibrate and shudder during heavy braking, or the vehicle feels like it is coming apart during a panic stop. Loose gravel, mud, wet leaves, or anything that makes the road surface slick can cause the ABS to function under light to moderate braking. A brake inspection and customer education may be the only repair needed.

Pedal pulsation is a rapid up-and-down movement of the brake pedal during brake application. The condition worsens the harder the brakes are applied.

Hot spots are formed by excessive heat on the weakest area of the disc or rotor. The spot may be crystallized, and the cutting tip on a lathe will tend to skip over that point. The spot will be very hard but brittle.

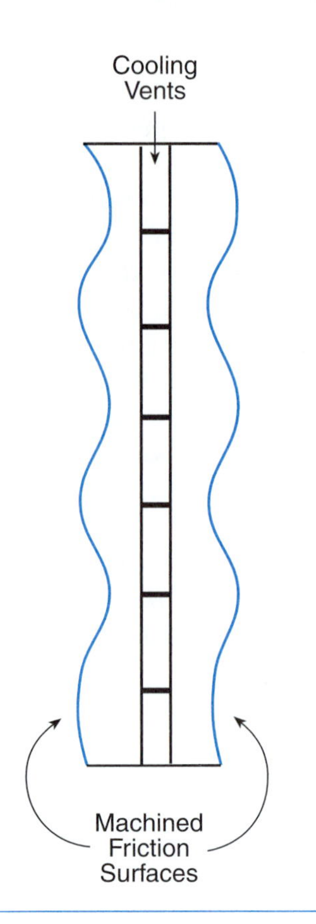

Figure 11-7 If the two-machined braking surfaces of a rotor are not parallel, the brake pedal will pulsate during brake application.

Spongy Pedals

A *spongy pedal* will feel like the driver is stepping on a light spring. The bounce may be reduced somewhat during heavy braking.

A driver may complain that his or her vehicle stops quickly and quietly, but the pedal feels soft or bouncy. This is known as a **spongy pedal** and is caused by air in the brake system. As discussed in the section on hydraulic theory, liquids cannot be compressed. However, air can be compressed and used to perform certain types of work even though it does not transmit force well. When air enters the brake system, it forms an air pocket and compresses or expands every time the brakes are applied and released. What the driver is feeling though the pedal is the compression of the air pocket as the brake fluid tries to transmit the driver's force to the wheels. It will take more driver force to achieve the same braking action. The only solution is to locate and repair the air leak and bleed the entire brake system.

Grinding and Scraping Noises

Grinding and scraping are classic symptoms of completely worn out brake shoes and pads. The metal of the shoe web or pad mount comes into direct contact with the drum or rotor and causes the noise (Figure 11-8). The sound will become louder until it occurs even when the brakes are not applied. It may also be caused by other problems such as a shoe return spring breaking and dropping loose from it mounts. However, the probable cause is lack of friction material on the pad or shoe.

Another condition that will cause loud grinding or scraping noises is installing the wrong type of pad or shoe. The friction material of the brake lining should always be suitable for the type of rotor used. With some composite rotors, a so-called "long life pad" may cause very loud grinding noises during braking because metal particles or compounds in the pad friction material dig into the rotor surface and create a metal-to-metal noise. On a few vehicles, it may be necessary to install factory parts instead of aftermarket items to retain quiet braking.

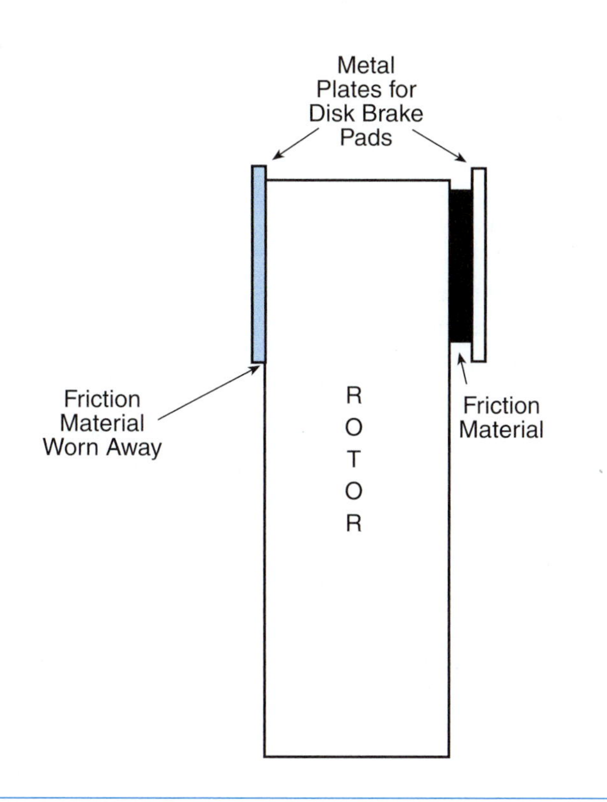

Figure 11-8 Metal-to-metal contact between the rotor and pad will occur when the friction material is missing. A similar problem can occur with shoes and drums.

Brake Booster Diagnosis

Classroom Manual
pages 316–319

WARNING: Before moving the vehicle after a brake repair, pump the pedal several times to test the brake. Failure to do so may cause an accident with damage to vehicles, the facility, or human beings.

WARNING: Before working on the brakes of a vehicle with ABS, consult the service manual for precautions and procedures. Failure to follow procedures to protect ABS components during routine brake work could damage the components and cause expensive repairs.

CAUTION: Do not work in or around the steering column without disarming the air bag systems. Serious injury could result from a deploring air bag.

Master cylinders and boosters are easy to diagnose and repair. Listen to the operator's complaint, understand how the system works, and use the information presented on the last few sections to determine which component failed.

Brake Booster Operational Checks

To check the brake's power booster, turn off the engine and press down hard on the brake pedal several times to bleed the booster's vacuum or pressure. The pedal will rise with each stroke until it is firm and much farther from the floorboard than normal. Switch the ignition to RUN. If PowerMaster or another system of that type is being used, the electric motor will switch on and a whining sound can usually be heard. At the same time, the pedal feel will become softer and the pedal will drop down as booster pressure builds. Engine-driven brake booster systems will not respond until the engine is started.

PowerMaster

If the PowerMaster motor runs but there is still no boost, the problem may be caused by the pump, the lines, a lack of brake fluid, or a problem with the power valve in the master cylinder. First, check the brake fluid. Assuming that the fluid level is correct, loosen the pressure line fitting at the master cylinder and accumulator and switch the pump on for a brief period (Figure 11-9). Use a wipe cloth to capture any fluid that may exit the line. If there is fluid and a reasonable amount of pressure behind it, the most probable failure is inside the master cylinder. The best option is to replace the master cylinder.

A bad pump will not pressurize the fluid. While fluid may still be present at the loosened fitting during motor operation, it may not be enough to operate the power valve. The problem is probably the motor/pump connection or a worn pump. Some pumps and motors can be separated and serviced as single items, but many require replacement as a unit.

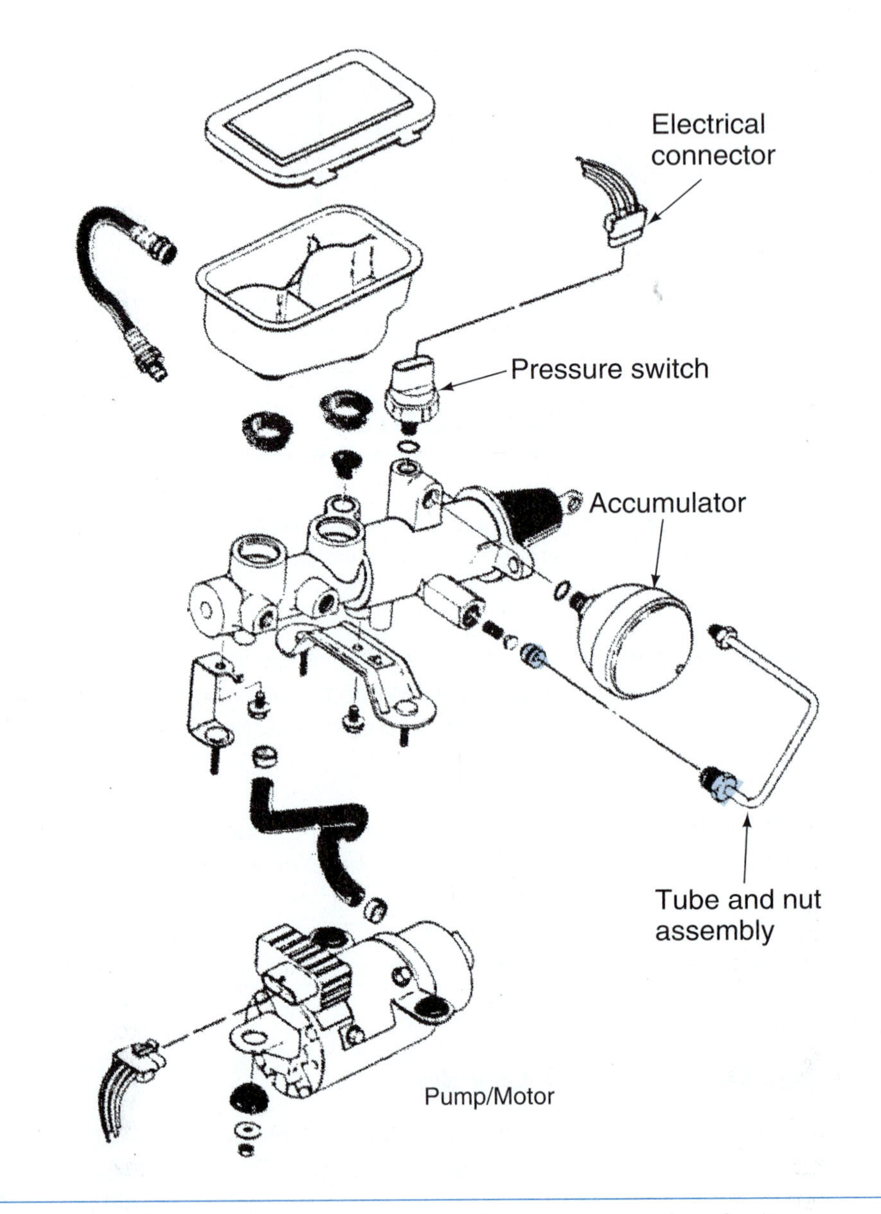

Figure 11-9 Slightly loosening the fitting will give an indication of the fluid's pressure. (Courtesy of General Motors Corporation, Service Operations)

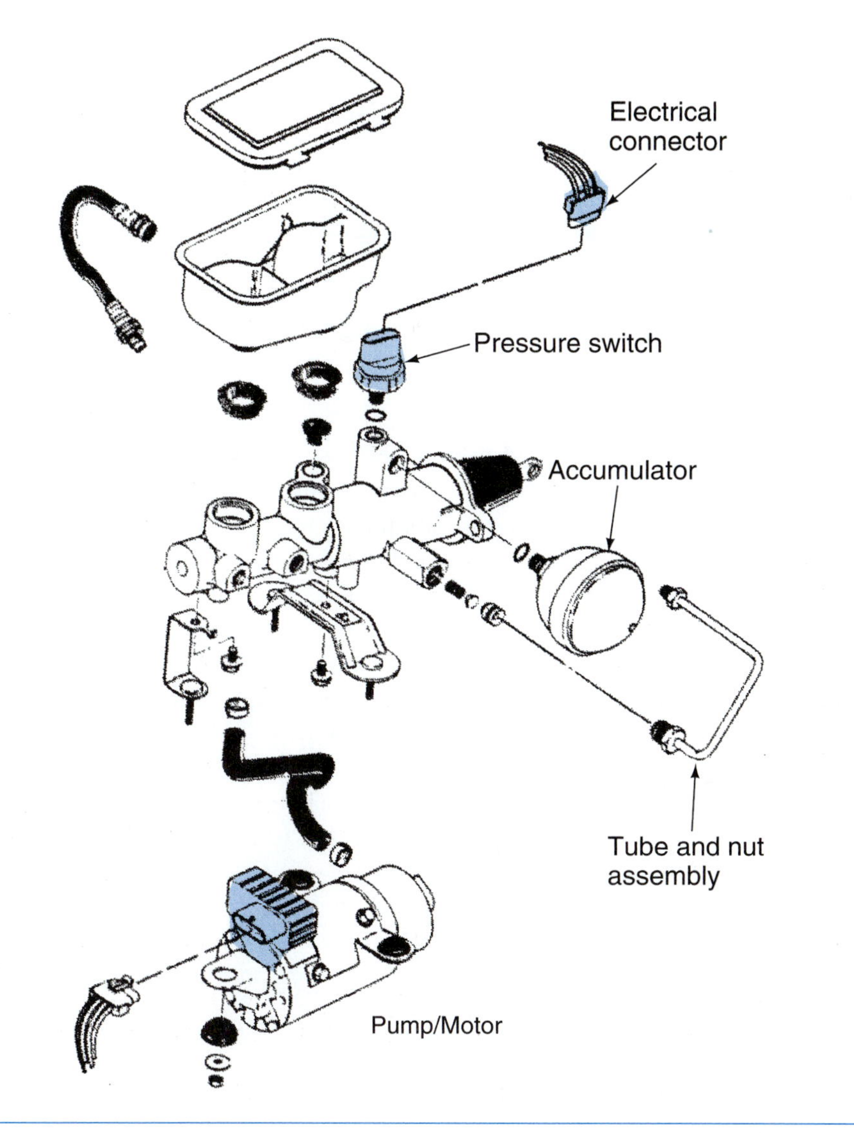

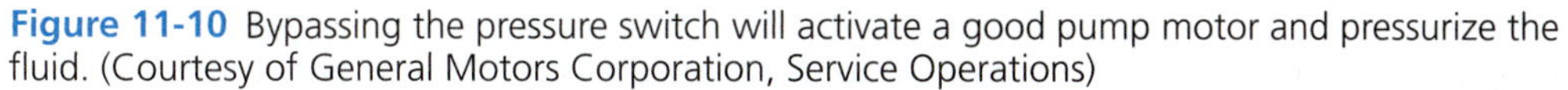

Figure 11-10 Bypassing the pressure switch will activate a good pump motor and pressurize the fluid. (Courtesy of General Motors Corporation, Service Operations)

Should the motor fail to switch on, perform electrical checks on the fuse, pressure switch, and motor (Figure 11-10). Before checking the pressure switch, determine if the switch grounds the motor control circuit or is in series with the motor control circuit. A grounding switch can be checked by:

1. Checking the brake fluid level. Top off as necessary.
2. Exhausting the boost pressure.
3. Turning the ignition to RUN.
4. Disconnecting the switch wire and grounding it to a metal component (Figure 11-11). If the switch is bad, the motor will switch on.

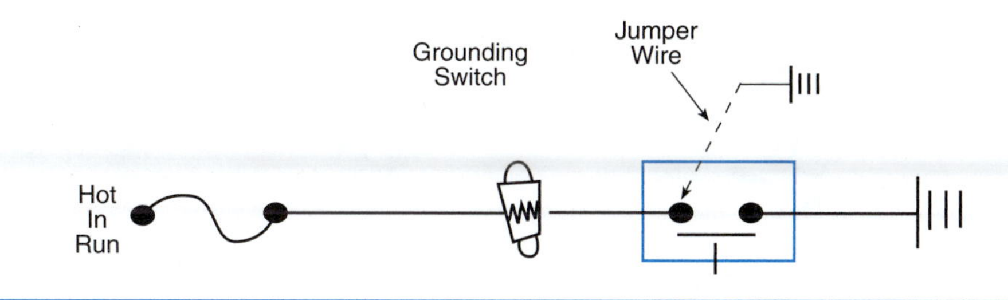

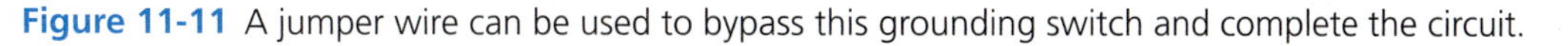

Figure 11-11 A jumper wire can be used to bypass this grounding switch and complete the circuit.

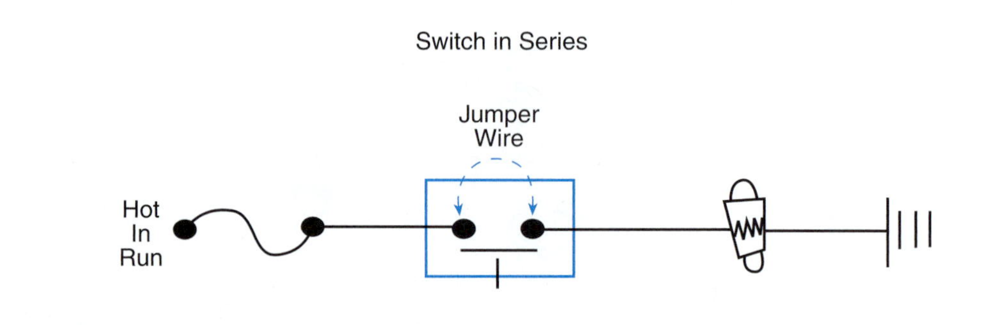

Figure 11-12 A jumper wire can be used to bypass this in-series switch to complete the circuit.

Steps 1 through 3 above are also used to test a switch wired in series with the motor. On step 4, disconnect the switch harness and install a jumper wire between the two terminals of the harness (Figure 11-12). If the switch is bad, the motor will switch on.

If the switch and fuse are good, do not condemn the motor until further checks are made. Use the service manual to determine if a relay is used to control current to the motor. Check the relay according to the manufacturer's instructions. If the relay is good or there is no relay, check for the proper voltage at the motor's electrical terminal. A voltage reading close to battery voltage should be enough to operate the motor. Again, before condemning the motor, check its ground and repair if necessary. Two or more volts below battery voltage normally means that there is excessive resistance in the circuit. Check the service manual for exact test specifications. A zero volt reading requires the technician to start probing for voltage at different points in the circuit starting near the motor. Voltage testing is covered in Chapter 6, "Making and Reading Measurements."

Removal and replacement of the PowerMaster requires disconnecting the pedal's pushrod, brakes lines, and removing the mounting bolts. The manufacturer's instructions should be followed to bleed the master cylinder, power valve chamber, and remounting the master cylinder. The brakes may also have to be bled.

Vacuum and Hydro-Boost

On vacuum and hydro-boost systems, exhaust the boost pressure as stated previously and start the engine (Figure 11-13). The pedal should immediately feel softer and drop lower. A firm, steady pedal after engine startup means there is no boost. On a hydro-boost system, attempt to turn the steering wheel. If power steering is present, the problem is in the valve mechanism mounted on the master cylinder. Consult the service manual to determine if the valve mecha-

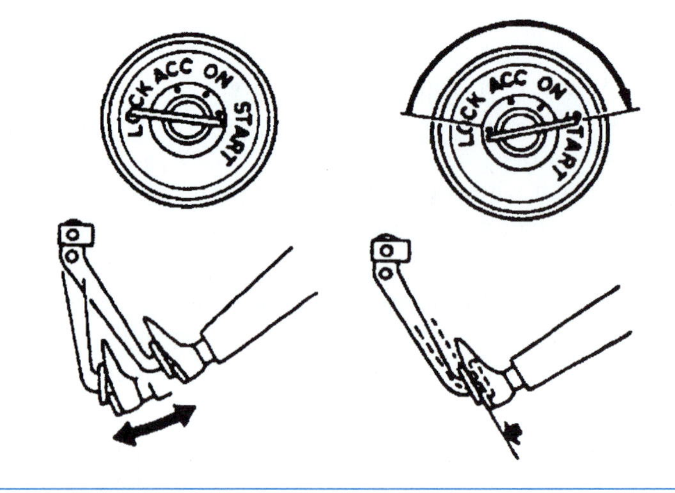

Figure 11-13 Exhaust the booster with the key off, hold the pedal down, and switch key to START. The pedal should fall away when the engine starts. (Reprinted with permission)

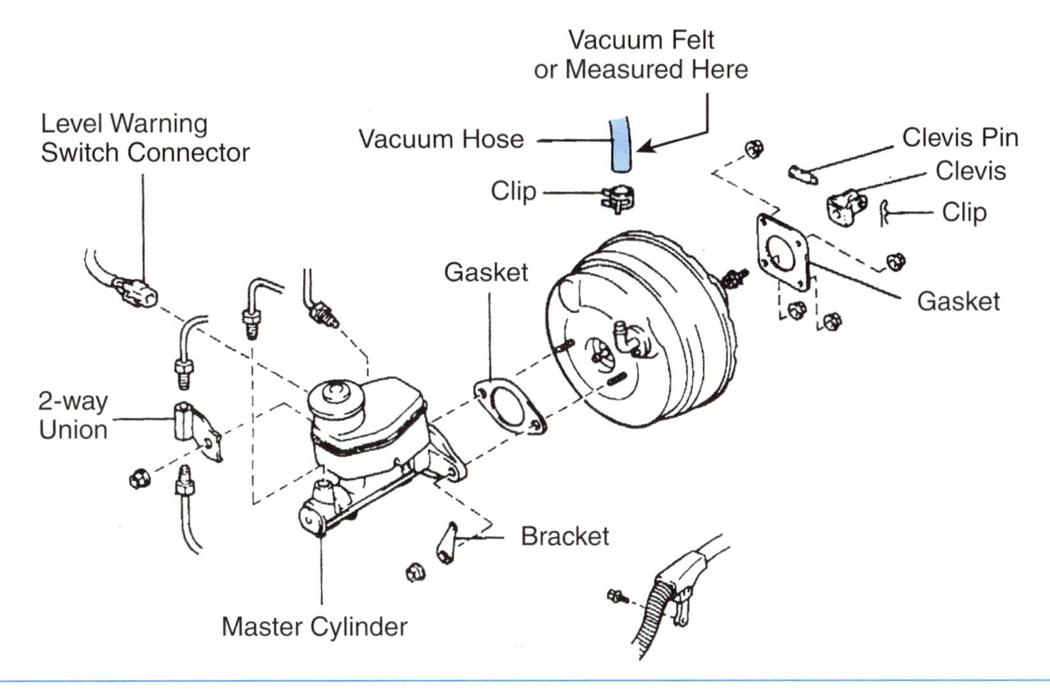

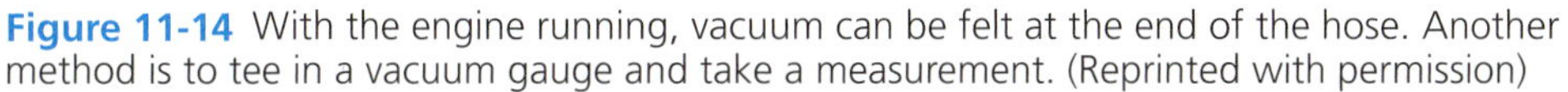

Figure 11-14 With the engine running, vacuum can be felt at the end of the hose. Another method is to tee in a vacuum gauge and take a measurement. (Reprinted with permission)

nism can be replaced or serviced. Many manufacturers require that the master cylinder and valve be replaced together. A lack of power steering indicates that the power-steering fluid is low, the power-steering pump is inoperative, or the power-steering belt is missing. Treat this condition as a steering component failure and perform the necessary repairs to that system.

A vacuum brake booster failure is caused by one of two things: booster failure or lack of vacuum. The first step is to disconnect the vacuum hose at the booster with the engine running (Figure 11-14). The engine will try to stall. Placing a thumb over the end of the hose will steady the engine and there should be a strong vacuum. A lack of vacuum usually means that the hose or check valve is closed or restricted. On a diesel engine, the **vacuum pump** and belt have to be checked if there is no vacuum at the brake booster hose.

A *vacuum pump* is required by diesel engines because this type of engine does not produced a vacuum within the intake manifold.

Vacuum Booster Replacement

Classroom Manual
pages 315–319

Special Tools

Service manual

Wheels blocks

■ **CAUTION:** Do not work in or around the steering column without disarming the air bag systems. Serious injury could result from a deploring air bag.

If there is vacuum to the booster, consult the service manual for instructions on removing the booster and disarming the air bags. The general method of removing the booster is to disarm the air bag, disconnect the pushrod, disconnect and move the master cylinder, and remove the booster mounting fasteners (Figure 11-15). Do not disconnect the brake lines from the master cylinder. The master cylinder can be moved to the side without disconnecting the lines. Reverse the procedures to install the booster, vacuum hose, master cylinder, and the push rod. Check the service manual to determine if the push rod's length has to be measured and adjusted. The stoplight switch should be checked after installing the pushrod to the pedal.

Test the new booster by starting the engine and applying the brakes. The pedal should be firm but easy to apply. The **engine** should not stumble or stall while the brakes are on. Stop the engine and pump the pedal to exhaust the vacuum from the booster. The pedal should rise in height above the floor and become very firm or hard to push. Hold the pedal down while starting the engine. The pedal should fall away as the engine is started. The brakes should be easy to apply with a soft but not spongy feel.

An *engine* that operates poorly usually does not cause a total loss of vacuum in the brake booster. However, a tear in the booster's diaphragm can cause a good engine to stall and run roughly.

Master Cylinder Replacement

Classroom Manual
pages 319–323

Special Tools

Wheel blocks

Vise

Master cylinder bleeder kit

Service manual

WARNING: Handle master cylinders carefully to prevent spilling the fluid. Brake fluid is very corrosive and will cause paint and finish damage.

The master cylinder creates and supplies the pressure for the brake system. An internal or external leak may affect both circuits of a dual system or just one circuit. In most cases, the master cylinder is fairly easy to replace. This discussion will cover the replacement of a master cylinder, with vacuum boost, on a non-ABS system. Consult the service manual before attempting to replace any brake components on an ABS system.

WARNING: Some ABS components may be located on or near the master cylinder. Consult the service manual on removing the master cylinder if any ABS or other system components present a problem to master cylinder replacement. Damage to the ABS or other systems may occur if proper procedures are not followed.

WARNING: Many ABS systems require special bleeding procedures. Do not remove a master cylinder on an ABS system before consulting the service manual. Damage to the brake system or loss of time could result if proper procedures are not followed.

Move the vehicle into the bay and block the wheels. Inspect the master cylinder mounting area. Remove the two or four metal brake lines attached to one side of the master cylinder body (Figure 11-16). An electrical connection for the fluid level sensor may be present. Four nuts should hold the master cylinder to the vacuum booster. Before loosening the nuts, place clean rags in the area under the master cylinder to capture any **brake fluid** that may leak.

Most manufacturers indicate that the *brake fluid* should be removed from the system and replaced with new fluid. This is usually referred to as "flushing the system."

Disconnect the electrical connection, if present. Select the proper line wrench and disconnect the brake lines. The flare nuts are loosened until the lines can be pulled out of the fitting. Bend the lines slightly so the master cylinder can be moved to the side.

Select the correct wrench to remove the nuts. Generally, a 3/8-inch drive ratchet, socket,

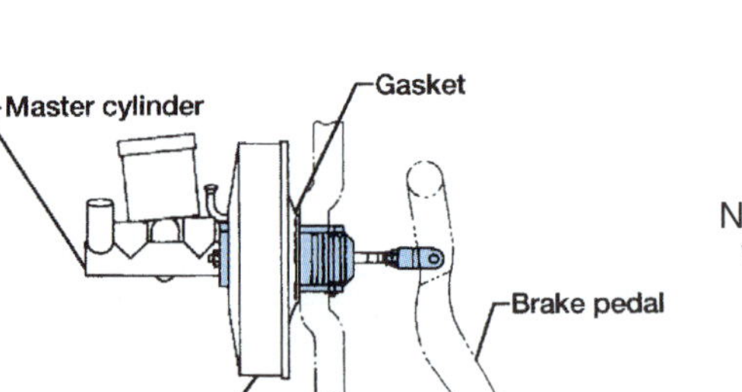

Figure 11-15 To remove most boosters, disconnect the push rod, remove the master cylinder fasteners, pull the master cylinder back, and remove the brake power boosters. (Courtesy of Nissan North America)

Figure 11-16 Place cloth under the master cylinder before disconnecting the brake lines. (Courtesy of General Motors Corporation, Service Operations)

and a three- or six-inch extension are the only tools needed to remove the cylinder's mounting nuts. With all of the nuts removed, slide the master cylinder toward the front of the vehicle until it clears the studs. Tilt the master cylinder slightly to one side with the line fittings up. This will reduce any leakage as the unit is moved.

Remove the reservoir cover and drain the brake fluid into a suitable container. If necessary, remove the reservoir and install it on the new master cylinder using the new O-rings supplied (Figure 11-17). Place the old master cylinder in the box that the new one came in. It will be sent back to the parts vendor as a core for rebuilding. Secure a master cylinder bleeder kit or make one using two, short pieces of brake line and two fittings.

The kit is basically two plastic lines with two plastic male fittings (Figure 11-18). The threads on the fittings are cut to fit several types of master cylinder threads and will flex or give to hold the fittings to prevent any leakage during braking. Usually, there are three types of threads on the fitting, each one a little larger then one before it. The fittings are screwed into the line ports on the master cylinder. They do not need to be very tight. Screw them in by hand and add two or three turns with a wrench. There will not be much pressure applied during bench bleeding. Route the lines from the fittings back into the cylinder's reservoir, one per chamber. The lines should nearly touch the bottom of the chamber. The kit should include a small holding device to keep the lines in place during this procedure. Some master cylinders are packaged with a kit included, but most kits can be bought separately and kept on hand.

A bleeder kit can be made if necessary, but it is time consuming and can only be used for one type of master cylinder fitting thread. A brake line must be cut for two six- to eight-inch pieces and bent to fit between the fitting and the bottom of the master cylinder's chamber. Select two male fittings that will slide over the line and screw easily into the master cylinder. The threads must be exact since the fitting is either brass or steel and the wrong threads will cause problems. The fitting slides over the line and one end of the line flares. The line is then bent to fit into the master cylinder chamber. Generally, a technician does not make a master cylinder bleeder kit unless one type of vehicle is being maintained or a special- purpose vehicle requires special adapters. For the most part, it not worth the trouble to build a bleeder kit for each vehicle that comes into the shop when an inexpensive kit can be bought at almost any parts vendor.

WARNING: Before adding brake fluid, consult the vehicle's service manual. Many manufacturers require a specific classification of brake fluid to be used.

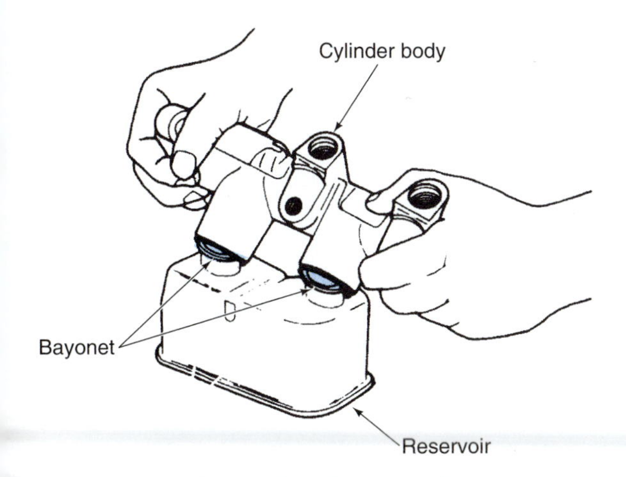

Figure 11-17 Lube the new O-rings with brake fluid before pressing on the reservoir. (Courtesy of General Motors Corporation, Service Operations)

Figure 11-18 This bleeder kit can fit several types and makes of master cylinders. (Courtesy of Snap-on Tools Company)

Photo Sequence 12
Typical Procedure for Bench Bleeding the Master Cylinder

P12-1 Mount the master cylinder firmly in a vise, being careful not to apply excessive pressure to the casting. Position the master cylinder so the bore is horizontal.

P12-2 Connect short lengths of tubing to the outlet ports, making sure the connections are tight.

P12-3 Bend the tubing lines so that the ends are in each chamber of the master cylinder reservoir.

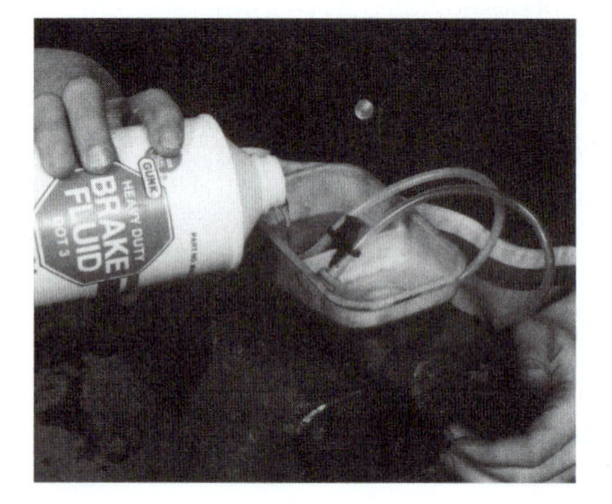

P12-4 Fill the reservoirs with fresh brake fluid until the level is above the ends of the tubes.

P12-5 Using a wooden dowel or the blunt end of a drift or punch, slowly push on the master cylinder's pistons until both are completely bottomed out in their bore.

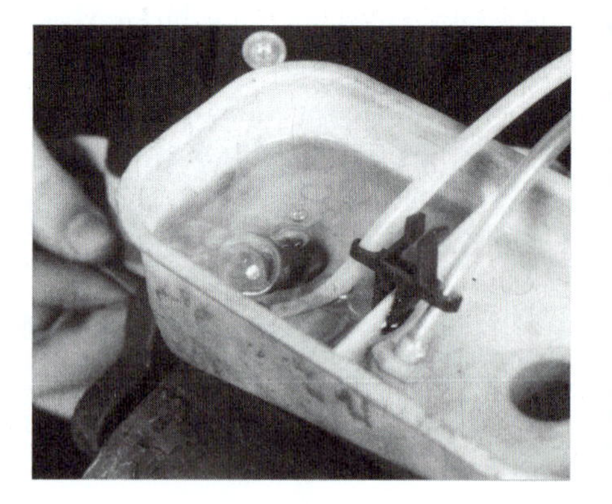

P12-6 Watch for bubbles to appear as the tube ends are immersed in the fluid. Slowly release the cylinder piston and allow it to return to its original position. On quick take-up master cylinders, wait 15 seconds before pushing in the piston again. On non-quick, take-up units, repeat the stroke as soon as the piston returns to its original position. Slow piston return is normal on some master cylinders.

Typical Procedure for Bench Bleeding the Master Cylinder (continued)

P12-7 Pump the cylinder piston until no bubbles appear in the fluid.

P12-8 Remove the tubes from the outlet port and plug the openings with temporary plugs or fingers. Keep the ports covered until the master cylinder on the vehicle is installed.

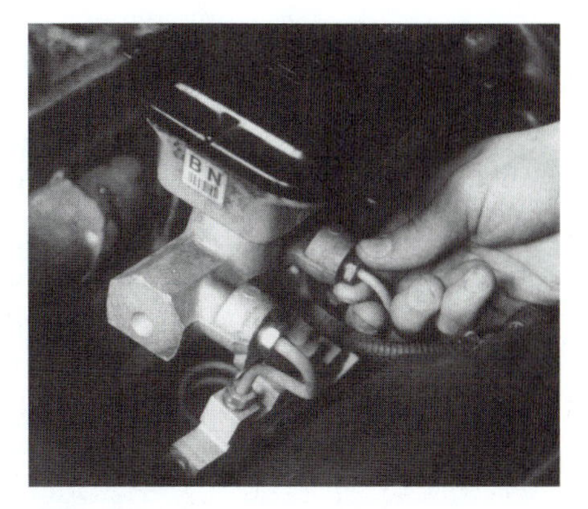

P12-9 Install the master cylinder on the vehicle. Attach the lines, but do not tighten the tube connections.

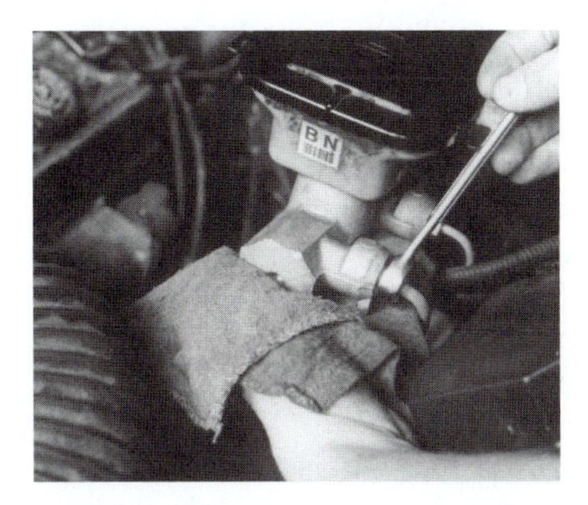

P12-10 Slowly depress the pedal several times to force out any air that might be trapped in the connections. Before releasing the pedal, tighten the nut slightly and loosen it before depressing the pedal each time. Soak up the fluid with a rag to avoid damaging the car finish.

P12-11 When there are no air bubbles in the fluid, tighten the connections to the manufacturer's specifications. Make sure the master cylinder reservoirs are adequately filled with brake fluid.

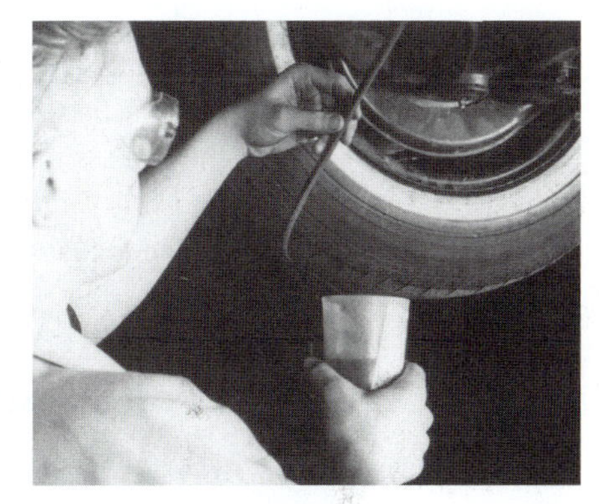

P12-12 After reinstalling the master cylinder, bleed the entire brake system on the vehicle.

Mount the master cylinder in a vise with the forward end elevated slightly. Install the bleeder kit and remove the reservoir cover. File each chamber about half full with brake fluid. Use a punch or other rod to slowly apply force to the master cylinder's primary piston (Figure 11-19). Continue the stroke until both pistons bottom out. As the pistons are pushed in, air should escape from the bleeder tubes as evidenced by bubbles in the reservoir. Slowly release the force from the pistons and allow them to return to their resting position. Repeat the bleeding procedure until there are no bubbles visible in the reservoir during the pistons' inward movement. The rear or secondary cylinder bore will probably bleed completely for one or two strokes before the secondary chamber. During the bleeding process, always push both pistons to the bottom of their bore. Before removing the cylinder from the vise, close the two bleeder tubes with small paper clips and install the reservoir cover. The clips will prevent leaks and they will prevent air from entering as the cylinder is installed on the vehicle.

Place the master cylinder in position on the brake booster and torque the fasteners. Position the two brake lines near their fittings. Remove one of the brake bleeder fittings and quickly insert the correct brake line into the master cylinder. Hand tighten the fitting enough to prevent leaking. Do not tighten completely at this time. Install the second brake line in the same manner. Drain and store the bleeder kit.

It is best to have a second person assist during the final stage of master cylinder bleeding. The next step is to remove any air that may have entered the line or cylinder fitting during the change over from bleeder to brake line. Ensure that enough rags are placed under the master cylinder brake lines. Ask the second person to press slowly on the brake pedal. Both brake lines will loosen and retighten on each stroke of the brake pedal. Loosen the rear line slightly as the pedal is being applied. Watch the small leak for signs of air. If no air is present, tighten the fitting and loosen the front line. Ensure that both fittings are tight before the brake pedal is released. Repeat this step until there is no air being expelled past the fittings. Generally, one or two strokes of the brake pedal are sufficient. Complete the installation by tightening the fittings and topping off the fluid level. Before moving the vehicle from the bay, apply the brakes and start the engine. If the vehicle has an automatic transmission, apply the brakes and put it in gear to see if the brakes will hold the vehicle. On a manual transmission, release the brakes and engage the clutch just enough to move the vehicle slowly. Apply the brakes to ensure that they will stop and hold the vehicle.

Brake Flushings

There are other facts that must be considered during master cylinder replacement. If the vehicle is over two years old, or it has been more than two years since the last *brake flushing,*

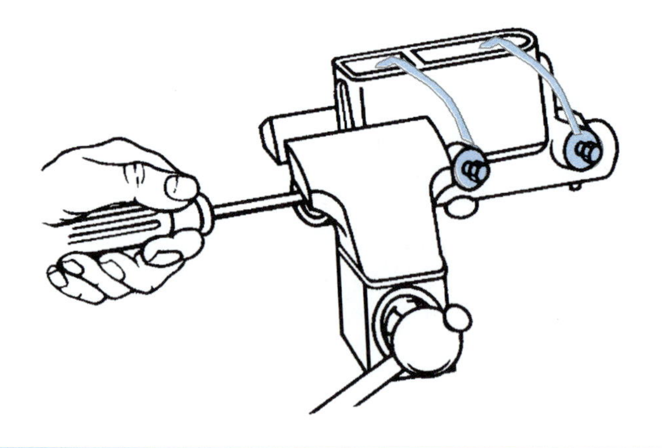

Figure 11-19 A screwdriver is being used to move the master cylinder pistons during bench bleeding.

the old fluid should be flushed out with new fluid. This is done by bleeding each wheel until new fluid is visible. When a master cylinder fails because of age, it is probable that other components may require replacement. At the minimum, a thorough inspection of the brake system should be conducted.

Disc Brake Pad Replacement

WARNING: Before moving the vehicle after a brake repair, pump the pedal several times to test the brake. Failure to do so may cause an accident which might damage the vehicle, the facility, or injure humans.

WARNING: Before working on the brakes of a vehicle with ABS, consult the service manual for precautions and procedures. Failure to follow procedures to protect ABS components during routine brake work could damage them and require expensive repairs.

WARNING: Failure to lower the reservoir brake fluid level may result in spillover during brake repair. Damage to the vehicle's finish can occur. The fluid will overflow as the caliper pistons are pushed back into their bores and the fluid behind them is forced back into the master cylinder.

Classroom Manual
pages 333–335

Special Tools

Lift or jack and jack stands

Impact wrench

Torque wrenches

C-clamp

Disc brake silencer

Disc brake lubricant

Disc brake pad replacement does not require any special tools except a C-clamp (Figure 11-20). Before lifting the vehicle, inspect the fluid level in the reservoir. If the reservoir is more than half full, some of the fluid must be removed. A suction gun *should not* be used because the gun's rubber hose and piston seal will probably be damaged. A small, inexpensive syringe is the best. It can be discarded after several uses. Use cloth to capture any leaking fluid.

Lift the vehicle to a good working height, set the locks, and remove the tire and wheel assembly. Inspect the disc brake area to determine the type of rotor/hub assembly and the method used to attach the caliper. Some systems require the adapter to be removed before the rotor is free. The rotor and hub may be one or two piece. The wheel bearings and seal have to be removed on a one-piece assembly before the rotor can be machined. For this example, the rotor is separate and the adapter must be removed.

Most rotors can be removed and turned on the brake lathe (Figure 11-21), but there are portable lathes that can machine the rotor on the vehicle (Figure 11-22). They were originally sold to service one-piece hub/rotors on FWD vehicles where the drive axle and the wheel bearing were press-fitted to the hub. This lathe required some special setup procedures but

Figure 11-20 A typical C-clamp. (Courtesy of MATCO Tool Company)

Photo Sequence 13
Typical Procedure for Replacing Brake Pads

P13-1 Begin front brake pad replacement by removing brake fluid from the master cylinder reservoir or reservoirs serving the disc brakes. Use a syphon to remove about two-thirds of the fluid.

P13-2 Raise the vehicle on the hoist, making certain it is positioned correctly. Remove the wheel assemblies.

P13-3 Inspect the brake assembly, including the caliper, brake lines, and hoses, and rotors. Look for signs of fluid leaks, broken or cracked lines or hoses, and a damaged brake rotor. Correct any problems found before replacing the pads.

P13-4 Loosen and remove the pad locator pins.

P13-5 Lift and rotate the caliper assembly up and off of the rotor.

P13-6 Remove the old pads from the caliper assembly.

P13-7 To reduce the chance of damaging the caliper, suspend the caliper assembly from the underbody using a strong piece of wire.

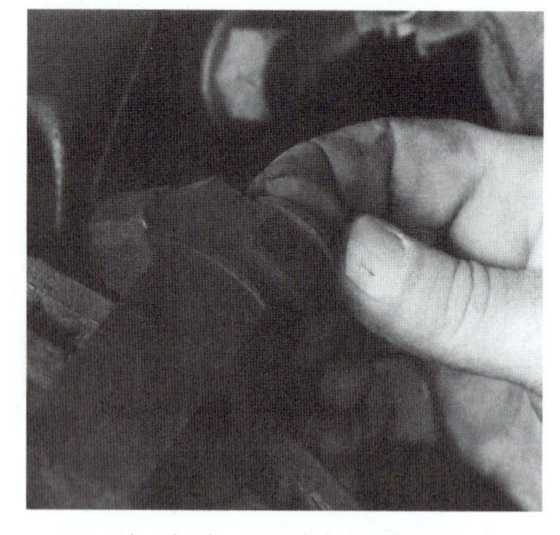

P13-8 Check the condition of the caliper locating pin insulators and sleeves.

P13-9 Place a block of wood over the caliper piston and install a C-clamp over the wood and caliper. Tighten the C-clamp to force the piston back into its bore.

Typical Procedure for Replacing Brake Pads (continued)

P13-10 Remove the C-clamp and check the piston boot, then install new locating pin insulators and sleeves if needed.

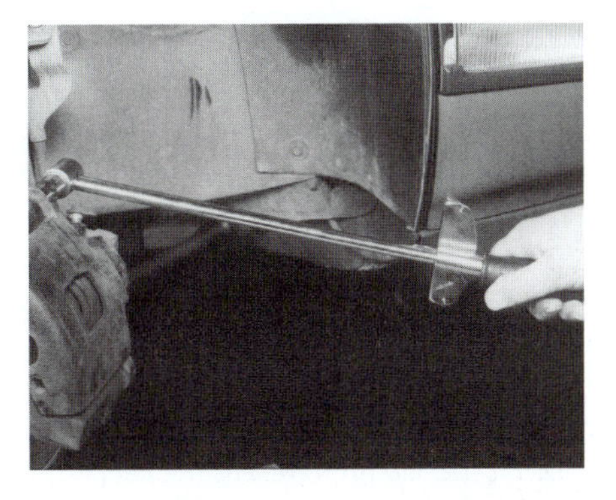

P13-11 Install the new pads in the caliper. Then set the caliper with its new pads onto the rotor and install the locating pins. Check the assembly for proper position. Torque the locator pins to proper specifications.

P13-12 Install the tire and wheel assembly and tighten to torque specifications. Then, press slowly on the brake pedal to set the brakes.

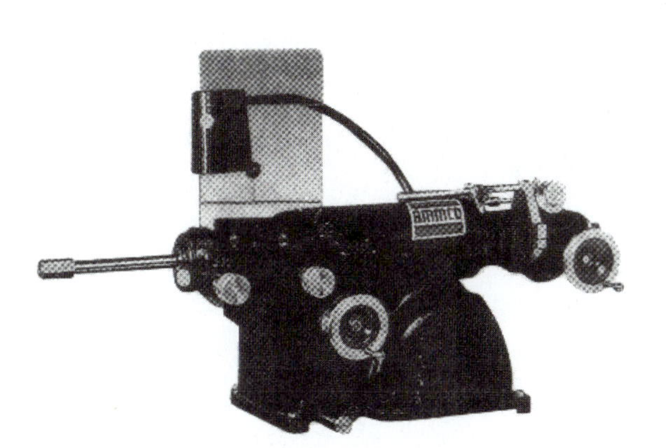

Figure 11-21 A typical bench-mounted brake lathe. (Courtesy of MATCO Tool Company)

Figure 11-22 This is one type of on-vehicle brake lathe. (Courtesy of MATCO Tool Company)

could be learned quickly. The on-vehicle brake lathe can be used for machining other rotor setups as long as the rotor, machining head, and hub can be clamped tightly together with the lug nuts. Machining a rotor or drum should be done by an experienced technician following the lathe operating instructions and the procedures set forth in the vehicle's service manual.

WARNING: Consult the service manual before pushing the caliper's piston into its bore. Some ABS systems require that the caliper's bleeder screw be opened or they require other special procedures to prevent damage.

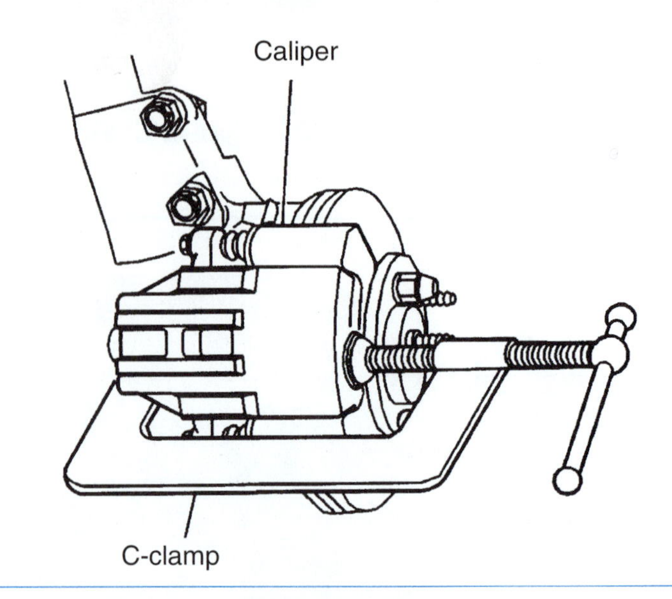

Figure 11-23 Using a C-clamp to reseat the caliper piston into the bore. (Courtesy of Chevrolet Motor Division, General Motors Corporation)

A *C-clamp* is so named because of its shape and function.

A **C-clamp** may be used to press the piston back into the caliper bore (Figure 11-23). The piston must be placed as far into the caliper as possible in order for the new pads to fit over the rotor.

Remove the fasteners holding the adapter to the steering knuckle. With the fasteners removed, the adapter, caliper, and pads will slide off the rotor (Figure 11-24). Be cautious at this point because many rotors can drop off the hub once the caliper is removed. Once the assem-

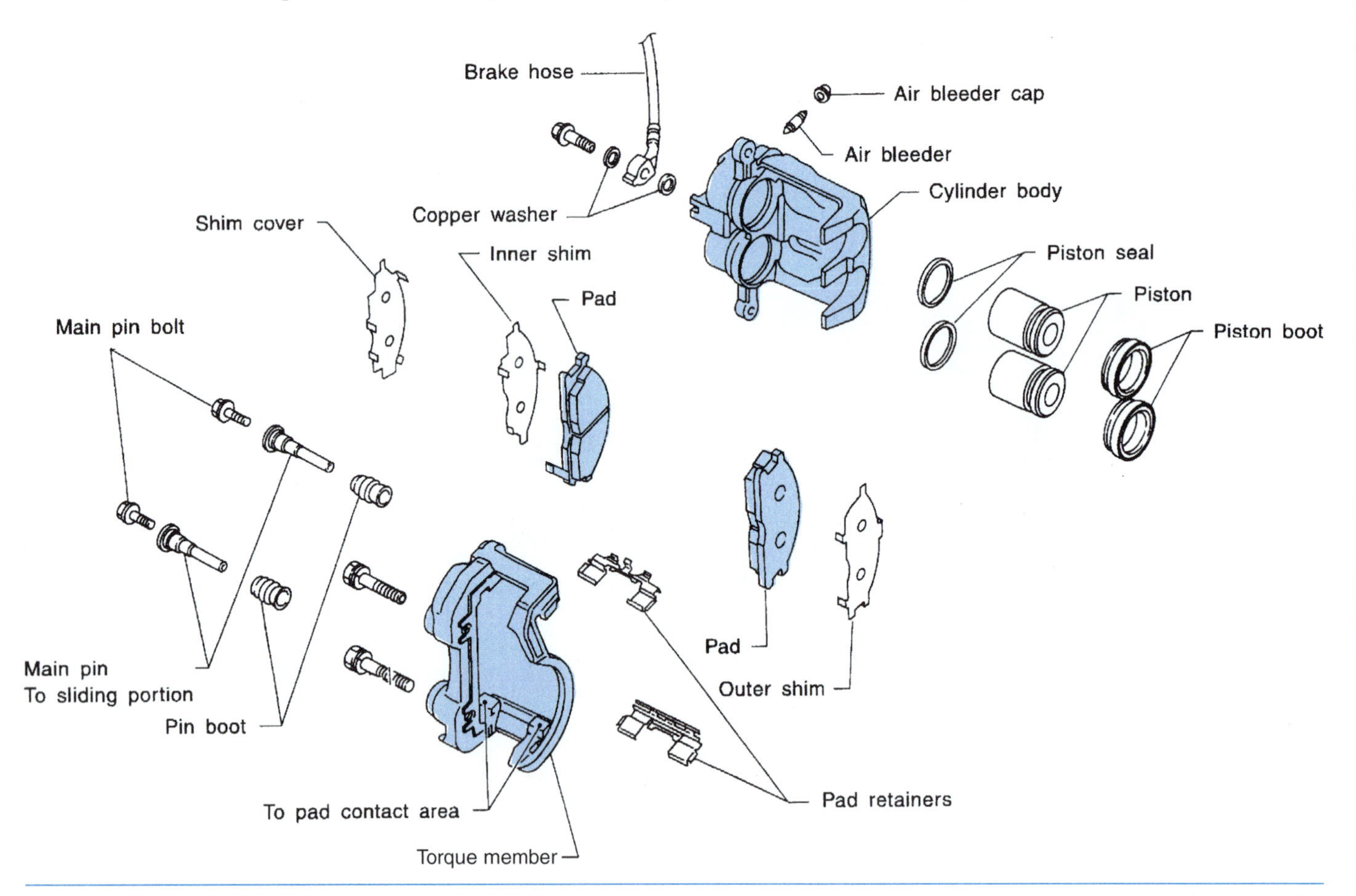

Figure 11-24 A typical dual piston caliper. Note the pad retainers that help quiet the noise of brake application. (Courtesy of Nissan North America)

bly is removed, hang it with **mechanic wire** or other strong wire so the caliper's weight does not hang on the flexible brake hose. Remove the pads and note where they are located and what is used to hold them in place. Normally, anti-rattle clips keep the pads from clicking in the caliper when the brakes are applied and released.

Mechanic wire is a generic term for very flexible, low-strength wire used by technicians to hold parts during repair work. Usually, it is sold in small rolls, but it is being replaced by plastic tie strips.

If there is brake fluid present on the outside of the caliper, the caliper must be rebuilt or replaced. Like the wheel caliper, it is usually best and cheapest to *replace* the caliper, but there are rebuilding kits on the market.

Before installing the pads, the slide pins or guides on the caliper and adapter must be lubed. Inspect the caliper and adapter to determine the slide points (Figure 11-25). Some systems have slide pins only while others slide the calipers along rails on the adapter. Once located, clean the slides and then lube with a disc brake lubricant.

Cover the pads' metal portion with **disc brake quieter** (Figure 11-26). The quieter changes the frequency of the sound made by the pads during braking. Brake application sounds should not be heard. Brake noises usually point to brake component failure or mismatched components. Assemble the pads into the caliper, inboard pad first. Ensure that the retaining clips or springs are in place before fitting the assembly over the machined rotor. Install and torque the adapter bolts, and install the wheel assembly.

A *disc brake quieter* changes the sound frequently of pads during braking.

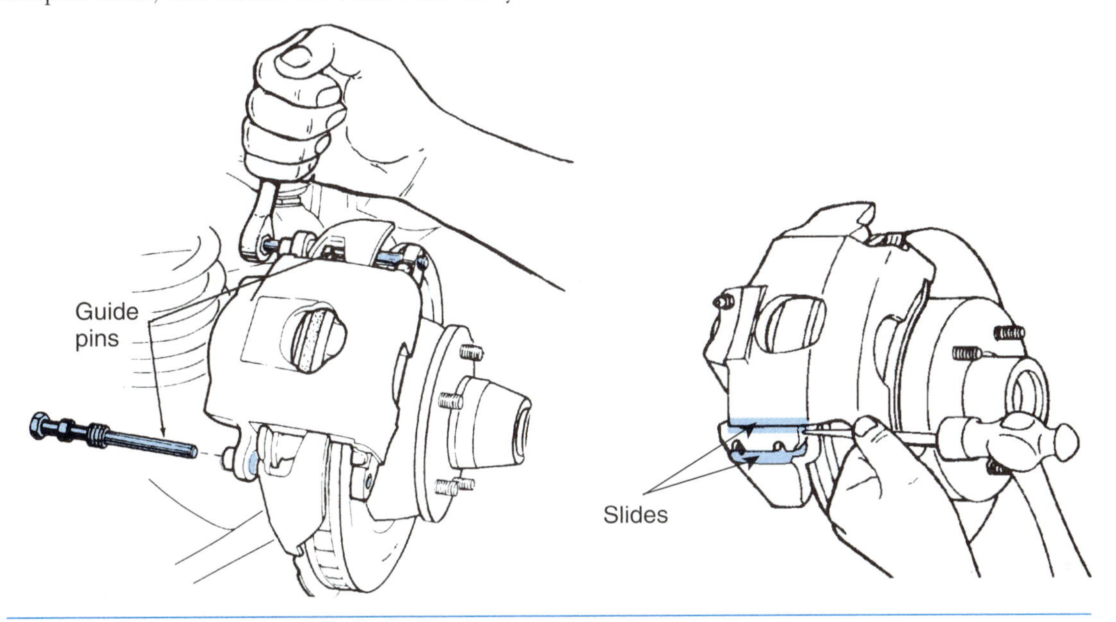

Figure 11-25 Note the caliper on the right has one wedge-type retainer on each end. They also act as the slide areas for the caliper. (Courtesy of EIS Brake Parts)

Figure 11-26 The small tube of silencer is sometimes supplied with a new set of disc pads.

Classroom Manual
pages 328–336

Special Tools

The tools required for other brake repairs depend on the repair, make, and vehicle model.

When the vehicle is lowered to the floor and before the vehicle is moved, slowly pump the brake pedal several times to seat the pads against the rotor. The pedal should be firm and at a normal height above the floor before the vehicle is moved. Top off the brake fluid level as necessary and inspect for any overflow around the master cylinder.

Other Brake Repairs

The repairs discussed in this chapter were selected for the relatively simple and straightforward mechanical procedures involved. As noted, even the simplest brake repair involves many things that can affect brake operation. This section will highlight other brake repairs and offer some hints that may help the new technician to gain experience without too many mistakes.

Drum Brakes

Drum brake shoe replacement is a fairly simple process of removing the old and installing the new. However, drum brake shoes are held in place with return springs, retainers, parking brake cable, and adjusters, most of which can easily be installed incorrectly (Figure 11-27). As always, the service manual contains instructions for the repair. A quick method of learning drum brake assemblies is to lay out all parts in the order in which they were removed and never

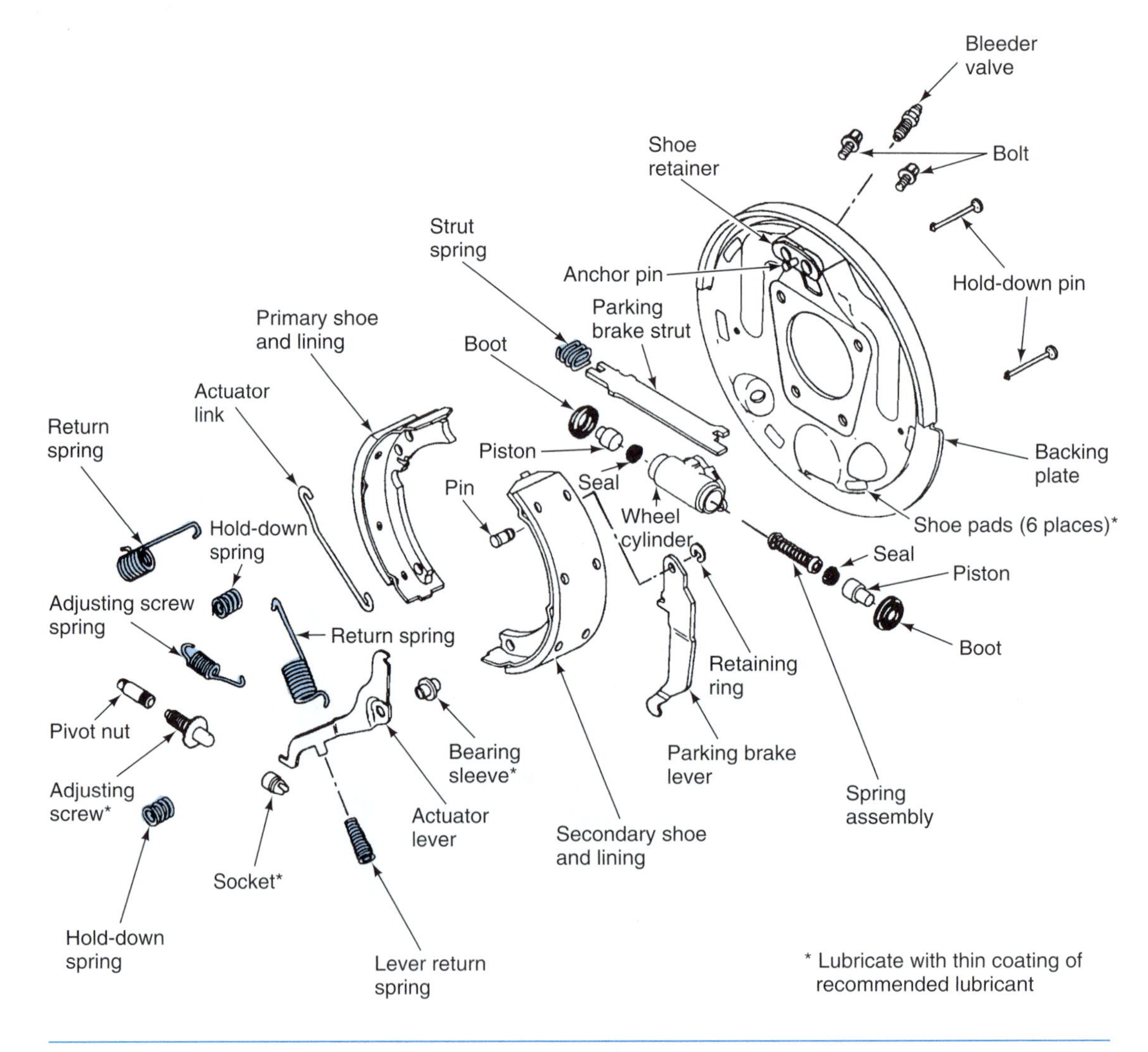

Figure 11-27 Strict attention must be given when reassembling some drum brakes. (Courtesy of Chevrolet Motor Division, General Motors Corporation)

remove both sets of brakes at the same time. Many new technicians lost time because they pulled both sides down at the same time and could not remember how the drum brakes were assembled. Keeping one set intact provides a model to reassemble the other set. Since the drum has to be machined, no time is lost by keeping one assembly intact. It is suggested that this practice be followed during most drum brake repairs.

Removing and installing drum brakes may also expose the technician to brake dust. The brakes should be cleaned with a low-pressure, aquatic cleaner to prevent dust. One way is to use a window-cleaner-style spray bottle or similar device filled with water. Position a catch basin under the brake assembly after the wheel assembly and drum are removed. Squirt the water over the entire brake assembly starting at the top. Use a great deal of water to flush all of the loose dust into the basin. A wet brush can be used to complete the task. This job will take only a few minutes and it provides a clean work area.

One step that sometimes gets missed is the lubrication of the rub pads on the backing plate. There are three rub pads under each shoe (Figure 11-28). They should be lubed lightly with brake lubricant before the new shoes are installed.

Parking Brakes

Parking brake repairs normally consist of cable replacements and adjustments or replacement of the ratchet locks on the pedal or lever (Figure 11-29). Most repairs are straightforward procedures laid out in the service manual. However, in most systems, some of the repairs require removal or some partial disassembly of the rear brakes.

Once the new parts have been installed, the rear brakes must be adjusted before the parking brake is adjusted. The parking brake should lock solidly about halfway along the pedal or lever's travel. Some manufacturers require that the parking braking locks after a certain number of clicks are heard or felt as the parking brake is applied. During the installation of any parking brake cable, ensure that the cable and sheaths are not clamped shut by fasteners.

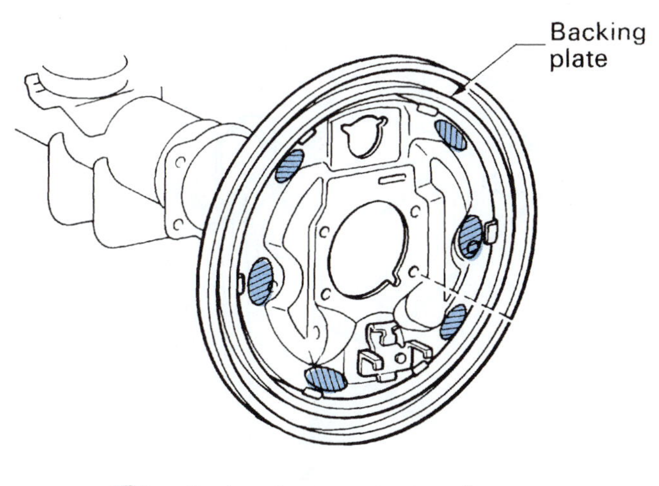

Figure 11-28 Failure to lube the shoe/backing plate rub points will result in noise during brake application.

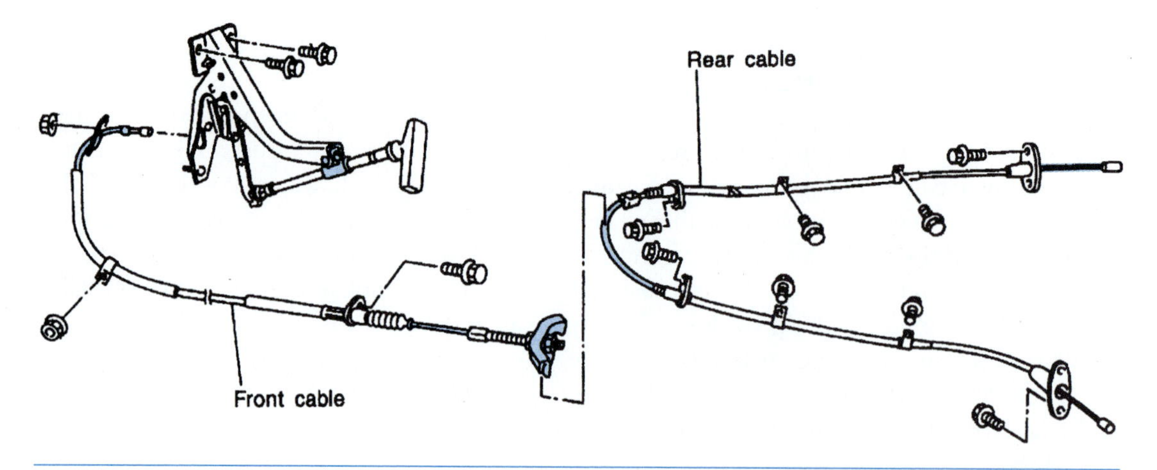

Figure 11-29 The highlighted points are the typical wear points on the parking brake actuation cable. They are also common lubrication points. (Reprinted with permission)

The parking brake warning light switch may need adjusting after work is performed on the parking brake lever or pedal mechanism (Figure 11-30). The light should come on as soon as the parking pedal or level is moved. It should not go off until the brake is completely released. After repairs are complete, turn the tires to make sure the parking brake is not causing the shoes or pads to drag.

Rear Discs

On vehicles with rear discs, the procedure for replacing the pads becomes a little more complicated. The rear disc may also be used as the parking brake or the disc rotor may have a small, internal drum brake whose sole function is for parking (Figure 11-31). In either case,

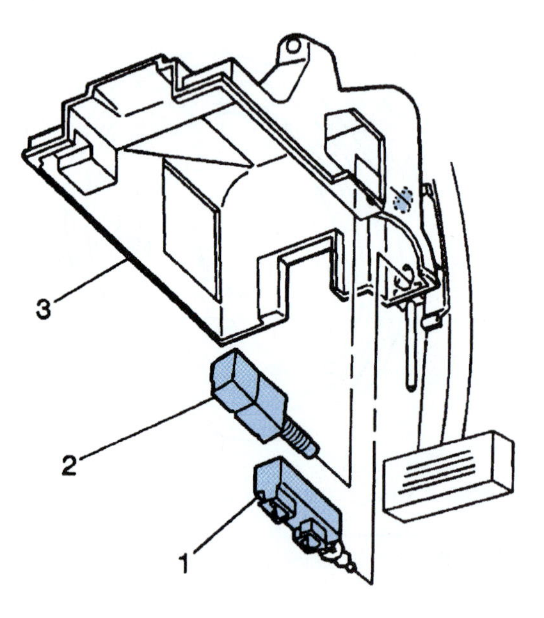

Figure 11-30 This type of parking brake warning light switch needs adjusting any time the pedal is removed for any reasons. (Courtesy of Chevrolet Motor Division, General Motors Corporation)

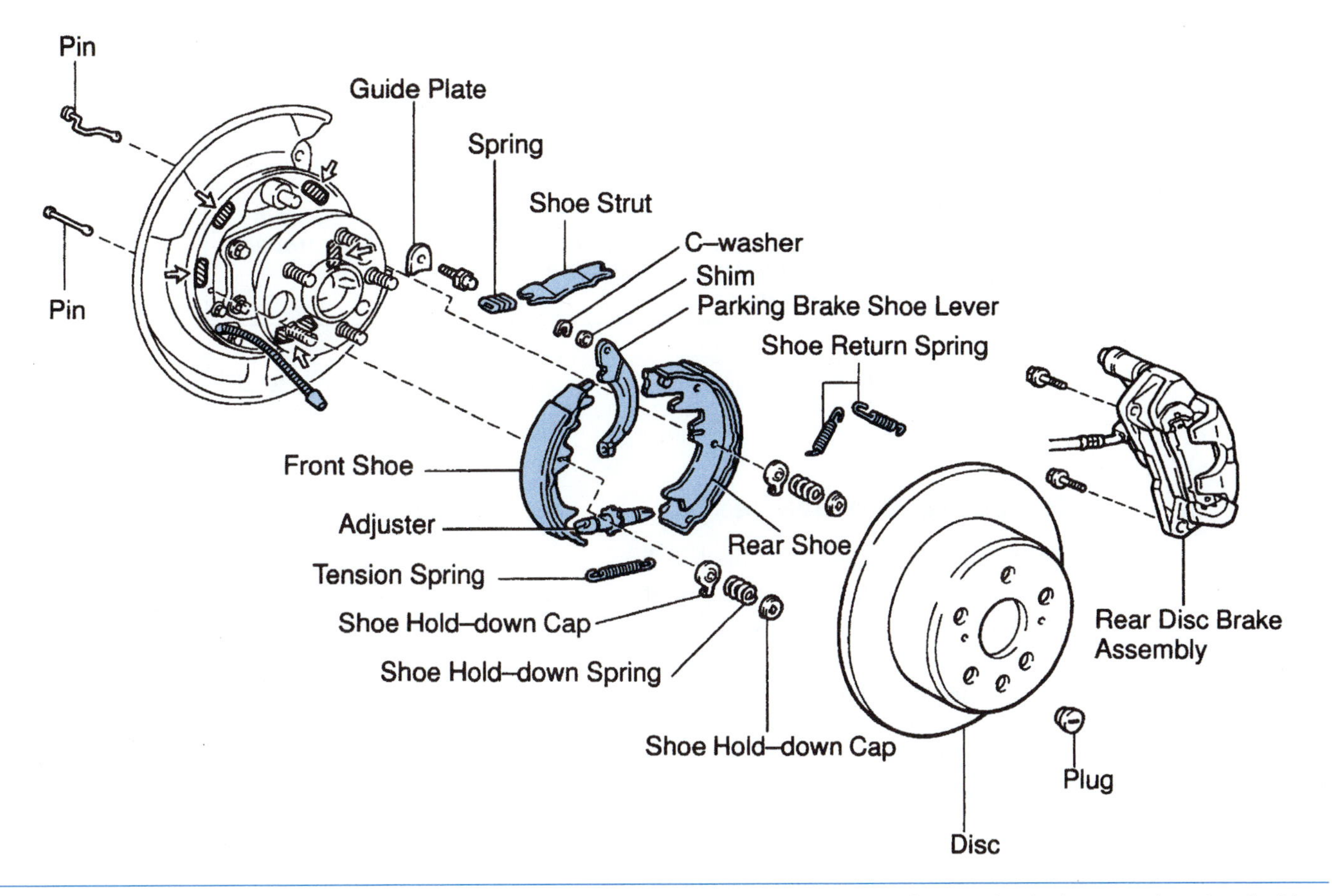

Figure 11-31 This parking brake is located within a small drum built into the back of a rear disc brake rotor. (Reprinted with permission)

compressing the piston back into the bore requires special tools and procedures. Normally, the piston is screwed into the bore with a special wrench (Figure 11-32). At times, even the correct procedures may not work properly and the technician has to spend time analyzing the problem. In addition, ABS presents special situations that must be dealt with before a proper repair can be completed.

Wheel Bearings

Wheel bearings are serviced as a part of a brake repair. This practice applies to systems where the drum and hub or the rotor and hub are one-piece assemblies where the bearings and seals are easily removed. This is a common setup on rear-wheel-drive vehicles. Most front-wheel-drive vehicles use press-in, permanently lubed, wheel bearings. Some FWD vehicles have wheel bearings that can be cleaned and repacked during a brake service. See Chapter 7, "Servicing Bearings and Installing Sealants," for bearing inspection and service.

Figure 11-32 It is not obvious in this picture, but this tool has six types of adapters to fit almost any rear-disc caliper piston. (Courtesy of MATCO Tool Company)

ABS Handling Caution

Caution: Certain components in the Antilock Brake System (ABS) are not intended to be serviced individually. Attempting to remove or disconnect certain system components may result in personal injury and/or improper system operation. Only those components with approved removal and installation procedures should be serviced.

Figure 11-33 This is a typical ABS warning or caution found in vehicle service manuals. (Courtesy of Chevrolet Motor Division, General Motors Corporation)

Bleeding the Brakes

The brake system must be bled if the system was exposed to air at any time. This includes disconnecting the lines from the master cylinder, disc calipers, or wheel cylinders and external leaks at any point in the system.

Bleeding can be done in several ways. Special suction equipment can be used as well as a **pressure bleeder** or two persons working together. In most systems, the wheels are bled starting with the one farthest from the master cylinder. The typical sequence is right rear, left rear, right front, and left front. Depending on the repairs accomplished, the master cylinder and some control valves may have to be bled before bleeding the wheels. Most ABS brakes require special bleeding procedures. Improper bleeding of the ABS may not lead to damage, but it could cost in labor time and frustration on the part of the customer, the technician, and the supervisor.

A *pressure bleeder* is a small holding tank with a diaphragm between a compressed air chamber and a fluid chamber. The bleeder uses adapters to attach itself to the master cylinder, and the compressed air (5 to 8 psi) forces the fluid through the brake system.

ABS and Air Bags

It is suggested that regular brake repairs never be performed on an ABS until the service manual has been consulted and any special procedures, cautions, and warnings have been noted (Figure 11-33). Air bags usually do not present a problem during braking repair unless work is being done under or near the steering column and dash. Repairs on the parking brake, warning light and switch, and stoplight switch place the technician in a position to be injured by a deploying air bag. Before starting any work within the passenger compartment, disarm the air bag system following the manufacturer's instruction (Figure 11-34).

CASE STUDY

The repair order on a light truck stated that the customer's complaint was a lowering of the brake pedal while at a stoplight or stop sign. A test drive confirmed the complaint. The technician replaced the master cylinder and conducted a test drive. The problem remained. The technician removed the brake lines at the master cylinder and plugged the ports. The pedal remained firm when tested. After reconnecting the lines, the technician disconnected the rear brake line from the isolation/dump valve of the RWAL system. The dropping-pedal symptom returned. The isolation/dump valve was replaced, the system bled, and the vehicle was road tested. The symptom did not return and the vehicle was released to the owner. The problem was a leaking dump valve within the assembly that allowed brake fluid to drain slowly from the rear brake lines.

SIR Caution

Caution: This vehicle has a Supplemental Inflatable Restraint (SIR) System. Refer to SIR Component Location View in order to determine whether you are performing service on or near the SIR components or the SIR wiring. When you are performing service on or near the SIR components or the SIR wiring, refer to SIR On-Vehicle Service information. Failure to follow the CAUTIONS could cause air bag deployment, personal injury, or unnecessary SIR system repairs.

SIR Handling Caution

Caution: When you are performing service on or near the SIR components or the SIR wiring, you must disable the SIR system. Use the following procedure to temporarily disable the SIR system. Failure to follow the correct procedure could cause air bag deployment, personal injury, or unnecessary SIR system repairs.

SIR Inflator Module Disposal Caution

Caution: In order to prevent accidental deployment of the air bag which could cause personal injury, do not dispose of an undeployed inflator module as normal shop waste. The undeployed inflator module contains substances that could cause severe illness or personal injury if the sealed container is damaged during disposal. Use the following deployment procedures to safely dispose of an undeployed inflator module. Failure to dispose of an inflator module as instructed may be a violation of federal, state, province, or local laws.

SIR Inflator Module Handling and Storage Caution

Caution: When you are carrying an undeployed inflator module:

- ***Do not carry the inflator module by the wires or connector on the inflator module***
- ***Make sure the bag opening points away from you***

When you are storing an undeployed inflator module, make sure the bag opening points away from the surface on which the inflator module rests. When you are storing a steering column, do not rest the column with the bag opening facing down and the column vertical. Provide free space for the air bag to expand in case of an accidental deployment. Otherwise, personal injury may result.

SIR Special Tool Caution

Caution: In order to avoid deploying the air bag when troubleshooting the SIR system, use only the equipment specified in this manual and the instructions given in this manual. Failure to use the specified equipment as instructed could cause air bag deployment, personal injury to you or someone else, or unnecessary SIR system repairs.

Figure 11-34 Typical air bag cautions/warnings. Note that the air bag system is referred to as the *supplemental restraint system*. (Courtesy of Chevrolet Motor Division, General Motors Corporation)

Terms to Know

Brake flushing
C-clamp
Disc brake quieter
DOT 3
Fading brakes
Glazing
Glycol-based
Grabbing brakes
Hard pedal
Hot spot
Locking brakes
Mechanic wire
Pedal pulsation
Pressure bleeder
Silicone-based
Spongy pedal

ASE Style Review Questions

1. Drum brake repairs are being discussed. *Technician A* says to keep one side intact while the other is disassembled to save some time. *Technician B* says wheel grease can be used to lube the rub points. Who is correct?
   A. A only
   B. B only
   C. Both A and B
   D. Neither A nor B

2. Disc brakes are being discussed. *Technician A* says a vibration during braking may be caused by the caliper's piston being pushed in and out of the caliper bore. *Technician B* says the piston may have to be screwed into the bore. Who is correct?
   A. A only
   B. B only
   C. Both A and B
   D. Neither A nor B

3. *Technician A* says a locking brake may be caused by a broken shoe retaining spring. *Technician B* says a leaking wheel cylinder or axle seal may cause a brake to grab. Who is correct?
   A. A only
   B. B only
   C. Both A and B
   D. Neither A nor B

4. The driver complains of the vehicle with front disc is pulling to the left during braking. *Technician A* says the left front brake is the cause. *Technician B* says the right front pads may be glazed from heat. Who is correct?
   A. A only
   B. B only
   C. Both A and B
   D. Neither A nor B

5. Diagnosing a disc/drum brake system is being discussed. *Technician A* says a bad hose on the rear brakes may cause the front wheels to lock too quickly. *Technician B* says pedal pulsation may be caused by incorrect pressure between the front and rear brakes. Who is correct?
   A. A only
   B. B only
   C. Both A and B
   D. Neither A nor B

6. Drum brake diagnosis is being discussed. *Technician A* says a grinding noise may be caused by broken hardware. *Technician B* says a firm pedal with little braking action could be caused by a parking brake misadjustment. Who is correct?
   A. A only
   B. B only
   C. Both A and B
   D. Neither A nor B

7. Rotor and drum machining is being discussed. *Technician A* says a drum can be machined on the vehicle. *Technician B* says a disc rotor must be removed from the vehicle for machining. Who is correct?
   A. A only
   B. B only
   C. Both A and B
   D. Neither A nor B

8. The customer complains of a slight pull to the left during braking. The technician finds the right inner pad worn out and the right outer pad with almost no wear. *Technician A* says the inner pad wear could be caused by dirty caliper slide pins. *Technician B* says the left pull could be a result of the right side brake failure. Who is correct?
   A. A only
   B. B only
   C. Both A and B
   D. Neither A nor B

9. The vehicle has very little brake boost and the engine stumbles severely during braking. *Technician A* says the hydro-boost power valve may be the problem. *Technician B* says the diaphragm in the booster is probably split. Who is correct?
   A. A only
   B. B only
   C. Both A and B
   D. Neither A nor B

10. A vehicle pulls slightly to one side during normal or slow braking. *Technician A* says a right rear drum brake that is too tight could be the cause. *Technician B* says brakes shoes soaked with brake fluid is the probable cause. Who is correct?
    A. A only
    B. B only
    C. Both A and B
    D. Neither A nor B

Table 11-1 ASE Task

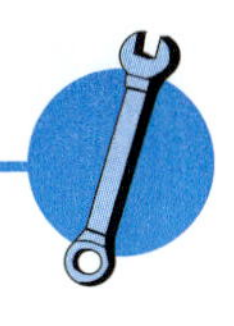

Measure and adjust pedal pushrod length.

Problem Area	Symptoms	Possible Causes	Classroom Manual	Shop Manual
Poor braking	High hard pedal or low hard pedal.	**1.** Improper pushrod length.	315, 316	299
		2. Glazed pads/shoes.	330, 331	299
		3. Improperly adjusted shoes.	332, 333	299

SAFETY

Wear safety glasses.
Disable air bag system.
Exercise ABS safety procedures.

Table 11-2 ASE Task

Diagnose the master cylinder for defects by depressing the brake pedal; determine needed repairs.

Problem Area	Symptoms	Possible Causes	Classroom Manual	Shop Manual
Poor braking	Hard or soft pedal.	**1.** Air in brake system.	319	300
		2. Blocked brake hoses.	325–326	297
		3. Leaking master cylinder seals.	319–323	295–296

SAFETY

Wear safety glasses.
Exercise ABS safety procedures.

Table 11-3 ASE Task

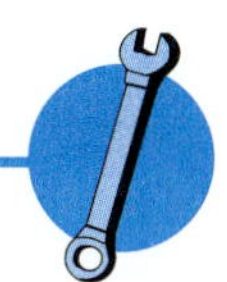

Diagnose the master cylinder for secondary cup defects by inspecting for external fluid leaks.

Problem Area	Symptoms	Possible Causes	Classroom Manual	Shop Manual
Low fluid level	Pedal travel excessive. Poor braking.	**1.** Fluid leaks.	319–323	295–296

SAFETY

Wear safety glasses.

Table 11-4 ASE Task

Remove the master cylinder from the vehicle.

Bench bleed (check for functions and remove air) on all non-ABS master cylinders.

Install the master cylinder in the vehicle; test the operation of the hydraulic system.

Problem Area	Symptoms	Possible Causes	Classroom Manual	Shop Manual
Poor braking	Pedal drops. Fluid low.	**1.** Leaks in master cylinder.	319–323	295–296
		2. Piston cups leaking.	319–323	296
		3. Secondary piston seal leaking.	319–323	296, 306–310

SAFETY

Wear safety glasses.
Block wheels.
Use catch basin and cloth.
Check brake fluid type.
Exercise ABS safety procedures.

Table 11-5 ASE Task

Inspect brake lines and fittings for leaks, dents, kinks, rust, cracks, or wear; tighten loose fittings and supports.

Inspect flexible brake hoses for leaks, kinks, cracks, bulging, or wear; tighten loose fittings and supports.

Problem Area	Symptoms	Possible Causes	Classroom Manual	Shop Manual
Brakes	Poor braking.	**1.** Brake lines cracked.	325–326	295
		2. Brake hoses cracked, split.	325–326	295, 297
		3. Low brake fluid.	320	293–294
		4. Loose brake line/hose fitting.	325–326	295–297

SAFETY

Wear safety glasses.
Use lift locks or jack stands.
Exercise ABS safety procedures.

Table 11-6 ASE Task

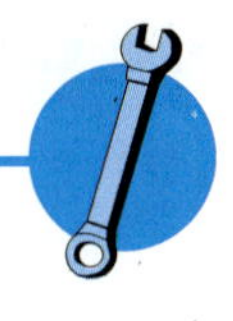

Select, handle, store, and install brake fluid.

Problem Area	Symptoms	Possible Causes	Classroom Manual	Shop Manual
Poor braking	Improper brake fluid. Dirty brake fluid.	**1.** Corrosion on container.	—	293–295
		2. Corrosion in system.	—	293–295
		3. Aged fluid.	—	293–295

SAFETY

Wear safety glasses.
Consult service manual.
Select tight, dry, clean storage containers.

Table 11-7 ASE Task

Bleed (manual, pressure, vacuum, or surge) and/or flush the hydraulic system.

Problem Area	Symptoms	Possible Causes	Classroom Manual	Shop Manual
Poor braking	Low spongy pedal. Aged fluid. Dirty fluid.	**1.** Air in system.	312–314	320
		2. System opened for repair.	312–314	320
		3. Fluid old, dirty, boils.	—	311

SAFETY

Wear safety glasses.
Use lift locks or jack stands.
Block wheels.
Catch and dispose used brake fluid.
Exercise ABS safety procedures.

Table 11-8 ASE Task

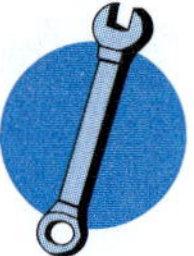

Diagnose poor stopping, noise, pulling, grabbing, dragging, or pedal pulsation caused by drum brake mechanical problems; determine needed repairs.

Lubricate brake shoe support pads on backing (support) plate, adjuster/self-adjuster mechanisms, and other brake hardware.

Problem Area	Symptoms	Possible Causes	Classroom Manual	Shop Manual
Brake noise	Noise. Grabbing. Vibration during braking.	**1.** Loose, missing retainers/ springs.	330	317
		2. Warped drums, shoes worn unevenly.	330	297, 299, 317
		3. Leaking axle/brake seals.	—	297

SAFETY

Wear safety glasses.
Block wheels.
Use lift locks or jack stands.
Exercise ABS safety procedures.

Table 11-9 ASE Task

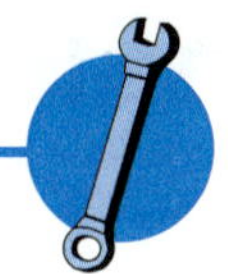

Diagnose poor stopping, noise, pulling, grabbing, dragging, or pedal pulsation caused by disc brake mechanical problems; determine needed repairs.

Remove the caliper assembly from the mountings; clean and inspect for leaks and damage to the caliper housing.

Clean and inspect caliper mountings and slides for wear and damage.

Remove, clean, and inspect pads and retaining hardware; determine needed repairs, adjustments, and replacements.

Clean, inspect, and measure the rotor with a dial indicator and a micrometer, follow manufacturers' recommendations in determining the need to machine or replace.

Remove and replace the rotor.

Install pads, calipers, and related attaching hardware; bleed the system.

Reinstall the wheel, torque lug nuts, and make final checks and adjustments.

Problem Area	Symptoms	Possible Causes	Classroom Manual	Shop Manual
Pads worn. Routine service.	Noise. Pedal pulsates.	**1.** Mileage.	—	300
		2. Abuse of brakes.	333	299, 300
		3. Warped rotors, pads worn.	334	299, 300, 311–316
		4. Stuck caliper/slides.	334	299, 311–316

SAFETY

Wear safety glasses.
Block wheels.
Use lift locks or jack stands.
Use catch basin and spray to clean brake components.
Exercise ABS safety procedures.

Table 11-10 ASE Task

Test pedal free travel with and without the engine running to check power booster operation. Check vacuum supply (manifold or auxiliary pump) to the vacuum-type power booster with a vacuum gauge.

Inspect the vacuum-type power booster unit for vacuum leaks; inspect the check valve for proper operation; repair, adjust, or replace parts as necessary.

Inspect and test the hydro-boost system and accumulator for leaks and proper operation-repair, adjust or replace parts as necessary.

Problem Area	Symptoms	Possible Causes	Classroom Manual	Shop Manual
No brake boost	Hard pedal. Poor braking effect.	**1.** Worn engine.	314	301
		2. Leaking check valve or hose.	314	301–305
		3. Damaged booster.	316–319	301–305

SAFETY

Wear safety glasses.
Block wheels.
Use lift locks or jack stands.
Exercise ABS safety procedures.
Disarm air bag.

Table 11-11 ASE Task

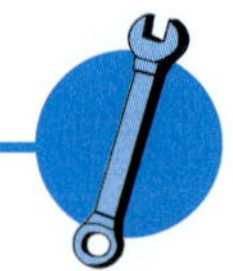

Check the parking brake system; inspect cables and parts for wear, rusting, and corrosion; clean or replace parts as necessary; lubricate assembly.

Adjust the parking brake assembly; check operation.

Test, adjust, repair or replace the brake stop light switch and wiring.

Test the parking brake indicator light(s), switch(es), and wiring.

Problem Area	Symptoms	Possible Causes	Classroom Manual	Shop Manual
Park brake	Brake will not hold. Vehicle noise.	**1.** Worn shoes.	330	317–318
		2. Misadjusted park brake.	336	318–319
		3. Worn, damage cable/lever.	336	318–319

SAFETY

Wear safety glasses.
Block wheels.
Use lift locks or jack stands.
Exercise ABS safety procedures.
Disarm air bag.

Job Sheet 21

21

Name ______________________________ Date ______________

Replacing a Master Cylinder

Upon completion of this job sheet, you should be to remove, bleed, and install a master cylinder.

Tools Needed

1/2-inch impact wrench
Vise
Oil drain pan
Brake fluid
Hand tools
Line wrenches
Fender cover
Service manual
Cloth
Master cylinder bleeder kit

Procedures

1. Determine the following information from the service manual.

   Vehicle Make___________ Model___________ Year___________

   Master cylinder fasteners torque___________

   Brake boost? _____Yes _____ No

   If yes, what type?_________________

   Brake fluid type and classification ____________________

   ABS cautions ____

   ____

   ____

   ____

   ____

   ▲ **WARNING:** Do not mix glycol-based and silicone-based brake fluids. The mixture could cause a loss of braking efficiency.

   ■ **CAUTION:** Wear eye protection when dealing with brake fluid. Brake fluid can cause permanent eye damage. If brake fluid gets in the eyes, flush them with cold running water and see a doctor immediately.

   ■ **CAUTION:** Do not work in or around the steering column without disarming the air bag systems. Serious injury could result from a deploring air bag.

▲ **WARNING:** Before working on the brakes of a vehicle with ABS, consult the service manual for precautions and procedures. Failure to follow procedures to protect ABS components during routine brake work could damage the components and cause expensive repairs.

2. Place thick wipe cloth under the master cylinder brake line connections.
3. Disconnect the fluid level sensor if equipped.
4. Loosen but do not remove the master cylinder fasteners.
5. Loosen and remove the brake lines.
6. Remove the master cylinder fasteners. Remove the master cylinder.
7. Drain the reservoir into an oil drain pan and dispose of the old fluid.
8. Place the master cylinder body in a vise.
9. Use a small pry bar to remove the reservoir from the master cylinder. Be careful not to crack the reservoir.
10. Lube the new O-rings with brake fluid before installing the reservoir on the new master cylinder. The new master cylinder should have new O-rings.
11. Install the reservoir to the master cylinder.
12. Install the bleeder kit.

▲ **WARNING:** Do not mix glycol-based and silicone-based brake fluids. The mixture could cause a loss of braking efficiency.

13. Fill the reservoir chambers about half full of brake fluid.
14. If necessary, reposition the master cylinder in the vise so the front (closed) end is higher than the rear.
15. Use a punch or cross-tip screwdriver to push in the secondary piston. Push it in enough to bottom both pistons.
16. Observe the reservoir for bubbles emerging from the bleeder hoses.
17. Repeat steps 15 and 16 until there is no air coming out of the brake bleeder lines. Ensure that the line ends are kept below the level of the fluid.
18. Close the lines with small paper clips.
19. Install the reservoir cover.
20. Install the master cylinder to the brake booster or firewall.
21. Install and torque the master cylinder's fasteners.

▲ **WARNING:** Protect the vehicle's paint surface from glycol-based brake fluids. This type of brake fluid can damage and remove the finish quickly. If brake fluid gets on the vehicle's finish, wash immediately with cold, running water.

22. Remove the bleeder lines and attach the vehicle's brake line to the master cylinder. Do one line at a time and do not completely tighten the fittings.

■ **NOTE:** Step 23 should be done on a *single* stroke of the brake pedal.

23. Ask a second person to slowly depress the pedal while the secondary line fitting is loosened slightly to allow brake fluid to escape. Observe the fluid for air. Loosely tighten the fitting. Repeat with the secondary line. Ensure that both fittings are tightened before the pedal is released.

24. Release the brake pedal and repeat step 23 as necessary to ensure that the lines are clear of air. If the procedure is not completed within two or three strokes of the pedal, the master cylinder is not bled properly and should be removed for bench bleeding. See steps 2 through 8 and 12 through 24.

25. Test the system by applying the brakes with and without the engine running. The pedal should be firm and there should be no spongy feel. If the pedal is spongy, the complete brake system will have to be bled. It is suggested that the system should be bled any time the system is exposed to air.

26. Remove the wipe cloth and ensure that the area is clean of any brake fluid.

▲ **WARNING:** Before moving the vehicle after a brake repair, pump the pedal several times to test the brake. Failure to do so may cause an accident with damage to vehicles, the facility, or it may cause personal injury.

27. Test the brakes before moving the vehicle from the bay.

28. Clean the area and complete the repair order.

____

____

____

Job Sheet 22

22

Name ______________________________ Date ________________

Replacing a Vacuum Booster

Upon completion and review of this job sheet, you should be able to replace a vacuum boost.

Tools Needed

Standard tool set

Procedures

1. Determine the following information:

   ABS equipped? _____ Yes _____ No

   Air bag equipped? _____ Yes _____ No

   Notes on ABS cautions ____

   ____

   ____

   ____

   ____

   Notes on air bag cautions ____

   ____

   ____

   ____

   ____

   Push rod adjustment ________________

   Booster fastener torque ________________

   Master cylinder fastener torque ________________

   ■ **CAUTION:** Do not work in or around the steering column without disarming the air bag systems. Serious injury could result from a deploring air bag.

   ▲ **WARNING:** Before working on the brakes of a vehicle with ABS, consult the service manual for precautions and procedures. Failure to follow procedures to protect ABS components during routine brake work could damage the components and cause expensive repairs.

2. Disarm the air bag if equipped.
3. Disconnect the brake pedal push rod from the pedal.
4. Remove booster fasteners from under the dash if necessary.

5. Remove the master cylinder and booster fasteners.
6. Disconnect the electrical connections and the vacuum hose as needed.
7. Move the master cylinder from the studs, if necessary, and position it to one side of the booster.
8. Remove the booster fasteners as needed.
9. Remove the booster from the vehicle.
10. Install the new booster and torque the fasteners.
11. Reinstall a master cylinder and torque the fasteners.
12. Connect electrical connections and the vacuum hose.
13. Connect the push rod to the brake pedal.
14. Measure and adjust the push rod to the manufacturer's specifications.
15. Start the engine and test the brake booster. The pedal should be firm and easy to apply.

    Notes ____

    ____

16. Stop the engine and pump the brake pedal to exhaust the booster.
17. Hold the brake pedal down and start the engine. The pedal should fall away but retain firmness without being spongy.

    Notes ____

    ____

    ▲ **WARNING:** Before moving the vehicle after a brake repair, pump the pedal several times to test the brake. Failure to do so may cause an accident with damage to vehicles and the facility, or cause personal injury.

18. Arm the air bag system as needed
19. If the brake pedal feels correct, ensure that the brakes will hold the vehicle before moving it from the bay.
20. Clean the area and complete the repair order.

Instructor's Check ____

____

____

Job Sheet 23

23

Name ______________________________ Date ______________

Replacing Disc Brake Pads

Upon completion and review of this job sheet, you should be able to replace the brake pads on a front-wheel disc brake and inspect the disc brake components.

Tools Needed

Lift or jack and jack stands
Impact wrench
C-clamp
Disc brake silencer
Mechanic wire

Procedures

1. Determine the following information.

   ABS? _____ Yes _____ No

   Caliper/adapter torque ____________

   Wheel nut torque ____________

   Notes on ABS cautions ____

   ____

   ____

   ____

   ____

   ▲ **WARNING:** Before working on the brakes of a vehicle with ABS, consult the service manual for precautions and procedures. Failure to follow procedures to protect ABS components during routine brake work could damage the components and cause expensive repairs.

2. Inspect and adjust the fluid level in the master cylinder so the reservoir is about half full.
3. Lift the vehicle and remove the wheel assembly.
4. Inspect the brake caliper mounting area for ABS components, caliper and adapter, and general condition of the steering, suspension, and brake components on each side of the vehicle.

   Notes ____

   ____

   ____

   ____

   ▲ **WARNING:** Do not use petroleum (gasoline, kerosene, motor oil, transmission fluid) or mineral-oil-based products to clean brake components. This type of liquid will cause damage to the seals and rubber cups and it decreases braking efficiency.

5. Position a catch basin and spray down the braking components to remove dust.
6. Select the correct wrench and remove the caliper or adapter mounting fasteners.
7. Slide the caliper/adapter from the rotor. Use a large, flat screwdriver, if necessary, to pry the pads far enough to clear any ridge.

8. Use mechanic wire to support caliper/adapter weight from a vehicle component.
9. Remove the pads and anti-rattle clips from the caliper.
10. Inspect the rotor for damage. If deemed necessary, measure the thickness of the rotor and compare it to specification. If the rotor requires machining, consult with an experienced technician.

    Inspection results and comments ______________________________

    ____
11. Use a C-clamp or similar tool to press the piston completely into the bore.
12. Inspect the caliper piston boot, slide pins, and slide areas on the adapter and caliper.

    Inspection results and comments ______________________________

    ____
13. Clean the slide pins or areas. Lube the pins or areas with disc brake lubricant.
14. Coat the metal portion of each of the new pads with disc brake silencer. Allow them to set for about five minutes.
15. Install the inner pad into the caliper. Ensure that the pad mates and snaps into the piston. Properly install the anti-rattle springs. Ensure that the friction material, not the metal, faces the rotor. Install the outer pad.
16. If necessary, install the slide pins into the rotor.
17. Slide the caliper/adapter with pads installed over the rotor and align the fastener holes.
18. Install and torque the caliper/adapter fasteners.
19. Install the wheel assembly and torque the lug nuts.
20. Repeat steps 2 through 18 for the opposite wheel.
21. Lower the vehicle when both wheel assemblies have been installed.
22. Press the brake pedal several times to seat the pads to the rotor.

    ▲ **WARNING:** Before adding brake fluid, consult the vehicle's service manual. Many manufacturers require a specific classification of brake fluid to be used.

23. Check the brake fluid level and top off as needed.

    ▲ **WARNING:** Before moving the vehicle after a brake repair, pump the pedal several times to test the brake. Failure to do so may cause an accident with damage to vehicles and the facility, or it may cause personal injury.

24. Perform a brake test to ensure that the brakes will stop and hold the vehicle. Do this test before moving the vehicle from the bay.
25. Clean the area and complete the work order.

☑ **Instructor's Check** ______________________________

____

Auxiliary System Service

Upon completion and review of this chapter, you should be able to:

- ❑ Diagnose and repair the warning light systems.
- ❑ Conduct a performance test of dash-mounted instrument circuits.
- ❑ Diagnose and repair or replace the interior lighting.
- ❑ Diagnose the power windows and door locks.
- ❑ Conduct a performance test of the climate control system.
- ❑ Perform a function test of the anti-theft system.
- ❑ Demonstrate the proper method to disarm the air bag system

Introduction

Most passenger comfort and convenience features will fail like any other component on the vehicle. Most of the failures will be electrical in nature, either a failure of the wiring or the electrical components. Since many of the accessories on today's vehicles require skills gained from many hours of experience and training, this chapter will cover some of the common tests and repair that could be performed by entry-level technicians.

Warning Lights

Classroom Manual
pages 347–350

Special Tools

Service manual

Vehicle's diagnostic routine

KOEO and *KOER* running are two test conditions that are associated with PCM test routines, but they appear in other system test procedures as well.

Many of the warning lights are simple, electrical circuits. A good understanding of their operation will give the technician a head start on basic repairs. Most tests have to be run when the key is on with engine off **(KOEO)** or key on with the engine running **(KOER).**

Older vehicles use a mechanical switch to turn the light on when certain conditions are present. Other vehicles use a sending unit for the warning light and gauge. One basic function occurs on both types: the ground to the warning light is completed.

WARNING: Do not grab hoses or other components under the hood if an overheated engine is suspected. Even an engine that is operating at normal temperature can cause serious burns.

WARNING: Never remove the radiator cap on a warm or hot engine. Always let the engine cool down. Extremely serious burns and other injuries can occur when removing a hot radiator cap.

If the red engine light is lit, the problem indicated is either an overheated engine or low oil pressure. On some vehicles, there is a light for oil and one for overheating (Figure 12-1). Before starting work on the electrical circuit, do the obvious and check the level and condition of the coolant and oil. For a cooling problem, look for missing or damaged fan belts. The coolant level on a hot engine can be checked in the overflow reservoir. The reservoir may not contain a true representation of the coolant level, but it will be an indication. Note that on some cooling systems, the reservoir may become pressurized. Notice the condition of the engine by checking for oil or coolant leaks. The hoses should be inspected for swelling. Do not touch hot hoses. In many cases of low oil pressure or overheating, the condition will be found during the inspection.

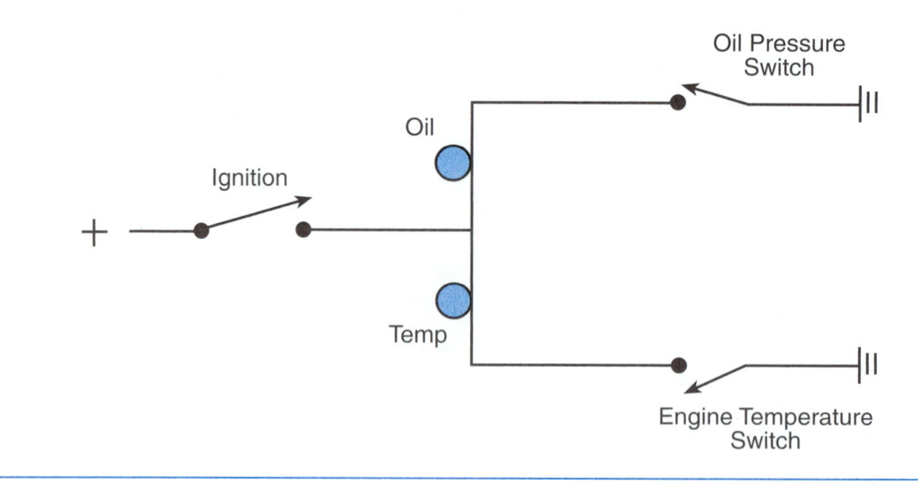

Figure 12-1 Two warning lights require two completely different circuits powered through the ignition switch.

WARNING: Do not perform an KOER test if there are any visible or audible signs that the engine is damaged or could possibly suffer damage when running. Doing so could cause more damage.

WARNING: Never apply 12 VDC to a sending unit unless directed to do so by manufacturers' test routines. Sending units are almost always on the negative side, may be electronic, and will be damaged by direct 12-VDC, positive current.

A bulb test or proof-out test of the warning light is done every time the ignition key is switched to *RUN*. Other warning lights are tested while the key is in the START position.

Assuming that the problem was not apparent under the hood and the engine operates normally, run the engine just long enough to confirm the customer's complaint and determine if there is more than one warning light for the engine. Before starting the engine, switch the ignition to run and observe the lights. There should be several red lights and possibly an amber light. If all of the red warning lights are lit, the circuit to each light is good.

Do not run the engine any longer than it takes to check the lights. If the warning light stays on during engine idle, speed up the engine to see if the light goes out. As the engine speed increases, the oil pump produces more flow and pressure. If the light goes out, the likely problem is in the oil system. In this case, the oil pump and engine could be worn or sludge is developing in the oil pickup or passageways. Generally, an overheated engine will not cool down enough to turn off the light by simply speeding up the engine. Note the matching gauges, if equipped. When a red light turns on but the gauge reads normal, the problem is usually in the light circuit.

For a vehicle with a shared light for oil pressure and temperature warning, shut down the engine and locate the oil pressure and temperature sending units (Figure 12-2). Perform a KOER

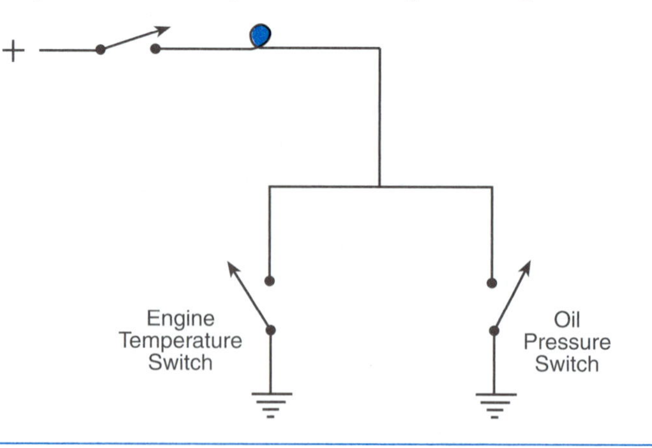

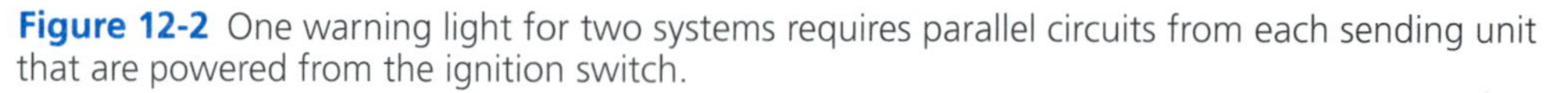

Figure 12-2 One warning light for two systems requires parallel circuits from each sending unit that are powered from the ignition switch.

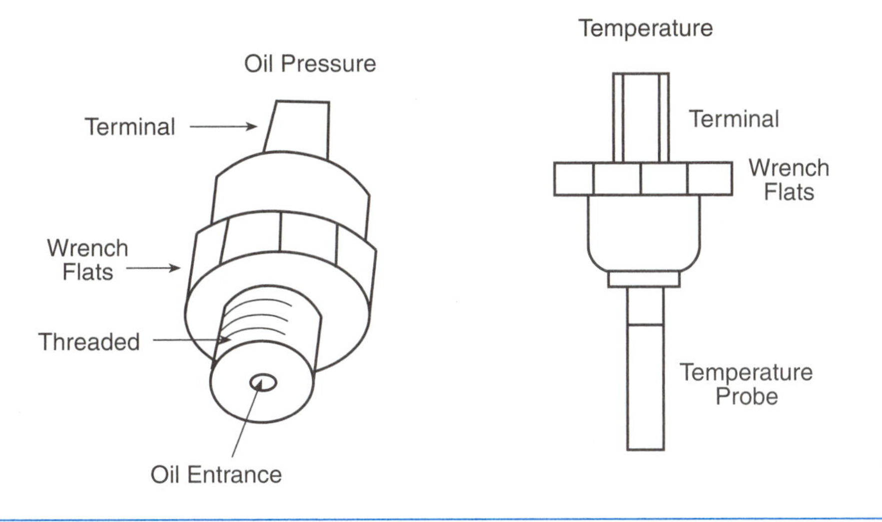

Figure 12-3 The oil pressure sending unit is usually different in appearance and operation from the temperature sending unit or switch.

test by disconnecting the temperature unit. It is the easiest unit to reach because it is on or near the thermostat housing in most vehicles. Start the engine. If the light goes off, the problem is either overheating or circuit trouble. A light that stays on may indicate an oil pressure or circuit problem. Repeat the test with the oil pressure sending unit. With the engine running and both circuits unplugged, the red light should go out. If it does not, there is a problem within the circuit. With the light out, connect one of the two sending units. Now, the appearance of a red light indicates system failure. Lack of a red light indicates that the problem is in the disconnected system.

If there are separate lights for temperature and oil, disconnect the sending unit for the affected circuit and test that circuit only (Figure 12-3).

On a dual light system, check the electrical circuit from the sending unit to the light by disconnecting the wire at the sending unit. Run a KOEO test. Use a jumper to ground the wire to a metal part with ignition switch in RUN (Figure 12-4). Each time the wire is grounded, the warning light should come on. If the light and circuit function properly, the sending unit is bad. On a temperature warning system, let the engine cool down, if necessary, and replace the sending unit. Do not remove the temperature sending unit on a hot engine. If the lubrication system is being checked, it is possible to manually check the oil pressure before replacing the sending unit. This would also be a direct test of the oil pump output.

Special Tools

Service manual

Jumper wire

Figure 12-4 This test will ground the circuit and it should turn on the warning light or make the gauge read at its maximum.

WARNING: Do not start the engine if the oil level is low or the engine is making unusual noises. Serious damage could result if there is not enough lubrication on the bearings and journals.

Special Tools

Manual oil pressure gauge

Remove the oil pressure sending unit and install a manual oil pressure gauge (Figure 12-5). Start the engine and record the oil pressure. Consult the service manual for exact specifications. Generally, pressures over 20 psi at idle are acceptable. Shut down the engine immediately if oil pressure does not reach 20 psi within a few seconds of engine startup. If the pressure is at least 20 psi, speed up the engine. There should be an increase in pressure. If the pressure is low or does not increase, the probable cause is a worn oil pump or a worn engine. However, each of these conditions occurs over a period of time, depending on the condition of the engine. A discussion with the operator can verify this. A broken oil pump will suddenly turn on the light, cause almost immediate engine failure, and the condition will be obvious to the technician.

Some vehicles have *cooling fan* switches that closely resemble sending units. These switches may be located near the sending unit.

CAUTION: Electric **cooling fans** may turn on at any time when the temperature-sending unit is connected. Observe safety procedures by remaining clear of an electrical cooling fan during testing.

CAUTION: Some electric cooling fans with a temperature switch (not to light or gauge) will turn on when the switch is disconnected with the ignition key on. Observe safety procedures by remaining clear of an electrical cooling fan during testing.

If the temperature light/gauge circuit proves correct during the test, the engine's cooling system must be checked. That includes the fan, water pump, thermostat, and coolant condition and level. A detailed diagnosis routine will be set forth in the service manual. This is particularly true of computer-controlled engines. Many computer-controlled systems use the PCM to turn on the warning lights based on input signals from its sensors. Diagnosing computer systems is best left to experienced technicians, but check the basics such as belts, oil and coolant level, leaks, or anything else that may point to a mechanical failure. If the top of the engine

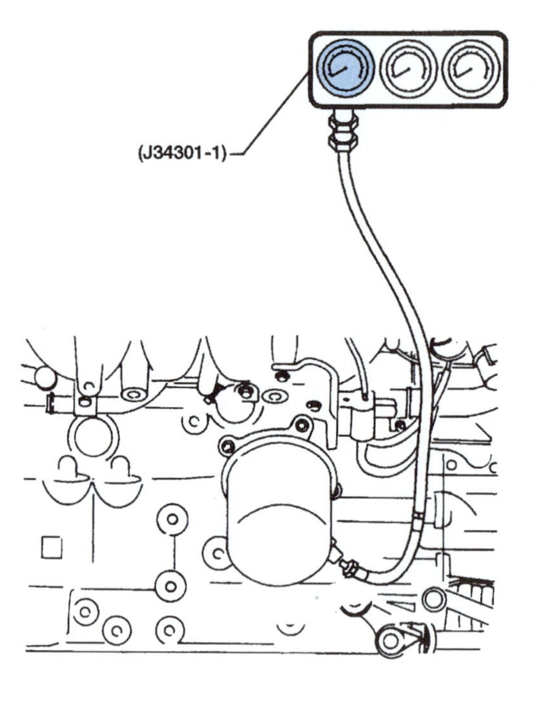

Figure 12-5 The gauge for this test is a dash-mounted aftermarket gauge fitted with a long tube/hose for attachment to the engine's block. (Courtesy of Nissan North America)

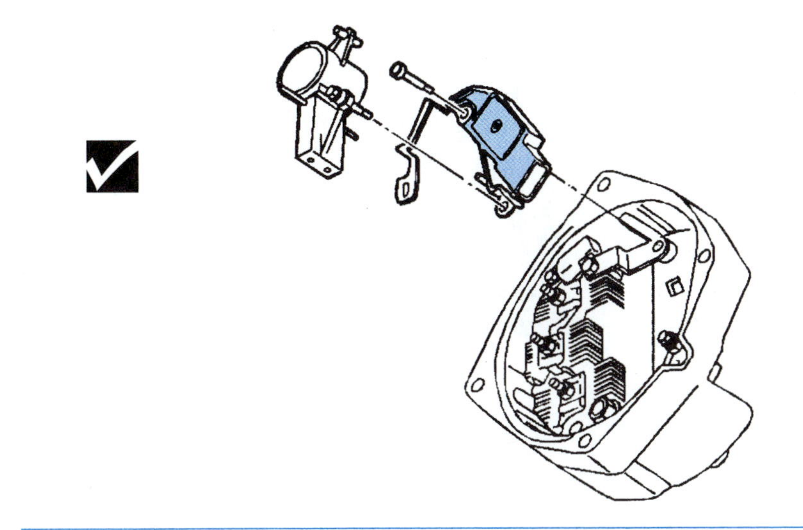

Figure 12-6 The regulator is set up to adjust alternator output based on the vehicle's electrical load. A voltage sensor is enclosed for the warning light. (Courtesy of Chevrolet Motor Division, General Motors Corporation)

shows signs of recent repairs, check to make sure all connectors are, in fact, properly connected and tight.

SERVICE TIP: If the alternator has been replaced, it is possible that the wrong alternator pulley has been installed. An oversized pulley will slow down the speed of the alternator, thereby reducing its output. It will be most notable at idle speeds.

Low voltage or current in the charging circuit will switch on the alternator warning lights. A loose alternator drive belt may slip and not allow the alternator to produce the required voltage or current. Always check the belt and all battery and alternator connections before testing the charging system or the battery. If the inspection does not indicate a mechanical problem, the battery and charging system can be checked using a VAT 40 or VAT 60 as outlined in Chapter 8, "Engine Maintenance." Do not use a VAT 40/60 on a computer-controlled alternator.

The voltage/current sensor is usually in the regulator and, depending on the system, the alternator may have to be replaced or rebuilt or an external regulator may need to be replaced (Figure 12-6).

Other warning light systems can be checked in the same manner. Before condemning the electrical part, always check the mechanical portion of the system. Most of the time, the warning system is doing what it was designed to do. The most common failure in a warning system is the switch or sending unit. Disconnecting the switch or sending unit should turn off the light.

Gauges

Classroom Manual
page 346

WARNING: Do not check electronic circuits using a test light, grounding a conductor, or jumping a connection unless specifically directed to do so by the service manual. Incorrect test equipment and procedures can damage electronic circuits and components.

Before proceeding, determine the type of gauges used on the vehicle. If the **instrumentation** is electronic and is operated by a computer module, do not use the test procedures suggested

Computerized *instrumentation* often runs through a proof-out mode, and various LEDs may flash during the test.

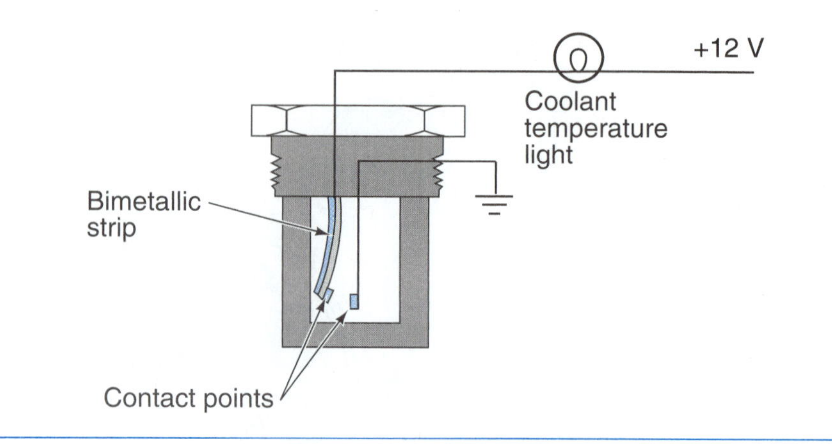

Figure 12-7 A cutaway view of a typical temperature warning light switch.

here. Most sending units with one wire are not considered to be electronic while two or more wires are generally part of an electronic circuit in some manner. Also consider the customer's complaint. If only *one gauge* is not working, then the problem is in that system. If *all gauges* are not working, it is probable that the fault lies in some shared component like a fuse or common conductor. Always consult the service manual to determine the system's operation and the basic diagnostic test procedures. The tests discussed here are very similar to or the same as the ones for warning lights. Some vehicles may use separate circuits for each but share a common sending unit.

Sending units for analog gauges use a material that changes resistance based on the amount of heat or pressure being applied or they use some variable resister (Figure 12-7). Both types can fail at OFF, ON, or any point in between. If there is a one-wire connection at the sensor, grounding the wire in the KOEO mode will cause a maximum gauge reading. A disconnected wire will allow the gauge to drop to its lowest limit. If the gauge works in this test, then the sending unit is bad. A gauge that does not move when the wire is disconnected or grounded (KOEO) indicates a bad circuit or gauge. Consult the service manual for test procedures for the circuit and gauge.

Special Tools

Service manual

Jumper wire

Variable resister sending units are used for fuel gauges. A float in the tank moves a wiper arm over a very thin resister wire (Figure 12-8). The position of the arm on the wire deter-

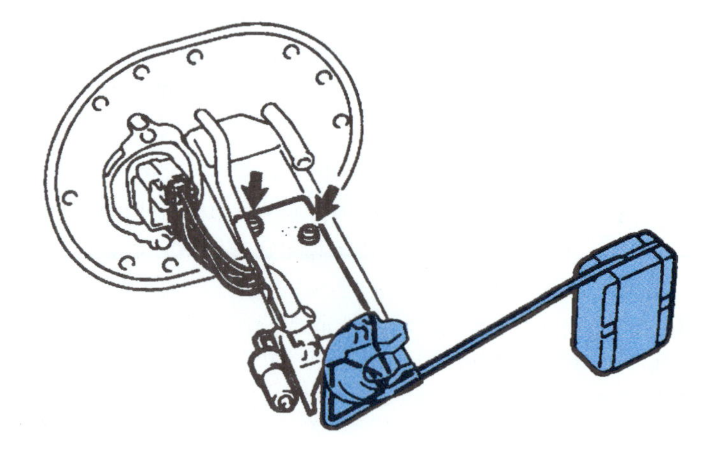

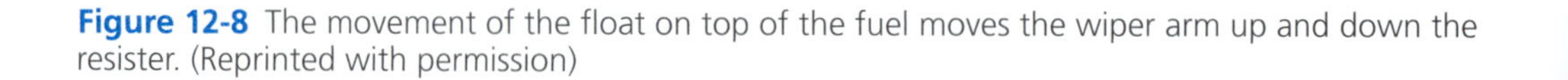

Figure 12-8 The movement of the float on top of the fuel moves the wiper arm up and down the resister. (Reprinted with permission)

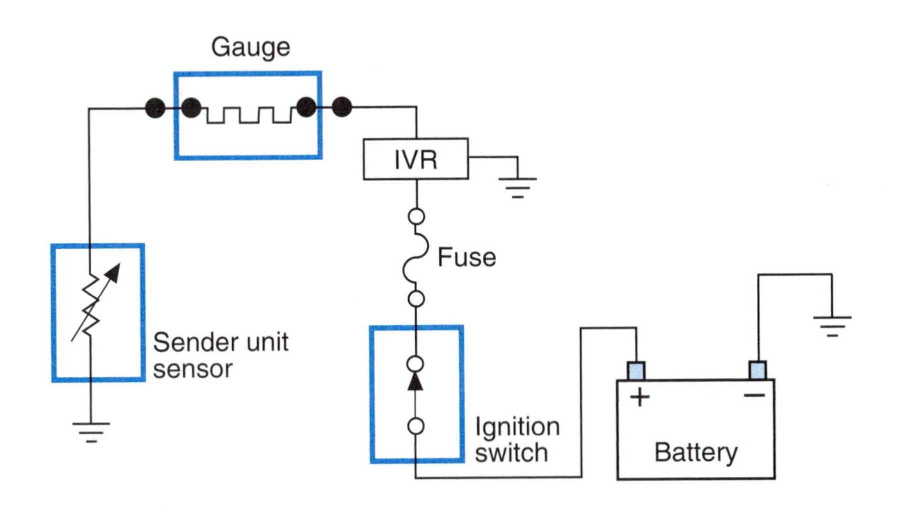

Figure 12-9 Most IVRs can be individually tested and replaced from the instrument panel or any gauge/light circuit.

mines the amount of resistance between the gauge mechanism and ground. The test for this circuit is similar to the one for one-wire sending units on the engine. Lift the vehicle and locate the connection for the sending unit. It will be near or on the tank. On older vehicles without in-tank electric fuel pumps, there is usually one wire leading to the tank. On systems with in-tank pumps, there may be several electrical connections. Before disconnecting any of them, consult the service manual. Once the sending unit wire is located and disconnected, perform a KOEO test. Turn the ignition key on while the wire is grounded. A good circuit will cause the gauge to read full. Removing the wire from ground should cause the gauge to drop to empty. If either condition is not met, then the circuit or gauge is bad. An operating circuit means the fuel tank has to be dropped and a new sending unit installed.

A word of caution before leaving this section. Most instrument panels use some type of voltage limiter to regulate the amount of voltage being supplied to the various gauges and sometimes to the entire gauge circuits (Figure 12-9). It is normally referred to as an *instrument voltage regulator (IVR)*. Jumping 12 VDC to a gauge or circuit may damage them and cover up the original problem. The tests listed above are simple and cover most initial testing of analog, non-electronic gauges, and circuits. If the test indicates that the circuit or gauge is inoperative, the test procedures to be followed from that point are found in the service manual. Testing of the actual instrument panel components should always be done according to the manufacturer's procedures and performed or supervised by an experienced technician.

Lighting Systems

Classroom Manual
pages 352–363

Automotive lighting systems are very simple electrical systems. They have a positive side and a negative side. Almost all systems are series-parallel circuits with the controls and protection devices located in the positive or hot side (Figure 12-10). After the last control, the circuit parallels out to two or more lamps. The initial diagnosis is easy. If one lamp on a system does not light but the others do, the circuit, controls, and protection devices are operating properly. The usual problem would be a burned out lamp. However, there are times when all the lamps on a system go out, indicating a control or protection device failure. It is possible, but not probable, that all the lamps on a system burn out at once. Even then the root cause is probably somewhere else in the circuit such as a protection device not opening when the circuit is overloaded with current or grounded. Diagnosing the headlight and marking light circuits and the turn signals will be discussed, covering the most probable causes of complete system failure.

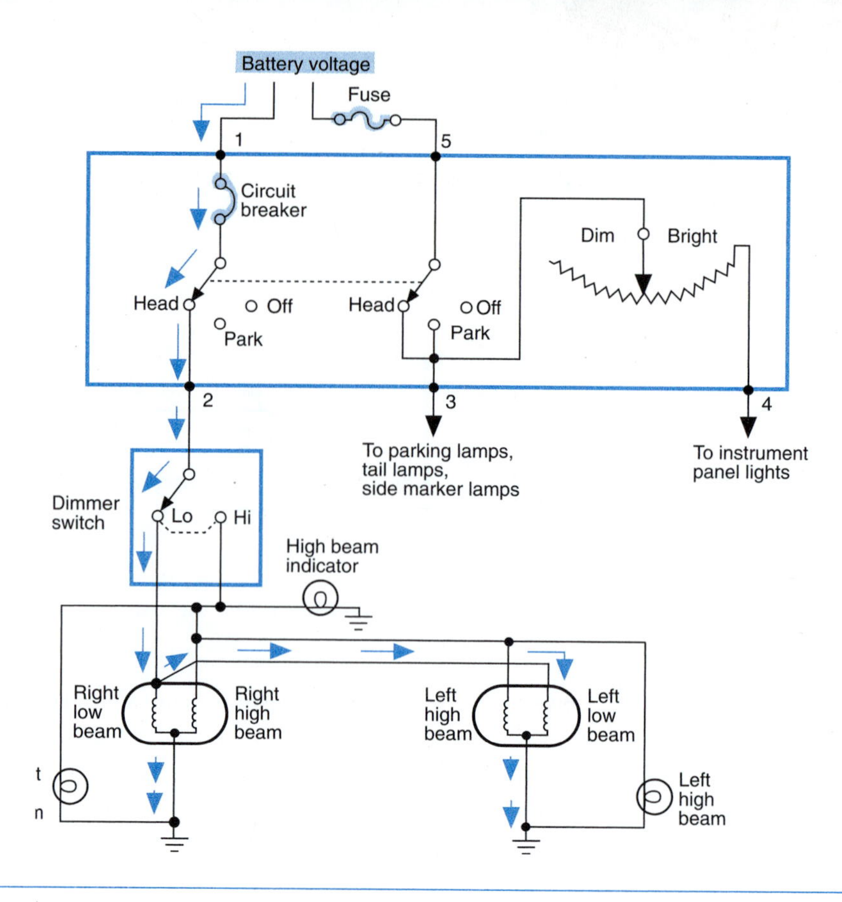

Figure 12-10 Note that the protection devices are on the positive side of this light system.

Headlights and Marking Lights

Daytime running lights are controlled by a different switch than the one that operates the headlights. One system may work, but not the other.

Special Tools

Multimeter or voltmeter

Service manual

Each of the two systems is turned on by different contacts within the same switch (Figure 12-11). This multi-position switch may be found on the dash or steering column and has OFF, PARK (marker), and headlight positions. Normally, the switch is connected directly to the battery with two positive conductors. A fuse protects power to the marking side while a circuit breaker is used on the headlight feed. With all of the marking lights (or **daytime running lights**) and headlights inoperative, the most probable cause is a bad headlight switch or total loss of power from the battery. However, always check the fuse and circuit breaker before proceeding further.

SERVICE TIP: A test light could be used in the following tests, but the amount of voltage will not be registered. It is best to use a multimeter when testing for voltage.

Remove the switch and use a multimeter to test for voltage at the two 12- or 10-gauge wires feeding into the switch (Figure 12-12). Usually, they are solid red or red/black in color. Each should register within one volt of battery voltage if a voltmeter is being used. On some vehicles, the ignition must be on for the marking lights, the headlights, or both to work. Turn the ignition key to RUN if there is no voltage found on the first test. This confirms the test result. If both conductors are within voltage limits, replace the switch. If neither is carrying voltage and the protection devices are good, the circuit from the switch back to the battery must be checked. Normally, in a case of battery loss, there will be other inoperative circuits.

In most cases, where only the headlights or marking lights are working, a portion of the headlight switch may be bad, but the most probable cause in this instance would be an open protection device for that circuit. Do not replace the protection device and return the vehicle to the customer. The protection device opened for a reason. Attempt to locate the root cause

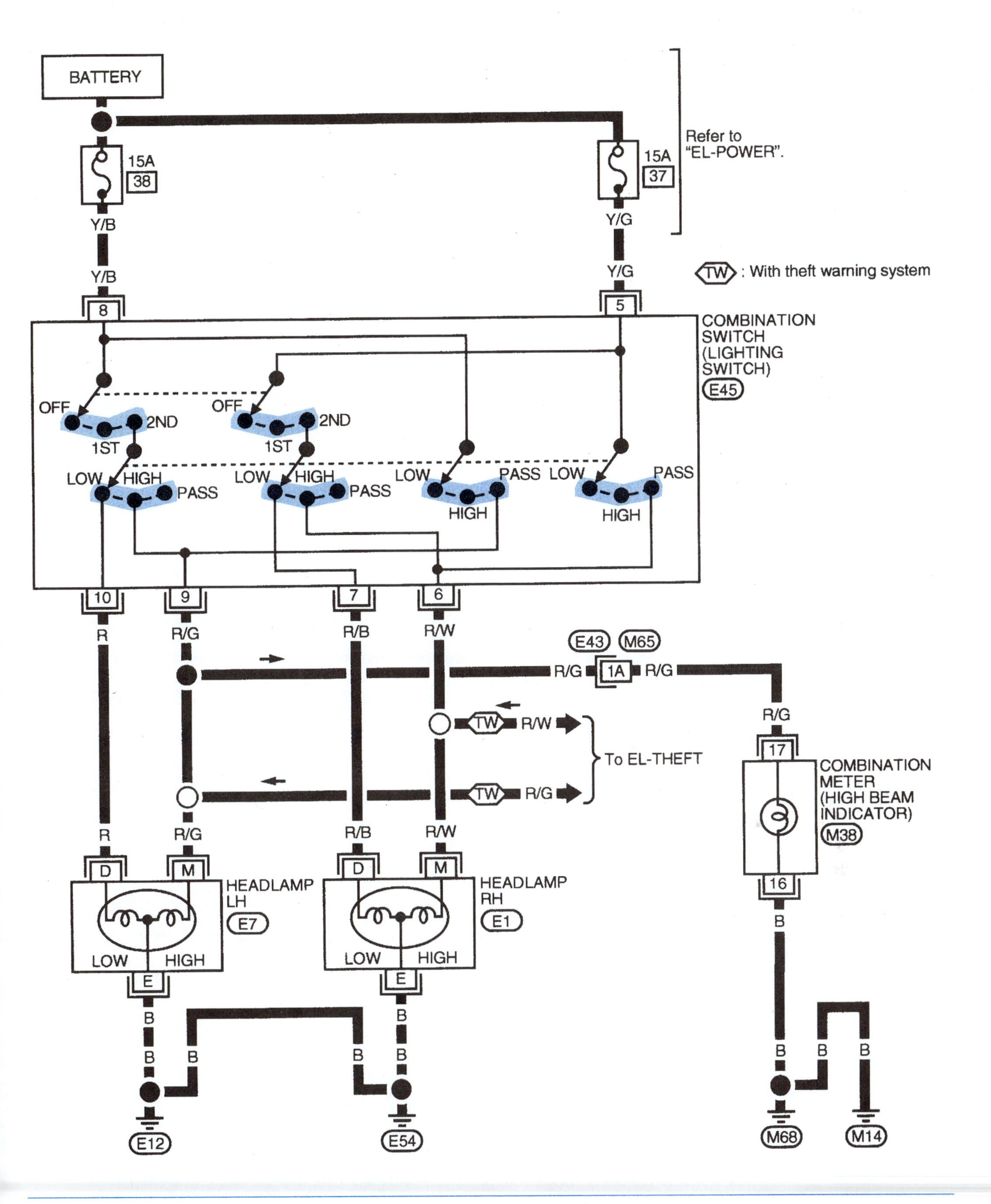

Figure 12-11 The headlight switch has several different contacts that feed parallel circuits. (Courtesy of Nissan North America)

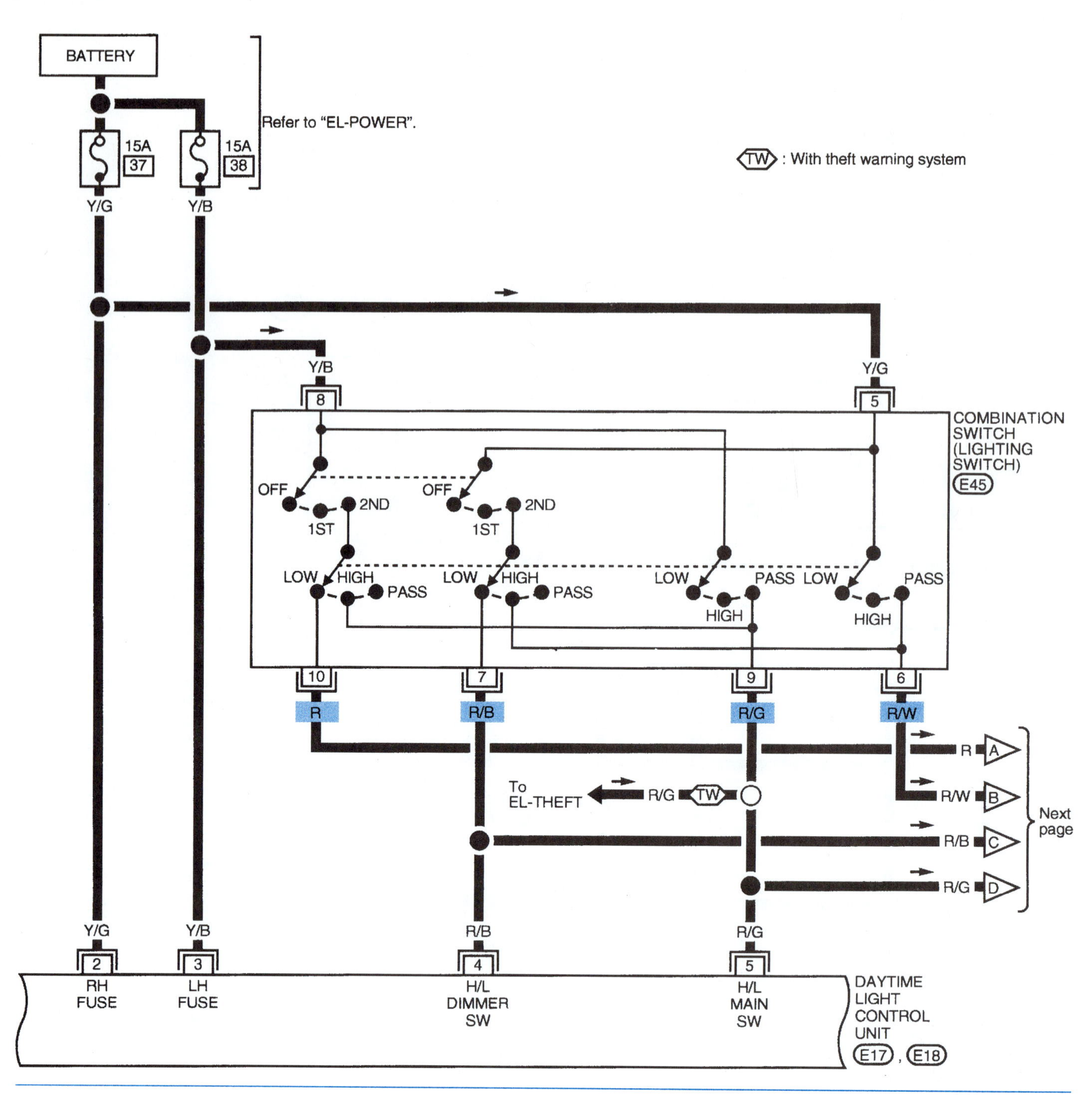

Figure 12-12 Most feed conductors are red in color. Another common color for feed is orange. (Courtesy of Nissan North America)

of the problem. The device itself may be old and failing or there may be a short or ground somewhere in the circuit. Replacing the device without checking the system may result in customer comeback and loss of time and profit.

Some vehicles, especially imports, have more than four terminals on the *relay*. Relays of this sort have more then one switch within them.

Most headlight systems have a **relay** that routes a large current flow directly from the battery to the headlamps (Figure 12-13). The relay is operated by a set of contacts in the headlight switch. If the relay, normally under the hood, is easier to access, it may be better and quicker to check it before checking the headlight switch. A typical simple relay has four terminals:

- One for battery feed
- One to feed the parallel circuits to the lamps
- One feed from the headlight switch
- One ground

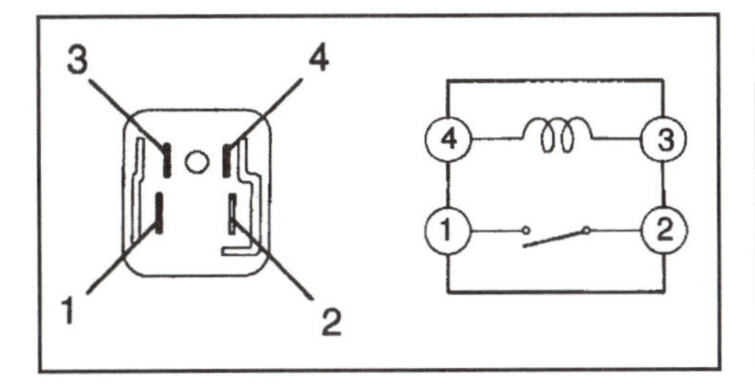

3. INSPECT HEADLIGHT RELAY CONTINUITY

Condition	Tester connection	Specified condition
Constant	3 – 4	Continuity
Apply B+ between terminal 3 and 4.	1 – 2	Continuity

If continuity is not as specified, replace the relay.

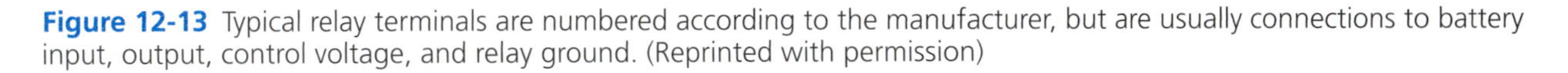

Figure 12-13 Typical relay terminals are numbered according to the manufacturer, but are usually connections to battery input, output, control voltage, and relay ground. (Reprinted with permission)

The most common failure in any electrical circuit is the loss of ground. Use a jumper wire to connect the ground terminal to a good metal ground. If the system works, replace the ground conductor or connector at the ground. A good ground is needed to check for voltage at the two **feed conductors.** The two largest conductors will be the *battery feed* and *lamp feed*. Use a multimeter to check for battery feed (Figure 12-14). If correct, turn on the headlight switch and check for voltage on the conductor at the relay connection (Figure 12-15). If one of the two is not within one volt of battery voltage, that **circuit** must be checked back to the switch and protection device or to the battery. If both feeds are correct, leave the switch on and check the conductor from the relay to the lamps (Figure 12-16). No voltage on the output indicates a bad relay if the two feeds and the ground are correct.

Almost all automotive *feed conductors* are red or predominately red in color.

Some more expensive vehicles have a *circuit* that measures the amount of current being used in a light system. If a lamp burns out and the current drops, a warning light is lit on the dash.

Special Tools

Jumper wire

Multimeter or voltmeter

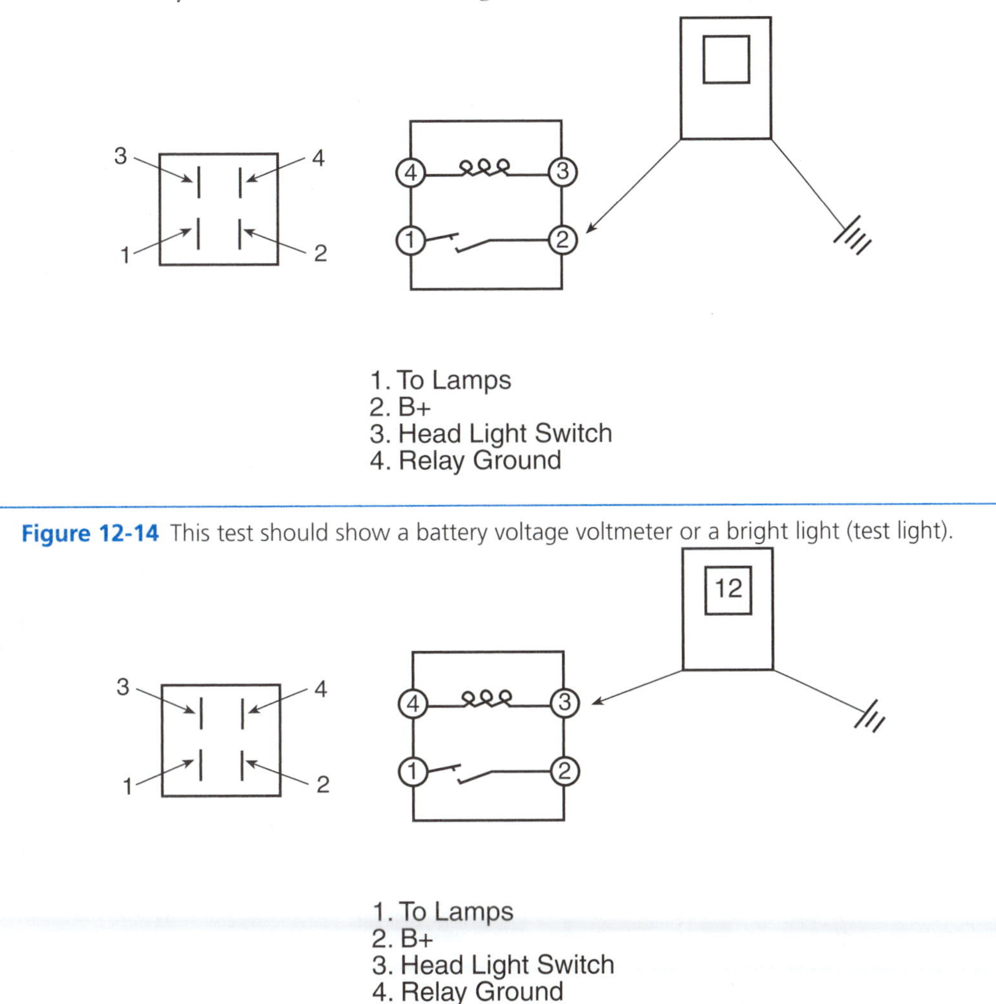

Figure 12-14 This test should show a battery voltage voltmeter or a bright light (test light).

Figure 12-15 Since the headlight switch receives battery voltage, this test should show the same results as the one.

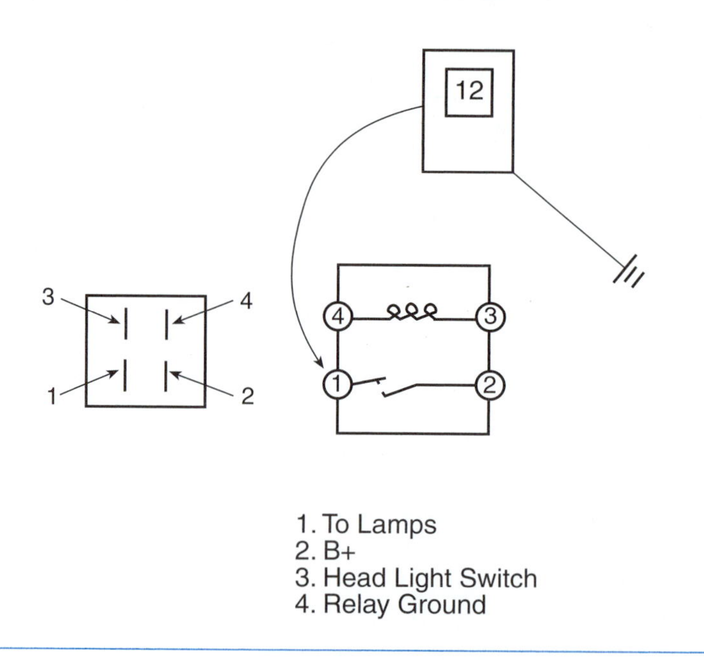

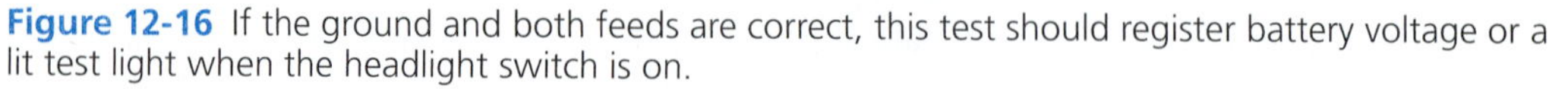

Figure 12-16 If the ground and both feeds are correct, this test should register battery voltage or a lit test light when the headlight switch is on.

The tests above seem to be more difficult than the ones discussed before. However, many times the most difficult part of electrical testing on a vehicle is locating and accessing the conductor or the component being tested. Experience, wiring diagrams, and electrical component locator manuals will show the technician how to conduct the same tests at different and easier-to-access points in the circuit (Figure 12-17). The preceding tests on the headlight circuit can be used to test almost any electrical, but not electronic, system on the vehicle by understanding the circuit's operation and tracking the circuit and components back to the voltage source.

There are two tools that can assist the technician in testing electrical circuits. One is known as a *short finder*. It is connected to the circuit and a power source. The short finder indicates the presence of a magnetic field. When electric current flows through a conductor, a magnetic field is established around that conductor. A needle on the face will move from side to side as the finder is moved over a conductor. When the needle stops moving, there is no current in the conductor at that point.

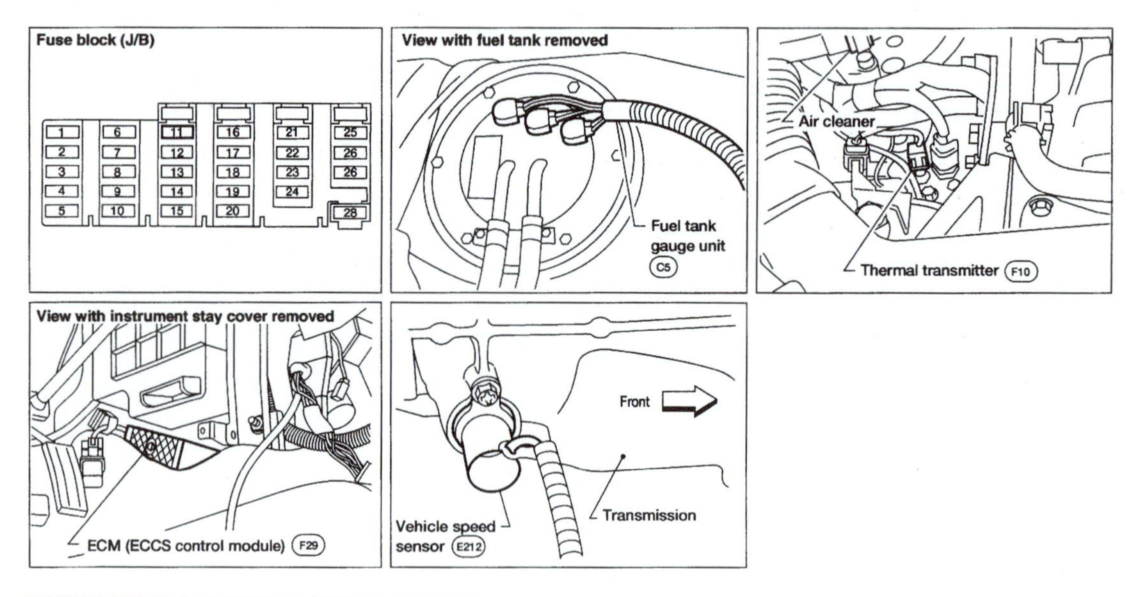

Figure 12-17 Component locators show wiring routing and components within a circuit. This can help make it easier to find an accessible test point. (Courtesy of North America)

A second tool is designed for testing relays. The tool has a female plug for connection to the suspected relay. A short *pigtail* is plugged into the connection where the relay is usually inserted. Operating the tester and observing the tester's meter and the relay operation can pinpoint the electrical or mechanical problem.

Turn Signals

The turn signal circuits work like most other lighting systems except they use a fast-acting circuit breaker called a *flasher* to blink the lamps. Testing this system in done a manner similar to other lighting systems.

Special Tools

Jumper wire

Multimeter or voltmeter

When the turn signal switch is moved to either the left or right turn position, current flows from a fuse through the flasher and switch to a front and rear lamp. When one lamp burns out, the flasher will do one of three things: blink the other lamp faster, blink the other lamp slower, or turn on the other lamp continuously. Replacing the non-operative lamp will repair the system.

A second problem may be indicated by signals working on one side that do not turn on or do not blink on the other side. Because of the design of the signal switch, one side of the turn signal system may work well, but the other side may not receive any power (Figure 12-18). Another condition that could cause only one side to work is the installation of the wrong lamp or excessive or extremely low resistance in the non-working side. Resistance determines the amount of current needed for operation, and the proper amount of current is needed to make the flasher open and close.

SERVICE TIP: The most common cause of brake, turn, and marking light failure on a trailer is its grounding wire. If all trailer lights blink or become very bright when the turn signal or brake is used, suspect the trailer's master ground. Usually, it is a white wire connected to the trailer frame near the hitch. The problem may also affect some or all of the towing vehicle lights.

The easiest method of checking a turn signal with one inoperative side is to try the hazard warning system. If all the lights work, the circuits are good and the turn signal switch is probably bad. Sometimes, replacing the turn signal flasher with a new one will make the system work, but the flasher will not be the cause of the failure. Corrosion or looseness of the switch contacts would be the most likely cause, and the switch must be replaced. The new flasher will temporarily make the circuit a little better and allow the current to flow.

When the turn signals, marking lights, or brakes work incorrectly, one thing to check is the addition of a trailer electrical connector on the vehicle (Figure 12-19). When additional lamps are added to the vehicle's lighting system, a heavy-duty flasher is required. Also, the connection between the trailer harness and the vehicle wiring could be faulty, thereby adding resistance to the circuits. A proper T-connection is shown in Figure 12-20.

Flasher failure is a common problem and does not mean there is any other problem within the circuit, but repeated failure of the flasher indicates a circuit problem. New electronic flashers are fairly expensive and require certain procedures and test equipment to properly diagnose them. The procedures in the service manual must be followed to successfully test and isolate the malfunction in electronic circuits.

Like all lighting systems, failure of all turn signal lamps at one time points to a breakdown of the protection device, switch, or some other common component of the system. Most turn signals will not work unless the ignition switch is in the RUN position. Four-way or hazard circuits can be checked in the same manner as the turn signals. A different switch and a heavy-duty flasher are used for hazard warning lights, but the same conductors and lamps are used after the switch. The hazard light circuit will work in any ignition switch position.

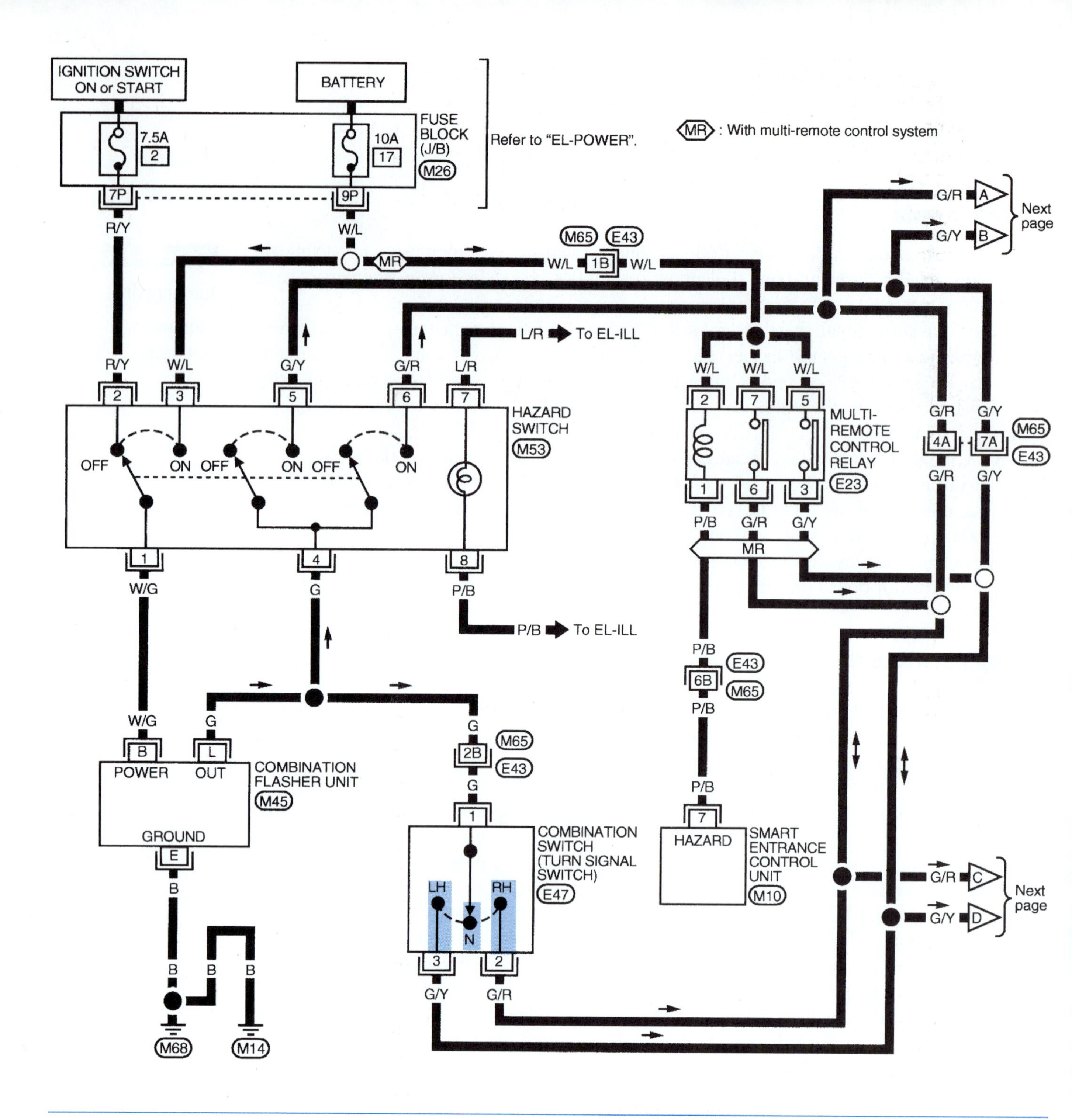

Figure 12-18 The operation of the turn signal switch sometimes results in excessive wear on one set of contacts, causing failure of that circuit. (Courtesy of North America)

Figure 12-19 A trailer's electrical connection can be easily damaged if it is left to hang loose, thereby causing shorts and grounds within the circuits.

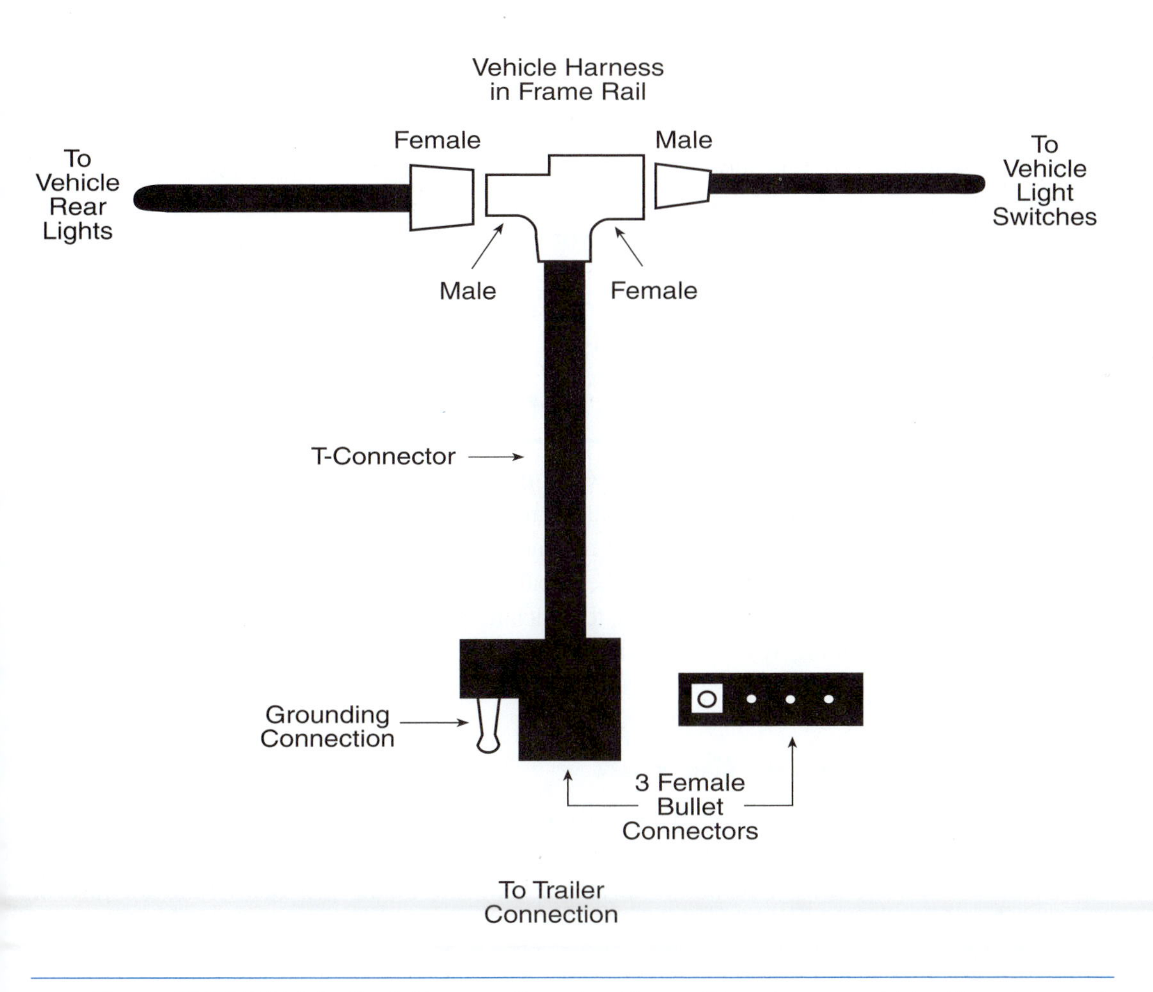

Figure 12-20 Vehicle/trailer light connections should use a T-type harness, which will provide good contacts and weatherproofing.

Power Windows, Locks, and Seats

Service of the power windows, door locks, and seats consists of checking operation, fuses, and some slight lubrication. Almost any repair to these systems requires disassembly of the inner door panels or seat removal. In this section, preliminary checks will be discussed.

The first step in troubleshooting is to check all operations and determine answers to the following questions:

1. Are all the devices inoperative?
2. Do some of the devices work with one switch but not with the others?
3. Do some or all of the devices work slowly or create noise during operation?
4. Do the noisy devices attempt to move the window, lock, or seat?

Remember from the classroom manual that switches for the power windows and door locks are wired parallel to the master switch with a common protection device. If the answer to question 1 is yes, the obvious problem is in the power feed to the master switch. Should the answer to question 1 be no, then the problem is not the power supply.

An example of the situation in question 2 is the operation of the power windows. If the master switch lowers all of the windows, then the power feed, the motors, and the conductors are good. The only thing left to check is the individual switch for each window. The situation is reversed if the individual switches work the windows and the master does not. The same theory of operation applies to the power door locks.

Questions 3 and 4 cover the same general problem. A motor that lowers and raises the window slowly could be worn or the window or regulator is binding. The same is true of power door locks. Worn solenoids or binding linkage may cause the solenoid to hum, hang, or refuse to work at all. A noisy window motor could indicate worn gears, a broken ribbon strip, or loose regulator fasteners. If the motor or solenoid does not attempt to move the glass or lock but continues to operate, the problem is usually stripped gears or ribbon.

Power seats use different motors and switches to move the seat, but a common fuse protects everything. A fuse or other power source failure will cause the failure of all operational modes. If any of the two or three switches work, the power source and fuse are good. The probable cause would be the switch or motor. Again, if the motor can be heard but the seat component does not move, the problem will probably be found in the cables or screwnut.

Climate Control Systems

Classroom Manual
pages 367–375

Special Tools

Jumper wire

Multimeter or voltmeter

Normally, an entry-level technician will not perform full repairs on climate control systems but could perform basic, troubleshooting procedures and possibly correct a minor problem. Since this system relies on the transfer of heat and the movement of air, many times the malfunction can be traced to a switch or component. The repair may not require the opening of the A/C system, which is done by an EPA-certified technician. Some common complaints of manual heating and air conditioning systems that do not require the opening of the A/C system and a full systems performance test will be discussed.

No Air from Vents at the Halt

The most common problem here is a failed blower motor or motor circuit. If the customer states that air comes from the vents while driving, the blower motor is inoperative for some reason. Airflow during driving results from air entering the intake just below and forward of the windshield. Any time the vehicle stops or slows, there is no intake. Perform the following test on the blower motor circuit.

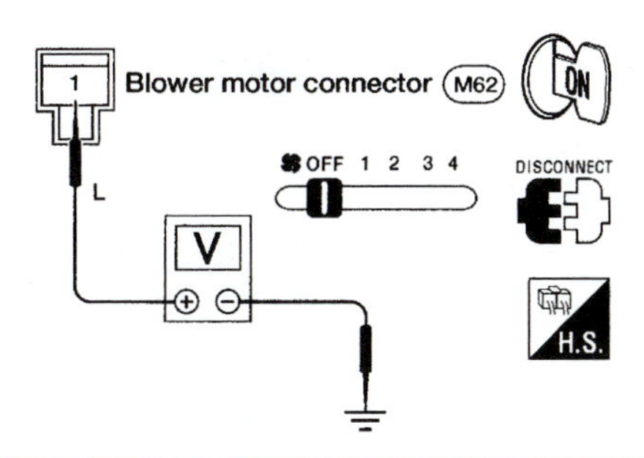

Figure 12-21 This motor has positive power even though the ignition key is on and the blower switch is off. The circuit uses a grounding switch that is typical of those found in electronic circuits. (Courtesy of Nissan North America)

WARNING: Do not use a test light on automatic climate control or electronic systems. The electronic circuits can be damaged using test lights. Consult the service manual for proper test procedures and equipment.

WARNING: Do not use a test light on blower motors that are on automatic climate control systems unless directed to do so by the service manual. The motor may not be considered electronic, but its controls will be, and most do not have a resister block for motor speed control. Damage to the circuit can result if test procedures are violated.

As usual, check the fuse first. If it is good, locate the feed wire to the blower motor and check for voltage with a multimeter or voltmeter. With the meter connected from the feed to a good ground, turn the ignition switch to RUN and place the blower control switch in each position (Figure 12-21). If there is voltage in one or more speeds but not on all, check the service manual to locate any resistors. Most vehicles have a *resistor block* between the switch and motor. The resistor block controls the amount of voltage reaching the motor, thus adjusting its speed (Figure 12-22). At the resistor block connection, use a wiring diagram to determine which of the three or four wires are coming from the switch. Test for voltage in each switch position. Voltage on each conductor indicates that the resister block is open. Voltage on some or none of them indicates there is a problem within the switch. The switch and block are not repairable and must be replaced individually.

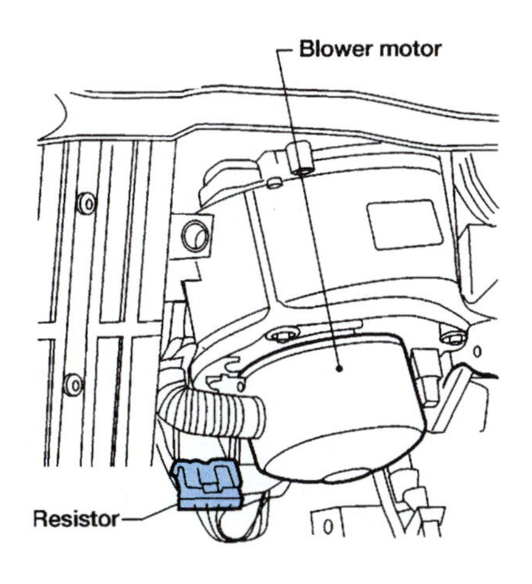

Figure 12-22 Typical location of a resister for manual heating and A/C systems. (Courtesy of Nissan North America)

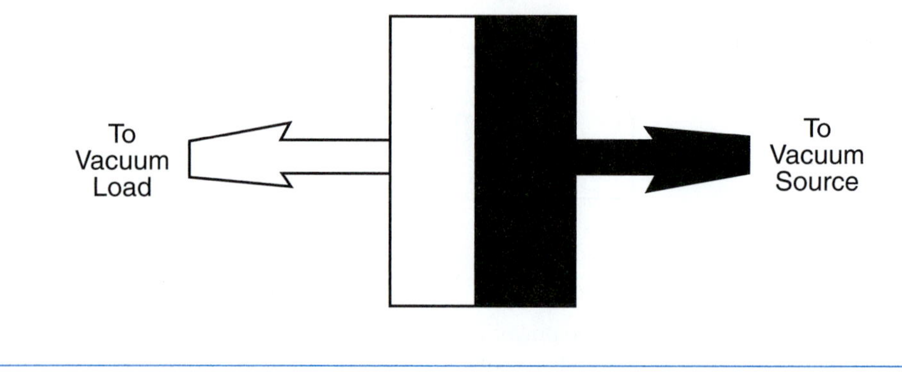

Figure 12-23 Vacuum check valves are used to maintain system vacuum when the engine is under load or shut down.

No voltage at the resistor block output or at the motor indicates an open in the circuit. Some manufacturers like Ford use a *thermal limiter* in the resistor block. The limiter acts like a circuit breaker and opens because of excessive current or heat. The thermal limiter is usually part of the resister block and the block has to be replaced.

If there is voltage at the motor in all speeds, check the motor's ground before replacing it. Connect the meter between the feed wire and the ground wire at the motor. Move the blower switch and ignition to ON. A voltage reading indicates that the power and ground are good and the motor must be replaced. No voltage at this point indicates the motor's ground is bad. Repair the ground wire or connection.

Air Comes from the Vent Only

Automatic climate control systems use small, electric motors to set any or all of the various *blend doors* to almost any position needed to raise or lower the passenger compartment temperature.

Special Tools

Vacuum gauge

The air is directed to selected vents by moving a lever to different modes: floor, panel, or defrost. The lever is attached to a vacuum switch or connected to the **blend doors** via cable. Before removing the climate control panel, consult the service manual to determine the type of controls. If vacuum operated, start your diagnosis by checking under the hood along the firewall. There should be one or two small vacuum hoses entering the passenger compartment. Locate the nearest vacuum connection in each hose found and check for vacuum with the engine running. It is best to use a vacuum gauge, but placing a finger over the hose end on the engine side will be a quick check. Vacuum will be felt if it is present. If there is no vacuum, follow the hose to its connection with the engine's intake and then repair the problem. Observe the vacuum for any check valves that may be used (Figure 12-23). Many times, a loss of heater or air controls within the passenger compartment is a result of rotten or disconnected vacuum hoses or bad check valves.

Assuming vacuum is present on the hoses or the system is cable operated, it may be necessary to remove a portion of the dash or dash-mounted components to correct the problem. It is best that an experienced technician either performs the repairs or provides supervision.

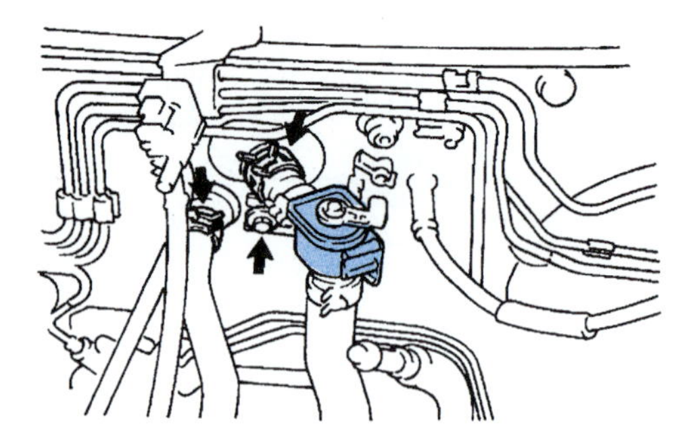

Figure 12-24 A typical water control valve for the heater. (Reprinted with permission)

Heater Does Not Work, A/C Is Good

CAUTION: Do not remove the radiator cap from a warm or hot engine. Extreme burns could be caused by pressurized, hot coolant. Always allow the engine to cool before opening the cooling system.

In this instance, always check the coolant level and engine temperature. The coolant level on most vehicles can be checked through the *translucent* overflow reservoir. An engine that is still cool or cold after a mile or two of driving indicates a stuck-open or missing thermostat. Low coolant level will cause poor flow through the heater core while an open thermostat will not allow the engine to warm. Both can cause heater problems.

The *water control valve* can cause a heater to stop producing warm air (Figure 12-24). The valve on manual systems is generally located in a heater hose on the engine side of the firewall. The valve may be cable operated, but most are vacuum controlled. Locate the valve and, if possible, operate the valve by hand and check the heater's operation. If the system works, inspect the valve and valve controls and repair as necessary. Note that the repair may require some portion of the dash to be removed.

Heater Works, but No A/C

WARNING: Do not open the A/C system before recovering the refrigerant using EPA equipment. Releasing refrigerant to the atmosphere could result in heavy fines or injury or damage to the vehicle.

CAUTION: Use eye and face protection while recovering or recharging A/C refrigerant. Refrigerant can cause eye and flesh injuries by freezing.

Most repairs on the A/C system are best left to EPA-certified and experienced technicians. However, there are a few checks that can be made quickly and safely by a new technician. Performing the following performance test will give the technician a better idea of what is happening within the system.

Performance Tests

A performance test involves operating the system through every mode. The following tests are set up using certain terms based on a specified control system. The system being tested may have controls that are labeled a little differently, but the test procedures are the same. Before starting the engine, ensure that the controls for the heater, air conditioning, and blower are set to OFF.

Special Tools

Service manual

Belt tension gauge

Start the engine and set the heater/A/C mode to VENT, the temperature to COOL, and switch the blower to LOW or 1 (Figure 12-25). Air about the same temperature as the outside air should be felt coming from the vents in the dash. Check each vent for airflow. If one vent does not have airflow, the ducting to that vent is disconnected or missing. Switch the blower to the other speeds, one at a time, and feel for an increase in airflow at each higher speed. Failure to obtain an airflow in any speed normally indicates there is a problem with the blower or blower switch. Loss of flow at one or more speed, but not all, indicates a problem with the blower switch. Leave the blower on HIGH for the next tests.

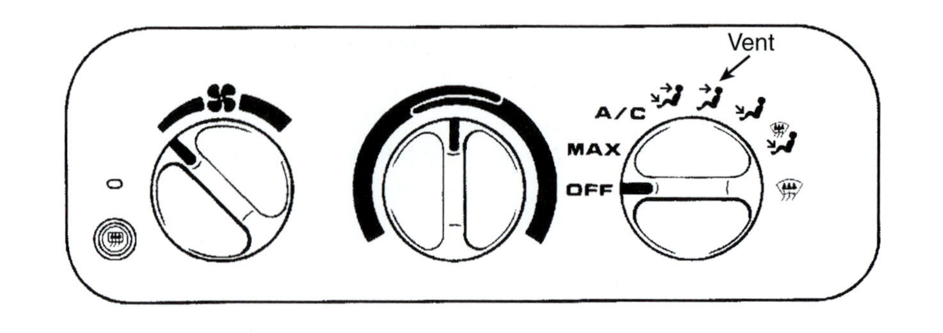

Figure 12-25 Set the blower (left) to LOW; temperature (center) to COOL, and mode (right) to VENT. (Courtesy of Chevrolet Motor Division, General Motors Corporation)

Move the mode lever (knob) to FLOOR and feel for airflow at the floor (Figure 12-26). Move the lever (knob) to DEFROST and feel for airflow at the bottom of the windshield. If there is a mode setting for HI/LO or mixed, move the lever to that position. The airflow should come from the floor and the vent or from the floor and the defrost. Failure to receive airflow at one of the modes indicates that the controlling cables or lever (knob) is broken or the blend door is broken. Move the mode to VENT for the following test. Keep the blower on HIGH.

Set the temperature lever (knob) to HOT. Within a few seconds, there should be warm or hot air (depending on engine temperature) from the vents. If not, then the heater control valve or control mechanism is damaged. Move the temperature lever (knob) back to COOL and leave the blower on HIGH.

Select the MAX position with the air conditioning controls (Figure 12-27). The sound of the blower should increase immediately and, within a few seconds, the A/C compressor should switch on, sometimes noticeable by a loud under-the-hood "click" and a slowing of the engine. A few seconds later, cool air should come from the vents. Switch the mode to NORMAL. The only noticeable difference will be less noise from the blower. Switch to DEFROST mode. There should be cold air at the bottom of the windshield. Failure to obtain cold air after a few minutes indicates that the A/C system is not functioning properly.

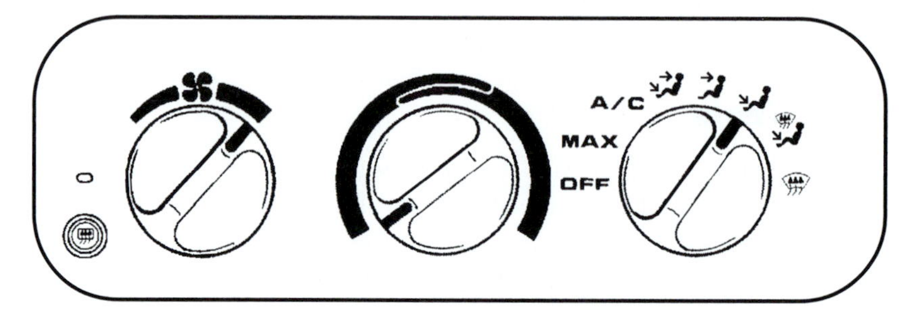

Figure 12-26 The mode is set to FLOOR and the blower is on HIGH. (Courtesy of Chevrolet Motor Division, General Motors Corporation)

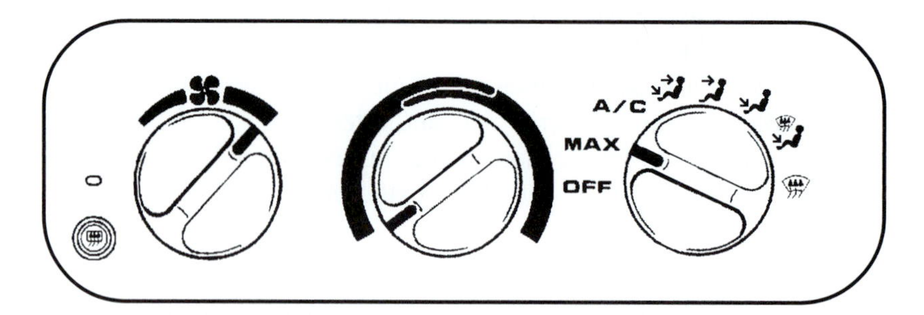

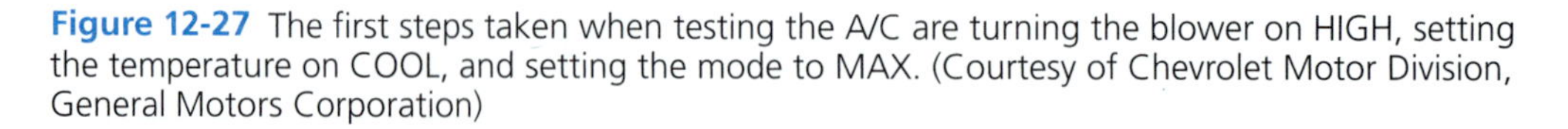

Figure 12-27 The first steps taken when testing the A/C are turning the blower on HIGH, setting the temperature on COOL, and setting the mode to MAX. (Courtesy of Chevrolet Motor Division, General Motors Corporation)

The following are some common faults found with climate control systems. In most cases, the suggested cause can occur on manual or ATC systems. However, the typical sensors, actuator, and program commands in an ATC control module can cause *what appears to be* a mechanical problem. Experienced technicians should perform the final diagnosis and test of ATC systems.

Mechanical Diagnosing

Special Tools

Belt tension gauge

Service manual

Loose or worn engine drive belts tend to slip and produce a squeal when placed under load. The A/C compressor is one of the heaviest belt-driven loads. The A/C drive belt may drive other components like the alternator, fan, and water pump. A seized compressor will cause the belt to slip or pull apart.

Check the tension for the belt by pressing down on it about halfway across its longest span or by using a belt tensioner tool. The belt should not deflect more than one-half inch on most systems. An oily or worn belt may slip even if the tension is correct. The belt must be replaced if it is oily or worn. The belt is replaced following similar procedures used on other drive belts. The most common problem with replacing A/C compressor drive belt is locating and accessing the fasteners. There may be up to four fasteners.

Special Tools

Small pry bar

1/2-inch breaker bar

Belt tension gauge

Loosen the pivot adjustment fasteners. Before forcing the component to move, check for a square, 1/2-inch hole somewhere at the front. The hole is used for inserting the drive end of 1/2-inch-drive breaker bar (Figure 12-28). If there is no hole, then the component must be forced to move with a small pry bar since the belt tension cannot be set using hand and arm power alone. Use extreme caution in the placement of the bar. One end of the bar must be placed on the engine block without catching any wires or hoses. The portion of the bar touching the compressor must be placed against a solid, metal portion. Ensure that the bar is not placed against the wires, hoses, or lines. In most cases when a pry bar must be used, the best place is beside the mounting bracket. This is the strongest part of the component outside the housing.

Force the component toward the engine and remove the belt. Install the new belt, ensuring that it is properly seated in each pulley. Apply outer force against the component until the correct belt tension is achieved. Hold the component in place while tightening the adjustment or the easiest fastener to access. Remove the bar and tighten all other fasteners. Sometimes, and whenever possible, it is necessary to tighten the adjustment and pivot fasteners before removing the bar to prevent the compressor from twisting and loosening the belt. Measure the belt tension after all fasteners have been tightened.

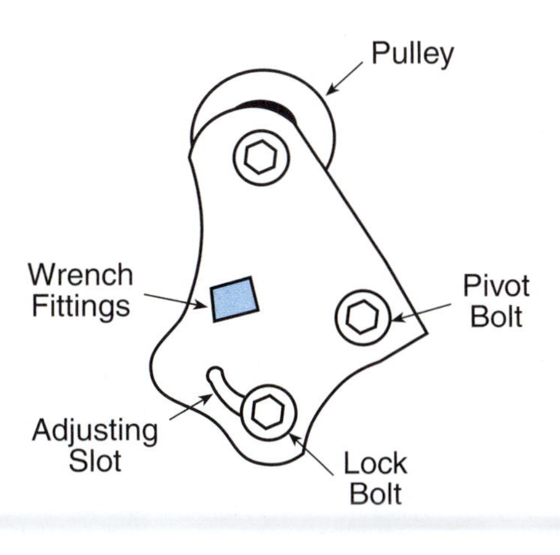

Figure 12-28 A 1/2-inch-square breaker bar is used to make tension adjustments in this unit.

Function Testing the Anti-Theft System

Special Tools

Service manual

The *panic* mode button is used to set off the vehicle's alarm in case of personal danger. It is designed to attract attention from anyone within hearing distance.

Before testing the anti-theft system, alert everyone in the shop. If possible, move the vehicle outside and away from the shop doors. A remote control that has lock, unlock, **panic**, and trunk buttons will be discussed.

Close and lock all vehicle doors using the remote-control button. An audible or a light should indicate that the system is armed and the alarms are set. Test the panic mode next. The vehicle's alarm should sound and some of the exterior lights should flash. Disarm the system by pressing the unlock button. Rearm the system and press the button to unlock the trunk. The trunk lock should function, but the alarm should not. Close the trunk and ensure that the system is armed. Next, test the ability of the system to recognize tampering. With the system armed, violently rock the vehicle. The alarms should sound and flash. Disarm the system.

Rearm the system and note if there is a red or green light on the instrument panel. The red light indicates that the vehicle has been tampered with while the green light indicates that the system is armed and no one tampered with it since the last time it was armed. Different systems may have different lights with different meanings. Consult the service manual.

CASE STUDY

A customer complains that the A/C air is directed to the floor vents every time the engine is accelerated, especially from the halt. It would eventually return to the panel vents. The technician confirmed the customer's complaint during a test drive. Knowing the blend doors were moved by vacuum motors and returned by springs to an OFF position, the technician began checking the vacuum delivery system to the switch. The supply hose led directly to a vacuum reservoir and then to the engine's intake manifold. Testing proved that every time the throttle was opened, the reservoir would empty and the door would spring back to the floor (OFF) position. When the engine load dropped, vacuum accumulated in the reservoir and the door returned to panel vents. A vacuum check valve was installed between the intake and the reservoir. It worked to a degree. The problem was completely solved by trying several different check valves until one allowed quick vacuum build-up on restart and closed the system when increased engine load was installed. A similar fix was listed later in the manufacturer's *technical bulletins* for engines that were approaching 80,000 miles of service.

Terms to Know

Blend door

KOEO

KOER

Resistor block

Technical bulletin

Translucent

Water control valve

ASE Style Review Questions

1. Climate control system diagnosis is being discussed. *Technician A* says an open blower motor fuse could cause a compressor clutch malfunction. *Technician B* says an inoperative blower motor may be caused by a poor ground connection. Who is correct?
   A. A only
   B. B only
   C. Both A and B
   D. Neither A nor B
2. Lighting systems are being discussed. *Technician A* says inoperative marking lights may be caused by an open circuit breaker. *Technician B* says testing the lighting system can normally be done with a test light. Who is correct?
   A. A only
   B. B only
   C. Both A and B
   D. Neither A nor B
3. A customer complains that the headlights turn on and off, generally when traveling on a rough road. *Technician A* says the circuit breaker may be weak or failing. *Technician B* says a loose ground on the relay may be the fault. Who is correct?
   A. A only
   B. B only
   C. Both A and B
   D. Neither A nor B
4. A power window stuck about half way up and will not move up or down. There is a "clunk" each time the switch is operated. *Technician A* says the drive gear is stripped. *Technician B* says the window regulator or glass is probably stuck. Who is correct?
   A. A only
   B. B only
   C. Both A and B
   D. Neither A nor B
5. Instrument gauge diagnosis is being discussed. *Technician A* says a test light can be used to test digital gauge systems. *Technician B* says supplying 12 VDC to a gauge will cause it to read full. Who is correct?
   A. A only
   B. B only
   C. Both A and B
   D. Neither A nor B
6. Warning light systems are being discussed. *Technician A* says grounding an oil pressure warning light wire at the sending unit will cause the gauge to read low. *Technician B* says grounding the fuel sending unit wire will cause the gauge to read full. Who is correct?
   A. A only
   B. B only
   C. Both A and B
   D. Neither A nor B
7. A vehicle with low engine oil pressure is being discussed. *Technician A* says using a manual pressure gauge can test the oil pump output. *Technician B* says the low oil pressure reading may be caused by a bad gauge. Who is correct?
   A. A only
   B. B only
   C. Both A and B
   D. Neither A nor B
8. The alternator warning light is on at idle and off at speeds above idle. *Technician A* says the problem may be a misadjusted throttle stop screw. *Technician B* says the battery connections could cause this problem. Who is correct?
   A. A only
   B. B only
   C. Both A and B
   D. Neither A nor B

9. Climate control is being discussed. *Technician A* says a non-certified technician can perform preliminary testing.
Technician B says an EPA-certified technician must perform any repairs requiring removal and replacing of the A/C refrigerant. Who is correct?
A. A only
B. B only
C. Both A and B
D. Neither A nor B

10. A vehicle power seat has forward and backward movement but not tilt. There is a low, "whirring" sound when the tilt switch is operated. *Technician A* says the motor drive gear may be stripped. *Technician B* says the forward and upward cable may be broken. Who is correct?
A. A only
B. B only
C. Both A and B
D. Neither A nor B

Table 12-1 ASE Task

Check voltages and voltage drops in electrical/electronic circuits with a voltmeter; determine needed repairs.

Problem Area	Symptoms	Possible Causes	Classroom Manual	Shop Manual
Poor system performance	Electrical device does not work.	**1.** Weak battery/alternator.	346	337, 341
		2. Corrosion.	346	341
		3. Loose/damaged connections.	346	339–340
		4. Worn/damaged device.	346	339–340

SAFETY

Wear safety glasses.
Follow electrical/electronic safety procedures.

Table 12-2 ASE Task

Find shorts, grounds, opens, and resistance problems in electrical/electronic circuits; determine needed repairs.

Inspect, test, and replace fusible links, circuit breakers, fuses, and other current-limiting devices.

Problem Area	Symptoms	Possible Causes	Classroom Manual	Shop Manual
Poor system performance	Electrical devices work poorly or not at all.	**1.** Short between two feed conductors.	348	338–339
		2. Connection between feed conductor and ground.	348	338–342
		3. Protection device open.	349	343
		4. Control device inoperative.	348–349	338–342

SAFETY

Wear safety glasses.
Follow electrical/electronic safety procedures.

Table 12-3 ASE Task

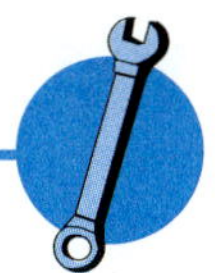

Inspect, test, and repair or replace headlight and dimmer switches, relays, control units, sensors, sockets, connectors, and wires of headlight circuits.

Inspect, test, and repair or replace switches, relays, bulbs, sockets, connectors, and wires of parking light and taillight circuits.

Inspect, test, and repair or replace switches, relays, bulbs, sockets, connectors, wires and controllers of printed circuit boards of instrument lighting circuits.

Inspect test, and repair or replace gauges, gauge sending units, connectors, wires, controllers, and printed circuit boards of gauge circuits.

Problem Area	Symptoms	Possible Causes	Classroom Manual	Shop Manual
Gauge circuits faulty. No light in instrument panel. Warning lights.	Gauges do not work or give wrong reading. No lights in instrument panel.	**1.** IVR inoperative.	352	343
		2. Sending unit, switch inoperative.	347–351	344–348
		3. Conductor damaged.	—	348–349
		4. Shorted, grounded, open circuits.	347–363	348–349

SAFETY

Wear safety glasses.
Follow electrical/electronic safety procedures.

Table 12-4 ASE Task

Inspect, test, and repair or replace switches, relays, bulbs, sockets, connectors, wires and controllers of courtesy lights (dome, map, vanity) circuits.

Diagnose the cause of intermittent, dim, or lack of operation of courtesy lights (dome, map, vanity).

Problem Area	Symptoms	Possible Causes	Classroom Manual	Shop Manual
Interior lights function improperly	Lights dim., Lights will not come on. Lights flicker.	**1.** Corrosion on switch, socket.	362–363	344–348
		2. Shorted conductor, controls.	362–363	344–348
		3. Loose connections.	362–363	344–348

SAFETY

Wear safety glasses.
Follow electrical/electronic safety procedures.

Table 12-5 ASE Task

Diagnose the cause of slow, intermittent, or lack of operation of power side windows.

Diagnose the cause of poor, intermittent, or lack of operation of electric door locks.

Diagnose the cause of poor, intermittent, or lack of operation of keyless and remote lock/unlock devices.

Problem Area	Symptoms	Possible Causes	Classroom Manual	Shop Manual
Windows will not lower or raise. Locks will not function.	Windows hang up or down. Windows will not move or move slowly. Locks will not function.	**1.** Linkage hanging.	363–366	352
		2. Gears stripped.	365–366	352
		3. Motors worn out.	365–366	352
		4. Controller damaged.	363–366	352
		5. Solenoid shorted, grounded, open.	364–365	352
		6. Dead battery in remote.	365	358

SAFETY

Wear safety glasses.
Follow electrical/electronic safety procedures.

Table 12-6 ASE Task

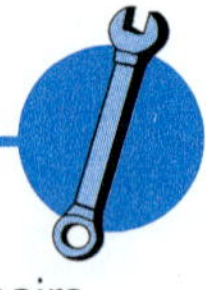

Identify system type and conduct performance test on the A/C system; determine needed repairs.

Problem Area	Symptoms	Possible Causes	Classroom Manual	Shop Manual
Poor A/C performance	Warm or cool air from A/C.	**1.** Blower inoperative.	368	352
		2. Duct work damaged.	368	342
		3. Low refrigerant.	369	355
		4. Weak compressor.	369	355
		5. System leaking refrigerant.	369	355

SAFETY

Wear safety glasses.
Follow electrical/electronic safety procedures.
Follow A/C safety procedures.

Job Sheet 24

24

Name ______________________________ Date ______________

Accessories Performance Tests

Upon completion of this job sheet, you should be able to conduct a performance test on selected accessories.

Tool Needed

Service manual

■ **CAUTION:** Before beginning a performance test, study the service or operator's manual and the system controls for familiarization and expected results.

Procedures

Remote Control Anti-Theft Systems

1. Confirm that the vehicle has a remote-control, anti-theft system.
2. Shut all doors and arm the system from the outside with the remote.

   Did the door locks function? ____________

   Did a horn or light function to indicate that the alarm is armed? ____________
   If not, check the remote's battery before proceeding.
3. Press the panic mode button. Did the alarms sound? ____________
4. Disarm and rearm the system.
5. Look through the driver's window without touching the vehicle. Is there a light in the dash indicating that the alarm is armed? ____________
6. Ensure that the remote is available to disarm the system.
7. Attempt to open each door. If all are locked, rock the vehicle from side to side. Disarm the system if the alarm sounds.
8. Did the alarm operate correctly? ____________
9. Explain. ____

   ____

   ____

Power Windows

10. Block the wheels and apply the parking brake.
11. Open all doors and switch the ignition to RUN.
12. Operate each window in turn using the master switch.

    Did each window work smoothly up and down? ____________

    Did each window stop smoothly at the top and bottom? ____________

13. Operate each window using the individual switches on the door.

Did each window work smoothly up and down? ______________

Did each window stop smoothly at the top and bottom? ______________

14. Set the window lock. Attempt to operate each window, first with the master switch and then with the individual switches.

Did each window work with the master switch? ______________

Did the right front or either of the rear windows work using the their individual switches? ______________

Comments ____

____

____

Power Seats

15. Sit in the driver's seat with the driver's door open. Switch the ignition to RUN.

16. Operate the switch to move the entire seat fully forward and fully backward.

Did the seat move smoothly and quietly? ______________

17. With the seat fully back, operate the switch for the backrest. Do not raise or lower the backrest completely.

Did the backrest perform as expected? ______________

18. Set the backrest to a comfortable position and operate the tilt switch.

Did the seat tilt up and down at the rear? ______________

Comments ____

____

____

19. If the other front seat is equipped with power, repeat steps 10 through 18 for that seat. Did the seat perform as expected? ______________

Instructor's Check ____

____

____

Job Sheet 25

25

Name ______________________________ Date ______________

Diagnosing an Oil Pressure Warning Light Circuit

Upon completion of this job sheet, you should be able to diagnose a non-electronic oil pressure warning light circuit.

Tools Needed

Hand tools
Manual oil pressure gauge
Jumper wire
Service manual

■ **CAUTION:** Do not use the following test procedures or equipment on electronic circuits. Damage to the circuit could result.

Procedures

1. Locate the service data.

   Make and model of vehicle ______________

   Engine ______________

   Type of instruments ______________

   Warning lights only or lights and gauges ______________

   Minimum and maximum oil pressures ______________

2. Is the oil level correct? ______________
3. Are there signs of oil leakage? ______________
4. Proceed with this step only after correcting the oil level. Turn the key to KOEO.
5. Did all the warning lights come on? ______________ If not, replace defective lamps and retest.
6. Start the engine and idle.
7. Is the oil light still on? ______________ If yes, slightly speed up the engine.
8. Did the light go out? ______________ If no, proceed with the circuit test. If yes, proceed with the manual oil pressure test starting with step 13.
9. Stop the engine, locate the oil pressure sending unit under the hood, and disconnect the wire from the sending unit.
10. Use jumper wire to ground the sending unit wire.
11. Set the ignition to RUN (KOEO). Is the light on? ______________If yes, go to the next step. If no, follow the circuit test routine in the service manual.

12. Remove the wire from the ground. Did the light go out? ______________ If yes, proceed with the next step. If no, follow the circuit test in the service manual.

13. Remove the oil pressure sending unit.

14. Install the manual oil pressure gauge.

15. Start the engine and idle. Record the oil pressure.

16. Does the oil pressure meet minimum specifications? ______________ If yes, proceed with the next step. If no, stop the engine and install the sending unit. Consult with the customer on all test results and recommended actions.

17. Remove the pressure gauge. Install a new oil pressure sending unit and connect its harness. Check the light operation. Return the vehicle to the customer.

Dual-Purpose Light-Oil and Temperature Combined

18. Locate the temperature sending unit under the hood.

19. Disconnect the wire. Set the ignition to RUN (KOEO).

20. Did the light go out? ______________ If yes, check the condition of the engine cooling system and sending unit. If no, check the oil pressure warning circuit following steps 8 through 16.

____

____

____

APPENDIX

Metric Conversions

	to convert these	to these,	multiply by:
TEMPERATURE	Centigrade Degrees	Fahrenheit Degrees	1.8 then + 32
	Fahrenheit Degrees	Centigrade Degrees	0.556 after − 32
LENGTH	Millimeters	Inches	0.03937
	Inches	Millimeters	25.4
	Meters	Feet	3.28084
	Feet	Meters	0.3048
	Kilometers	Miles	0.62137
	Miles	Kilometers	1.60935
AREA	Square Centimeters	Square Inches	0.155
	Square Inches	Square Centimeters	6.45159
VOLUME	Cubic Centimeters	Cubic Inches	0.06103
	Cubic Inches	Cubic Centimeters	16.38703
	Cubic Centimeters	Liters	0.001
	Liters	Cubic Centimeters	1000
	Liters	Cubic Inches	61.025
	Cubic Inches	Liters	0.01639
	Liters	Quarts	1.05672
	Quarts	Liters	0.94633
	Liters	Pints	2.11344
	Pints	Liters	0.47317
	Liters	Ounces	33.81497
	Ounces	Liters	0.02957
WEIGHT	Grams	Ounces	0.03527
	Ounces	Grams	28.34953
	Kilograms	Pounds	2.20462
	Pounds	Kilograms	0.45359
WORK	Centimeter Kilograms	Inch Pounds	0.8676
	Inch Pounds	Centimeter Kilograms	1.15262
	Meter Kilograms	Foot Pounds	7.23301
	Foot Pounds	Newton Meters	1.3558
PRESSURE	Kilograms/Sq. Cm	Pounds/Sq. Inch	14.22334
	Pounds/Sq. Inch	Kilograms/Sq. Cm	0.07031
	Bar	Pounds/Sq. Inch	14.504
	Pounds/Sq. Inch	Bar	0.06895

GLOSSARY

Note: Terms are highlighted in color, followed by Spanish translations in bold.

Accountant: The person or persons who receive payments, bill for payment, and manage the financial statement for the department or business.

Contador: La persona o las personas que reciben los pagos, mandan las cuentas, y administran los resumenes financieras del departamento o el negocio.

Antifreeze: An ethylene glycol-based fluid added to the cooling system to prevent freezing and to raise the boiling point.

Anticongelante: Fluido con una base de glicol de elileno que se mezcla al agua del radiador de un motor con el fin de evitar su congelación y elevar el punto de ebullición.

Baking soda: A base chemical used to neutralize acids. A cooking product.

Bicarbonato sódico: Un químico base que se usa para neutralizar los ácidos. Un producto para hornear.

Ball joint: A flexible joint used to connect the steering knuckle to the suspension system. So named because the ball end fits permanently into a closed socket.

Junta esférica: Una junta flexible que se usa para conectar el muñon de dirección al sistema de suspensión. Se nombra así porque la bola esférica queda permanentemente en un casquillo.

Bleeder screw: A holed screw fitted into the highest part of the wheel cylinder or brake caliper to remove air from the brake system in a process known as "bleeding the brakes."

Tornillo de descarga: Un tornillo agujerado puesto en la parte más alta del cilíndro de la rueda o del calíbre del freno para remover el aire del sistema de frenos en un proceso conocido como "purgar los frenos."

Blind hole: A hole that does not go all the way through a part.

Agujero ciego: Agujero que no atraviesa completamente una pieza.

Brake flushing: The process of removing all old brake fluid from the brake system. A form of brake bleeding.

Descargar los frenos: Un proceso de remover todo el fluido de freno viejo del sistema de freno. Una forma de purgar los frenos.

Castellated nut: A nut with castle-like turrets on top for insertion of a cotter key as a locking device.

Tuerca de corona: Una tuerca con las torrecitas parecidas a un castillo en la parte superior para insertar un pasador de chaveta como un dispositivo de cerrojo.

C-clamp: A small portable clamping tool so named because of its shape.

Presilla: Una pequeña herramienta de apriete que se llama así en inglés por su forma de C.

Corrosion: Deterioration caused by acid or other acidic material on metal. Commonly associated with electrical conductor, but includes rusting of metal.

Corrosión: La deterioración causado por el ácido u otra materia o materiales acídicas. Comunmente asociado con los conductores eléctricos, pero incluye la oxidación del metal.

Cotter key: A folded piece of metal used to lock nuts in place. *See* castellated nut.

Pasador de chaveta: Un pedazo de metal plegado que se usa para enclavar las tuercas. *Ver tambien* tuerca de corona (castellated nut).

Cover letter: A letter of introduction to a possible employer with the intention that the letter writer will be hired.

Carta de introducción: Una carta de introducción a un patrón potential con la intención de que él que lo escribe recibirá un puesto.

Crimping: A procedure used to join a wire to a solderless connector.

Engarzado: Procedimiento a través del cual se une un alambre a un conectador sin soldadura.

Database: A computerized information file.

Datos: Un archivo de información computerizado.

Dipstick: A flat, metal part with a handle and gauge marks used to measure the oil level in an engine.

Varilla de nivel: Pieza metálica plana con un mango y una escala graduada utilizada para medir el nivel del aciete en un motor.

Disc brake quieter: A chemical placed on the metal portion of disc brake pads to reduce squealing during brake application.

Silenciador de disco de freno: Una química puesta en la porción metálica de las balatas de freno para disminuir los chillidos durante la aplicación de los frenos.

Diverter valve: A valve used with Secondary Air Injection systems. *See* Secondary Air Injection in the *Classroom Manual.*

Válvula desviador: Una válvula que se usa con los sistemas de inyección de aire secundaria. *Ver tambien* la sección de inyección de aire secundaria (Secondary Air Injection) en *La Manual de Clase.*

DOT 3: A grade of brake fluid under the Federal Department of Transportation (DOT).

DOT 3: Un grado de fluido de freno según el departamento federal de transportación (DOT).

Drain pan: A pan used to capture liquids being removed from or leaking from vehicle components.

Cárter de aceite: Una charola que sirve para guardar los líquidos que se remuevan o que gotean de los componentes de un vehículo.

Drain plug: The plug removed at the bottom of the oil pan to drain the oil.

Tapón de vaciado del aceite: Tapón localizado en la parte inferior del colector de aceite que se remueve para vaciar el aceite.

Drive axle: A shaft that delivers torque from the differential to the rear wheels in a rear-wheel-drive vehicle. A shaft that delivers torque from the final drive assembly to the front wheels in a front-wheel-drive vehicle.

Eje de propulsión: Un eje que entrega la torsión del diferential a las ruedas traseras en un vehículo de tracción trasera. Un eje que entrega la torsión de la ensamblaje de engrane final a las ruedas delanteras en un vehículo de tracción delantera.

Electrodes: The ends of electrical conductors. Usually much larger that regular terminals and designed to carry larger current or voltage.

Electrodos: Las extremidades de los conductores eléctricos. Suelen ser mucha más grandes que los terminales regulares y se diseñan para tomar una corriente o un voltaje más grande.

Fading brakes: A brake condition resulting from overheated brake pads or shoes.

Desvanecimiento de los frenos: Una condición de los frenos que resultar del sobrecalentamiento de las balatas o los zapatos de freno.

Flux: A material made from acid or rosin used when soldering to clean and remove oxides from the solder joint.

Fundente: Material fabricado de ácido o resina utilizado durante la soldadura con el propósito de limpiar y remover óxidos de la junta del soldador.

Free travel: The amount of clutch pedal movement until the clutch release bearing contacts the release levers of the clutch's pressure plate.

Juego libre: La cantidad de movimiento del pedal de embrague hasta cuando el cojinete de desconexión del embrague toca las palancas de reposición de la placa de presión del embrague.

Friction area: The area where components move against each other. Normally referring to the area of the brake rotor or brake drums where the pads (rotor) or shoes (drum) contact for braking action.

área de fricción: El área en donde dos componentes muevan uno contra el otro. Normalmente refiera al área del rotor de frenos o los tambores del freno en dónde las balatas (rotor) o los zapatos (tambor) hacen contacto para efectuar el acción del frenado.

Gap: The air gap between two electrical or electronic devices. Usually the spark plug gap or an air gap between a sensor and its trigger.

Separación: La abertura de aire entre dos dispositivos eléctricos o electrónicos. Suele ser la separación de las puntas o una separación de aire entre un sensor y su desconectador.

Glazing: A condition that occurs when a drive belt slips and is polished by excessive rubbing on the pulleys.

Pulido: Condición que ocurre como consecuencia del deslizamiento de una correa de transmisión que es pulida por la fricción excesiva de las poleas.

Glycol-based: A type of commonly used brake fluid.

A base de glicol: Un tipo de fluido de freno muy común.

Grabbing brakes: A brake failure when the pads or shoes grab and release the rotor or drum without releasing or applying extra force to the brake pedal. May be caused by brake fluid on the friction material.

Frenos agarrantes: Un fallo de frenos cuando las balatas o los zapatos agarran y sueltan el rotor o el tambor sin soltar o aplicar una fuerza extra al pedal de frenos. Puede causarse por el fluido del freno en la materia de fricción.

Grease fitting: A part installed on a ball joint or tie-rod end that allows grease from the lube gun nozzle to enter the part.

Accesorio para la lubrificación: Pieza instalada en una juna esférica o en un extremo de la barra de acoplamiento que permite que la grasa en la boquilla de la pistola de lubrificación entre en la pieza.

Hard pedal: A brake failure where the brake pedal is hard to push, but there is little braking effect at the wheels. Usually caused by overheated brakes.

Pedal duro: Un fallo de los frenos en el cual el pedal de frenos es muy difícil de oprimir, pero hay muy poco efecto del frenado en las ruedas. Suele ser causado por el sobrecalentamiento de los frenos.

Isolators: Rubber bushings or pads between components used to reduce component vibration from reaching the body. Used under the body and on most suspension components.

Aisladores: Las bornas o arandelas entre los componentes usados para disminuir la vibración de los componentes que llega al chasis. Se usan bajo el chasis y en la mayoría de los componentes de suspensión.

Joint stud: The threaded stud of a ball joint or steering linkage that fits tightly into a mated bore on adjacent components.

Espiga de la junta: El poste enroscado de una junta esférica o una biela de dirección que se ajusta apretadamente en un agujero emparejado en los componentes adyacentes.

Labor time: The amount of labor used to complete a repair. May be actual time or the time shown in a labor guide.

Tiempo de trabajo: La cantidad de trabajo para completar una reparación. Puede ser tiempo actual o el tiempo indicado en una guía de trabajo.

Limited slip: A type of differential or final drive where the power is automatically shifted from a spinning drive wheel to the other drive wheels.

Deslizamiento limitado: Un tipo de diferencial o manejo final en que la fuerza se desvia automáticamente de una rueda propulsor girando a la otra rueda de propulsión.

Load test: A test of the battery to determine the amount of current available at 1/2 battery capacity.

Prueba de carga: Una prueba de la batería para determinar la cantidad de corriente disponible en un 1/2 de capacidad de la batería.

Locking brakes: A brake failure where the brakes lock the wheel with less than full brake pedal force. May be caused by misadjusted brakes or brake fluid on the friction material.

Frenos bloqueantes: Un fallo de los frenos en que los frenos enclavan la rueda con menos del total de la fuerza del pedal de frenos. Puede causarse por los frenos fuera de ajuste o el fluido de freno en la material de fricción.

Mechanic wire: A light-strength wire technicians use to tie components out of the way during repairs. Commonly used to hang brake calipers so their weight is not hanging on the flexible hose during brake repairs.

Alambre de mecánicos: Un alambre de poca fuerza que usan los técnicos para amarrar y apartar los componentes mientras que se efectuan las reparaciones. Suelen usarse para colgar los calibres de freno para que su peso no este colgando en el tubo flexible durante la reparación de los frenos.

Parts manager: The person responsible for managing the parts department. Most commonly found in dealerships and reports to service manager.

Gerente de partes: La persona encargada para la administración del departamento de refacciones. Se encuentran comunmente en los distribuidores y se comunican con el gerente de servicio.

Peacock: A manual draining valve for the radiator or engine block. Used to drain coolant.

Válvula de purga: Una válvula de purga manual del radiador o del bloque del motor. Sirve para purgar el fluido refrigerante.

Pedal pulsation: A brake failure where the pedal tries to move up and down while the driver is applying force to the pedal. Usually caused when the brake rotors' friction areas are not parallel or when a rotor is warped.

Pulsación del pedal: Un fallo del freno en el cual el pedal se mueva verticalmente mientras que el conductor aplica la fuerza al pedal. Suele ser causado cuando las áreas de fricción de un rotor no son paralelas o cuando un rotor es alabeado.

Pinch bolt: A high-strength bolt used to clamp or pinch a clamping type mount. Usually found on ball joint to steering knuckle connection.

Perno de anclaje: Un perno de alta resistencia que sirve para sujetar o apretar un montaje de tipo anclaje. Suele encontrarse en una junta esférica en una conexión de muñón de dirección.

Power pack: Usually refers to the assembly of the engine, transaxle, sub-frame, and front suspension components of a front-wheel-drive vehicle.

Batería común: Suele referirse a la ensambladura del motor, el transeje, el bastidor auxiliar, y los componentes delanteros de suspensión de un vehículo de tracción delantera.

Pressure bleeder: A tool used to pressure bleed a brake system by forcing brake fluid through the master cylinder into the brake lines.

Purgador de presión: Una herramienta que se usa para purgar un sistema de frenos bajo presión por medio de forzar el fluido de freno através del cilindro maestro a las lineas de freno.

Ratio: The proportional amount of one item to another. Commonly used to express gear relationship in size or number of teeth.

Relación: La cantidad proporcional de una cosa a otra. Suele usarse para expresar la relación de tamaño o número de dientes de los engranajes.

Recall: A manufacturer's term used when certain vehicles must be returned to a service center, usually a dealership, for correction of some sort. It is not a warranty and usually involves all of certain brand, model, and year of vehicle. It is paid by the manufacturer and may be ordered by the federal government based on safety data.

Retiro: Un término del fabricante que se usa cuando un cierto vehículo debe regresarse al centro de servicio, tipicamente al distribuidor, para algún reparación. No esta bajo garantía, y normalmente involucra todos los vehículos de cierta marca, modelo y año. Las refacciones se pagan por el fabricante y pueden ser por orden del gobierno federal basado en los datos de seguridad.

Resumé: A short education and work history submitted to perspective employer.

Resumen: Una historia corta de educación y de trabajo que se entrega a un patrón prospectivo.

Ring travel area: The area of the cylinder where the piston rings normally move.

Área de carrera del anillo: El área del cilíndro en donde suelen moverse los anillos del pistón.

Repair order: A form that is filled out to identify the car, customer, and complaint on a car to be serviced.

Solicitud de reparación: Formulario que se llena para identificar el vehículo y el cliente, donde se detalla el problema de un vehículo que necesita servicio.

Runout: The measurable side-to-side variation or wobble in a part as it rotates.

Desviación: Cantidad de desalineación o bamboleo de un lado al otro de una pieza mientras ésta gira.

Safety inspection: An inspection of a workplace to find safety hazards and to check for safety equipment such as fire extinguishers or hazardous material storage.

Inspección de seguridad: Una inspección de un taller para encontrar los peligros a la seguridad y para verificar la presencia del equipo de seguridad tal como los extinguidores o revisar el almacenaje de las materiales nocivas.

Seepage: A slow leak that does not produce a liquid drop or a run. Most seals require seepage to function properly. Should not be considered a leak at seals and gaskets.

Infiltración: Un goteo despacio que no produce gotas ni chorro líquido. La mayoría de los sellos requieren una infiltración para funcionar correctamente. No se debe considerar como un fuga en los sellos y empaques.

Serpentine belt: A single drive belt used to drive all engine-mounted accessories.

Correa serpentina: Una correa de un sólo mando que sirve para mandar todos los accesorios montados en el motor.

Signage: The signs in and around a business naming and listing the business' work.

Letreros: Los letreros en y alrededor de un negocio que detallan y nombran el trabajo del negocio.

Silicone-based: A type of brake fluid. Not used often and will caused damage if mixed with other types of brake fluid.

A base de silicona: Un tipo de fluido de frenos. No se usa frecuentamente y puede causar daños si se mezcla con otros tipos de fluido de freno.

Soldering gun: A tool with an electric heating element used to melt solder.

Pistola de soldadura: Herramienta con un elemento eléctrico de calentamiento utilizada para fundir la soldadura.

Soldering iron: A tool with an electric heating element used to melt solder.

Soldador: Herramienta con un elemento eléctrico de calentamiento utilizada para fundir la soldadura.

Span: The distance between two pulleys. The belt tension is taken at the longest span.

Abertura: La distancia entre dos poleas. La tensión de la correa se calcula en la abertura más grande.

Splicing: A term used to describe the joining of two or more electrical wires.

Empalme: Término utilizado para describir la unión de dos a más alambres eléctricos.

Spongy pedal: A brake failure where the pedal feels bouncy when the brakes are applied. Caused by air trapped in the brake system.

Pedal esponjosa: Un fallo de frenos en el cual el pedal tiene una sensación de brincar cuando se aplican los frenos. Causado por el aire atrapado en el sistema de frenos.

Steering knuckle: A major component of the suspension and steering systems. Anchored to the front suspension and steering linkage. Allows the wheel to be steered left and right while the suspension is absorbing road conditions.

Muñón de dirección: Un componente mayor de los sistemas de suspensión y dirección. Anclado a la suspensión delantera y a la biela de dirección. Permite girar la rueda a la izquierda y a la derecha mientras que la suspensión absorba las condiciones del camino.

Stripped threads: Threads that have been damaged by cross threading.

Filetes roidos: Filetes que han sido averiados debido al cruzamiento de roscas.

Stripping: The procedure used to remove insulation from an electrical wire.

Desaislación: Procedimiento a través del cual se remueve el aislamiento de un alambre eléctrico.

Stripping and crimping pliers: Pliers used to strip insulation from wire and crimp a solderless terminal in wire.

Alicates para la desaislación y el engarzado: Alicates utilizados para remover el aislamiento de un alambre ye engarzar un borne sin soldadura en un alambre.

Stud remover: A tool that grips a stud and can be driven by a wrench to unscrew a stud.

Removedor de espárragos: Herramienta que prende un espárrago y que puede ser accionado por una llave para destornillar el mismo.

Sub-frame: A small frame used to support the power pack in a unibody-constructed vehicle. May be used on some vehicles to support the suspension or drive components.

Bastidor auxiliar: Una pequeña armazón que se usa para soportar la batería común en un vehículo de construcción monocasco. Puede usarse en algunos vehículos para soportar los componentes de suspensión o tracción trasera.

Suction pump: Any pump use to suck liquid or other material to move it to a different location. Normally not a pressure pump.

Aspirador: Cualquier bomba que sirve para aspira al líquido u otra material moviendolo a otro lugar. Normalmente no es bomba de presión.

Tachometer: A gauge that translates electronic pulses from an engine sensor into data that is read as revolution per minutes (rpm). Usually scaled in 100 rpm.

Taquímetro: Un medidor que traduce los impulsos electrónicos de un sensor de motor a los datos que se interpretan como revoluciones por minuto (rpm). Suele tener la escala de 100 rpm.

Tap wrench: A tool used to drive the square shank of a tap.

Giramacho: Herramienta utilizada para apretar la espinilla cuadrada de un macho de roscar.

Technical bulletin: Repair instructions from a vehicle manufacturer detailing how certain repairs or corrections are to be made.

Boletín técnico: Las instrucciones de reparaciones de un fabricante de vehículos detallando como se deben efectuar ciertas reparaciones o correcciones.

Threadlocking compound: A material used on threads to prevent them from unscrewing from vibration.

Compuesto de sujeción para fieltes: Material utilizado en los filetes para prevenir que se aflojen debido a la vibración.

Thread sealing compound: A material used on threads to prevent liquid-like coolant from damaging the threads.

Compuesto sellador para filetes: Material utilizado en los filetes para prevenir que un líquido, como por ejemplo un refrigerante, averie los filetes.

Through hole: A hole that goes all the way through a part.

Agujero pasante: Agujero que atraviesa completamente una pieza.

Throttle body: The part of the intake system where the amount of air entering the intake is controlled through the throttle linkage. Usually houses the throttle valve, idle speed control, and may have other electronic or mechanical component.

Cuerpo de acceleración: La parte del sistema de entrada en donde la cantidad del aire entrando a la entrada se controla por medio de la biela de acceleración. Suele incluir la válvula de acceleración, el control de marcha mínima, y puede incluir otro componente electrónico o mecánico.

Thrust wear: Wear created by two components rubbing together as a result of some force being applied. Normally associated with cylinder wear when the piston is thrust against the cylinder wall during the power stroke.

Desgaste del empuje: El desgaste creado por dos componentes que se frotan como resultado de una fuerza aplicada. Normalmente se asocia con el desgaste del cilindro cuando el piston se oprima contra el muro del piston durante la carrera de potencia.

Toe wear: Tire wear created by improperly set toe. Usually causes a sawtooth pattern across the tread, and is worse on one side of the tread.

Desgaste de divergencia o convergencia: El desgaste del neumático debido a una divergencia o convergencia indebido. Suele causar un daño de aparencia de un diseño parecido a dientes de sierra en la banda del rodadura del neumático, y es peor de un lado.

Top off: The process of adding liquid to bring the level to specifications.

Llenar: El proceso de añadir el líquido para que el nivel queda a las especificaciones.

Transaxle: The gear system for a front-wheel-drive vehicle. Similar to the transmission for a rear-wheel-drive vehicle.

Transeje: El sistema de engranaje de un vehículo de tracción delantera. Parecido a la transmisión de un vehículo de tracción trasera.

Translucent: A semi-clear container. The contents of the container can be seen through the container without removing a cap.

Translucente: Un recipiente casi transparente. Los contenidos del recipiente se puede ver a través del recipiente sin quitar la tapa.

Transversal: Placing an object at angles to the major assembly as in the way an engine is set into the body of a front-wheel-drive vehicle.

Transversal: Ubicando un objeto a en ángulo al ensamblaje principal tal como se coloca un motor en el armazón de un vehículo de tracción delantera.

U-bolts: High-strength bolts with threads at both ends used to hold round components and flat components together in one assembly. Normally associated with leaf springs and solid axle housing. So named because of their U shape.

Perno en forma de U: Los pernos de alta resistencia enroscados de las dos extremidades para juntar los componentes redondos y los componentes planos en un ensamblaje. Normalmente se asocia con los resortes de hoja y la caja del eje sólido. Se nombran así por su forma de U.

Undercarriage: Under the vehicle body. Includes mounting attachments for components.

Aterrizador: Debajo de la carrocería del vehículo. Incluye los retenes de montaje de los componentes.

Vehicle identification number (VIN): A set of numbers and letters on a plate used to identify information about a car.

Número de identificación del vehículo: Juego de números y letras en una placa utilizado para identificar un vehículo.

Voltage drop: The amount of voltage used by a circuit or part of a circuit.

Caída de voltaje: La cantidad del voltaje que usa un circuito o parte de un circuito.

Water control valve: The valve used to control the flow of coolant into the heater core of a climate control system.

Válvula de control de agua: La válvula que sirve para controlar el flujo del fluido refrigerante al núcleo de calefacción de un sistema de control de climatizaje.

Windshield washer: A small motor and pump system used to spray a cleaner onto the windshield. The motor may be electric or vacuum operated.

Limpiaparabrisa: Un motor pequeño y el sistema de bombas que sirven para rociar un líquido limpiador en la parabrisa. El motor puede ser operado electricamente o por vacío.

Work apron: A type of heavy material used to protect the worker. In the automotive field, most commonly used during welding or cleaning parts.

Mandil: Un tipo de material muy grueso que se usa para protejer al empleado. En el campo automotivo, se usa más frecuentamente durante la soldadura o en la limpieza de las partes.

INDEX